CELL SIGNALING

T0314529

Cell Signaling provides undergraduate and graduate students with the conceptual tools needed to make sense of the dizzying array of pathways that cells use to detect, process, and respond to signals from the environment. By emphasizing the common design principles and molecular processes that underlie all signaling mechanisms, this book develops a broad conceptual framework through which students can understand diverse signaling pathways and networks. This book first examines the common currencies of cellular information processing and the core components of the signaling machinery. It then shows how these individual components link together into networks and pathways to perform more sophisticated tasks. Many specific examples are provided throughout to illustrate common principles and to provide a comprehensive overview of major signaling pathways.

Thoroughly revised, this second edition includes two new chapters and substantial updates to the text and figures throughout the book.

Key features:

- This book provides a conceptual framework through which all signaling pathways can be understood without memorization of details

- It is extensively illustrated, including high-quality diagrams and schematics to elucidate important concepts and processes

- Each chapter concludes with a useful summary section that brings together key concepts

- End-of-chapter review questions test the reader's understanding of the material covered

- Two new chapters have been written especially for this edition: "Signaling and Disease" and "Diversity in Signaling across Phylogeny"

Wendell A. Lim is Professor of Cellular and Molecular Pharmacology at the University of California, San Francisco. The principal focus of his research is the structure and mechanism of protein interaction domains and the logic by which these components are used to build complex cellular signaling systems.

Bruce J. Mayer is Professor of Genetics and Genome Sciences at the University of Connecticut, School of Medicine. Current work of his group focuses on characterizing and manipulating tyrosine kinase–mediated signal transduction pathways.

CELL SIGNALING

Principles and Mechanisms

Second Edition

Wendell A. Lim
and Bruce J. Mayer

CRC Press
Taylor & Francis Group
Boca Raton London New York

CRC Press is an imprint of the
Taylor & Francis Group, an **informa** business

A GARLAND SCIENCE BOOK

Designed cover image: Original artwork for cover and chapter openings by Wendell Lim.

Second edition published 2025
by CRC Press
2385 NW Executive Center Drive, Suite 320, Boca Raton FL 33431

and by CRC Press
4 Park Square, Milton Park, Abingdon, Oxon, OX14 4RN

CRC Press is an imprint of Taylor & Francis Group, LLC

Library of Congress Cataloging-in-Publication Data
Names: Lim, Wendell, author. | Mayer, Bruce, author.
Title: Cell signaling : principles and mechanisms / Wendell Lim, Bruce Mayer.
Description: Second edition. | Boca Raton : CRC Press, 2024. |
Includes bibliographical references and index.
Identifiers: LCCN 2023050318 (print) | LCCN 2023050319 (ebook) |
ISBN 9780367279424 (hbk) | ISBN 9780367279370 (pbk) | ISBN 9780429298844 (ebk)
Subjects: LCSH: Cell interaction. | Cellular signal transduction. | BISAC:
SCIENCE / General. | SCIENCE / Life Sciences / Biology / General. |
SCIENCE / Life Sciences / Cytology.
Classification: LCC QH604.2 .L56 2024 (print) | LCC QH604.2 (ebook) |
DDC 571.7/4–dc23/eng/20240307
LC record available at https://lccn.loc.gov/2023050318
LC ebook record available at https://lccn.loc.gov/2023050319

ISBN: 978-0-367-27942-4 (hbk)
ISBN: 978-0-367-27937-0 (pbk)
ISBN: 978-0-429-29884-4 (ebk)

DOI: 10.1201/9780429298844

Typeset in Times
by codeMantra

This book is dedicated to the memory of Tony Pawson (1952–2013). Tony was a towering figure in the field of cell signaling, and he will be sorely missed. The photo was taken at the Gerstle Park Inn in San Rafael, CA, where we often met to work on the book under the spreading branches of this ancient live oak tree.

Contents

Preface to the Second Edition xv

Preface to the First Edition xvii

Acknowledgments to the First Edition xix

Acknowledgments to the Second Edition xxi

Chapter 1
Introduction to Cell Signaling 1

WHAT IS CELL SIGNALING? 1

All cells have the ability to respond to their environment 2

Cells must perceive and respond to a wide range of signals 3

Signaling systems need to solve a number of
common problems 4

THE FUNDAMENTAL ROLE OF SIGNALING IN
BIOLOGICAL PROCESSES 6

Work in many different fields converged to reveal the
underlying mechanisms of signaling 6

Despite the diversity of signaling pathways and mechanisms,
fundamental commonalities have emerged 7

Signaling must operate at multiple scales in space and time 9

THE MOLECULAR CURRENCIES OF INFORMATION
PROCESSING 10

Information is transferred by changes in the state of proteins 10

There is a limited number of ways in which the state
of proteins can change 11

Most changes in state involve simultaneous changes in several
different currencies 13

LINKING SIGNALING NODES INTO PATHWAYS AND
NETWORKS 14

Information transfer involves linking different changes
of state together 15

Multiple state changes are linked together to generate
pathways and networks 15

Cellular information-processing systems have a
hierarchical architecture 16

SUMMARY 17

QUESTIONS 17

BIBLIOGRAPHY 18

Chapter 2
Principles and Mechanisms of Protein Interactions 19

PROPERTIES OF PROTEIN–PROTEIN INTERACTIONS 20

Changes in protein binding have both direct and
indirect functional consequences 20

Protein binding can be mediated by broad interaction
surfaces or by short, linear peptides 21

The affinity and specificity of an interaction determine
how likely it is to occur in the cell 22

The strength of a binding interaction is defined by the
dissociation constant (K_d) 23

The dissociation constant is related to the binding
energy of the interaction 25

The dissociation constant is also related to rates of
binding and dissociation 26

PROTEIN INTERACTIONS IN THEIR CELLULAR AND
MOLECULAR CONTEXT 28

The apparent dissociation constant can be strongly
affected by the local cellular environment and other
binding partners 28

Ideal affinity and specificity depend on biological
function and ligand concentrations 29

There are functional constraints on interaction
affinities and specificities 29

Interaction affinity and specificity can be independently
modulated 31

Cooperativity involves the coupled binding of
multiple ligands 32

Diverse molecular mechanisms underlie cooperativity 33

Cooperative binding has a variety of functional consequences 33

Protein assemblies differ in their stability and homogeneity 34

SUMMARY 36

QUESTIONS 36

BIBLIOGRAPHY 38

Chapter 3
Signaling Enzymes and Their Allosteric Regulation 39

PRINCIPLES OF ENZYME CATALYSIS 40

Enzymes have a number of properties that make them
useful for transmitting signals in the cell 40

Enzymes use a variety of mechanisms to enhance the rate of
chemical reactions 41

Enzymes can drive reactions in one direction by
energetic coupling 42

ALLOSTERIC CONFORMATIONAL CHANGES 43

Conformational flexibility of proteins enables
allosteric control 43

Signaling proteins employ diverse classes of
conformational rearrangements 44

PROTEIN PHOSPHORYLATION AS A REGULATORY
MECHANISM 45

Phosphorylation can act as a regulatory mark 45

Phosphorylation can either disrupt or induce protein structure 45

PROTEIN KINASES 47

The structure and catalytic mechanism of protein kinases are
conserved 47

The activation loop and C-helix are conserved molecular
levers that conformationally control kinase activity 48

Insulin receptor kinase activity is controlled via
activation-loop phosphorylation 49

Phosphorylation mediates long-range conformational
regulation of Src family kinases 50

Multiple binding interactions regulate protein kinase
substrate specificity 51

Protein kinases can be divided into nine families 52

PROTEIN PHOSPHATASES 52

Serine/threonine phosphatases are metalloenzymes 54

Most tyrosine phosphatases utilize a catalytic cysteine residue 54

Tyrosine phosphatases are regulated by modular domains while
serine/threonine phosphatases often associate with
regulatory accessory subunits 58

G PROTEIN SIGNALING 58

G proteins are conformational switches controlled by two
opposing enzymes 59

The presence of the GTP γ-phosphate determines the
structure of G protein switch I and II regions 60

There are two major classes of signaling G proteins 60

Subfamilies of small G proteins regulate diverse biological
functions 61

Many upstream receptors feed into a small set of common
heterotrimeric G proteins 62

REGULATORY ENZYMES FOR G PROTEIN SIGNALING 63

G-protein-coupled receptors act as GEFs for heterotrimeric G
proteins 64

Distinct GEF and GAP domains regulate specific small
G protein families 64

GEFs catalyze GDP/GTP exchange by deforming the
nucleotide-binding pocket 66

GAPs order the catalytic machinery for hydrolysis 66

Regulators of G protein signaling (RGS) proteins act as
GAPs for heterotrimeric G proteins 67

Additional mechanisms are used to fine-tune the activity
of G proteins 68

SIGNALING ENZYME CASCADES 68

The three-tiered MAP kinase cascade forms a signaling
module in all eukaryotes 68

Scaffold proteins often organize MAPK cascades 70

G protein activity can also be regulated by signaling cascades 71

SUMMARY 72

QUESTIONS 72

BIBLIOGRAPHY 74

Chapter 4
Role of Post-Translational Modifications in Signaling 77

THE LOGIC OF POST-TRANSLATIONAL REGULATION 77

Proteins can be covalently modified by the addition of simple
functional groups 78

Proteins can also be covalently modified by the addition of
sugars, lipids, and even proteins 79

Post-translational modifications can alter protein structure,
localization, and stability 81

Post-translational control machinery often works as part of
"writer/eraser/reader" systems 83

Post-translational modifications allow very rapid signaling
and transmission of spatial information 84

INTERPLAY BETWEEN POST-TRANSLATIONAL
MODIFICATIONS 84

A post-translational modification can promote or
antagonize other modifications 85

p53 is tightly regulated by a wide variety of post-translational
modifications 86

The level and activity of p53 are regulated by
ubiquitylation and acetylation 87

Additional modifications further fine-tune
p53 activity 88

PROTEIN PHOSPHORYLATION 88

Phosphorylation is often coupled with protein interactions 89

Kinases and phosphatases vary in their substrate specificity 91

Multiple phosphorylation of proteins can arise by different
mechanisms 92

Histidine and other amino acids can be phosphorylated,
especially in prokaryotes 93

Two-component systems and histidine phosphorylation
are also present in eukaryotes 94

ADDITION OF UBIQUITIN AND RELATED PROTEINS 95

Specialized enzymes mediate the addition and removal of
ubiquitin 95

E3 ubiquitin ligases determine which proteins will be
ubiquitylated 96

Ubiquitin-binding domains read ubiquitin-mediated signals in
diverse cellular activities 97

HISTONE ACETYLATION AND METHYLATION 98

Chromatin structure is regulated by post-translational
modification of histones and associated proteins 98

Two writer/eraser/reader systems are based on protein
methylation and acetylation 99

Chromatin modification in transcription is dynamic and
leads to highly cooperative interactions 102

SUMMARY 103

QUESTIONS 103

BIBLIOGRAPHY 104

Chapter 5
Subcellular Localization of Signaling Molecules 107

LOCALIZATION AS A SIGNALING CURRENCY 107

Changes in subcellular localization can transmit information 108

Subcellular localization can be regulated by a
variety of mechanisms 109

CONTROL OF NUCLEAR LOCALIZATION 109

Short, modular peptide motifs direct nuclear import
and export 110

Nuclear transport is controlled by shuttle proteins and the
G protein Ran 110

Phosphorylation of transcription factor Pho4 regulates
nuclear import and export 111

Nuclear import of STATs is regulated by phosphorylation and
conformational change 112

Localization of MAP kinases is regulated by association with
nuclear and cytosolic binding partners 112

Notch nuclear localization is regulated by proteolytic
cleavage 113

CONTROL OF MEMBRANE LOCALIZATION 114

Proteins can span the membrane or be associated with it
peripherally 114

Proteins can be covalently modified with lipids
after translation 114

Modular lipid-binding domains are important for regulated
association of proteins with membranes 115

Some lipid-modified proteins can reversibly associate with
membranes 116

Coupling effector protein activation to membrane
recruitment is a common theme in signaling 117

Akt kinase is regulated by membrane recruitment and
phosphorylation 117

MODULATION OF SIGNALING BY MEMBRANE
TRAFFICKING 118

Proteins can be internalized by a variety of mechanisms 119

Internalization of receptors can modulate signal transduction 119

TGFβ signaling output depends on the mechanism of
receptor internalization 120

Retrograde signaling allows effects distant from the
site of ligand binding 121

Ras isoforms in distinct subcellular locations have different
signaling outputs 121

BIOMOLECULAR CONDENSATES 122

Condensate formation is driven by multivalent interactions 122

Biomolecular condensates can make reactions more
efficient and specific 124

SUMMARY 126

QUESTIONS 126

BIBLIOGRAPHY 127

Chapter 6
Second Messengers: Small Signaling Mediators 129

PROPERTIES OF SMALL SIGNALING MEDIATORS 129

Small signaling mediators are controlled by an
interplay of their production and elimination 130

Small signaling mediators exert their effects by binding
downstream effectors 130

Small signaling mediators can lead to fast, distant,
and amplified signal transmission 131

Small signaling mediators can generate complex
temporal and spatial patterns 132

CLASSES OF SMALL SIGNALING MEDIATORS 133

Small signaling mediators have a wide range of
physical properties 133

The cyclic nucleotides cAMP and cGMP are produced
by cyclase enzymes and destroyed by phosphodiesterases 134

Cyclic nucleotides regulate diverse cellular activities 135

The regulatory (R) subunit of protein kinase A is a
conformational sensor of cAMP binding 136

Some small signaling mediators are derived from
membrane lipids 137

PLC generates two signaling mediators,
IP_3 and DAG 138

Activation of protein kinase C is regulated by IP_3 and DAG 138

CALCIUM SIGNALING 139

Activation of Ca^{2+} channels is a common means
of regulation 139

Ca^{2+} influx is rapid and local 140

Calmodulin is a conformational sensor of intracellular
calcium levels 141

Signaling can lead to propagating Ca^{2+} waves 142

SPECIFICITY AND REGULATION 143

Scaffold proteins can increase input and output specificity of small-molecule signaling 143

AKAP scaffold proteins can also regulate dynamics of cAMP signaling 145

SUMMARY 145

QUESTIONS 146

BIBLIOGRAPHY 146

Chapter 7
Membranes, Lipids, and Enzymes That Modify Them **147**

BIOLOGICAL MEMBRANES AND THEIR PROPERTIES 147

Biological membranes consist of a variety of polar lipids 148

Structural properties of membrane lipids favor the formation of bilayers 150

The composition of the membrane determines its physical properties 150

There are fundamental differences between biochemistry in solution and on the membrane 151

LIPID-MODIFYING ENZYMES USED IN SIGNALING 153

Cleavage of membrane lipids by phospholipases generates a variety of bioactive products 153

A variety of lipid kinases and phosphatases are involved in signaling 154

EXAMPLES OF MAJOR LIPID SIGNALING PATHWAYS 155

Phosphoinositides can serve as membrane binding sites and as a source of signaling mediators 156

Phosphoinositide species provide a set of membrane binding signals 157

Phospholipase D generates the important signaling mediator, phosphatidic acid (PA) 159

Phospholipase D plays a role in mTOR signaling 159

The metabolism of sphingomyelin generates a host of signaling mediators 161

Phospholipase A_2 generates the precursor for a family of potent inflammatory mediators 163

SUMMARY 165

QUESTIONS 165

BIBLIOGRAPHY 165

Chapter 8
Information Transfer across the Membrane **167**

PRINCIPLES OF TRANSMEMBRANE SIGNALING 167

The cell must process and respond to a diversity of environmental cues 168

Three general strategies are used to transfer information across the membrane 169

Many drugs target receptors 169

TRANSDUCTION STRATEGIES USED BY TRANSMEMBRANE RECEPTORS 170

Receptors with multiple membrane-spanning segments undergo conformational changes upon ligand binding 170

Receptors with a single membrane-spanning segment form higher-order assemblies upon ligand binding 171

Receptor clustering confers advantages for signal propagation 172

G-PROTEIN-COUPLED RECEPTORS 174

G-protein-coupled receptors have intrinsic enzymatic activity 174

Signaling by GPCRs can be very fast and lead to enormous signal amplification 176

TRANSMEMBRANE RECEPTORS ASSOCIATED WITH ENZYMATIC ACTIVITY 176

Receptor tyrosine kinases control important cell fate decisions in multicellular eukaryotes 176

TGFβ receptors are serine/threonine kinases that activate transcription factors 177

Some receptors have intrinsic protein phosphatase or guanylyl cyclase activity 178

Noncovalent coupling of receptors to protein kinases is a common signaling strategy 179

Some receptors use complex activation pathways that involve both kinase activation and proteolytic processing 183

Wnt and Hedgehog are two important signaling pathways in development 184

A variety of receptors couple to proteolytic activities 185

GATED CHANNELS 188

Gated channels share a similar overall structure 188

The voltage-gated potassium channel provides clues to mechanisms of gating and ion specificity 189

Ligand-gated ion channels play a central role in neurotransmission 191

MEMBRANE-PERMEABLE SIGNALING 193

Nitric oxide mediates short-range signaling in the vascular system 193

O_2 binding regulates the response to hypoxia 194

The receptors for steroid hormones are transcription factors 195

DOWN-REGULATION OF RECEPTOR SIGNALING 196

Ubiquitylation regulates the endocytosis, recycling, and degradation of cell-surface receptors 197

G-protein-coupled receptors are desensitized by phosphorylation and adaptor binding 199

SUMMARY 201

QUESTIONS 202

BIBLIOGRAPHY 203

Chapter 9
Regulated Protein Degradation   **205**

GENERAL PROPERTIES AND EXAMPLES OF
SIGNAL-REGULATED PROTEOLYSIS   205

Proteases are a diverse group of enzymes   206

Blood coagulation is regulated by a cascade of proteases   206

Regulated proteolysis by metalloproteases can generate
signaling molecules and alter the extracellular
environment   208

ADAMs regulate signaling pathways by cleaving
membrane-associated proteins   208

MMPs participate in remodeling the extracellular
environment   210

Proteolysis activates the thrombin receptor   211

Regulated intramembrane proteolysis (RIP) is an
essential step in signaling by some receptors   211

UBIQUITIN AND THE PROTEASOME DEGRADATION
PATHWAY   212

The proteasome is a specialized molecular machine that
degrades intracellular proteins   212

The cell cycle is controlled by two large
ubiquitin-conjugating complexes   214

SCF recognizes specific phosphorylated proteins,
targeting them for destruction   215

Two APC species act at distinct points
in the cell cycle   215

NF-κB is controlled by regulated degradation of
its inhibitor   217

PROTEASE-MEDIATED CELL DEATH PATHWAYS   219

Apoptosis is an orderly and highly regulated form of
cell death   219

The activity of caspases is tightly regulated   220

The extrinsic pathway links cell death receptors to caspase
activation   222

Mitochondria orchestrate the intrinsic
cell death pathway   225

Autophagy is a mechanism used by cells to digest
themselves, which can lead to cell death   228

SUMMARY   229

QUESTIONS   230

BIBLIOGRAPHY   230

Chapter 10
**The Modular Architecture and Evolution of
Signaling Proteins**   **231**

MODULAR PROTEIN DOMAINS   232

Protein domains usually have a globular structure   232

Bioinformatic approaches can identify
protein domains   232

Domains can be composed of several
smaller repeats   233

Protein domains often act as recognition modules   234

INTERACTION DOMAINS THAT RECOGNIZE
POST-TRANSLATIONAL MODIFICATIONS   237

SH2 domains bind phosphotyrosine-containing sites   238

Some SH2 domains are elements of larger
binding structures   239

Several different types of interaction domains recognize
phosphotyrosine   241

Multiple domains recognize motifs phosphorylated on serine/
threonine   242

14-3-3 proteins recognize specific phosphoserine/
phosphothreonine motifs   242

Interaction domains recognize acetylated and
methylated sites   243

Ubiquitylation regulates protein-protein interactions   244

INTERACTION DOMAINS THAT RECOGNIZE
UNMODIFIED PEPTIDE MOTIFS OR PROTEINS   245

Proline-rich sequences are favorable recognition motifs   245

SH3 domains bind proline-rich motifs   246

PDZ domains recognize C-terminal
peptide motifs   246

Protein interaction domains can form dimers or oligomers   247

INTERACTION DOMAINS THAT RECOGNIZE
PHOSPHOLIPIDS   247

PH domains form a major class of phosphoinositide-binding
domains   248

FYVE domains are phospholipid-binding domains found in
endocytic proteins   249

BAR domains bind and stabilize curved membranes   249

CREATING COMPLEX FUNCTIONS BY COMBINING
INTERACTION DOMAINS   250

Recombination of domains occurs through evolution   250

Combinations of interaction domains or motifs can be
used as a scaffold for the assembly of
signaling complexes   251

Scaffold proteins containing PDZ domains organize
cell–cell signaling complexes such as the
postsynaptic density   252

Proteins with multiple phosphotyrosine motifs function as
dynamically regulated scaffolds   253

RECOMBINING INTERACTION AND CATALYTIC
DOMAINS TO BUILD COMPLEX ALLOSTERIC
SWITCH PROTEINS   253

Many signaling enzymes are allosteric switches   254

14-3-3 Protein regulates the Raf kinase by coordinately
binding two phosphorylation sites   254

Certain plant protein kinases are regulated by modular light-gated domains 255

Regulation of the neutrophil NADPH oxidase by modular interactions 255

CREATING NEW FUNCTIONS THROUGH DOMAIN RECOMBINATION 256

Some modular domain rearrangements can lead to cancer 257

Modules can be recombined experimentally to engineer new signaling behaviors 258

SUMMARY 259

QUESTIONS 260

BIBLIOGRAPHY 261

Chapter 11
Information Processing by Signaling Devices and Networks 263

SIGNALING SYSTEMS AS INFORMATION-PROCESSING DEVICES 264

Signaling devices can be considered as state machines 264

Signaling devices are organized in a hierarchical fashion 264

Signaling devices face a variety of challenges in input detection 266

Proteins can function as simple signaling devices 267

INTEGRATING MULTIPLE SIGNALING INPUTS 269

Logic gates process information from multiple inputs 269

Simple peptide motifs can integrate multiple post-translational modification inputs 270

Cyclin-dependent kinase is an allosteric signal-integrating device 271

Modular signaling proteins can integrate multiple inputs 271

Transcriptional promoters can integrate input from multiple signaling pathways 273

RESPONDING TO THE STRENGTH OR DURATION OF AN INPUT 273

Signaling systems can respond to signal amplitude in a graded or a digital manner 275

An enzyme can behave as a switch through cooperativity 277

Networks can also yield switchlike activation 277

Signaling systems can distinguish between transient and sustained input 279

MODIFYING THE STRENGTH OR DURATION OF OUTPUT 281

Signaling pathways often amplify signals as they are transmitted 281

Negative feedback allows fine-tuning of output 281

Adaptation allows cells to control output duration 283

Feedback can cause output levels to oscillate between two stable states 286

Bistable responses also underlie more permanent outputs 288

SUMMARY 290

QUESTIONS 290

BIBLIOGRAPHY 290

Chapter 12
Cell Signaling and Disease 293

INTRODUCTION 293

SIGNALING IN CANCER AND CANCER THERAPY 294

Malfunctions in signaling contribute to the genesis of cancer 294

Mitogenic and growth stimulating signaling proteins can drive cancer 296

Cell cycle regulators can drive cancer 298

Cancers are associated with loss of tumor suppressors—signaling molecules that induce apoptosis or suppress growth 300

Immune evasion by cancers: downregulating antigen presentation 301

Cancers can avoid immune surveillance by immune checkpoint activation 301

Targeted cancer therapies inhibit oncogenic signaling proteins 302

Immune checkpoint inhibitors can be effective cancer therapies 304

Engineered T cells can recognize and kill cancers 305

PATHOGENIC MICROBES THAT HIJACK HOST CELL SIGNALING 307

Yersinia pestis provides an example of the diverse ways in which pathogens can hijack host signaling 307

Many pathogens manipulate the host actin cytoskeleton to facilitate adhesion, internalization, and cell-to-cell spread 309

DNA tumor viruses manipulate the host cell to support their replication 314

SUMMARY 316

QUESTIONS 317

BIBLIOGRAPHY 318

Chapter 13
Diversity of Signaling across Phylogeny 321

THE CONSTRAINTS ON SIGNALING IN DIFFERENT ORGANISMS 321

Differences in the physical properties of different cells constrain the signaling machinery 321

The lifestyle of an organism dictates what type of signaling mechanisms are needed 323

Comparative genomics provides insight into the evolution of signaling mechanisms 324

SIGNALING THEMES IN PROKARYOTIC ORGANISMS 325

Bacterial operons couple transcriptional regulation with metabolic activity 326

Bacterial populations can exchange information and coordinate their activities 329

SIGNALING MECHANISMS IN MULTICELLULAR
PLANTS 330

Signaling needs of plants are affected by their structure 330

Plant hormones (phytohormones) use a variety of
mechanisms to modify transcriptional programs 332

Plants have extensive stress response and innate immunity
systems to respond to pathogens and predation 336

THE TOOLKIT NEEDED FOR THE TRANSITION TO
MULTICELLULAR LIFE 338

New signaling pathways emerge through a combination
of novel elements and the repurposing of
existing elements 339

Eight signaling pathways are closely associated with the
emergence of animals 340

SUMMARY 342

QUESTIONS 343

BIBLIOGRAPHY 343

Chapter 14
How Cells Make Decisions **345**

VERTEBRATE VISION—how photoreceptor cells
sense and amplify light inputs 347

PDGF SIGNALING—triggering controlled cell
proliferation during wound healing 356

THE CELL CYCLE—how cells control their replication and
division 363

T LYMPHOCYTE ACTIVATION—how key cells in
our immune system are mobilized to fight infection 374

Chapter 15
Methods for Studying Signaling Proteins
and Networks **389**

BIOCHEMICAL AND BIOPHYSICAL ANALYSIS OF
PROTEINS 389

Analytical methods can determine quantitative
binding parameters 389

Michaelis–Menten analysis provides a way to measure the
catalytic power of enzymes 391

Methods to determine and analyze protein conformation are
central to the study of signaling 392

X-ray crystallography provides high-resolution
protein structures 395

Nuclear magnetic resonance (NMR) can reveal the
dynamic structure of small proteins 396

Electron microscopy can map the shape of large proteins and
complexes 396

Specialized spectroscopic methods can be used to
study protein dynamics 397

MAPPING PROTEIN INTERACTIONS AND
LOCALIZATION 398

Interacting proteins can be identified by isolating protein
complexes from cell extracts 398

Binding partners can be identified by screening large
libraries of genes 399

Direct protein–protein interactions can be detected by
solid-phase screening 399

Fluorescent protein tags are used to locate and track
proteins in living cells 400

Protein–protein interactions can be visualized
directly in living cells 401

METHODS TO PERTURB CELL SIGNALING
NETWORKS AND MONITOR CELLULAR RESPONSES 402

Genetic and pharmacological methods can be used to perturb
networks 403

Chemical dimerizers and optogenetic proteins provide a
dynamic way to artificially activate pathways 404

High-throughput sequencing is used to monitor the
transcriptional state of a cell 405

Modification-specific antibodies provide a method to track
post-translational changes 405

Mass spectrometry is the workhorse for identification of
proteins and their modifications 408

Live-cell time-lapse microscopy provides a way to track the
dynamics of single-cell responses 410

Biosensors allow signaling activity to be monitored in
living cells 411

Flow cytometry provides a method to analyze single-cell
responses rapidly 413

QUESTIONS 414

BIBLIOGRAPHY 415

Glossary **417**

Index **427**

Preface to the Second Edition

One of our main goals in writing the first edition of *Cell Signaling* was to emphasize the common design principles and molecular components that provide the basis for all cell signaling mechanisms. We reasoned that such an approach would provide readers with the conceptual tools to understand any signaling pathway, despite the enormous diversity in specific details. We also thought an advantage of this bottom-up approach would be that it would shield the book from quickly becoming outdated, a serious concern in a field as fast-moving as cell signaling. While we knew some of the details would inevitably change over time, we hoped the broader principles that serve as the foundation of the book would be more enduring.

It has now been close to 10 years since the first edition was published. As we had hoped, the conceptual framework we laid out has stood the test of time. So why a second edition? First, technical and conceptual advances in a number of areas have provided new insights that we felt needed to be included. A few examples include the increasing availability via cryo-electron microscopy of high-resolution molecular structures for large protein complexes, and the role of phase-separated biomolecular condensates in signaling. Therefore, we have updated the previous text and figures to highlight such advances where possible. Second, readers pointed out several areas that were not covered in sufficient depth in the first edition. For example, our focus on signaling in multicellular animals did not do justice to the diversity of signaling mechanisms used by a wider spectrum of organisms, such as prokaryotes and plants. We also heard that the central role of cell signaling in human health and disease was given short shrift in the first edition. To address these gaps, we have included two entirely new chapters in the second edition. *Diversity of Signaling across Phylogeny* explores the variety of signaling mechanisms found in different organisms, and how these signaling mechanisms might have evolved over time. We hope that researchers working on non-vertebrate systems will find this helpful in highlighting both the commonalities and differences in signaling across living systems. The second new chapter, *Cell Signaling and Disease*, explores how dysregulated signaling lies at the root of most human diseases, and how many drugs and treatments work by manipulating cell signaling mechanisms. In this exciting time where advances in cell signaling are quickly translated into the clinic, we hope this chapter will serve as an entryway to the field for medical students and other health care professionals.

Finally, a brief comment on our dear colleague and co-author Tony Pawson, who sadly passed away shortly before the publication of the first edition. Although he could not help us with the second edition, many of his words and ideas remain and shine through from the first edition. We never could have written this book without his wisdom and inspiration, and we can only hope that he would be pleased with this updated edition.

ORGANIZATION OF THIS BOOK

The first major section of this book (Chapters 2–9) focuses on the molecular parts and molecular principles used in cell signaling. These chapters introduce the types of molecular players, including catalytic domains, protein interaction domains, and receptor molecules. They also introduce the core currencies that are used to store and transmit signaling information at the molecular level, including molecular conformation, interaction, localization, modification, and degradation. In this section, we hope the reader will gain a deep understanding of how these molecular players can be used to conditionally modulate the core information-storage currencies. The second major section of this book (Chapters 10–14) moves gradually toward understanding cell signaling systems—how these components and currencies can be used to make more complex devices and networks that are capable of executing higher-level physiological decisions. The last chapter of this section (Chapter 14) is organized as a set of exploratory, visual panels. Here, we return to a more classical view of signaling—focusing on four physiologically important pathways (vision, mitogen response, the cell cycle, and T cell activation)—and describe their molecular components from top to bottom. We attempt to illustrate how the components discussed in previous sections of this book are used to solve key physiological problems in these model pathways. The goal of this unique chapter is to integrate a top-down physiological perspective with a bottom-up perspective of central principles and molecular components. Finally, the last chapter of this book (Chapter 15) is a compendium of experimental methods and approaches used to study cell signaling. It provides an overview of "how we know" what we know and of important tools of the trade.

Although our aim in organizing this book is to provide a conceptual framework for understanding signaling mechanisms, we understand that some readers may also wish to use it to provide an introduction to specific signaling pathways. For this reason, we use many specific examples throughout this book (mostly but not exclusively from multicellular animals) to illustrate the broader principles. In this way, most major signaling pathways and families of signaling molecules are explained and illustrated in some detail. However, we imagine that many instructors using this book may choose to supplement it with readings selected from recent experimental papers, in order to provide greater depth in specific areas of interest.

The end of each chapter includes a set of study questions. These questions are designed to encourage students to think more concretely about the concepts covered in each chapter, to extract and synthesize this information in new ways, and to think about signaling from an experimental perspective. Some questions have one clear answer, whereas others are more open-ended and have multiple possible answers. Answers to the questions are available online from the publisher's website. Chapter 14 does not contain end-of-chapter questions, but it rather integrates exploratory questions throughout each panel set.

We also include a short list of references at the end of the chapters. The intent is to provide a starting point for readers who wish to delve more deeply into the topics covered. These references are by no means comprehensive, and in a field as rapidly moving as cell signaling, it would be impossible to cover fully the most recent developments. Therefore, we sincerely apologize to the many valued colleagues whose publications deserved to be included, but were omitted due to space constraints or inadvertent oversight.

SUPPORT MATERIALS

For qualified instructors

Figure slides – All the figures from this book can be downloaded for use in lectures and presentations. These images are available in two convenient formats: JPEG and PowerPoint®.

To access these, please visit this book's product page at www.routledge.com/9780367279370 and click 'Instructor Resources Download Hub'. There you'll be asked to fill out a short form concerning the course you teach, before being granted access to these figure slides.

For students

Answers to the end-of-chapter questions – These are available as a downloadable PDF. To download them, please visit this book's product page at www.routledge.com/9780367279370.

Preface to the First Edition

The ability to sense one's environment and respond to it is one of the most fundamental properties common to all living organisms. Single-celled organisms are able to seek out nutrients, avoid toxins, and change their shape, gene expression, and metabolism depending on conditions. In multicellular animals and plants, even more subtle and sophisticated interactions between cells and the outside environment are needed. Individual cells are able to sense and integrate vast amounts of information and use it to make decisions about whether to grow, divide, migrate, adopt a particular shape, or even die. Without these decisions, a multicellular organism could neither develop nor maintain its integrity as a coherent living entity. Cell signaling is everywhere we look.

For most of its relatively brief history, cell signaling (or signal transduction, as it is often termed) was typically taught from the perspective of individual, top-down physiological stories—for example, understanding how a cell responds to a particular type of hormone, such as insulin or epinephrine. However, this approach has become increasingly untenable as we attempt to understand and explain how thousands of signaling proteins interact with each other in vast interconnected networks. Fortunately, it is now apparent that diverse signaling processes are often executed by very similar sets of molecular components, which in turn often interact in common network patterns. We can now consider a more bottom-up view that there are hierarchical building blocks for signaling often used in archetypical ways to achieve particular classes of cellular behaviors. The overarching theme of this book is that signaling is best understood by focusing on the common design principles, components, and logic that drive all signaling.

We have written this book embracing this new, bottom-up view of cell signaling and focus on the principles and mechanisms that underlie many, if not all, cell signaling processes. Therefore, we emphasize the commonality of underlying mechanisms rather than the diversity of individual examples. As we enter the postgenomic era, research is shifting away from the discovery of specific signaling molecules toward understanding how such cellular components can execute nuanced, complex regulatory behaviors. Understanding how simple molecules specifically interact to form a system with emergent response properties is very much at the heart of cellular signaling, as is understanding how such a diversity of response behaviors has evolved. Signaling is a field in which our understanding is growing exponentially and our viewpoint is constantly changing. While the rapid growth and change in this field make writing a textbook on cell signaling challenging, it is also a particularly exciting time to try to lay out the core molecular concepts and principles that have emerged in this young field.

Memorizing the details of "who did what to whom and when" for a number of specific pathways, especially where most of the players have similar-sounding three-letter acronyms, can be a recipe for confusion. It is our hope that by focusing on the common themes and concepts that underlie all signaling pathways, students will better see the beauty and logic of the system, and be able to apply this understanding to any signaling problem they encounter. This is particularly relevant in our era of superabundant information where manifold details are readily available at a few keystrokes. We hope this book conveys the deeper principles of the subject that cannot be encapsulated in a Wikipedia entry.

Acknowledgments to the First Edition

This book is the product of nearly a decade of thinking, prodding, procrastination, and hard work. It owes its inception to Miranda Robertson, who was our first editor, and originally proposed a cell signaling textbook as part of the New Science Press "Primer" series. Miranda recruited and cajoled this team of authors and guided them until the "Primer" project was discontinued. Without Miranda's keen insight and shopping bags full of ginger biscuits, this book would never have taken flight. At that point, the project was brought to Garland Science thanks to Denise Schanck and Adam Sendroff. Here, our second editor, Janet Foltin, deftly shepherded us through this transition and the writing of the core chapters. Finally, our last editor, Mike Morales, took over the immense challenge of pulling us over the finish line and keeping us on task. We would also like to thank Natasha Wolfe, our production editor, for guiding us through the copyediting and typesetting processes. We are indebted to Mary Purton, our developmental editor, who read, edited, and formatted every word of this book. Special thanks for keeping track of constantly shifting figures while we rearranged the order of chapters. Our illustrator, Matt McClements, has been with us since the beginning, and has been invaluable in translating our ideas to pictures, while at the same time developing a coherent and consistent style for this book that fits perfectly with its conceptual organization. We also thank Kenneth Xavier Probst who generated the molecular structure images in this book, along with Lore Leighton and Tiago Barros. Many thanks to other Garland Science staff that worked on this project, including Monica Toledo, Alina Yurova, and Lamia Harik, as well as Joanna Miles from New Science Press.

We also want to thank Jesse Zalatan and Brian Yeh, who contributed to several chapters in this book, as well as colleagues who read chapters at various stages, including Steve Harrison, Henry Bourne, Jim Ferrell, David Foster, David Morgan, Chao Tang, and Art Weiss. We are grateful for the instructors and content experts who read draft chapters of the manuscript and provided detailed, formal reviews: Johannes L. Bos, University Medical Center Utrecht; Andrew Bradford, University of Colorado School of Medicine; Adrienne D. Cox, The University of North Carolina at Chapel Hill; Madhusudan Dey, University of Wisconsin, Milwaukee; Julian Downward, London Research Institute of Cancer Research UK; Yanlin Guo, The University of Southern Mississippi; Tony Hunter, Salk Institute; Do-Hyung Kim, University of Minnesota; Low Boon Chuan, National University of Singapore; Shigeki Miyamoto, University of Wisconsin, Madison; and Henry Hamilton Roehl, The University of Sheffield. We would also like to thank the many scientists who made a special effort to provide us with high-resolution figures for this book, including Susumu Antoku, Jonathon Ditlev, Vsevolod V. Gurevich, Mark Hollywood, Evi Kostenis, Marco Magalhaes, Michiyuki Matsuda, Holger Stark, and Yi Wu.

Through these past several years, much of the writing and discussion occurred at retreats held at the Gerstle Park Inn, San Rafael, California, and we thank Jim and Judy Dowling for their warm hospitality and delicious omelets. We would also like to thank the current owners, Gail S. Jones and David W. Pettus, for allowing us to take the dedication photograph at their beautiful home. Thanks also to Mary Collins for welcoming us into her San Francisco home for writing sessions, and for supplying us with delicious scones and biscotti. This book also owes much of its inception to discussions held at the FEBS Protein Modules meeting at the Hotel Veronika in Seefeld im Tirol, Austria, where the authors would gather every other year to meet with our wonderful colleagues.

WAL: I am deeply grateful to my wife, Karen Earle-Lim, and my children, Emilia, Nadia, and Jasper, for their constant love and support, and their patience in allowing me to undertake this important project. I am also grateful to my parents, Ramon and Victoria Lim, for their encouragement and support. I would also like to thank my mentors, Jeremy Knowles, Bob Sauer, and Fred Richards, as well as the many members of my laboratory, who have consistently filled my life with interesting ideas.

BJM: I will always be thankful to my wife, Rita Malenczyk, who has been unfailingly supportive throughout this seemingly interminable project, despite a busy career of her own and three growing boys at home. I am also grateful for the many colleagues and members of my lab who helped when needed and didn't complain when this book pushed other obligations aside. And finally, I owe an enormous debt to Jim Donady, who introduced me to the sheer joy of research in biology, and especially to Saburo Hanafusa and David Baltimore, who presided over the wonderful playpens where I found my love for cell signaling.

Acknowledgments to the Second Edition

We are very grateful to colleagues from around the world who provided feedback on the first edition and made valuable suggestions for improvements. We also greatly appreciate the many colleagues who read drafts of new or revised material and provided helpful comments, including Jake Brunkard, Jon Ditlev, Timothy DuBuc, Les Loew, Mike Morales, Iñaki Ruiz-Trillo, Martin Schwartz, and Peter Setlow. We are also grateful for those who provided new images for the second edition, including Zeynep Baltaci, Angika Basant, Ken Campellone, Pascale Cossart, Jon Ditlev, John Janetzko, Brian Kobilka, Robert Lefkowitz, Aashish Manglik, Andrew Nguyen, Georgios Skiniotis, Katrina Velle, and Michael Way.

WAL: I am grateful for the love and support from my wife, Karen Earle; my children, Emilia, Nadia and Jasper; sibling, Jennifer Lim Dunham and Caroline Starbird, and my parents, Ramon and Victoria Lim. I thank the current and former members of my lab for their support and constant stimulation, and my trusty assistant, Noleine Blizzard, for keeping the ship in order. Finally, I am indebted to my steady partner and friend in this endeavour, Bruce Mayer.

BJM: I will always be grateful for the continuing support of lab members, friends and colleagues, and my family. The past decade has been challenging, marked by devastating and unexpected loss. The kindness, support, and love of the community, and especially of my wife, Rita Malenczyk, made it possible to carry on.

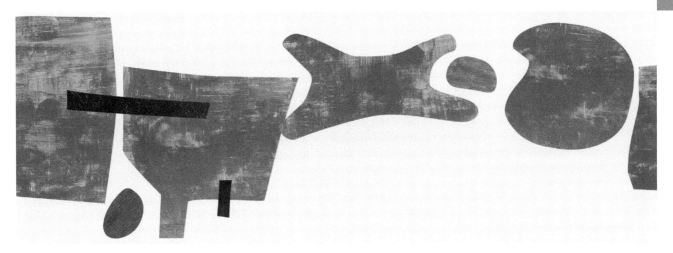

Introduction to Cell Signaling

All living cells perceive signals from the outside environment and adjust their behavior accordingly. If you think back to the earliest living cells, it is easy to imagine the incredible pressure they were under to evolve the ability to sense features of the environment and to change in response to these signals. The ability to sense and move toward nutrients, or to sense and avoid stresses and toxins, would give a unicellular organism a huge competitive advantage. This ability to respond to environmental cues is not only important for single cells, but it is also absolutely essential for the normal development and functioning of multicellular organisms, which depend on a continuous and extensive exchange of information to coordinate the activities of many individual cells. Furthermore, when this cellular communication goes awry, it can result in diseases such as cancer. In this chapter, we will introduce the basic principles of cell signaling and the molecular mechanisms that underlie it.

WHAT IS CELL SIGNALING?

Cells are the smallest fundamental units of life. Part of what makes them so distinctly "living" is their remarkable ability to sense stimuli and to respond to them in a dynamic manner. This ability of cells to detect or receive information and process it to make decisions can also be considered from the broader perspective of information processing. Here, we can draw analogies to the engineering and design principles of other, more familiar information-processing systems, such as human-made electronic devices. It is this interface between the unique properties of living systems and the more universal properties of any system that processes information that makes the study of cellular signaling mechanisms so compelling.

DOI: 10.1201/9780429298844-1

All cells have the ability to respond to their environment

While biologists and philosophers may not all agree on the precise definition of life, most definitions include a number of common properties, such as autonomy, the ability to generate energy, and the ability to reproduce. One of these common properties is adaptability—the ability to respond to changes in the environment. We all understand that one way to test whether a thing is animate or inanimate, or living or dead, is to poke it and see if it responds.

Single-celled organisms display the ability to detect diverse molecular species and stresses in their environment, and are able to change aspects of their gene expression, growth, structure, and metabolism in response, usually to improve their ability to survive under changing conditions. With the emergence of multicellular organisms, individual cells within the organism evolved the highly specialized ability to sense specific signals transmitted from other cells in the organism, allowing for extraordinary levels of communication within the organism. The coordinated regulation of growth, death, morphology, and metabolism is absolutely essential for many individual cells to function in concert as an integrated organism. Moreover, cells have the ability to monitor aspects of their own internal state, and to respond in a self-correcting way—the foundation of cellular homeostasis and repair. Thus, cell signaling, which encompasses the study of this wide range of stimulus–response behaviors observed in cells, is central to all of biology.

Today, as our knowledge of biological systems increases rapidly, we have begun to view cell signaling from the perspective of a more general question: how do cells process information? How a cell receives diverse signaling inputs, processes and integrates these signals, and converts them into responses is in many ways analogous to how other systems, at widely varying scales, process information. We can think about information processing and storage in the cell, just as we focus on these concepts when we consider a brain or a computer (**Figure 1.1a**). At the level of the organism, for example, we can marvel at the ability of an athlete to detect the movement of a ball, calculate its trajectory, and mount an effective response to intercept it that involves the coordinated action of hundreds of different muscles, all within a fraction of a second. Individual cells perform similarly remarkable feats, such as detecting the presence of just a few specific molecules in the outside environment, and responding by setting in motion elaborate cellular programs that culminate in behaviors such as proliferation, directed migration, or cell death. At each of these scales, a common series of challenges must be overcome. Just a few examples include detecting, amplifying, and robustly responding to a faint incoming signal; integrating and responding coherently to diverse and contradictory signals; and adapting to the strength or duration of a signal and shutting down the response when appropriate. These represent universal information-processing tasks that a cell or an organism must be able to perform.

In this book, we will focus on information processing by the smallest unit of life, the cell. It is particularly interesting to consider how such tasks are accomplished by the molecules that make up a cell. Unlike the tidy systems of wires, transistors, and other components that make up human-made signaling devices

Figure 1.1

Cell signaling systems process information. (a) The role of cell signaling systems is to receive input from the environment and, on the basis of that input, generate an appropriate output response. Information processing performed by the cell is conceptually similar to that of other, familiar information-processing systems, such as the brain or a computer. (b) Cellular information processing must be accomplished by a densely packed and diverse collection of biomolecules. (b, Adapted from D. S. Goodsell, *Trends Biochem. Sci.* 16: 203–206, 1991. With permission from Elsevier.)

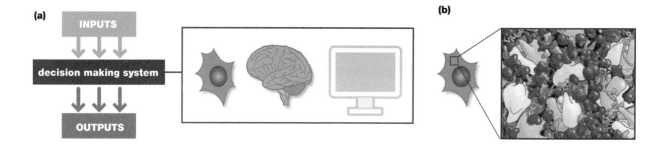

like mobile phones or computers, the cellular signal-processing machinery consists of a densely packed mixture of rapidly diffusing proteins, lipids, nucleic acids, and other biomolecules, all surrounded by a water-impermeable membrane (**Figure 1.1b**). How this genetically encoded set of molecules can perform complex information-processing behaviors is one of the most exciting and fundamental questions in modern biology.

Cells must perceive and respond to a wide range of signals

To get a better sense of the scope of the challenges faced by cellular signaling systems, we will briefly consider the types of inputs to which they must react. Some of these inputs, along with the types of responses that can be evoked, are illustrated in **Figure 1.2**.

The most basic inputs, those common both to free-living unicellular organisms and to the cells of multicellular organisms, are nutrients and other raw materials useful to the cell, along with various environmental stresses. In the case of nutrients, the cell is likely either to take advantage of a hospitable environment by staying in place, or to move to a place where resources are more plentiful. On the other hand, noxious physical or chemical stimuli may cause the cell to migrate away, or otherwise adapt to endure the hard times until conditions improve. A yeast cell that is starved for phosphate, for example, mounts a complex response in which phosphate utilization is minimized, phosphate transport is increased, and enzymes (known as phosphatases) are secreted by the cell to release phosphate from environmental sources.

In multicellular organisms, a whole new set of signals needs to be detected by each of the many cells that make up the various tissues and organs. For example, cells signal to nearby or adjacent cells to regulate the development and function of organs and tissues. This permits the cells to work together as an integrated unit, rather than a collection of individual cells, each marching to its own drummer. These localized signals include soluble signals that diffuse only a short distance, signals that are attached to the surface of the cell and thus can be detected only by cells they directly contact, and signals from the *extracellular matrix* that provides the physical framework to which the cells are attached.

Longer-range signals of various types are also secreted and transported throughout the organism, allowing spatially separated cells to act collectively. A familiar example is the *hormone* insulin, which is secreted by the pancreatic islet cells

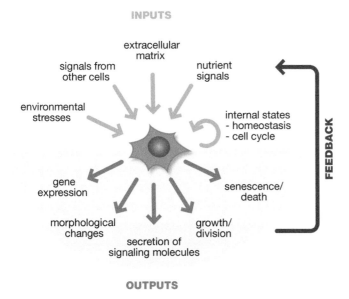

Figure 1.2

Cell signaling involves a diversity of inputs and outputs. Cells must read diverse external and intracellular signals, and use this information to generate many output responses. Some common inputs (*green arrows*) and outputs (*orange arrows*) are indicated. In many cases, the output of the system will change the response to future inputs (*feedback*).

and circulates through the bloodstream to regulate metabolic activity in tissues throughout the body. Other examples include epinephrine secreted by the adrenal gland, which induces an organism-wide and almost immediate "fight or flight" response, and sex hormones secreted by the gonads, which orchestrate the many physical changes in the body during puberty.

Finally, cells must continuously monitor their internal state to adjust to changing circumstances and respond to damage. The term **homeostasis** refers to the ability of living systems to adjust their behavior spontaneously to maintain a stable intracellular environment, despite varying environmental conditions. While this may sound like the antithesis of adaptability, it is really just another manifestation of the ability to detect changes in conditions and to modify cellular activities in concert with those changes. Typically, these homeostatic mechanisms involve *feedback*, which is the ability of the output from a system to adjust the incoming signal. For example, high levels of a certain cellular metabolite might cause the biosynthetic enzymes that generate the metabolite to shut down, or might close channels that regulate transport of the metabolite or its precursor into the cell. Feedback is an important component of almost all cell signaling systems.

Examples of the internal states that cells must monitor include physical or chemical damage (for example, to the genomic DNA) and the progression of cell division (the cell cycle). For example, cells that are preparing to go into **mitosis** (the process by which the cell, including its genomic DNA, is divided into two daughter cells) must ensure that DNA replication is completed before the physical events of mitosis are initiated.

The ways in which the cell can respond to changes in its environment (that is, the inputs discussed above) are equally varied. Change in gene expression or **transcription**, the production of mRNAs that will be translated into proteins, represents one of the most important long-term responses. Transcriptional changes alter the behavior of the cell in a fundamental way, because a cell's properties are almost entirely dictated by its protein constituents. Other important responses include whether the cell will grow in size or undergo cell division, or whether it will permanently stop proliferating or even undergo cell death. More short-term changes include alterations in cell shape or motility, or in the secretion of signaling molecules that can communicate with other nearby or distant cells. Of course, such changes may in turn serve as inputs for other cells in a multicellular organism or community.

Signaling systems need to solve a number of common problems

Cellular signaling systems must be able to overcome a number of challenges in order to reliably process information and mount an appropriate response. Some of these are listed in **Table 1.1**. While our understanding of the details of how this is accomplished remains far from complete, in this book we will discuss a number of solutions to each of these problems. Here, we briefly provide a sense of the magnitude and complexity of just a few of the challenges that must be overcome in cell signaling.

The issue of *specificity* is of particular importance considering the intracellular environment where cellular information processing must occur. The **cytosol** (and the extracellular environment for most of the cells in multicellular organisms) is a highly concentrated solution of organic molecules, in which the molar concentrations of individual components range over many orders of magnitude. Being able to pick out a specific signal in such a densely packed environment is a formidable challenge. The sheer variety of different molecules that are present is enormous: the protein products of many thousands of genes, along with the numerous variants of each that might be generated by alternative mRNA splicing or by various types

Challenges
Molecules must be able to sense the presence of stimuli, both inside and outside of the cell
This sensed information must be stored and transmitted, so that it can eventually alter the core physiological processes in the cell, such as gene expression, cell morphology, and metabolism
Information must be amplified so that small inputs can yield major changes in output, while random fluctuations (noise) are filtered out
Information must be integrated, so that outputs can be shaped by multiple inputs
Output information must be transmitted and processed to allow feedback control of behavior
Information must be transmitted through the impermeable plasma membrane that surrounds the cell
Information processing must be coordinated to give responses that are correctly shaped over both time and space
Signaling responses must be robust to a variety of conditions, such as changing ambient conditions or variations in the concentration of molecular components
The cell must achieve specificity in signaling, despite the many behaviors it must simultaneously coordinate and the vast number and variety of molecules diffusing about in the cell
Evolution must be able to give rise to new response behaviors and optimize/tune existing ones

Table 1.1

Fundamental challenges in cellular information processing

of post-translational modification and processing, plus numerous lipids, carbohydrates, nucleic acids, and simpler organic molecules. Yet cellular signaling systems can easily discriminate between two biomolecules that differ from each other in the minutest of details—a kink in a polypeptide chain, the addition of a phosphate group—and reliably detect slight differences in their concentration.

Within this complex molecular milieu, cellular signaling systems must also be able to respond strongly to the faintest of incoming signals. Sometimes just one or a few input molecules, such as of a hormone, are sufficient to induce wholesale changes in the cell receiving the signal, tipping the balance between life and death, or between quiescence and a round of cell division. This ability to amplify an incoming signal (signal amplification) is even more remarkable when one considers that the system must at the same time be resistant to various sorts of background noise, such as the random fluctuations in the conformation, activity, and local concentration of cellular components as they jostle about in the cell.

Another challenge faced by cell signaling systems is **signal integration**. A typical cell is simultaneously subjected to an enormous variety of different inputs, and often the response depends not on just one but many of these signals. For example, a cell may make the decision to proliferate only when nutrients are plentiful and specific pro-proliferative signaling molecules are present, when the cell is firmly attached to a particular type of extracellular matrix, when cell volume is sufficiently large to support two daughter cells, and when the genomic DNA is intact. Each of these individual conditions may be necessary but not sufficient on its own to switch on the proliferation program. Thus, cellular information processing is much more complicated than the simple conversion of one kind of input into another kind of output.

Finally, cells must modulate their output over time in various ways. For example, it is often useful for a cell to respond to a constant stimulus with an initial burst of output activity, and then to turn off the response. This down-regulation is a specialized kind of feedback, in which output from the system suppresses future output. More complex kinds of output modulation allow a system to adjust its sensitivity to respond to widely different levels of input (*adaptation*), or to generate waves or oscillations of output activity.

THE FUNDAMENTAL ROLE OF SIGNALING IN BIOLOGICAL PROCESSES

Our current understanding of signaling mechanisms is the result of many years of research in seemingly unrelated fields, using an array of different experimental approaches. It is really only recently, with the remarkable advances in our ability to identify, clone, and sequence key genes involved in a process, that we have discovered that the same or closely related types of communication molecules are utilized in a wide range of physiological information-processing functions. This synthesis, which has led to the emergence of the field of cell signaling, is one of the major scientific accomplishments of the last few decades.

Work in many different fields converged to reveal the underlying mechanisms of signaling

The field of cell signaling emerged from a number of disciplines that have historically been considered distinct (**Figure 1.3**). Indeed, the diversity of areas of inquiry that ultimately led to the field of cell signaling underlines the central role of signaling across biology. For example, because signaling is so important to normal physiology, the disruption or misregulation of signaling mechanisms is the basis for many human diseases, and thus these mechanisms are of interest in the areas of medicine and human health. Similarly, because normal development depends on the precise coordination of cell behaviors such as differentiation and movement, research on developmental events necessarily sheds light on the underlying signaling mechanisms. And because the signaling apparatus is comprised of biomolecules such as proteins, which are encoded by the genetic material, signaling mechanisms are amenable to the experimental approaches and analytic tools of biochemistry and genetics.

Cancer biology played a particularly important role in the emergence of the field of cell signaling. Our understanding of the molecular basis of cancer was revolutionized by the discovery of **oncogenes**, genes that when mutated or overexpressed

Figure 1.3

Many different fields of research contributed to the current understanding of cell signaling. Research in a wide variety of different disciplines revealed a common set of mechanisms and pathways that provide the basis for diverse biological activities.

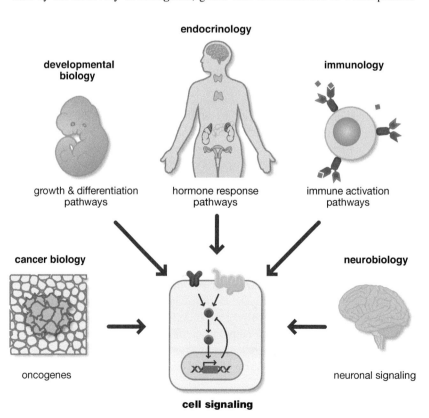

induce cells to respond inappropriately to normal signals and therefore proliferate uncontrollably, potentially leading to the formation of a malignant tumor (cancer). These oncogenes, once cloned and biochemically characterized, were in most cases found to be constitutively active or otherwise misregulated forms of signaling proteins. By understanding how signaling mistakes could lead to cancer, we learned a great deal about the normal signaling mechanisms that regulate cell proliferation and differentiation.

The field of endocrinology focuses on how hormones secreted into the blood, such as insulin, coordinate physiological communication between the different organs and glands that comprise an organism. As researchers delved into the biochemical basis by which target cells respond to hormones, a similar group of signaling molecules to those found in cancer biology was uncovered. For example, the receptor for insulin, because of its enormous significance to the common human disease diabetes, was one of the first hormone receptors to be cloned. The fact that it was highly related to some of the oncogenic proteins that cause cancer, and had the same biochemical activity (the ability to phosphorylate tyrosine residues on its target substrates), underscored the commonality of signaling mechanisms.

In the field of developmental biology, powerful genetic methods were used to identify mutations that disrupt the patterns of distinct cell fates that emerge within a multicellular organism. Model organisms such as the fruit fly *Drosophila melanogaster* and the roundworm *Caenorhabditis elegans* were screened for genetic mutations that affected normal development. As in cancer research, identifying genes whose mutation disrupted normal physiology was a key tool for understanding the normal signaling mechanisms. Embryologists using other approaches, such as manipulating frog or chick embryos, also zeroed in on regulatory molecules involved in cell fate determination and differentiation. As the genes that regulated development were identified and characterized, many were found to be similar or identical to signaling molecules implicated in oncogenesis.

Important contributions to our understanding of signaling were also provided by other specialized fields, such as neurobiology and immunology. The nervous system and the immune system are both clearly involved in information processing at the organismal and physiological levels. Research in these fields demonstrated that the development and proper function of these systems involve, at the molecular level, similar sets of signaling molecules and modules. Thus, despite the fact that one system is embodied by neurons that communicate to process higher-order cognitive function, and the other is embodied by immune cells that communicate to detect and mobilize a response to foreign invading pathogens, at the molecular level they share many common types of molecular parts and molecular network architectures.

Despite the diversity of signaling pathways and mechanisms, fundamental commonalities have emerged

In the years around 1990, studies in *Drosophila* and *C. elegans* converged with ongoing work in cancer biology to identify a key signaling pathway used in many different cellular contexts (**Figure 1.4**). In *Drosophila*, developmental biologists had used genetic screens to identify a number of genes important for the development of a specific retinal cell, the R7 photoreceptor. This particular cell was a good choice because it was not essential for viability, simplifying the design of genetic screens, and its presence or absence could be detected relatively easily in the compound eye of the fly. These genetic studies began to sketch out the signaling pathway that specified the R7 cell fate. This pathway began with a cell-surface receptor with tyrosine kinase activity (a *receptor tyrosine kinase*), led to a *G protein* closely related to *Ras*, which was already known as an oncogene in vertebrates, and then moved on to a cascade of three serine/threonine protein kinases (termed the *MAP*

Figure 1.4

A common signaling pathway in three different cell systems. Research on the misregulated signaling that drives cancer in humans (left), on the determination of retinal cell fate in the fruit fly (middle), and on the determination of vulval cell fate in roundworms (right) converged on the same signaling pathway. The names of the individual genes or gene products are given for each organism, with the function of each indicated on the far left. In humans, the epidermal growth factor receptor (EGFR), Ras, and Raf have all been shown to function as oncogenes (*pink*). GEF guanine nucleotide exchange factor; MAPK, mitogen-activated protein kinase; MAPKK, MAPK kinase; MAPKKK, MAPK kinase kinase.

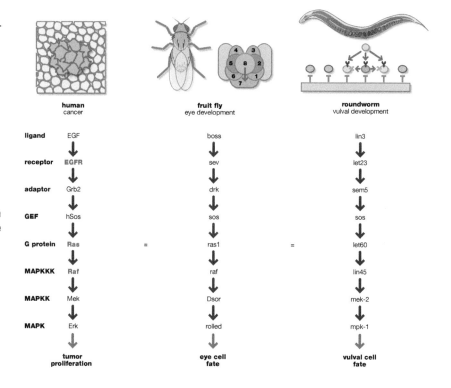

	human cancer		fruit fly eye development		roundworm vulval development
ligand	EGF		boss		lin3
receptor	EGFR		sev		let23
adaptor	Grb2		drk		sem5
GEF	hSos		sos		sos
G protein	Ras	=	ras1	=	let60
MAPKKK	Raf		raf		lin45
MAPKK	Mek		Dsor		mek-2
MAPK	Erk		rolled		mpk-1
	tumor proliferation		eye cell fate		vulval cell fate

kinase cascade). The vertebrate homolog of one of these kinases, Raf, was also a known oncogene. At the same time, investigators were using similar genetic approaches in *C. elegans* to identify the genes involved in specifying the cells that would develop into the vulva. Again, this particular system offered advantages for genetic screens, as defects were easily visualized and were not lethal. These experiments identified components of a signaling pathway very similar to that found in *Drosophila*, involving a receptor tyrosine kinase, a Ras-like G protein, and the MAP kinase cascade.

A key question in the field then became how the receptor might activate Ras, as was predicted by genetic experiments. In flies, a protein called Sos was discovered that seemed to function downstream of the receptor and upstream of Ras. Furthermore, it had homology to a protein that was thought to be a direct G protein activator in yeast. Meanwhile, in *C. elegans*, genetic screens uncovered another gene, *Sem-5*, that also seemed to act between the receptor and Ras. The Sem-5 protein was unrelated to any known catalytic domains, but did contain small regions of sequence similarity to a number of other signaling proteins including several vertebrate oncogenes.

Biochemical studies were then able to flesh out the physical and mechanistic details of the interactions that were implied by the genetic relationships. Sos was found to activate Ras by causing it to release the nucleotide GDP and bind GTP instead (Sos is an enzyme referred to as a *guanine nucleotide exchange factor* or GEF). The GTP-bound state of Ras that was generated was found to bind to and activate Raf, the first kinase in the MAP kinase cascade. Sem-5 (and its *Drosophila* and human homologs, Drk and Grb2, respectively) was found to function as an *adaptor*, binding to the activated receptor tyrosine kinase and bringing along Sos, which then activated Ras. Thus, a complete step-by-step pathway from external ligand to receptor to Ras and the MAP kinase pathway was elucidated, the end result of which was MAP kinase–mediated phosphorylation of nuclear transcription factors and changes in the genes transcribed in the cell. Remarkably, this same basic signaling framework was used to determine cell fate in fruit flies and roundworms, and to stimulate proliferation of human cells.

Mapping of this pathway was a beautiful convergence of developmental genetics and cancer biology. This convergence has led to a new appreciation of signaling, one that emphasizes common elements and strategies that are used in many different systems. Signaling is no longer simply a description of one specific physiological pathway (or set of pathways), nor is it associated with only one particular class of physiological processes. Rather, it can be viewed from the perspective of a more general set of information-processing tasks and solutions. In this book, we emphasize the common features of signaling mechanisms: how the same kinds of strategies and molecular components are used over and over to solve similar biological problems.

Signaling must operate at multiple scales in space and time

We can roughly divide signaling pathways into the events that occur mostly on the plasma membrane and in the cytosol, and the nuclear events that lead to increased or decreased transcription of specific genes (Figure 1.5). Changes in gene expression are the end result of many signaling pathways. This makes sense, as stable, wholesale changes in cell behavior are likely to require the synthesis of new sets of proteins that allow the cell to adapt to its changing environment. Those new proteins can only be generated by changes in the transcription of the messenger RNAs (mRNAs) that encode them. These transcriptional responses, however, are rather different in character than the information transfer and processing events that cause them. Here, we consider some of these differences in terms of timing, spatial control, and energetic cost to the cell.

Let us first consider the consequences of changes in gene expression. The process of transcribing a message into mRNA and translating it into protein is relatively slow, occurring on the time scale of minutes to hours. The effects are slower still, as it takes considerable time for new gene products to accumulate and for existing proteins to disappear. Furthermore, changes in transcriptional patterns are often highly stable, as when a cell terminally differentiates into a specialized cell type such as a muscle cell or neuron. In most cases, the effects of transcription are cell-wide, not limited spatially to specific sites in the cell. Finally, transcriptional changes require significant investment of cell resources, both in energy and in raw materials, to transcribe new mRNAs and to translate new proteins. Such investments are not to be undertaken lightly by a cell.

By contrast, the events that precipitate transcriptional changes are usually faster, more spatially organized, and less costly in terms of energy expenditure. First, unlike transcription, which is limited to the nucleus where the genomic DNA resides, the cell signaling machinery spans the cell from the external surface of the cell membrane into the nucleus. Indeed, one of the major tasks of many signaling systems is to transduce signals from outside the cell to the interior, where they can modulate intracellular processes such as transcription. A second major difference is in the speed of the response. Many cell signaling events are extremely rapid,

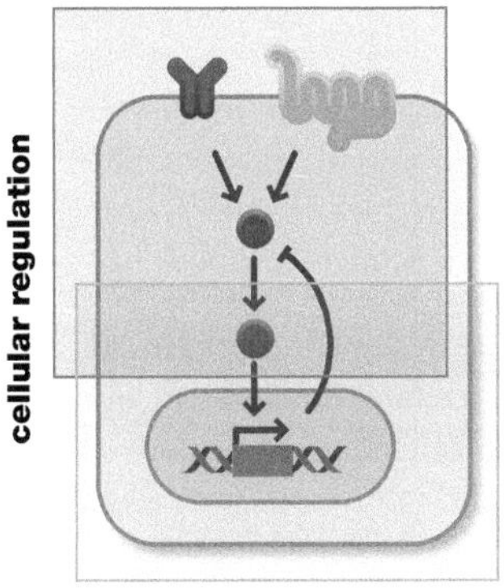

cell signaling

- enables transmission from outside of cell to nucleus
- fast ON and OFF (seconds to minutes)
- transient changes (minutes to hours)
- spatial/directional responses & organization
- energetically cheap (no protein synthesis)

gene expression

- slow ON and OFF (minutes to hours)
- stable changes (days to years)
- limited spatial responses
- energetically costly (transcription & translation)

Figure 1.5

Cellular regulation integrates signaling and gene expression networks. A typical cell signaling system is illustrated, in which binding of ligands to cell-surface receptors ultimately leads to changes in gene expression in the nucleus. Key properties of the cell signaling and gene expressions systems are compared to highlight important differences.

capable of switching states in fractions of a second, and the changes induced are often transient and reversible. Multiple time scales are often involved. For example, activation of a receptor may occur virtually instantaneously upon binding of its ligand; the activated receptor may then lead to the activation of a second intracellular enzyme, which may occur within a few seconds. This second enzyme may remain active for a few minutes, modifying downstream effectors, which may include transcription factors (there can be many intermediate steps, depending on the particular pathway). The modified transcription factors then become active in inducing or repressing transcription of certain genes, ultimately changing protein constituents of the cell at even longer time scales.

Another difference between cell signaling mechanisms and gene expression is that signaling can be spatially organized and localized. Again, this organization can occur over different scales, from the distance of a few molecules to the diameter of the cell and beyond. For example, certain receptors can cluster together when activated and dramatically change the protein and lipid composition in the immediate vicinity of the cluster, compared to the rest of the membrane. On a larger scale, one end of a cell may respond to an environmental cue by extending actin-rich protrusions (lamellipodia) that move the front edge of the cell forward, while at the other end, the cytosol retracts to push the cell body ahead, leading to directed motion. Overall, the complex morphologies that we observe in cells are usually organized or directed by signaling proteins and circuits that regulate components of the cytoskeleton.

Finally, most signaling transactions do not expend anywhere near the resources needed to make new proteins, usually requiring the equivalent of a molecule or two of ATP to transmit information from one molecule to another. Because signaling reactions are relatively cheap, it is not a serious disadvantage for the cell to continuously monitor the environment. In this book, we concentrate on the relatively fast, spatially organized, and cheap mechanisms that allow a cell to respond nimbly to environmental cues. We will devote much less attention to the ultimate long-term consequences of such signaling, such as gene expression.

THE MOLECULAR CURRENCIES OF INFORMATION PROCESSING

How are molecules in the cell used to store and transmit information? At its most fundamental level, the transfer of information requires some sort of change in state in the components of the signaling apparatus. In the cell, which is limited in the molecular raw materials from which the signaling apparatus can be built, a relatively small number of state changes are used over and over for signaling. In this section, we will consider these basic "currencies" of signal transduction, and how they are often combined in signaling mechanisms.

Information is transferred by changes in the state of proteins

We can think of information transfer as a series of switches or *nodes* that can change state in response to input signals; when a node changes state, a signaling output is generated (**Figure 1.6a**). When multiple nodes are linked together, information can be processed in a sophisticated fashion. Indeed, this kind of simple architecture provides the basis for electronic information-processing devices like a computer, which may contain millions of transistor-based switches linked together.

The idea that some kind of change is essential for information processing cannot be emphasized too strongly. We intuitively understand this concept from our own experience. For example, we notice and react to changes in sound—a loud noise breaking the silence, or a change in pitch and rhythm—whereas a sound of a constant volume and frequency rapidly fades into the background. Similarly,

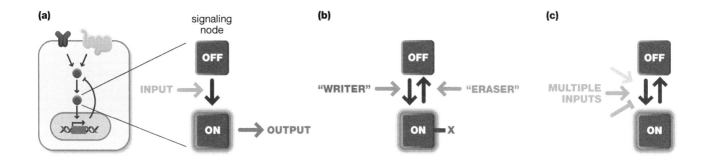

a constant radio signal has little information content, but if the amplitude of the radio waves changes over time, or their frequency, then an enormous amount of information can be conveyed (these correspond to amplitude modulation, or AM, and frequency modulation, or FM, on the radio dial).

What is the molecular basis for cellular information-processing machines? This is where cellular systems diverge significantly from the example of electronic circuits. In electronic circuits, there is one universal currency for information transmission, which is the electrons flowing through the circuit. There are relatively few devices that process information, and their inputs and outputs all involve this single currency. In cell signaling, there is a wider variety of signaling currencies, and therefore a wider variety of molecular devices or systems that are able to read one currency and convert it to other output currencies.

In the cell, proteins are the workhorses of signal processing. Proteins are enormously versatile in the types of physical structures that they can form and in the chemical reactions that they can perform. One class of proteins, the *enzymes*, act as catalysts that can enhance the rates of specific, useful biochemical reactions enormously, providing the basis for energy metabolism, replication, motility, and other behaviors associated with life. It is certainly then no surprise that proteins are essential for cell signaling mechanisms. Other biomolecules such as lipids, nucleic acids, and small molecules such as ions and nucleotides can play a supporting role, as we shall see, but for the most part, these molecules participate in signaling by changing the properties of proteins. Thus, if we want to understand the changes that underlie cell signaling, we must look to the properties of proteins.

In cell signaling, the most basic units of information are changes in the state of proteins. For example, a protein may have very different activities depending on whether or not it has a particular chemical modification (such as phosphorylation of a certain side chain). Perhaps the unmodified protein is inactive (off) and the phosphorylated form is active (on). The state of this protein is changed by input from other proteins that can add the phosphate group or remove it. In this example, we can think of the enzymes that add phosphate (termed *protein kinases*) as "writers," in that they add information in the form of post-translational marks, and the enzymes that remove those marks (*protein phosphatases* in this example) as "erasers" (**Figure 1.6b**). This kind of reversible change is essential to any kind of information-processing scheme. Furthermore, signaling proteins are often subject to regulation by multiple inputs, both positive and negative (**Figure 1.6c**). This allows the state or activity of the protein to depend in a relatively complex way on the specific combination of inputs at any particular time.

There is a limited number of ways in which the state of proteins can change

If information transfer requires change, and proteins lie at the heart of signaling, it follows that changes in the properties of proteins will provide the basis for cell signaling mechanisms. It turns out that the number of such changes that are used is really quite limited. These basic currencies of cell signaling are listed in

Figure 1.6

Changes in state link signal inputs to outputs. (a) A generic cell signaling system is illustrated on the left, and a single node of that system on the right. The node can be thought of as a switch that can exist in two states, OFF and ON. Signal inputs (*green arrow*) can switch the state, leading to a change in output (*orange arrow*). (b) In cases where the change of state is induced by a post-translational modification (*pink "X"*), enzymes that add or remove the modification are inputs that can be thought of as "writers" and "erasers." (c) Often, changes in state are affected by multiple different inputs, some of which promote switching (arrows) while others inhibit it (T-shaped arrow).

Table 1.2

The currencies of cell signaling

Currencies of cell signaling	Book chapter
Protein–protein interactions	Chapter 2
Conformation	Chapter 3
Enzymatic activity	Chapter 3
Post-translational modification	Chapter 4
Subcellular localization; concentration	Chapter 5
Small-molecule signaling mediators	Chapter 6

Figure 1.7

Different signaling currencies. Different ways that the state of proteins can be changed are illustrated. In all cases, the ON state (activation) is indicated by a *pink* halo.

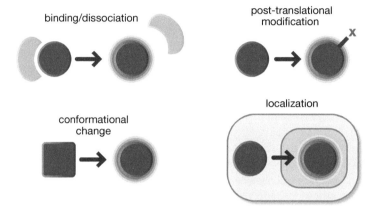

Table 1.2—and a subset is illustrated in **Figure 1.7**—and each is described briefly below. The next five chapters of this book deal with each of these currencies in greater detail.

Interactions between proteins, or between proteins and other biomolecules, can dramatically affect many aspects of their behavior, such as their activity and localization. The assembly or disassembly of multiprotein complexes is frequently a key step in transmitting cellular signals. Protein–protein interactions and their role in signaling are the topic of Chapter 2.

Proteins are not fixed in their three-dimensional shape or *conformation*; instead, they can often adopt multiple conformations that differ greatly in their activity. Protein conformation switches such as G proteins, which switch between active and inactive conformations depending on whether they are bound to GDP or GTP, are central to many signaling pathways.

Many signaling proteins are enzymes, which catalyze specific reactions such as the chemical modification of other proteins, lipids, or other biomolecules. The activation (or inactivation) of such enzymes by upstream signals can lead to widespread and massive downstream effects, due to the remarkable efficiency and specificity of enzymes as catalysts. Because changes in enzymatic activity are often tightly coupled to conformational changes, these two currencies and their interrelationship are discussed together in Chapter 3.

Proteins are subject to a number of different types of chemical modifications after they have been synthesized (these are collectively termed *post-translational modifications*). These modifications are performed by signaling enzymes, and include the addition or removal of small chemical groups such as phosphate, or of more substantial structures such as the small protein ubiquitin. *Proteolysis* (breaking the polypeptide backbone of the protein) is a particularly extreme form of post-translational modification. Chapter 4 examines the post-translational modifications used in signaling, how they are regulated, and the downstream consequences of modification.

Signaling also depends on the abundance (concentration) of signaling proteins. The rates at which biological reactions occur and the steady-state levels of their products are proportional to the concentrations of the reactants; the higher the concentration, the more likely two components are to interact with each other. The balance between the rates of synthesis and degradation of various components sets their overall concentration in the cell. Thus, a change in the concentration of a particular protein can itself represent a way to store information.

Even under conditions where the total amount of a molecule in the cell does not change, its *local concentration* can change dramatically upon changes in subcellular localization. Because the cell is not a well-mixed and homogeneous solution, changes in the localization of proteins within the cell can affect their activities. For example, the relocalization of a protein from the cytosol to the nucleus allows it to interact with the genomic DNA, which is entirely restricted to the nucleus. Similarly, many important targets of signaling enzymes are confined to the plasma membrane, and thus recruitment of proteins to the membrane can be a critical step in propagating signals. The regulation of subcellular localization and the role of changes of localization in signaling are discussed in Chapter 5.

A number of important signaling pathways involve changes in the abundance of small molecules such as calcium ions or cyclic AMP (cAMP). These small-molecule signaling mediators, often termed *second messengers*, are created and destroyed by upstream synthetic and degradative enzymes, and they exert their effects by binding to and altering the activities of downstream target proteins. Signaling that involves small-molecule mediators can have rather distinct properties due to the potential of these molecules to diffuse rapidly throughout the cell. Small signaling mediators are the topic of Chapter 6.

Most changes in state involve simultaneous changes in several different currencies

Although there is a limited number of ways in which proteins can change to transmit information, a change in one of these currencies is often intimately linked to change in one or more of the others. For example, phosphorylation of an enzyme (a change in post-translational modification) may induce a conformational change in the modified enzyme, which in turn changes its catalytic activity (**Figure 1.8**). The switch from one state to the other (inactive to active) involves changes in multiple properties that cannot easily be disentangled from each other. For the sake of simplicity, many of the following chapters discuss each type of change in isolation, but this way of organizing the book deliberately de-emphasizes the functional linkages and interrelationships among them.

To illustrate this point, we will consider a rather typical example of a switch between states that involves many different changes in protein properties. **Src family kinases** are enzymes that catalyze the phosphorylation of tyrosine side chains on target proteins in response to various extracellular signals. They play an important role in diverse physiological pathways, such as cell adhesion and lymphocyte activation. The namesake of the family is the Src oncogene, originally identified in a chicken tumor virus. Src was the first oncogene to be cloned and sequenced, and encodes the first tyrosine kinase to be identified, and so it holds an important place in the history of signal transduction research.

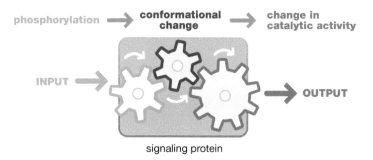

Figure 1.8

Different changes in protein state can be coupled. In this example, phosphorylation of a hypothetical protein is tightly coupled with conformational change and change in catalytic activity.

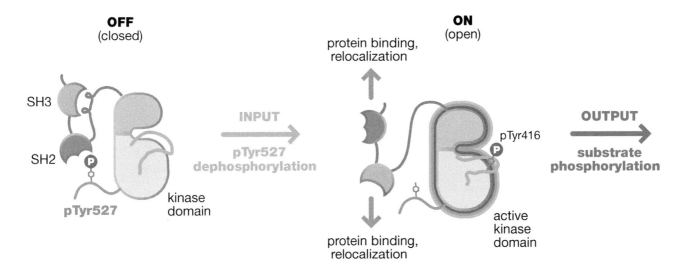

Figure 1.9

Inactive and active states of a Src family tyrosine kinase. In the inactive state (left), Tyr527 is phosphorylated (pTyr527) and the protein is folded in a compact, "closed" conformation via intramolecular interactions of the SH2 and SH3 domains. Upon activation (induced by dephosphorylation of Tyr527), the protein adopts a more open conformation in which catalytic activity is high, Tyr416 is phosphorylated, and SH2 and SH3 domains can bind to other proteins.

The two predominant states for Src family kinases are a "closed" catalytically inactive state that is tightly folded together, and a more "open" catalytically active state (**Figure 1.9**). The inactive state is stabilized by phosphorylation of a negative regulatory tyrosine at the C-terminus, tyrosine 527 (Tyr527). In the active state, Tyr527 is unphosphorylated, while a second site, tyrosine 416 (Tyr416), is phosphorylated. In this conformation, two protein-binding segments (the SH2 and SH3 domains) that had been involved in intramolecular interactions in the inactive state are now accessible to bind to other proteins. These domains are also important to localize the activated kinase to specific sites in the cell, for example focal adhesion complexes.

Let's go through in a little more detail the sequence of events leading from the "off" state to the "on" state. We will assume the initial input is dephosphorylation of Tyr527 by a phosphatase (a change in post-translational modification). The immediate consequence of dephosphorylation is to destabilize the closed conformation, which allows the SH2 and SH3 domains to dissociate from the catalytic domain (a change in conformation). This conformational change increases the activity of the catalytic domain (a change in enzymatic activity), which makes it more likely that the activating site, Tyr416, will become phosphorylated (a change in post-translational modification). In turn, Tyr416 phosphorylation stabilizes the active conformation of the catalytic domain. The SH2 and SH3 domains, released from their intramolecular interactions with the catalytic domain, are now free to bind to other cellular proteins (a change in protein–protein interactions). These interactions result in localization of Src to specific sites in the cell where its substrates are found (a change in subcellular localization).

In this rather typical example, the switch from the inactive to active state involves the concerted change of at least six distinct yet interrelated properties of the protein. Note also that such a regulatory scheme can accommodate multiple inputs that could impact the final state, including the levels of kinases that phosphorylate Tyr527, of phosphatases that dephosphorylate Tyr527, and of kinases and phosphatases that act on Tyr416, and the local concentrations of binding partners for the SH2 and SH3 domains.

LINKING SIGNALING NODES INTO PATHWAYS AND NETWORKS

In the previous section, we discussed how changes in the state of proteins provide the fundamental currencies of cellular signal processing. However, most signal transduction tasks do not involve a change in state of a single protein or node, but instead involve many different changes linked together in long chains (pathways) or interconnected networks. The increased complexity afforded by linking together individual signaling nodes is what permits more sophisticated and complex information processing.

Information transfer involves linking different changes of state together

In general, information transfer involves the conversion of one type of change into another type. We have already seen examples of this within a single signaling protein, as above, where dephosphorylation of the regulatory site on Src led to conformational changes, which in turn led to increased catalytic activity. But this is true in an even broader sense, in that virtually all signaling inputs must be converted into another form, often many times, in order to generate the appropriate cellular response. It is this concept of conversion of one type of signal into another that led to the widespread use of the term "signal transduction" to describe the field of cell signaling.

In addition to the conversion of different currencies within the same signaling protein, multiple proteins are often functionally linked together by their ability to change each other's state (**Figure 1.10**). For example, activation of an enzyme is linked to the post-translational modification of its substrate; modification of a protein may be linked to its ability to bind to a second protein; relocalization of an enzyme may be linked to modification of a substrate that is present only in the new location; conformational change in one protein may be linked to its dissociation from a second protein; and so on. These physical or functional linkages *between* proteins provide a way for the output of one signaling node to serve as the input for another. Much of the actual research in signal transduction has involved the discovery of such connections between signaling proteins and the demonstration of their functional importance to particular cell behaviors.

Multiple state changes are linked together to generate pathways and networks

Links between individual signaling nodes are generally assembled together into much larger architectures—pathways and networks. The term **pathway** is typically used to describe a linear chain of interactions where the output of each node serves as the input for the next downstream node. A specific example was provided above in the case of the signaling from the EGF receptor through Ras, Raf, and the MAP kinase cascade (see **Figure 1.4**). Like a row of falling dominoes, a long series of changes of state are linked together so that the initial input (high

transmission between signaling proteins

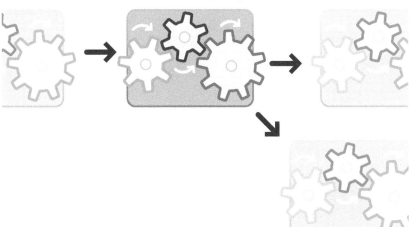

Figure 1.10

Linking signaling nodes together. Change in one signaling node can be induced by input from upstream nodes, and its output can, in turn, induce changes in downstream nodes. In this way, multiple nodes can be linked together to form pathways and networks.

concentrations of the activating ligand for the EGF receptor in this case) is tightly linked to the ultimate output (phosphorylation of substrates such as transcription factors by the MAP kinase Erk). Thus, a pathway defines the chain of interconnections that link a specific input to a specific output.

In fact, in most cases, such a linear pathway is a gross oversimplification of the actual relationships between the individual proteins involved. This is because at each step, it is more common for each node to receive inputs from multiple nodes, and for its output to affect the activity of multiple nodes. We have already discussed above how the Src family kinases can receive multiple activating and inactivating inputs. The same example also illustrates how the output of one node can provide inputs into multiple downstream nodes, as Src kinases can phosphorylate diverse substrates, serving to activate, inhibit, or otherwise modulate the activity of many different proteins in the cell.

It is this branched **network** of functional interactions among signaling proteins that allows more complicated information processing beyond the simple cause-and-effect relationships implied by a linear pathway. These interrelationships can generate behaviors such as feedback regulation, integration, adaptation, and complex spatiotemporally controlled output patterns. In Chapter 11, we will discuss in much more detail how these systematic behaviors can be built by linking together simple signaling nodes in various ways.

Cellular information-processing systems have a hierarchical architecture

In this chapter, we have discussed cellular information processing on a variety of different scales—from changes in the state of individual signaling proteins, all the way up to communication between distant organs in a multicellular animal. While it may seem at first glance that these mechanisms of information transfer are very different at these different scales, if we consider more closely, we can discern a common logic that allows ever-more complex systems to be built out of relatively simple components and rules.

This idea is illustrated in **Figure 1.11**. At the smallest scale, individual proteins act as molecular signaling devices, receiving inputs from the environment or upstream components and changing state in response, thereby generating an output signal. A Src family kinase is such a molecular signaling device. At a larger scale, a series of molecular devices are linked together into a network signaling device, in which the interrelationships can generate more complex and sophisticated information-processing behaviors. The EGF receptor–Ras–MAP-kinase pathway is an example

Figure 1.11

Signaling machines at different scales.
Cell signaling involves information-processing devices that operate at several scales, from individual proteins, to networks, to whole cells. Molecular signaling devices are linked together to make network signaling devices, which, in turn, are linked to regulate behaviors at the cellular level. These devices all share a common set of generic signal-processing tasks that they must accomplish, by receiving and integrating inputs and converting these to outputs.

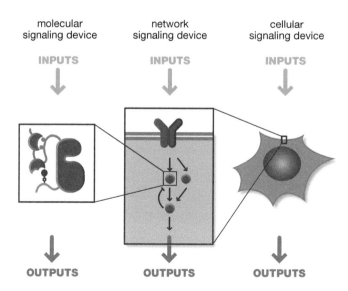

of such a network signaling device. Finally, at the scale of the cell, the outputs of various network signaling devices are harnessed together into a larger network that regulates complex cellular behaviors such as proliferation, differentiation, and directed movement.

We have organized most of this book around core molecular modules and mechanisms, and focused on explaining how they are commonly used to build generic information-processing systems within cells. This modular and conceptual treatment of signaling emphasizes how certain molecular components cannot be viewed as uniquely belonging to any one specific process; rather, in most cases, they have been incorporated in diverse processes through evolution because they are well suited to perform a particular class of information-processing tasks. Moreover, this view allows us to see the forest from the trees and to understand the logic of the higher-order architecture and organization of signaling systems.

While this approach highlights the core fundamentals underlying cell signaling, it poses a danger that one might also lose touch with the broader physiological context in which a particular signaling mechanism operates. To balance this, we have tried to organize this book in a hierarchical way that matches the hierarchical organization of signaling systems: to focus many of the earlier chapters on fundamental molecular mechanisms for transmitting information, but to transition in later chapters to how these molecular systems are used in a hierarchical manner to build networks that solve larger cellular and physiological problems.

SUMMARY

Cell signaling encompasses the ability of cells to detect changes in the environment and respond to those changes—a fundamental property of living systems. Our current understanding of cell signaling has resulted from a convergence of research in a variety of fields including cancer biology, developmental biology, endocrinology, biochemistry, and genetics. It is now apparent that common strategies and cellular components are used over and over to solve various cell signaling tasks. To perform this information processing, cells use genetically encoded biomolecules, predominantly proteins. Switching between protein states is the most fundamental building block for cell signaling mechanisms, and the number of ways that the state of proteins can be changed as a result of signaling inputs—the currencies of signaling—is quite limited. Such changes of state are often linked together, either within the same molecule (a molecular signaling device) or between different proteins (to generate signaling pathways and networks). By understanding how different signaling proteins are functionally linked together, we can understand the behavior of the signaling pathway or network as a whole.

QUESTIONS

1. Imagine that you are a primitive single-cell organism. What type of new signaling response could provide you with a potential evolutionary advantage over the other competitor organisms in the primordial soup? When might the same signaling response provide a fitness disadvantage? What general principles govern whether a new input–output response would be advantageous?

2. What are some of the common signals and stresses that must be detected by both single-cell organisms and individual cells that are part of multicellular organisms? What are some of the different signals that they must detect? What types of output response might be unique to cells from multicellular organisms?

3. Multicellular organisms often use complex organ systems to sense and process external information. For example, an animal will use its visual and nervous systems to detect many external stimuli. What types of inputs, decisions, and

Answers to these questions can be found online at www.routledge.com/9780367279370

outputs must be made by the individual cells that make up the visual and nervous systems?

4. Why are signaling proteins involved in cancer (such as oncogene products) also often found to be involved in the process of development?

5. What is the relationship between signaling and transcriptional regulation? What are the commonalities and differences between these two systems? What types of cellular response can only be generated by signaling systems?

6. Does phosphorylation of a signaling protein by an upstream kinase represent a signaling input or output? Explain.

7. How can the localization of a signaling protein be considered to be a currency for storing information? Describe how inputs can modify localization. Describe how localization can change downstream physiological outputs.

8. What are the other molecular currencies that signaling proteins can use to store and transmit information? How can upstream inputs alter these currencies? And how can changes in these currencies be converted into downstream responses?

BIBLIOGRAPHY

Alon U (2019) *An Introduction to Systems Biology: Design Principles of Biological Circuits* (2nd ed.). New York: Chapman & Hall/CRC.

Cox AD & Der CJ (2010) Ras history: The saga continues. *Small GTPases* 1, 2–27.

Koshland Jr. DE (2002) Special essay. The seven pillars of life. *Science* 295, 2215–2216.

Lee MJ & Yaffe MB (2016) Protein regulation in signal transduction. *Cold Spring Harb Perspect Biol.* 8(6), a005918. doi: 10.1101/cshperspect.a005918.

Lim WA, Lee CM & Tang C (2013) Design principles of regulatory networks: Searching for the molecular algorithms of the cell. *Mol. Cell* 49, 202–212.

Margolis B & Skolnik EY (1994) Activation of Ras by receptor tyrosine kinases. *J. Am. Soc. Nephrol.* 5, 1288–1299.

Martin GS (2001) The hunting of the SRC. *Nat. Rev. Mol. Cell Biol.* 2, 467–475.

Sternberg PW (2006) Pathway to RAS. *Genetics* 172, 727–731.

Tyson JJ, Chen KC & Novak B (2003) Sniffers, buzzers, toggles and blinkers: Dynamics of regulatory and signaling pathways in the cell. *Curr. Opin. Cell Biol.* 15, 221–231.

Principles and Mechanisms of Protein Interactions

2

The proteins that participate in signaling, like the components of any complex machine, must interact with each other in order to perform their function. Furthermore, changes in these interactions between signaling components can convey information. In this chapter, we will consider non-covalent physical interactions between signaling proteins and their binding partners, or **ligands**.

The cytosol and other biological fluids are very highly concentrated solutions of proteins and other components, in which molecules are continuously jostling and colliding with each other. In some cases, these collisions lead to **binding**—the relatively stable association of the two components. The likelihood that a particular interaction will occur at various concentrations of components, and the dynamics of how rapidly such interactions form and dissociate, all play fundamental roles in defining the behavior of signaling mechanisms. Here, we will examine the principles that govern the physical interactions of signaling molecules, and how these interactions are used to transmit information.

Protein–protein interactions will dominate our discussion in this chapter, but it is important to realize that the interactions of proteins with other cell components are also important for signal transduction. In particular, the binding of proteins to specific lipids is important for localizing signaling proteins to defined membrane compartments or for modulating their activity. And, of course, the specific binding of proteins to nucleic acids underlies the regulation of processes such as DNA replication and the transcription and processing of mRNA. Most of the concepts that we will introduce in discussing protein–protein interactions also underlie other macromolecular binding interactions.

DOI: 10.1201/9780429298844-2

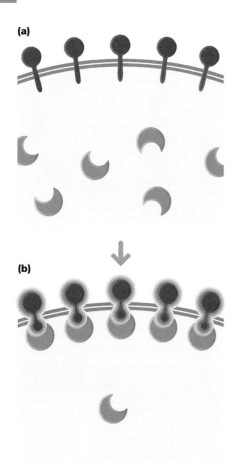

Figure 2.1

Binding can lead to relocation of proteins. (a) A transmembrane receptor (*brown*) resides on the surface of the cell, and its potential binding partner resides in the cytosol. In its inactivated state, the receptor cannot bind the cytosolic protein. (b) Upon activation, the receptor can now bind to its partner. Most of the previously cytosolic protein is now relocated to the membrane, in complex with activated receptor.

Scaffolds and MAP kinase cascades are discussed in Chapter 3.

PROPERTIES OF PROTEIN–PROTEIN INTERACTIONS

In this section, we will discuss the factors that determine whether or not two proteins will interact with each other in the cell. In particular, we will focus on two critical parameters of binding interactions, *affinity* and *specificity*, and address the quantitative description of binding interactions and their dynamic behavior.

Changes in protein binding have both direct and indirect functional consequences

A variety of signaling inputs can alter the binding of proteins with each other. These include changes in a protein's abundance or its distribution throughout the cell, changes in post-translational modifications such as phosphorylation, and changes in protein shape or *conformation*. In turn, protein–protein interactions can dramatically affect the behavior of the binding partners involved; this is why changes in binding are capable of transmitting information during signaling.

One of the most straightforward consequences of binding is the assembly of more complicated multifunctional structures from simpler component proteins. By combining the catalytic activity, binding activity, and other functions from several proteins, the resulting assemblies can perform more complex functions than the individual components can do. For example, many different inputs can be integrated into one output pathway, or one input signal can be diverted down many different output pathways, or different steps in a process can be coordinated.

Another common result of binding is to change the subcellular localization of one of the binding partners. For example, most cell-surface receptors undergo conformational changes or post-translational modification of their cytosolic domains upon activation. These changes create new binding sites for cytosolic proteins. Since the receptors are confined to the plasma membrane, by binding to the receptor, a previously soluble protein can be effectively localized to the membrane (**Figure 2.1**). Many enzymes, such as phospholipases and other lipid-modifying enzymes whose substrates reside in the membrane, are regulated by recruitment to the membrane in this way.

Another important consequence of binding is to bring enzymes into contact with potential substrates. For example, many protein kinases bind stably to their substrates via specific docking sites or substrate-binding domains. Such binding may be important to ensure that only a very specific substrate is phosphorylated, and that this phosphorylation is highly efficient. Such interactions may also be mediated by specialized proteins termed *scaffolds*. Scaffolds bind multiple proteins, such as enzymes and their substrates, involved in a single process. For example, in the *MAP kinase cascade*, a conserved signaling pathway that couples upstream signals to the phosphorylation of substrates such as transcription factors, a scaffold protein simultaneously binds three different protein kinases. These three kinases function in a cascade, in which they sequentially phosphorylate and thereby activate each other. The fact that they are all assembled into a single complex by the scaffold makes the overall reaction faster than if the three kinases needed to encounter each other separately through random collisions in solution. Furthermore, specificity and efficiency are increased because interaction with other competing substrates is prevented (**Figure 2.2**).

The interaction of proteins with one another can also affect their biological activities quite directly. For an enzyme, binding of another protein can induce changes in its shape *(allosteric* changes) that increase or decrease its catalytic activity or other properties (**Figure 2.3**). This type of regulated conformational change will be considered in more detail in **Chapter 3**. Even in cases where binding does not result in dramatic changes in the conformation of a protein, it may otherwise affect activity in a number of ways, for example, by blocking access to substrates or to other binding partners.

Protein binding can be mediated by broad interaction surfaces or by short, linear peptides

The basis for protein–protein binding (or the binding of proteins to other macromolecules) is a specific fit, or complementarity, between the two interacting surfaces. The shape of the two surfaces largely determines this fit. Like pieces of a puzzle, two surfaces are more likely to bind if they have complementary shapes, with bumps and ridges on one surface fitting into holes and grooves on the other (**Figure 2.4**). Binding is also stabilized by hydrophobic and electrostatic interactions and hydrogen bonds between the interacting surfaces. As one would intuitively expect from simple rules of chemistry, hydrophobic patches on one surface generally interact with hydrophobic patches on the other, negatively charged side chains are likely to interact with positively charged side chains (**Figure 2.4d**), and the formation of hydrogen bonds is favored.

One way to describe protein–protein interactions is by their **buried surface area**. This is the region that is exposed to solvent in the uncomplexed proteins, but is buried in the protein–protein interface in the complex (**Figure 2.4c**). Typical protein–protein interactions among signaling proteins have a buried surface area of ~1200–2000Å^2; in general, the greater the buried surface area, the stronger the interaction. In many cases, this binding surface is relatively hydrophobic, and thus binding to another protein is favored because it shields the hydrophobic region from the polar solvent. Individual amino acids do not contribute equally to binding; instead, some amino acids within the interface are absolutely critical for binding, while others can be altered by mutation without affecting binding.

Protein–protein interaction surfaces can be roughly divided into two categories. In the first, relatively broad surfaces of the two binding partners interact, with each binding surface composed of amino acids that may be widely separated in the linear sequence of the protein (see Figure 2.4). In the second, one of the binding surfaces consists of a relatively short, linear peptide sequence (usually four to eight amino acids in length) that fits into a corresponding groove on the surface of its binding partner (**Figure 2.5**). Both types of interactions are seen in signaling proteins, and there are important differences in their physical and evolutionary properties.

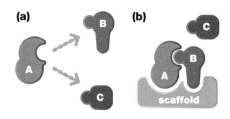

Figure 2.2

Protein binding can make reactions more efficient and specific. (a) Enzyme A and its two potential substrates B and C are free in solution; the reaction is relatively inefficient and targets both B and C. (b) A scaffold protein binds specifically to both enzyme A and substrate B, but not to substrate C. The enzymatic reaction is highly efficient and specific for substrate B.

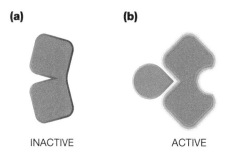

Figure 2.3

Binding can directly alter the activity of proteins. (a) In its unbound state, a protein adopts an inactive conformation. (b) Upon binding of a second protein (*orange*), changes in the conformation of the first protein result in its activation.

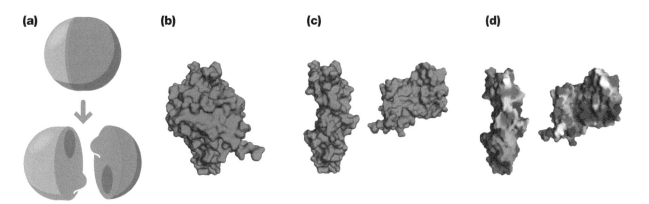

Figure 2.4

The nature of protein-binding interfaces. (a) A schematic representation of a protein–protein interaction, illustrating the complementary shapes of the interacting surfaces (*orange*). (b) Top, a surface representation of human growth hormone receptor (*green*) bound to growth hormone (*blue*), based on the X-ray crystal structure. Below, the binding surfaces of the receptor and growth factor are shown by peeling away the hormone from the receptor surface and rotating it 180°. The buried surface area of the interface is shaded *orange.* (c) The surface electrostatic potential of the two binding surfaces (*red*, acidic; *blue*, basic; *light gray*, hydrophobic); the orientation is the same as in the lower part of panel (b).

Modular binding domains are discussed in more detail in Chapter 10.

(a)

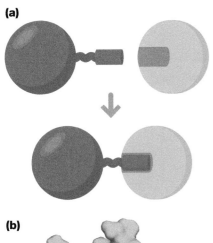

(b)

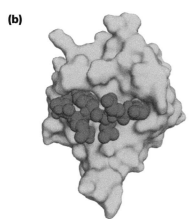

(c)

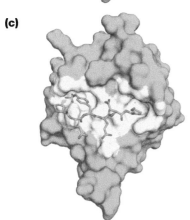

(d)

pTyr

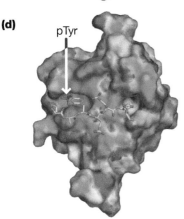

Peptide ligands are short and defined by only a few key amino acids, so such binding sites can be created or destroyed rather frequently by random mutation of protein-coding genes. By contrast, surface–surface interactions are less likely to arise by chance, because many amino acids spread over distant regions of different proteins play a part in the binding. Such interaction pairs are likely to evolve relatively slowly to fulfill a specific physiological function. Binding between two surfaces, because of the extensive contact area, is potentially stronger than surface–peptide binding.

Many of the protein interactions in signal transduction involve compact, structurally discrete modules or *domains*, whose sole known function is to confer on proteins containing them the ability to bind to specific types of ligands. Each type of modular protein-binding domain binds a characteristic linear peptide motif. Two specific examples are *SH3 domains*, most of which bind to proline-rich peptides that adopt a specific helical secondary structure, and *SH2 domains*, which bind to tyrosine-phosphorylated peptides. Examples of lipid-binding domains are the PH domains, many of which bind to specific phosphoinositol-derived lipids. Signaling proteins often contain more than one of these binding modules, each conferring specific binding properties. The presence of modular domains is often obvious from the protein sequence, and because such domains have relatively predictable binding partners, they provide important clues to the binding properties, and even the function, of the proteins that contain them. The presence of an SH2 domain in a protein, for example, strongly suggests that it functions in signaling pathways regulated by tyrosine phosphorylation.

The affinity and specificity of an interaction determine how likely it is to occur in the cell

Signaling proteins nearly all bind and interact with other molecules during the process of signal transduction. Thus it is critical to understand the likelihood that any particular signaling molecule will interact with another under specific conditions. We shall now look at the qualitative and quantitative aspects of binding. First, we will consider affinity and specificity, two important parameters of binding interactions. The **affinity** of an interaction is a measure of the intrinsic strength of the binding interaction. Simply put, if the affinity of the interaction between two species is high, then there is a higher probability of finding the molecules in complex at equilibrium.

In contrast to affinity, the **specificity** of an interaction is not an absolute measure but a relative one—it reflects the relative affinity of a particular interaction (between protein P and ligand A) with respect to other possible interactions (between protein P and all other molecules in the cell). Unlike affinity, which is an intrinsic property of two interacting molecules, specificity can only be defined in relation to a particular set of other potential interactions.

Because the cellular environment in which signaling occurs is both extremely concentrated (roughly 20% protein by weight) and diverse (containing many thousands

Figure 2.5

A peptide-protein interaction. (a) A schematic representation of a short, linear peptide (*pink*) binding to a corresponding groove on the surface of its binding partner (*green*). For clarity, the peptide is shown at the end of a protein; in reality, most binding peptides are located at internal sites, in loops or relatively unstructured regions of proteins. (b) A surface representation of an SH2 domain (*green*) binding to a tyrosine-phosphorylated peptide (*pink*). (c) The area (*yellow*) on the SH2 domain surface is buried by the bound peptide. The peptide is depicted in backbone format. (d) the electrostatic potential of the SH2 domain surface (*light pink*, acidic; *blue*, basic; *light gray*, hydrophobic). Note the concentrated positive charge (*blue*) in the area that binds the negatively charged phosphotyrosine (pTyr; indicated by arrow).

of distinct molecular species, varying in concentration over many orders of magnitude), any potential binding reaction is in constant competition with an enormous number of alternative interactions. Despite this competition, if an interaction is highly specific, it can predominate because it has a higher affinity compared with those of competing interactions. The specificity of a particular interaction can vary, however, depending on the exact intracellular context.

The relative nature of specificity is underscored by the fact that a single interaction between two molecules can be accurately described as either highly specific or nonspecific, depending on the context. For example, antibodies that bind to phosphotyrosine (**Figure 2.6**) are widely used reagents in signaling research. Such antibodies are considered to be highly specific for tyrosine-phosphorylated proteins, binding with much higher affinity to these targets than to the corresponding unphosphorylated proteins or the larger set of proteins that contain tyrosine. Yet in another context, these antibodies can be considered to be highly nonspecific, in that they bind with similar affinities to virtually all tyrosine-phosphorylated proteins, independent of the protein sequence context surrounding the phosphorylated tyrosine (Figure 2.6). It is possible to generate more specific antibodies that bind a particular phosphorylated site in a protein; such antibodies have much higher affinity for that site compared to all other phosphotyrosine-containing sequences.

Another example of relative specificity is provided by the lectins, which are proteins that bind to the sugar groups that make up the complex carbohydrates found on cell-surface glycoproteins. A lectin may be highly specific for a particular sugar group, failing to bind appreciably to other, closely related compounds. If the particular sugar recognized is widely distributed on a large number of different proteins in the cell, however, the lectin could also be described as nonspecific, in the sense that it binds with similar affinity to many different glycoprotein targets. These examples show how specificity is a relative quantity that is dependent on the definition of other interactions that are considered as competitive.

The strength of a binding interaction is defined by the dissociation constant (K_d)

Biologically important binding reactions can vary widely in their strength, ranging from weak and transient interactions that may be difficult to detect to extremely strong interactions that are nearly as stable as covalent bonds. Thus binding is not an all-or-none phenomenon, and a quantitative measure of the strength of binding interactions is needed in order to estimate whether they are likely to occur in the cell. The most commonly used measure of affinity is the **dissociation constant (K_d)** of the interaction—the ratio of unbound and bound species observed at equilibrium.

The definition of the dissociation constant is based on the law of mass action, which describes the relationship between the concentrations of reactants and products in a chemical reaction. In the case of a binding reaction between two molecules, the reactants are the two individual binding partners, and the product is the bound complex. Thus for the reaction A + B ↔ AB, the rate of formation of AB is given by a rate constant (on-rate or k_{on}) times the concentration of the binding partners: $k_{on}[A][B]$. Similarly, the rate of dissociation of the AB complex is given by another rate constant (off-rate or k_{off}) times the concentration of the complex: $k_{off}[AB]$. At equilibrium, the rate of formation of the complex equals the rate of dissociation of the complex, so $k_{on}[A][B] = k_{off}[AB]$. Rearranging, we obtain $k_{off}/k_{on} = [A][B]/[AB]$. The equilibrium constant k_{off}/k_{on} is defined as the dissociation constant, or K_d.

$$K_d = \frac{k_{off}}{k_{on}} = \frac{[A][B]}{[AB]} \qquad (2.1)$$

Phospho-specific antibodies are discussed in Chapter 15.

(a) context 1: all proteins

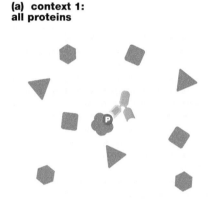

SPECIFIC

(b) context 2: all phosphotyrosine-containing proteins

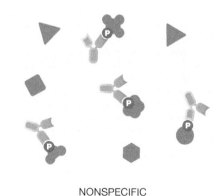

NONSPECIFIC

Figure 2.6

Specificity is a relative quantity that depends on context. (a) An antibody (*blue*) that recognizes phosphotyrosine would be considered specific in its ability to bind the relatively rare phosphotyrosine residues (*pink circles*) among all proteins (context 1). (b) However, if one is trying to distinguish a specific phosphotyrosine-containing protein from all other phosphotyrosine-containing proteins (context 2), then the antibody would be considered nonspecific.

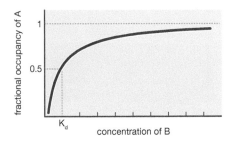

Figure 2.7

A binding isotherm. For a simple binding reaction A + B ↔ AB, the fractional occupancy of A is plotted as a function of the concentration of its binding partner B (under conditions where B is in excess, so unbound B is approximately equal to total B). The concentration of B where 50% of A is bound (fractional occupancy of A = 0.5) corresponds to the dissociation constant (K_d) for the binding reaction.

The dissociation constant has units of concentration (moles per liter, or M) and is quite useful because it tells us what fraction will be complexed at a given concentration of individual components. Consider a situation where there is a relatively small amount of A, so B is in large excess. When half of the total amount of A is complexed with B, [A] = [AB]. Then, according to Equation 2.1, under these conditions, K_d = [B]. Thus, the dissociation constant is the concentration of B at which half of A is free, and half is complexed with B. We can rearrange Equation 2.1 to obtain an equation that describes the fraction of the total amount of A that is complexed with B, or its **fractional occupancy**:

$$\text{Fractional occupancy of A} = \frac{[AB]}{[A]+[AB]} = \frac{[B]}{K_d+[B]} \tag{2.2}$$

If we plot the fractional occupancy of A as a function of [B] (again we assume that the total amount of A is low compared with that of B, so [B] is very nearly equal to the total concentration of B), we get a hyperbolic plot, often called a **binding isotherm** (**Figure 2.7**). It is obvious from this curve and Equation 2.2 that when the concentration of B is much higher than the K_d, almost all of A is complexed with B, while at concentrations of B much lower than the K_d, almost all of A is free (uncomplexed). Putting actual numbers into Equation 2.2, we see, for instance, that if [B] is nine times higher than the K_d, then 90% of A will be complexed with B; on the other hand, only 1% of A will be complexed when the concentration of B is 1/99 of the K_d. Another very useful way to think about fractional occupancy and the dissociation constant is in terms of binding probability. For any individual molecule of A, when the concentration of B is equal to the K_d, there is a 50% chance it will be bound to B; if the concentration of B is 1/99 the K_d, there is only a 1% chance that the molecule of A will be bound to B. The dissociation constants for some representative biological interactions are provided in **Table 2.1**. Note that because the dissociation constant is an equilibrium constant, it does not tell us the rates of complex formation or dissociation; the kinetics of binding are discussed later in this chapter.

Table 2.1

Physiological dissociation constants

Dissociation constant	Biological interaction	Notes
10^{-15} M	Avidin–biotin	Extremely high-affinity protein–small molecule interaction
10^{-14} M	Thrombin–hirudin (leech anticoagulant peptide)	Hirudin blocks blood clotting by binding and inhibiting thrombin
10^{-11} M	Methotrexate–dihydrofolate reductase (DHFR); platelet-derived growth factor (PDGF)-PDGF receptor	Methotrexate is a small-molecule drug that inhibits DHFR
10^{-9} M	Restriction enzyme *Eco*RI-DNA site; catalytic and regulatory subunits of protein kinase A (PKA)	Restriction enzymes cleave specific sites in DNA
10^{-7} M	SH2 domain–phosphotyrosine (pTyr) site	
10^{-6} M	Ca²⁺–Ca²⁺-activated enzymes	
10^{-5} M	SH3 domain–binding site; ATP–kinase	Kinases use ATP as a substrate in phosphotransfer reactions
10^{-4} M	Glutathione–glutathione S-transferase (GST)	

Source: Courtesy of C.T. Walsh.

The dissociation constant is related to the binding energy of the interaction

The dissociation constant also has a thermodynamic meaning, because the relative amounts of free components and of the complex present at equilibrium are directly related to differences in the **free energy** of the two states. Simply put, if the complex has lower free energy than the individual components, formation of the complex will be favored at equilibrium; conversely, if the complex has higher free energy than the components, little complex will be present at equilibrium. For the binding reaction $A + B \leftrightarrow AB$, this thermodynamic relationship is given by the equation:

$$\Delta G^\circ = -RT\ln \frac{[AB]}{[A][B]} \tag{2.3}$$

where, ΔG° is the **standard free energy** for the binding reaction, R is the gas constant, and T is the temperature in degrees K.

The term ($[AB]/[A][B]$) is the equilibrium constant (K_{eq}) for the reaction. For a binding reaction, this is termed the association constant (K_a) and, as can be seen from Equation 2.1, it is equal to $1/K_d$. Thus we can write:

$$\Delta G^\circ = -RT\ln K_a = -RT\ln \left(\frac{1}{K_d} \right) = RT\ln K_d \tag{2.4}$$

If we put in values for the gas constant and standard temperature, and convert from the natural logarithm, then $\Delta G^\circ = 1364 \log K_d$ (in calories) or $5707 \log K_d$ (in joules). So for a biological interaction of moderate affinity (assume $K_d = 10^{-8}$ M), then ΔG° is roughly $-11,000$ cal/mol, or -11 kcal/mol (roughly -46 kJ/mol). The fact that ΔG° is negative for the binding reaction means that under standard conditions, binding is favored; at equilibrium, most of the components will exist in a complex. From Equation 2.4, we can also see that relatively small differences in the binding energy will necessarily cause large differences in K_d. For example, the favorable free-energy contribution of a hydrogen bond in a protein is typically 1–1.5 kcal/mol (~4–6 kJ/mol), depending on context; thus adding or subtracting a single hydrogen bond to a binding interface between two proteins has the potential to change the K_d by more than tenfold. This fundamental relationship between binding energy and the dissociation constant is the basis of biological specificity, where quite subtle changes in the surface of a protein (changes in amino acid side chains or in overall conformation) can lead to enormous differences in the likelihood of two proteins binding to each other.

Another important thermodynamic aspect of binding is that the free energy of binding, like any change in free energy, can be described in terms of two components: the change in the disorder or **entropy** (S) of the system, and the change in **enthalpy** (H), or the heat given off or absorbed upon binding:

$$\Delta G = \Delta H - T\Delta S \tag{2.5}$$

In this equation, ΔH is the change in enthalpy and ΔS is the change in the entropy of the system. Binding is favored (ΔG is negative) when it results in a decrease in enthalpy (heat is given off on binding) or an increase in entropy (the disorder of the system increases). In general, one might expect that entropy would decrease after binding, owing to the decreased disorder of the components. But it is the entropy of the entire system, including the solvent, that must be considered, and the effect of binding on the entropy of the many ordered water molecules that form a shell around the surface of the protein can be quite large. Thus, overall entropy can increase upon binding. In practice, some binding interactions are enthalpically driven, and others are entropically driven.

How are the enthalpy and entropy of binding relevant to signal transduction? Understanding the thermodynamic basis of binding is critical for the ability to

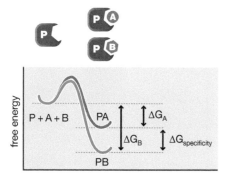

Figure 2.8

The quantitative definition of specificity.
In this example, protein P interacts
preferentially with ligand B compared with
ligand A. Plots of the change in free energy
as the binding reactions progress are shown
for binding to A (*pink*) and B (*blue*); for both
reactions, the free energy of the complex is
lower than the free energy of the unbound
components, indicating that binding is
favored at equilibrium. The difference
between the free energy before and after
binding (*ΔG*) is the free energy of binding.
The free energy of specificity of protein P for
the interaction with ligand B versus ligand A
is equal to the difference in the free energy
of binding of the two ligands under standard
conditions.

predict the affinity of a biological interaction. As an example, computer algorithms can simulate the docking of a virtual library of compounds onto a protein surface, and predict which are likely to bind with high affinity; this can help to identify those compounds that may ultimately be developed into new pharmaceuticals. However, such *in silico* predictions are still quite challenging. As we have seen, even relatively small errors in calculated binding energies propagate into large errors in K_d. Furthermore, the overall binding energy is the sum of many distinct and counterbalancing changes in enthalpy and entropy, both of the interacting proteins and the surrounding solvent, many of which are difficult to quantify precisely on a theoretical basis.

The free energy of binding also gives us a way to quantify the specificity of interactions. **Figure 2.8** shows a free-energy diagram illustrating the specificity of protein P for interaction with two competing ligand molecules, A and B. The difference in free energy between the binding of P to A and P to B is the free energy of specificity. Specificity can also be quantitatively described as the ratio of the dissociation constants for the two ligands (K_B/K_A). However, the specificity described here only reflects the preference of protein P for binding ligand B over ligand A, and does not reflect the preference of protein P for binding ligand B over any other competing ligands.

The dissociation constant is also related to rates of binding and dissociation

The dissociation constant describes the fraction of components bound to each other at equilibrium. In biological systems, however, binding reactions very seldom achieve equilibrium. In fact, it is changes in protein binding over relatively short periods of time that are usually most important for transmitting information during signaling. How can we describe the rates at which multicomponent systems assemble and disassemble in the cell, and what is the relationship of these rates to the dissociation constant?

As described before, for the bimolecular reaction $A + B \leftrightarrow AB$, the rate of formation of AB is given by a rate constant (on-rate or k_{on}) times the concentration of the binding partners, $k_{on}[A][B]$, while the rate of dissociation is given by another rate constant (off-rate or k_{off}) times the concentration of the complex, $k_{off}[AB]$. Let us consider each of these in a bit more detail.

From the rate term $k_{on}[A][B]$, we see that the actual rate of formation of the complex (in moles per liter per second: $M^{-1}s^{-1}$) depends on the concentration of each of the free components A and B. This makes intuitive sense, as binding can occur only when a molecule of A collides with a molecule of B, and the likelihood of such a collision is proportional to their concentrations. We can think of k_{on} as a combination of the rate of diffusion of A and B (which governs the rate of collision) and the likelihood that two molecules will stick to each other once they collide. In this case, k_{on} is a bimolecular rate constant and has units of $M^{-1}s^{-1}$. There is a limit to how large k_{on} can be, however: with rare exceptions, the rate of binding cannot be higher than the random rate of collision of A and B. For typical proteins in aqueous solution, the diffusion-limited rate of collision is on the order of 10^8–10^9 $M^{-1}s^{-1}$. If k_{on} approaches this rate, then a very high percentage of colliding molecules bind to each other, at least momentarily. The actual on-rates for biological binding reactions are usually considerably lower than this limit, typically 10^5–10^6 $M^{-1}s^{-1}$.

From the rate term $k_{off}[AB]$, we see that the rate of dissociation (again, in $M^{-1}s^{-1}$) is entirely dependent on the concentration of the AB complex. For any individual complex, the off-rate (k_{off}) describes the probability of dissociation as a function of time. This is a unimolecular rate constant, with units of s^{-1}. In fact, from k_{off}, we can calculate the **half-life** of the complex, or the amount of time it will take for half of the complex to dissociate (or, for an individual complex, the time at which there is a 50% likelihood it has dissociated). The general first-order rate equation describing such unimolecular reactions is $kt = \ln[\text{initial amount}]/[\text{amount at time } t]$. When time t is equal to the half-life, [initial amount]/[amount at time t] is equal to 2; thus under these conditions $t = \ln 2/k$, or $0.693/k$,

$$\text{Half-life of complex} = 0.693 / k_{off} \qquad\qquad (2.6)$$

Thus if k_{off} for a particular reaction is $10^{-2}\,\text{s}^{-1}$ (in the range typically seen for physiological interactions), then the half-life of the complex is $0.693/10^{-2}$ seconds, or 69.3 seconds (about a minute).

The dissociation constant is the ratio of two rate constants (k_{off}/k_{on}), and so the affinity of a binding reaction is inextricably linked to both the rate of formation and the rate of dissociation of the complex. Thus, two different complexes can have the same dissociation constant and thus the same overall affinity, yet have entirely different kinetic behavior as a result of compensating differences in on-rate and off-rate. Consider a situation where both the on-rate and the off-rate are relatively high. In this case, complex formation is rapid even at low levels of the individual components, but the half-life of those complexes is very short. Compare this with a second reaction with the same dissociation constant, where both the on-rate and the off-rate are relatively low. Now complex formation is quite slow unless the concentrations of components are high, but the complexes are relatively stable once formed (**Figure 2.9**). Such considerations are crucial to the dynamic behavior of signaling mechanisms, as they govern how rapidly existing complexes can be remodeled and how quickly the system can respond to changes.

We stated above that in most cases, the on-rate cannot be faster than the diffusion-limited rate of collision. This has profound implications for the half-life of biological complexes. Consider a relatively high-affinity interaction, for example, between a receptor and its ligand ($K_d = 10^{-11}$ M). Using the definition of the dissociation constant, we can write: $K_d = 10^{-11} = k_{off}/k_{on}$. Since k_{on} can be no greater than $\sim 10^{8}\,\text{M}^{-1}\text{s}^{-1}$, then k_{off} must be less than $10^{-3}\,\text{s}^{-1}$. From Equation 2.6, we can then calculate that the half-life of the receptor–ligand complex must exceed $0.693/10^{-3}$ seconds, or 693 seconds (about 12 minutes). If a more typical value for k_{on} is used ($10^{6}\,\text{M}^{-1}\text{s}^{-1}$), the calculated half-life is nearly a day. Thus a signal generated by binding of the ligand to the receptor will be quite stable and sustained, unless other mechanisms (such as degradation or other modification of the ligand or receptor) are provided to down-regulate it more rapidly. In the case of very high-affinity interactions, such as that between biotin and avidin ($K_d = 10^{-15}$ M), the half-life of the complex must be many days, making such interactions essentially irreversible. This is consistent with the biological role of avidin, which is found in chicken eggs. By binding and irreversibly sequestering biotin, an essential microbial nutrient, avidin helps prevent the growth of bacteria in the egg.

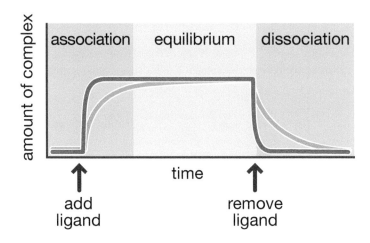

Figure 2.9

Interactions with the same affinity can have different rates of binding and dissociation. The time course of binding is shown for two interactions with the same dissociation constant (K_d), one with relatively fast on-rate and off-rate (*pink* line), and the other with relatively slow on-rate and off-rate (*green* line). Although the two reactions reach the same level of binding at equilibrium, the rates of binding and dissociation after ligand addition or removal are very different.

PROTEIN INTERACTIONS IN THEIR CELLULAR AND MOLECULAR CONTEXT

The dissociation constant provides essential information about the intrinsic properties of a protein–protein interaction and places physical limits on how it can behave in a cell. However, the likelihood that the interaction actually takes place *in vivo* is highly dependent on context. Taking the most obvious case, a particular interaction may be very strong *in vitro,* but if the two potential partner proteins are not expressed at the same time in the same cell, or in the same subcellular compartment, then the interaction will not occur. On the other hand, an interaction that appears relatively nonspecific *in vitro* may actually be more specific *in vivo.* For example, if a protein has several partners that all bind with the same affinity *in vitro,* but only one of these is coexpressed with the protein in a particular cell, then the *in vivo* binding specificity of that interaction will be high.

Furthermore, our discussions of affinity so far have involved some assumptions that may not be valid in biological systems. For example, simple binding equations assume that all components are well mixed and freely diffusing in solution, conditions that may not be met in the cell. We have also assumed that only one molecule of A binds to one molecule of B. In this section, we will explore a few situations where these assumptions are violated in biological systems, and how this can affect binding in the cell.

The apparent dissociation constant can be strongly affected by the local cellular environment and other binding partners

One way in which the apparent affinity of biological interactions can be significantly increased is through the presence of multiple binding sites on the two interaction partners. This is known as the **avidity** effect. The most prominent example of avidity, and where the effect was first noted, is in the binding of antibodies to ligands with multiple binding sites (known as polymeric ligands) such as bacterial cell walls, which consist of many identical subunits. Antibodies are dimeric, Y-shaped molecules possessing two identical binding sites for an **antigen** (loosely defined as a molecule recognized by an antibody). Thus when an antibody binds an antigen that is present in many copies on a surface, both of its antigen-binding sites are likely to be bound to the surface simultaneously. The fact that the antibody is held onto the surface by two points of contact instead of one makes dissociation much less likely, because this would require that both sites dissociate at the same time. If one of the binding sites momentarily dissociates, the other is likely to remain bound, and since the concentration of the antigen on the surface is very high, the first site will rebind rapidly (remember, the rate of association is directly proportional to the concentration of the binding partner) (**Figure 2.10a**). The effective off-rate is therefore much slower than the off-rate for an individual binding site, and the apparent dissociation constant is correspondingly much lower. In the case of antibodies, the off-rate can be essentially zero, and binding is thus irreversible.

Note that avidity depends on there being multiple binding sites on each partner. Thus an intact antibody (with two antigen-binding sites) binds to a polymeric ligand

Figure 2.10

Avidity of antibody binding. (a) Antibody molecules (*blue*) have two identical antigen-binding sites. When an antibody binds an antigen present in many copies on a surface, such as a bacterial cell wall, both antigen-binding sites are bound simultaneously. If one of the sites dissociates, the antibody is still held by the second site and the first site rapidly rebinds, and the off-rate is low. (b) If an antibody fragment containing a single antigen-binding site (an Fab fragment) is bound to the surface, the off-rate is relatively high. (c) If the antibody binds the same antigen in a soluble, monomeric form, the off-rate is also relatively high.

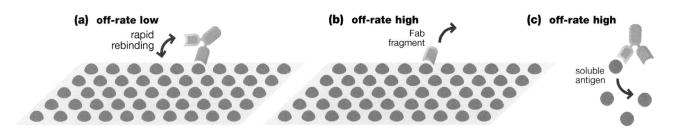

(a) off-rate low
rapid rebinding

(b) off-rate high
Fab fragment

(c) off-rate high
soluble antigen

such as a bacterium much better than does a proteolytic fragment of the antibody containing only a single binding site, termed an Fab fragment (**Figure 2.10b**). However, when an antibody recognizes a soluble, monomeric antigen instead of a surface with multiple binding sites, there is no avidity effect and an Fab would bind just as well as the intact immunoglobulin (**Figure 2.10c**).

The above example illustrates how the effective concentration of a ligand in the immediate vicinity of its binding partner, or its **local concentration**, influences binding. The local concentration of a ligand may be very different from its overall (or average) concentration in the cell. For example, in the case of an intramolecular binding interaction (between two parts of the same protein), the local concentration of the reactants is extremely high and thus binding can be favored even when the isolated interaction (intermolecular) is relatively weak. Local concentration also has an important role in facilitating binding to structures with a high density of binding sites, such as membranes or other surfaces. Under such conditions, it may be quite difficult for a molecule bound to such a surface to escape into solution; if it should dissociate, the very high local concentration of binding sites on the surface makes it likely that it will rebind before it can diffuse away (**Figure 2.11**).

Ideal affinity and specificity depend on biological function and ligand concentrations

Most binding interactions in signaling might be assumed to have high affinity and specificity, to ensure efficient and unambiguous transfer of information. However, this is not the case—binding interactions are observed that span a wide range of both affinities and specificities. What affinity and specificity would be most appropriate for a given interaction depends strongly on the function of the interaction and the endogenous concentrations of the partners. For example, many complexes are preassembled and need to be long-lived for their function; in such complexes, the affinities of the partners are likely to be high. In particular, it is important for the dissociation constant to be lower than the endogenous concentrations of the components so that binding will be close to saturation. However, there is a limit to how high an affinity is required—once below the endogenous ligand concentration, decreasing the dissociation constant by many orders of magnitude would provide little improvement in the degree of binding, and might be less likely to evolve.

By contrast, if an interaction is dynamically regulated, as are many signaling interactions, then it is important that the dissociation constant be roughly equal to or slightly higher than the endogenous ligand concentration. This weaker affinity (higher K_d) is important for two reasons. First, it allows small changes in either ligand concentration or other modulating factors (for example, local concentration or allosteric changes) to lead to large changes in the fraction of ligand bound—that is, it allows the interaction to be regulated. Second, weaker affinities allow the interaction to be potentially more dynamic. As described above, the dissociation constant (K_d) of an interaction is defined by its off-rate (k_{off}) divided by its on-rate (k_{on}). Since the on-rate is essentially limited to the rate of diffusion ($\approx 10^8$–10^9 $M^{-1}s^{-1}$), then the off-rate is limited by the affinity. A higher off-rate (shorter half-life) may be required for interactions that must be rapidly disassembled as part of their biological function.

There are functional constraints on interaction affinities and specificities

Because endogenous protein concentrations are so closely tied to binding affinities, it is useful to survey the range of individual protein concentrations observed in biological signaling systems. Overall, the concentrations of intracellular signaling proteins tend to be significantly higher than those of extracellular signaling proteins, such as hormones. As we shall see below, such concentrations constrain the range of observed affinities and specificities.

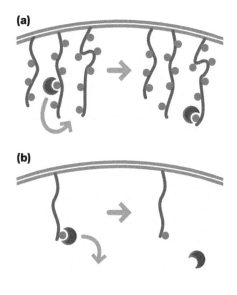

(a)

(b)

Figure 2.11

Effect of surface density on binding.
(a) When there is a high local concentration of a binding site on a surface, a ligand dissociating from one site is likely to rebind immediately, and so dissociation of the ligand from the surface is a rare event. (b) If the density of binding sites is low, however, rebinding is much less likely, and dissociation is favored.

Signaling interactions can be divided into broad classes on the basis of affinity and specificity (**Figure 2.12**). Interactions that have both high affinity and high specificity include hormone–receptor interactions; for example, human growth hormone (hGH) and its receptor interact with $K_d = 3 \times 10^{-10}$ M. Such receptors must recognize low concentrations of hormone in the bloodstream and have affinities that allow binding in the range of the signaling concentrations of their ligands. For example, under basal conditions, the concentration of hGH is $<10^{-10}$ M, while during activation, the concentrations reach ~10^{-9} to 10^{-8} M. Thus, the dissociation constant is closely matched with the change in concentration that must be physiologically detected—resulting in a significant signal-dependent change in the fraction of bound receptor. These types of interactions must also be specific enough to prevent cross-talk between related hormones and receptor sets. Interactions that have low affinity (here arbitrarily defined as interactions with $K_d > 10^{-7}$ M) but high specificity include the interactions of modular peptide-recognition domains, such as SH3 domains, with their partners. These interactions function in the intracellular environment, where protein concentrations are much higher. Low affinities are probably necessary to allow for subtle, dynamic regulation of transient signaling complex assembly and disassembly in response to changing input. High specificity, however, is still essential to achieve proper flow of information.

An extreme example of an interaction that has high affinity but relatively low specificity is the interaction of **major histocompatibility complex (MHC)** molecules with peptide antigens. MHC molecules bind peptides derived from proteins degraded inside cells, and carry them to the cell surface where they can be surveyed by lymphocytes of the immune system. In this way, protein components of intracellular infectious agents such as viruses, or from bacteria that have been engulfed by phagocytes, can be detected and elicit an immune response. To ensure detection of microbial intruders, each MHC molecule must be able to bind a wide range of peptides, and this is achieved by combined recognition of general features of the peptide backbone with recognition of specific amino acid side chains that vary between peptides recognized by any given MHC molecule. Despite the resulting promiscuity of this interaction, binding is quite tight—estimated K_d values for such complexes are between 10^{-6} and 10^{-9} M. These unusual binding properties

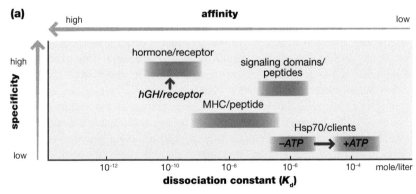

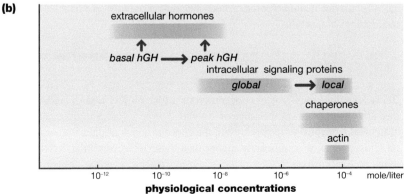

Figure 2.12

Range of biological affinities and concentrations. (a) The general range of affinities and specificities observed for a particular class of interactions. (b) The endogenous concentrations of these classes of proteins. Overall, the K_d values of signaling interactions are matched to endogenous concentrations so that the extent of binding can be easily regulated by small environmental changes. hGH, human growth hormone; MHC, major histocompatibility complex molecule; hsp70/clients, chaperone binding to newly synthesized client proteins during folding.

are observed because the MHC molecule folds around the peptide as an integrated unit. The intimate engagement of MHC–peptide leads to a very slow off-rate—approximately $10^{-6}\,s^{-1}$, equivalent to a half-life of more than 100 hours. This extremely slow off-rate helps to ensure that the peptide–MHC complex remains at the cell surface long enough for sustained interaction with the antigen receptors of lymphocytes, a requirement for immune activation.

By contrast, the housekeeping role of chaperone proteins, which bind transiently to hydrophobic regions on newly synthesized proteins during folding to prevent their misaggregation, requires interactions that have low affinity and low specificity. Chaperone interactions must be nonspecific enough to allow binding to a wide range of ligands, and allow rapid dissociation. Chaperones like hsp70 bind a wide variety of hydrophobic peptides with K_d values ranging from 10^{-7} to $10^{-5}\,M$ when bound to ADP. When ATP is bound, the affinity of hsp70 for the peptide is further reduced by between 10- and 100-fold. This enables ATP hydrolysis and exchange of ADP for ATP to drive cycles of peptide binding and release. Chaperones are present at high concentrations in the cell, ensuring that binding is favored despite their relatively low affinity for their ligands.

Interaction affinity and specificity can be independently modulated

One might intuitively assume that specificity and affinity are well correlated: the tighter the binding of a protein to its correct partner, the less likely it is to cross-react with others. As suggested above, however, this is not always the case. In fact, affinity and specificity can be tuned independently via several different mechanisms. We will consider here the interaction of a protein with a series of related competing ligands. If the interaction of the protein with all its ligands is initially both relatively weak and nonspecific, there are, in principle, several ways in which improved affinity and specificity for one of the ligands can evolve.

In a process of positive discrimination, the affinity of the protein for one of its ligands can be increased by the formation of additional favorable stereochemical interactions. Whether these new interactions increase specificity, however, depends on their nature. Favorable interactions that recognize chemical features common to all members of a competing ligand family will increase the affinity, but not the specificity, for the given ligand. For example, many relatively nonspecific peptide-binding proteins, such as the MHC molecules described previously, recognize their peptide ligands partly through extensive hydrogen-bonding to the peptide backbone—which, unlike bonding to side chains, is not specific to a particular sequence of amino acids. Positive discrimination will generally increase specificity only if the increase in affinity involves the formation of specific side-chain interactions—elements that are unique to the correct ligand compared with its related competitors (**Figure 2.13a–c**).

Structural studies have shown that for most proteins that recognize specific peptide ligands, the binding interface consists of a combination of specificity-determining contacts to specific side chains of the target, and contacts to generic features that are common to all peptides.

Specificity can also be increased via a process of negative discrimination. In this case, affinity for related competitor ligands is decreased, while the affinity for one ligand remains unchanged (**Figure 2.13d** and **2.13e**). Negative discrimination often involves the presence of interactions that would be unfavorable if the protein were to form a complex with the incorrect competing ligand. Where many closely related protein interactions occur in the same cell, such as those involving families of interaction domains, there is evidence that negative discrimination prevents incorrect cross-interactions in some cases.

Figure 2.13

Increasing specificity through positive and negative discrimination. (a) Binding of a receptor to two possible ligands, A and B, is initially weak and nonspecific. (b) Modification of the receptor to increase complementarity to both ligands results in higher affinity but no increase in specificity. (c) Modification of the receptor to increase complementarity only to ligand B results in an increase in specificity for ligand B versus A. (d) Modification of the receptor to decrease complementarity to ligand A results in reduced affinity for ligand A and thus an increase in specificity for B versus A. (e) The receptor could also be altered to simultaneously decrease complementarity to ligand A and increase complementarity to ligand B, resulting in an even greater increase in specificity. Free-energy diagrams on the right show how the changes alter the difference in free energy of binding between the two ligands ($\Delta G_{specificity}$).

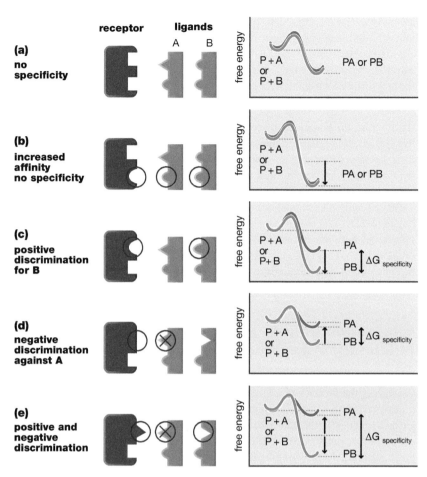

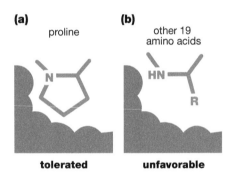

Figure 2.14

Negative discrimination enables SH3 domains to specifically recognize proline. The binding pocket of SH3 domains tolerates an imino group (a), but poorly recognizes the standard amide group found in most peptide linkages (b). Because proline is the only amino acid with an imino backbone group, this is the only one of the 20 natural amino acids tolerated at this site, and thus the binding by the SH3 domain is highly specific for ligands containing proline.

Negative discrimination can also be observed in the specific binding of SH3 domains to proline-containing peptide motifs (**Figure 2.14**). Pockets on the SH3 domain recognize proline via the imino acid backbone group. This interaction is not highly optimized, but is highly discriminatory against peptides lacking an imino group at these positions. Because proline is the only one of the 20 natural amino acids with an imino backbone group (all others have an amide backbone group), the selection for proline by this binding site is absolute. This strong negative selectivity would not be observed if there were other imino acids among the constituents of natural proteins, illustrating how the effectiveness of negative discrimination as a mechanism for specificity is highly dependent on the nature and range of potentially competing ligands present in the cell.

In most known interactions, affinity and specificity are tuned by both positive and negative discrimination. For example, charge–charge (salt bridge) interactions are observed at the interfaces in many protein–protein complexes. These not only provide a favorable energy for binding when they are complementary within the correct complex, but they often also repel related ligands that lack charge at the appropriate position, thus preventing the formation of incorrect complexes. The negative discrimination at such surfaces is strong because it is highly energetically unfavorable to bury a charged residue at an interface without a compensating interaction with an oppositely charged residue.

Cooperativity involves the coupled binding of multiple ligands

Many macromolecular complexes in cells are composed of multiple interacting partners, and the higher-order interplay between these partners can yield interesting and functionally important binding behaviors. One of these higher-order effects is cooperative binding, which is yet another mechanism by which the affinity and specificity of an interaction can be increased.

Cooperativity refers to an interaction in which binding of one ligand enhances the binding of an additional ligand(s). In thermodynamic terms, cooperativity is observed when the free energy of two ligands binding simultaneously is different from the sum of the free energies of the two ligands binding individually. If the binding of one ligand increases the affinity for an additional ligand, then positive cooperativity is said to have occurred. Here, we will focus only on positive cooperativity, although negative cooperativity—when the binding of one ligand decreases the affinity for an additional ligand—does occur. An important consequence of positive cooperativity is that assembly of a complex occurs in more of an all-or-none fashion: the multiple ligands cooperate to assemble the complete complex, and intermediate states of assembly are poorly populated (**Figure 2.15**).

Diverse molecular mechanisms underlie cooperativity

Cooperativity can be produced by many different mechanisms, which differ in the source of the additional interaction energy (**Figure 2.16**). For example, if two ligands directly interact with one another when they are bound to the same receptor protein, they will enhance each other's binding in a cooperative manner—the two binding events will be dependent, with the second binding event being more favorable than the first. Here, the source of the energy for cooperativity is the ligand–ligand interaction. Cooperativity is also observed if binding of one ligand leads to a conformational change in the receptor protein that in turn increases its affinity for a second ligand; for example, it alters the conformation of the second binding site. In this case, the extra energy of cooperativity comes from the reorganization of the receptor protein structure.

Cooperativity can also occur when two interaction domains in a protein bind to two distinct sites on the same ligand. When one domain binds to the ligand, the increased effective concentration of the second binding site further enhances its binding; here, the covalent tethering of the two binding sites reduces the entropic cost of the second binding event (avidity, described above, is an example of this type of cooperativity). A common example of cooperativity in intracellular signaling is when one protein is recruited to the membrane by a protein–lipid interaction, thereby increasing its local concentration (see Chapter 5, Figure 5.10). This cooperatively increases its effective affinity for another membrane-localized partner. In addition, cooperativity can be observed if two binding sites on the protein are occluded by a single inhibitory element—binding of one ligand displaces the inhibitory element, thus increasing affinity for the second ligand.

Cooperative binding has a variety of functional consequences

One outcome of cooperativity can be to increase the biological specificity of an interaction, restricting it to a particular time and place. For proteins participating in multiple cooperative interactions, for example, the likelihood of interaction depends on the concentration of a specific set of ligands, rather than of a single ligand. The correct combination of ligands may only be present at a specific time or at a specific location in a cell.

Cooperativity between multiple interaction domains within a protein often serves to increase specificity. For example, the intracellular protein kinase ZAP-70 in the T lymphocytes of the immune system has two phosphotyrosine-binding SH2 domains (**Figure 2.17**). On its own, each domain has only modest specificity. When linked in the intact protein, however, they cooperate to recognize particular tandem arrangements of phosphotyrosine motifs known as *immunoreceptor*

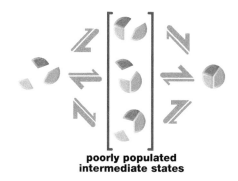

**poorly populated
intermediate states**

Figure 2.15

Positive cooperativity leads to all-or-none assembly. An example of a three-component assembly in which binding is cooperative. The binding of any pair of the three components shown on the left facilitates the binding of a third component. Thus, intermediate two-component states are rare, and the components are said to cooperate to favor assembly of the complete complex.

 The role of ZAP-70 in T cell signaling is discussed further in Chapter 14.

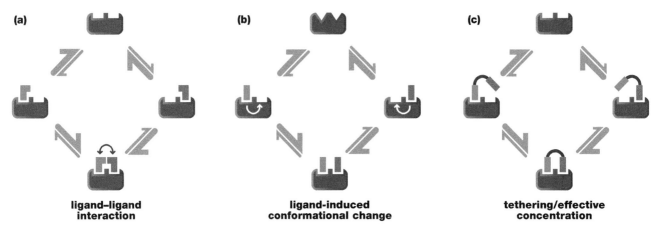

(a) ligand–ligand interaction

(b) ligand-induced conformational change

(c) tethering/effective concentration

Figure 2.16

Diverse mechanisms for cooperative binding of two ligands to one receptor. (a) The binding of the second ligand to the receptor is enhanced by ligand–ligand interactions. (b) Conformational changes in the receptor induced by the binding of the first ligand increase the affinity for the second ligand. (c) If ligands are tethered, then binding of one ligand by the first binding site increases the local concentration at the second binding site, enhancing binding.

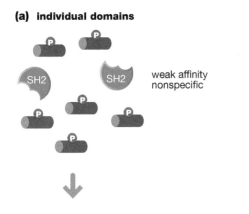

(a) individual domains

weak affinity nonspecific

(b) linked domains

high affinity and specificity for tandem phosphorylated motifs

cooperativity:
effective concentration and domain–domain interactions

Figure 2.17

Cooperative recognition of a tandem phosphotyrosine motif by coupled SH2 domains. (a) Individual SH2 domains show weak affinity and specificity for a single phosphotyrosine motif. (b) In the protein kinase ZAP-70, linked SH2 domains show high affinity and specificity for phosphotyrosine motifs linked in tandem. This is the result of cooperativity due to interactions between the SH2 domains and also to ligand tethering and increased local concentration.

tyrosine-based activating motifs (ITAMs), which are present in the cytoplasmic tails of important immune-system receptor proteins.

When positive cooperativity involves multiple ligands of the same species, it is known as homotypic cooperativity; cooperativity involving two or more different ligands is referred to as heterotypic cooperativity. As well as making binding interactions more specific, homotypic cooperativity can make them act more like a binary switch, which is either in the "on" or "off" state. In such cases, the binding curve is no longer a simple hyperbola, but assumes a sigmoidal shape (**Figure 2.18**). Within a narrow critical range of ligand concentration, the protein shifts from a state in which almost all sites are unbound to one where almost all sites are bound; thus, at this threshold range of concentration, very small changes in concentration can have disproportionately large effects on fractional occupancy. By contrast, for single-site (noncooperative) binding, a greater than 80-fold increase in ligand concentration is required to increase the fractional occupancy from 10% to 90%. Because binding is often linked to protein activity, many signaling pathways use cooperativity to sculpt the response of the system so that activation of a protein only occurs at a precise concentration threshold of a ligand.

Protein assemblies differ in their stability and homogeneity

There is wide variation in the size, complexity, and uniformity of multiprotein complexes involved in signaling. At one extreme of this spectrum are very stable structures that have a precise and invariant molecular architecture and **stoichiometry** (that is, the numbers of the different types of subunits). Examples include ribosomes, proteasomes, and nuclear pores. The assembly of such complexes is often highly cooperative, and partially assembled structures are unstable. Such macromolecular machines execute important functions (such as making, destroying, or transporting proteins) in the cellular response to a signal, but their immutability makes them poorly suited for the business of sensing, integrating, and transmitting signals that are rapidly changing over time.

The protein complexes that perform more dynamic roles in signaling are often much less stable and homogeneous. Interactions among the components are likely to be relatively weak, and there can be many possible binding partners for a given binding site. Thus, many different possible combinations of interactions are possible. An important property of such structures is that their interactions are likely to change and reorganize relatively quickly. A good illustration is provided by the

Figure 2.18

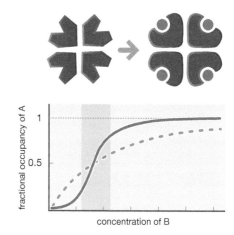

The effect of positive cooperativity on binding curves. Binding of one molecule of ligand B (*orange circle*) to one subunit of protein A (*brown*) increases the affinity of the other subunits for ligand B. The *pink* solid line shows the sigmoidal binding curve characteristic of such positive homotypic cooperativity. Note that within a critical range of concentration (*shaded area*), relatively small differences in the concentration of B have large effects on binding. For comparison, the *blue* dotted line shows a simple binding curve with no cooperativity.

complexes induced by the stimulation of receptors coupled to tyrosine kinases (see Figure 4.12 in Chapter 4). Activation of such receptors induces the phosphorylation of many different sites on the receptor itself, which then serve as docking sites to recruit SH2-domain-containing proteins. Receptor phosphorylation is relatively slow and inefficient, however, and each site is subject to dephosphorylation by phosphatases. Thus, individual receptor molecules are likely to bear different and constantly changing combinations of phosphorylated sites. Furthermore, each site may bind to several different SH2-containing proteins with similar affinity, so which one, if any, binds will depend on chance and the local availability of binding partners. This situation leads to a combinatorial explosion of possible states for the receptor; millions of distinct receptor complexes can be described for even a relatively small number of phosphorylation sites and binding partners.

Such complexes, which can be termed dynamic molecular assemblies, differ greatly from more stable macromolecular machines such as the ribosome. This is an important concept to keep in mind when thinking about signaling pathways: although we often loosely refer to the signaling machinery, this analogy should not be taken too literally. The actual complexes that transmit signals can be difficult (if not impossible) to define precisely, and the contributions of a single interaction among many possible interactions can be difficult to tease out. What advantages might there be to signaling mechanisms based on such dynamic molecular assemblies? It may be that because such complexes exist in a wide range of possible states, depending on the specific combinations of binding partners, they are better able to respond to and transmit complex, finely graded signals. The relative instability of such assemblies also allows them to be highly sensitive to rapid and subtle changes in the environment. And since they function in a combinatorial fashion, a relatively limited number of components can participate in an almost infinite variety of pathways having different inputs and outputs.

A particular type of dynamic molecular assembly results in the separation of the components into two phases—a dense phase with a high concentrations of interacting components, surrounded by a dilute phase where the components are present at much lower concentrations. Such *biomolecular condensates* are typically formed by the interaction of multivalent partners, each of which contains multiple binding sites for the other (**Figure 2.19**). While the affinity of each individual interaction is typically weak, collectively they can mediate formation of highly dynamic, macroscopic complexes in the cell that have properties of liquid droplets (sometimes called "membraneless organelles"). The special properties of biomolecular condensates are further discussed in Chapter 5.

(a)

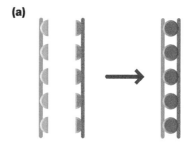

(b)

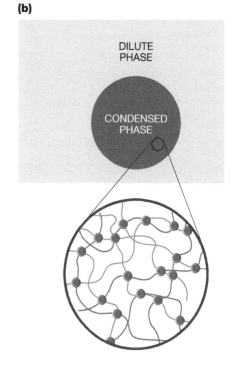

Figure 2.19

Biomolecular condensates. (a) Multivalent low-affinity interactions mediate the binding of two molecules, represented by the red and blue lines. In this example, each molecule contains five binding sites for the other. (b) When the two molecules are present in the same solution, under certain conditions, the mixture will separate into two phases: a dilute phase and a condensed phase. Inset of the condensed phase shows that it consists of a dense network of the two molecules held together by multiple weak interactions.

SUMMARY

Protein–protein interactions play a part in virtually all signaling events in the cell. Changes in these interactions, in turn, drive many other changes of importance for signal transmission, as we will see in coming chapters. The dissociation constant (K_d) of an interaction is an intrinsic, quantitative measure of the likelihood that the interaction will occur, whereas the specificity of an interaction is determined by the relative affinities of competing interactions. The affinity and specificity of physiological binding interactions vary widely, but are tuned to the specific circumstances so that information can be processed reliably.

QUESTIONS

Answers to these questions can be found online at www. routledge.com/9780367279370

1. What is the concentration of a protein if it is present at one molecule per cell? Assume the volume of the cell is $\sim 10^{-12}$ L.

2. If a cell is $\sim 25\%$ protein (mass/vol), then what is the approximate total number of protein molecules in a cell? Assume an average protein molecular mass of 100 kD.

3. You measure that there are approximately 10,000 copies of protein X in the cell. Again, assuming that the volume of a mammalian cell is $\sim 10^{-12}$ L, what is the approximate concentration of this protein when distributed throughout the whole cell? What happens to the concentration if all of protein X is translocated to the nucleus (use an estimated nuclear volume of $\sim 10^{-13}$ L)?

4. Protein X described in Question 3 has a physiological binding partner Y in the nucleus that is present at a concentration of $\sim 10^{-10}$ M. You observe a significant interaction between molecules X and Y ($>50\%$ of Y is bound to X) only after all of molecule X is concentrated in the nucleus (which we estimate is one-tenth the volume of the entire cell). Can you estimate the K_d for the interaction of X and Y?

5. The observed extent of an interaction between two molecules in a living cell is often different from what might be expected from the simple K_d of the interaction measured *in vitro*. What factors might lead to a higher fraction of binding than expected? What factors might lead to a lower fraction of binding than expected?

6. Proteins X and Y interact with each other, both *in vivo* and *in vitro*. You identify a disease-causing mutation in protein X that leads to disruption of the X–Y interaction *in vivo*. Surprisingly, when you purify proteins X and Y, the *in vitro* binding affinity is unchanged by the mutation. What is a simple hypothesis for how the mutation in X might change the extent of binding to Y *in vivo*?

7. A protein domain recognizes a peptide with the following sequence profile: RxxF/L/VxF (the residues F, L, and V at position 4 are recognized with equal affinity to each other; x denotes any amino acid is tolerated at that position). A homologous protein domain recognizes peptides with the profile: RxxVxF. Which domain has higher specificity? Which domain will bind with higher affinity? Discuss your answers in terms of how positive or negative discrimination could be involved in these two interactions and how these mechanisms might correlate with affinities.

8. Glutathione S-transferase (GST) has often been used as a "tag" for recombinant proteins, allowing the GST-tagged proteins to be purified away from other proteins using small beads coated with a high surface density of glutathione. Table 2.1 shows that the affinity of the GST–glutathione interaction is relatively low ($K_d \sim 10^{-4}$ M). What are the advantages and disadvantages of the GST–glutathione system as a purification tag, compared to one with

much higher affinity, such as biotin–avidin? GST-tagged proteins can often be purified efficiently even when working at concentrations lower than 10^{-4} M, at which one might not expect to observe a high fraction of binding. What features of the GST interaction with glutathione-coated beads might account for this?

9. The emergence of new interactions between signaling proteins is thought to be a key driver of the evolution of new signaling pathways and behaviors. Describe different mechanisms whereby a new protein–protein interaction could evolve.

10. Most protein–protein interactions involved in signaling are dynamic (that is, the kinetics of the interaction—on-rates and off-rates—are fast). Why might this be important? What other types of function might require protein interactions that are less dynamic? How are the dynamics of protein interactions related to their thermodynamic affinities?

11. A number of proteomic databases are now available that enumerate all known protein interactions in a given system. For example, over 20,000 human protein–protein interactions have been identified in one such dataset. Generally, these datasets provide little if any information about affinity. What are the limitations of such databases? Are there conditions where a predicted interaction might occur only rarely in a cell? When might a biologically important interaction not be predicted by such databases?

12. Scaffold proteins bind to multiple binding partners simultaneously. You discover a new scaffold protein that binds both molecules A and B, with no independent interaction between A and B. Signal transmission depends on the concentration of the trimolecular complex (scaffold plus A and B). How will signal output (proportional to the amount of the trimolecular complex) depend on the concentration of the scaffold protein? What are the conditions under which signal output will be maximal? How would you expect signal output to be affected if the scaffold is experimentally overexpressed to a level much higher than normal?

BIBLIOGRAPHY

PROPERTIES OF PROTEIN–PROTEIN INTERACTIONS

Hammes GG (2000) *Thermodynamics and Kinetics for the Biological Sciences.* New York: John Wiley & Sons.

Harrison SC (1996) Peptide-surface association: The case of PDZ and PTB domains. *Cell* 86, 341–343.

Jones S & Thornton JM (1996) Principles of protein-protein interactions. *Proc. Natl. Acad. Sci. USA.* 93, 13–20.

Kastritis PL & Bonvin AM (2012) On the binding affinity of macromolecular interactions: Daring to ask why proteins interact. *J. R. Soc. Interface* 10, 20120835.

Lo Conte L, Chothia C & Janin J (1999) The atomic structure of protein-protein recognition sites. *J. Mol. Biol.* 285, 2177–2198.

Moreira IS, Fernandes PA & Ramos MJ (2007) Hot spots-a review of the protein-protein interface determinant amino-acid residues. *Proteins* 68, 803–812.

Nooren IM & Thornton JM (2003) Diversity of protein-protein interactions. *EMBO J.* 22, 3486–3492.

Siebenmorgen T & Zacharias M (2020) Computational prediction of protein–protein binding affinities. *WIREs Comput. Mol. Sci.* 10, e1448. doi: 10.1002/wcms.1448.

Van Roey K, Uyar B, Weatheritt RJ et al. (2014) Short linear motifs: ubiquitous and functionally diverse protein interaction modules directing cell regulation. *Chem Rev.* 114(13), 6733–6778. doi: 10.1021/cr400585q.

Wells JA (1996) Binding in the growth hormone receptor complex. *Proc. Natl Acad. Sci. USA.* 93, 1–6.

Winzor DJ & Sawyer WH (1995) *Quantitative Characterization of Ligand Binding.* New York: John Wiley & Sons.

Wyman J & Gill SJ (1990) *Binding and Linkage: Functional Chemistry of Biological Macromolecules.* Mill Valley, CA: University Science Books.

PROTEIN INTERACTIONS IN THEIR CELLULAR AND MOLECULAR CONTEXT

Bhattacharyya RP, Reményi A, Yeh BJ & Lim WA (2006) Domains, motifs, and scaffolds: The role of modular interactions in the evolution and wiring of cell signaling circuits. *Annu. Rev. Biochem.* 75, 655–680.

Bukau B, Weissman J & Horwich A (2006) Molecular chaperones and protein quality control. *Cell* 125, 443–451.

Capra EJ, Perchuk BS, Skerker JM & Laub MT (2012) Adaptive mutations that prevent crosstalk enable the expansion of paralogous signaling protein families. *Cell* 150, 222–232.

Darzacq X, Tjian R (2022) Weak multivalent biomolecular interactions: a strength versus numbers tug of war with implications for phase partitioning. *RNA* 2022, 28(1), 48–51. doi: 10.1261/rna.079004.121.

Edwards LJ & Evavold BD (2011) T cell recognition of weak ligands: Roles of signaling, receptor number, and affinity. *Immunol. Res.* 50, 39–48.

Gibson TJ (2009) Cell regulation: Determined to signal discrete cooperation. *Trends Biochem. Sci.* 34, 471–482.

Hatada MH, Lu X, Laird ER, et al. (1995) Molecular basis for interaction of the protein tyrosine kinase ZAP-70 with the T-cell receptor. *Nature* 377, 32–38.

Mayer BJ, Blinov ML & Loew LM (2009) Molecular machines or pleiomorphic ensembles: Signaling complexes revisited. *J. Biol.* 8, 81.

Müller KM, Arndt KM & Plückthun A (1998) Model and simulation of multivalent binding to fixed ligands. *Anal. Biochem.* 261, 149–158.

Nguyen JT, Turck CW, Cohen FE, et al. (1998) Exploiting the basis of proline recognition by SH3 and WW domains: Design of N-substituted inhibitors. *Science* 282, 2088–2092.

Pawson T & Nash P (2003) Assembly of cell regulatory systems through protein interaction domains. *Science* 300, 445–452.

Sadegh-Nasseri S, Stern LJ, Wiley DC & Germain RN (1994) MHC class II function preserved by low-affinity peptide interactions preceding stable binding. *Nature* 370, 647–650.

Szwajkajzer D & Carey J (1997) Molecular and biological constraints on ligand-binding affinity and specificity. *Biopolymers* 44, 181–198.

Zarrinpar A, Park SH & Lim WA (2003) Optimization of specificity in a cellular protein interaction network by negative selection. *Nature* 426, 676–680.

Signaling Enzymes and Their Allosteric Regulation

3

Many of the key steps in cellular signal transduction are controlled by specific chemical reactions, ranging from covalent modifications like phosphorylation to the generation of small diffusible mediators such as cAMP. In most cases, these chemical reactions are intrinsically slow, and thus specific enzymes are required for them to occur at reasonable rates. Cells use enzymes like protein kinases to catalyze phosphorylation reactions, and enzymes like adenylyl cyclase to catalyze the formation of cAMP from ATP. These enzymes are required for signaling processes to occur at the time scales (fractions of a second to minutes) necessary for cellular sensing and responses.

In this chapter, we will discuss the basic mechanism of signaling enzymes, focusing on a few canonical examples: kinases and phosphatases, which control protein phosphorylation, and G proteins and the enzymes that regulate them. These represent two of the most important enzymatic control systems used throughout eukaryotic signaling pathways. Phosphorylation is the protein modification most frequently used in signaling, while G proteins use conformational changes to transmit signaling information. In this chapter, we will review the detailed chemical mechanisms by which these common classes of enzymes catalyze their reactions. Other classes of signaling enzymes are discussed throughout this book, including those that create or destroy other types of post-translational modifications (Chapter 4), those that produce or destroy second messengers such as cAMP (Chapter 6), lipid-modifying enzymes (Chapter 7), and proteases (Chapter 9), though their detailed molecular mechanisms will not be covered.

When discussing signaling enzymes, it is impossible to separate their catalytic mechanisms from their mechanisms of regulation. The central role of signaling enzymes is, after all, to relay information. Thus, their absolute catalytic efficiency is often less important than the capacity of their catalytic activity to be differentially regulated in response to inputs such as ligand binding or covalent modification (unlike, for example, a metabolic enzyme whose function might be to produce

DOI: 10.1201/9780429298844-3

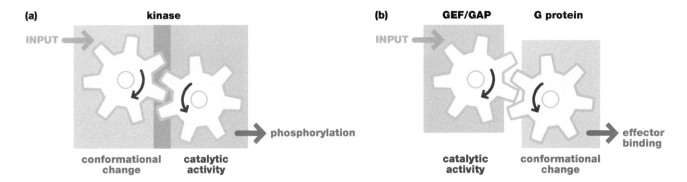

Figure 3.1

Signaling proteins convert conformational changes into catalytic activity changes, and vice versa.

(a) Kinases undergo conformational changes in response to diverse inputs, which in turn regulate the kinase catalytic function (phosphorylation). (b) G proteins are controlled by opposing enzymes—guanine nucleotide exchange factors (GEFs) and GTPase-activator proteins (GAPs)—which regulate the conformation of the G protein and, in turn, its interactions with downstream effector molecules.

the maximal amount of a chemical product). A hallmark of signaling proteins, such as kinases and G proteins, is their ability to convert conformational changes into catalytic activity changes, or vice versa (**Figure 3.1**). The mechanisms by which these canonical classes of signaling proteins can couple conformation change with catalytic activity are a major focus of this chapter.

PRINCIPLES OF ENZYME CATALYSIS

Enzymes are biological macromolecules that catalyze chemical reactions in living systems. They are remarkable for their ability to promote specific chemical reactions. They can greatly increase the rates of reactions that occur spontaneously but slowly. In addition, they can promote unfavorable reactions that would not occur spontaneously by coupling them to energetically more favorable reactions (such as ATP or GTP hydrolysis).

Enzymes have a number of properties that make them useful for transmitting signals in the cell

Signaling enzymes provide a flexible way to relay cellular information—they are capable of both receiving inputs and transmitting outputs. The output of a signaling enzyme is to catalyze a functional change in downstream targets. A substrate that is phosphorylated by a kinase enzyme might undergo a change in conformation and activity, while a small-molecule mediator like cAMP that is produced by the enzyme adenylyl cyclase might change the function of a number of downstream effectors. Thus, the output from the enzyme is information that is stored and passed downstream via the states of the downstream targets of the enzyme.

In turn, the activity of these enzymes themselves is often subject to regulation by upstream inputs. Signaling enzymes often show low levels of activity under basal conditions, but large increases in activity upon regulatory changes such as ligand binding (conversely, active enzymes can be inhibited by such inputs). Because signaling proteins act as relay switches, they are often selected not for optimum catalytic efficiency, but rather for the ability to be regulated.

At the molecular level, many of these individual signal-transducing events involve some sort of change in the **conformation** in the enzyme (broadly speaking, a change in its three-dimensional arrangement or shape), which, in turn, is coupled to a change in its activity. Conformational changes that are induced by upstream inputs such as ligand binding or post-translational modification, and which result in changes in the protein activity, are referred to as **allosteric** changes. This conformational coupling between an upstream input and a change in protein activity provides a basic mechanism for the protein to propagate signals. In this chapter, we discuss the principles that drive allosteric changes, and show examples of how these changes can be used to propagate signals.

Another common feature of signaling enzymes is that they often occur as complementary pairs. For example, a specific phosphorylation reaction might be catalyzed by a **protein kinase**, while the reversal of this modification—the removal of the phosphate group—is catalyzed by a **protein phosphatase**. The kinase would essentially act as a "writer" element that drives the substrate to a new state, while the phosphatase acts as an "eraser" that returns the substrate to the original state (**Figure 3.2**). All information-storage and transmission systems, be they natural cellular systems or human-made electronic ones, require some sort of mechanism like this to write and erase information in a controllable fashion.

Enzymes have another useful property for transmitting information: the potential to amplify a signal. Because they are catalysts, enzymes remain unchanged while promoting their target reaction, and thus they are capable of carrying out many rounds of the reaction. This catalytic behavior can result in *signal amplification*: a single activated enzyme molecule can, under the right conditions, generate many molecules of product in a short period of time.

Enzymes use a variety of mechanisms to enhance the rate of chemical reactions

Enzymes can typically increase the rate of a specific reaction by several orders of magnitude relative to the uncatalyzed reaction in water. In some cases, including the phosphoryl transfer reactions discussed below, rate enhancements can be as large as 10^{20}-fold or more. Here, we will consider the ways that enzymes can achieve such remarkable rate enhancement and specificity.

In any reaction, the reactants and the products have defined **ground-state energies**, and the conversion of reactants to products requires the molecules to pass over a **transition state** of much higher energy. Enzymes catalyze reactions by stabilizing the transition state relative to the ground state of the reactants (**Figure 3.3**). Lowering the transition-state energy reduces the **free-energy barrier** for the reaction and increases the probability that reactants will pass over the barrier and be converted into the products. While enzymes can alter the relative free-energy difference between ground states and transition states, the thermodynamic equilibrium between free reactants and products is unchanged. In other words, at infinitely long times, all chemical reactions will reach an equilibrium based on the free energies of the reactants and products; this equilibrium is not affected by the presence of an enzyme. What is affected is how quickly this equilibrium is approached. As mentioned above, enzymes can accelerate reaction rates by many orders of magnitude.

There are several general mechanisms that enzymes can use to stabilize the transition state of a chemical reaction, all of which are observed in various signaling enzymes. First, enzymes can lower the transition-state energy by binding and orienting key reactive groups within a substrate (or two different substrates) such that they react with one another in a more favorable fashion. Second, enzymes can provide general acids and general bases to donate or accept protons that are transferred to and from the substrate during the reaction. The use of general acids, bases, and metal ions to activate attacking groups and stabilize charge development can be particularly important for nucleophilic displacement reactions (such as phosphoryl transfer reactions) that require activation of an attacking nucleophile and stabilization of a leaving group. Third, enzymes can provide a binding site that is more complementary to the electrostatic or geometric properties of the transition state than the ground state; in this case, some of the binding energy of the transition state is used to reduce the free-energy barrier for the reaction. This type of electrostatic or geometric catalysis can be accomplished by the precise positioning of active-site functional groups and cofactors such as metal ions.

Finally, in some cases, enzymes alter the path or mechanism of a reaction, causing the reaction to proceed through a reaction intermediate that does not normally

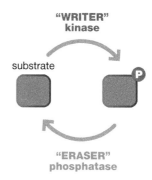

Figure 3.2

Signaling enzymes act as "writers" and "erasers" of information-carrying marks. Phosphorylation of a protein acts as a mark that can control its activity. Kinases catalyze the writing of this mark (phosphorylation reaction), while opposing enzymes—phosphatases—catalyze the erasure of this mark (dephosphorylation). Many other cellular signal-processing systems are also jointly controlled by analogous opposing regulatory enzymes.

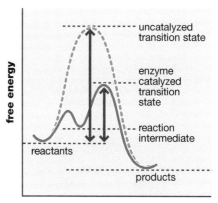

reaction coordinate

Figure 3.3

Enzymes catalyze chemical reactions by lowering the free energy of the transition-state barrier. Free-energy diagram illustrating the effect of an enzyme on reaction energetics. An uncatalyzed reaction *(blue dotted line)* is slow because the conversion of reactants to products must normally pass through a high-energy transition state. An enzyme does not change the ground-state energies of the reactants and products, but increases the rate of conversion by reducing the free-energy barrier to the reaction, sometimes introducing a new reaction intermediate, as shown here by the *pink* line. This and other mechanisms by which enzymes alter the reaction energetics are discussed in the text.

occur in the uncatalyzed reaction (see Figure 3.3). If the free-energy barriers are lower along this new reaction path, then the rate will be faster (this is akin to finding an alternative path through a mountain range, which, though less direct, requires less cumulative ascent). In the context of signaling enzymes, a typical example of this type of catalysis involves the use of enzyme functional groups as nucleophiles, leading to the formation of covalent intermediates. Several phosphatase enzymes that remove phosphate groups from protein residues form enzyme-bound covalent intermediates as part of their catalytic cycle.

Methods to analyze the catalytic power of enzymes quantitatively (Michaelis–Menten analysis) are discussed in Chapter 15. However, we will briefly introduce here two terms that are often used to describe enzymatic properties. The **catalytic rate constant (k_{cat})** describes the maximum rate of reaction that can be achieved by the enzyme (that is, in the presence of saturating substrate concentration). The **Michaelis–Menten constant (K_m)** is a measure of the affinity of the enzyme for its substrate. The K_m is the substrate concentration at which half-maximal reaction velocity is achieved; if the K_m is relatively low, the reaction will proceed efficiently, even at relatively low concentrations of substrate.

Enzymes can drive reactions in one direction by energetic coupling

Enzymes do not alter the thermodynamic equilibrium between reactants and products; rather, enzymes accelerate the approach to equilibrium. Many biologically important signaling reactions are thermodynamically unfavorable, however. That is, the free energy of the products is greater than that of the reactants, so at equilibrium reactants would predominate over product. For example, protein phosphorylation is thermodynamically unfavorable (for tyrosine phosphate) or approximately neutral (for serine/threonine phosphates) relative to inorganic phosphate and the unphosphorylated protein. To overcome this fundamental problem, signaling reactions are often coupled to highly energetically favorable reactions that drive the overall process in only one possible direction. For example, phosphorylation of protein targets is coupled with cleavage of the high-energy β–γ phosphodiester bond of ATP. Thus, a kinase reaction will be essentially irreversible as long as the cell supplies a sufficiently high concentration of ATP, which is normally the case.

By the same logic, removal of phosphate from proteins cannot be accomplished through the exact reverse reaction (that is, regeneration of ATP), because this would be thermodynamically unfavorable. Instead, dephosphorylation proceeds through an alternative reaction: hydrolysis of the phosphoester linkage to the protein by water. This reaction produces inorganic phosphate and the free, unphosphorylated protein. This process is thermodynamically favorable for tyrosine phosphates and approximately neutral for serine/threonine phosphates. Nevertheless, under physiological conditions, hydrolysis of serine/threonine phosphates is driven to completion by the large excess of water relative to the concentration of protein (remember that the rates of chemical reactions are proportional to the concentrations of the reactants—the molar concentration of water is ~55 M). Thus, both phosphorylation and dephosphorylation are energetically driven in one direction, forming a unidirectional reaction cycle (**Figure 3.4**), ultimately powered by the cell's constant supply of ATP. The energy provided by ATP functions analogously to the electric power that is required to drive an electronic circuit.

Because uncatalyzed phosphorylation and dephosphorylation reactions are incredibly slow, the degree to which a substrate is phosphorylated will be determined by the kinetics of the opposing enzymatic phosphorylation and dephosphorylation reactions and not by the thermodynamic stability of one state over the other. An important consequence of this behavior is that the output of the system (the degree

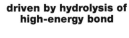

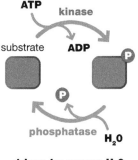

Figure 3.4

Phosphorylation/dephosphorylation reactions form a unidirectional, energetically driven cycle. Both the kinase-catalyzed phosphorylation reaction and the phosphatase-catalyzed dephosphorylation reaction are energetically favorable, allowing the reactions to form a unidirectional cycle that can alternate between the phosphorylated and dephosphorylated state. Energy provided by cellular ATP synthesis drives this cycle.

of substrate phosphorylation) will essentially be a readout of the relative activities of upstream enzymes—a kinase and a phosphatase. These are ideal properties for a dynamic system that must transmit both positive and negative signals rapidly. We will see many times that signal transduction mechanisms exploit and manipulate reaction kinetics to transmit information.

ALLOSTERIC CONFORMATIONAL CHANGES

The ability of upstream signaling inputs to induce changes in protein conformation, which in turn lead to changes in protein activity, is fundamental to many signaling mechanisms. This coupling allows enzymes to act as nodes that can relay and process information. In this section, we will consider the basis for conformational change in proteins, and the different types of rearrangements that are commonly seen in signaling enzymes.

Conformational flexibility of proteins enables allosteric control

Signaling mechanisms take advantage of the intrinsic conformational flexibility of proteins, which provides a means by which regulatory inputs can modulate protein function. When looking at the crystal structure of a protein, it is easy to think of the molecule as a rigid, static structure. This impression is quite misleading, however. Although most proteins adopt a stably folded state, they are also highly dynamic, constantly sampling a range of slightly different conformational substates (**Figure 3.5**). This continuous jostling motion increases with temperature (it is sometimes called thermal "breathing" of the structure), but occurs even at physiological temperature. Some of these conformational substates will be more stable than others (will have lower free energy), and thus will be adopted by a higher fraction of molecules in the population. Other, less stable substates will be more poorly populated. The relative free energies of these conformational substates, however, can be altered by the binding of other molecules (ligands) or by post-translational modifications. Thus, a previously unstable conformational substate may be stabilized through events such as ligand binding or phosphorylation (**Figure 3.6**).

An allosteric conformational change refers to a ligand- or modification-induced change in the secondary, tertiary, or quaternary structure of a protein. **Figure 3.7** shows schematically how changes such as ligand binding can be associated with stabilization of distinct folded states of proteins. In this illustration, the protein exists in two conformational substates in the absence of ligand, one with low activity and another with high activity. In the case of an enzyme, the active-site residues might be in the correct position to promote catalysis in the active state, but not in the inactive state. In this example, the inactive conformation is intrinsically more stable; thus, in the absence of ligand, the protein has low activity. However, ligand binding alters the relative free energy of the two substates, leading to a relative stabilization of the active conformation over the inactive conformation. Thus, increasing ligand concentration has two energetically linked effects on the

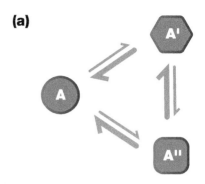
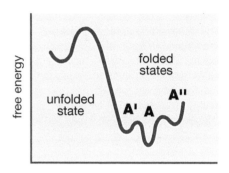

Figure 3.5

A folded protein can exist in multiple conformational substates of slightly differing stabilities. In the cell, multiple conformational states of a protein exist, and their relative abundance at equilibrium is determined by their relative free energy. (a) Conformation A (circle) predominates over conformation A′ (hexagon) or A″ (square). (b) Conformation A has a lower relative free energy than either A′ or A″. Note that all of these conformations have much lower free energy than the unfolded state.

 See Chapter 15 for a review of protein structure.

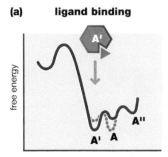
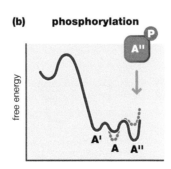

Figure 3.6

Inputs can change the relative stability of distinct conformational substates. (a) Binding of ligand *(orange triangle)* stabilizes conformation A′, lowering its free energy relative to conformations A and A″. (b) Phosphorylation makes conformation A″ the most stable conformation. Dotted lines indicate relative free energies in the absence of ligand binding or phosphorylation.

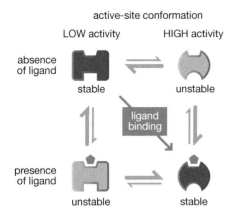

active-site conformation

LOW activity   HIGH activity

absence of ligand — stable / unstable

ligand binding

presence of ligand — unstable / stable

Figure 3.7

The general principle of allosteric regulation. An input such as ligand binding can stabilize an otherwise unstable protein conformation. If this unstable conformation has higher activity, ligand binding increases activity by increasing the fraction of the population in the high-activity conformation.

Figure 3.8

Many kinds of conformational changes are observed in signaling proteins. Regions in the protein can undergo disorder-to-order transitions (or vice versa) leading to changes in secondary structure (illustrated here by formation of an α helix). Proteins could undergo changes in tertiary structure, such as hinge-bending between two subdomains, or the formation of new interdomain interactions. Finally, proteins can undergo changes in quaternary structure, including changes in the molecule's oligomeric state, or reorganization of the monomers within a preformed oligomer. Areas undergoing significant changes in conformation or interaction are highlighted in *light pink*.

protein. First, it leads to an increase in the fraction of proteins with bound ligand. Second, because ligand binding stabilizes the active state, it leads to an increase in the proportion of proteins with high activity. Thus, the protein behaves as a sensor that can be activated in response to increasing ligand concentration.

There are a number of other specific ways in which signaling inputs can affect output by changing the stability of different protein conformations. For example, similar effects can be mediated by covalent modifications such as phosphorylation. Furthermore, such allosteric changes need not be activating—they can also inhibit or otherwise alter protein activity. For example, ligand binding can preferentially stabilize a conformation with lower activity, or one that binds to a different set of interaction partners. Many signaling proteins can respond to multiple allosteric inputs, some positive and some negative. In this way, a protein can integrate many different environmental signals through changes in its conformation.

Signaling proteins employ diverse classes of conformational rearrangements

From a thermodynamic perspective, inputs such as ligand binding or phosphorylation can modify the energies of different conformational substates by either disrupting interactions unique to one state, or introducing new, favorable interactions that stabilize another state. There are many diverse structural mechanisms for these types of conformational changes, some of which are summarized in **Figure 3.8**. Some changes involve transitions between order and disorder, whereby a region of a protein may be unstructured and flexible (adopting no defined secondary structure) in one substate, but structured in another. Indeed, many key regulatory proteins appear to have important **intrinsically disordered regions (IDRs)** that adopt a defined structure upon ligand binding or post-translational modification. Other changes involve tertiary rearrangements. These may involve relative repositioning of atoms within a single folded unit (intradomain changes in tertiary structure). An example of such a change is a hinge-bending motion in an enzyme. In proteins composed of multiple domains, regulatory conformational changes might involve changes in the relative positions of structurally independent modular domains (interdomain changes in tertiary structure). In proteins capable of oligomerization, regulatory changes may be mediated through quaternary changes including oligomerization or a change in the relative orientation of monomers within an oligomer. This type of quaternary structural change is observed in hemoglobin, a classic example of an allosterically regulated system. The binding of oxygen to one subunit in the hemoglobin tetramer induces a change in quaternary structure that increases the oxygen-binding affinity of the other subunits.

disorder/order transitions (secondary structure)

tertiary structure transitions

(e.g., hinge bending)

inter-domain rearrangements

quaternary structure transitions

oligomerization

reorganization of monomers within an oligomer

PROTEIN PHOSPHORYLATION AS A REGULATORY MECHANISM

The addition of the phosphate group to proteins by protein kinases and its removal by protein phosphatases is a frequently used mechanism to regulate protein activity. In this section, we discuss the properties of phosphorylation that lend itself to this role, and the molecular consequences of phosphorylation on protein conformation.

Phosphorylation can act as a regulatory mark

Protein phosphorylation represents a key mechanism for altering the structure and function of target substrate proteins. Several reasons have been suggested to explain why phosphorylation is so commonly used for biological regulation. First, phosphorylation reactions can be coupled to ATP hydrolysis to drive the reaction to completion. Second, although hydrolysis of ATP or of a phosphorylated residue may be highly energetically favorable, these reactions are, by themselves, kinetically unfavorable—hydrolysis is extremely slow in the absence of a catalyst. Thus, these reactions provide excellent points for kinetic control via enzymes that are required to add and remove phosphate groups (kinases and phosphatases). If, for example, ATP or a phosphorylated residue were kinetically less stable (that is, spontaneously hydrolyzed rapidly), then kinases and phosphatases would not be able to provide significant rate enhancement over the uncatalyzed reactions, and could serve little regulatory function. This combination of thermodynamic instability but kinetic stability may help explain why phosphate ester bonds drive so many biological processes.

In eukaryotes, proteins are most commonly phosphorylated (marked) on the three hydroxylated amino acid residues: serine, threonine, and tyrosine (**Figure 3.9**). Phosphorylation on histidine and aspartate commonly occurs in prokaryotes. Serine and threonine phosphorylation is controlled by **serine/threonine kinases** (Ser/Thr kinases), which write these marks, and by **serine/threonine phosphatases** (Ser/Thr phosphatases), which erase these marks. Modification of serine and threonine is controlled by the same class of enzymes, because both amino acids have short side chains that are stereochemically identical except for a single methyl group. By contrast, the bulky aromatic tyrosine side chain sterically requires a larger active site and therefore a distinct class of enzyme in most cases. Tyrosine phosphorylation is controlled by **tyrosine kinases** (Tyr kinases) and **tyrosine phosphatases** (Tyr phosphatases). However, there are a few kinases that can phosphorylate serine and tyrosine residues, as well as dual-specificity phosphatases (DSPs) that have the flexibility to accommodate both phosphotyrosine and phosphoserine/threonine substrates.

Phosphorylation can either disrupt or induce protein structure

Protein phosphorylation is a common mechanism for regulating protein structure and function. What explains the ability of phosphoryl modification to promote conformational changes? The phosphate group is relatively compact, but it represents a major chemical change in the protein. A newly introduced phosphate moiety (pK$_a$ 6.7) is most likely to carry a double negative charge at neutral pH. The introduction of this large electrostatic perturbation (relative to the unmodified residue, which is uncharged) at specific sites in a protein can have dramatic conformational effects, both disrupting previously existing interactions and generating new ones.

From structural analysis of proteins that are phosphorylated, we know that phosphorylation can alter both local structure (regions near the site of phosphorylation) and long-range (tertiary or quaternary) structure. At the local level, introduction of a new phosphate group on an amino acid can lead to dramatic steric or electrostatic effects. Such effects can disrupt an otherwise structured part of the

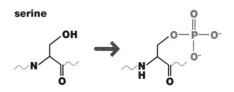

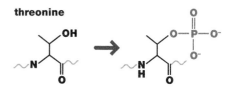

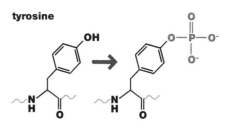

Figure 3.9

The chemical structures of common protein phosphorylation modifications. In eukaryotes, the most common phosphorylated residues are serine and threonine. Tyrosine phosphorylation is less common, but also plays a central role in signaling in multicellular organisms. The phosphate group is highlighted in *pink*. At neutral pH, the phosphate modification is likely to introduce two new negative charges.

See Chapter 4 for a more detailed discussion of different types of protein post-translational modifications

Figure 3.10

Protein phosphorylation can both disrupt and induce structure. (a) Introduction of a phosphate modification can sterically or electrostatically disrupt interactions previously made by the modified residue (or nearby residues). Here, phosphorylation disrupts a hydrogen bond (pink dotted line) between an α helix and an adjacent strand of a β sheet. As a consequence, the α helix unfolds. (b) A new phosphate group can also form new electrostatic or hydrogen-bonding interactions with other parts of the protein, inducing new structure. Here, an introduced phosphate hydrogen-bonds with two positively charged residues, allowing a previously unstructured region to fold into an α helix.

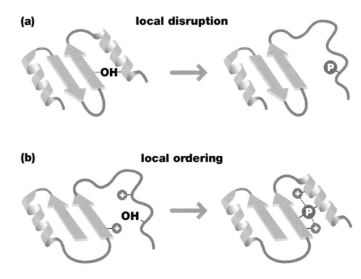

(a) local disruption

(b) local ordering

protein (for example, an α helix) in the region bearing the phosphorylated residue (**Figure 3.10a**). Disruption of local structure is observed if phosphorylation results in repulsion of another nearby negative charge, or disruption of a hydrogen bond made by the nonphosphorylated form of the side chain. The resulting conformational changes could, for example, move active-site residues of an enzyme out of position, resulting in a loss of activity.

In other cases, introduction of a phosphate group can lead to the formation of new local structures (**Figure 3.10b**). In such cases, a newly introduced phosphate group often participates in new interactions with nearby positively charged moieties. Structural analysis has revealed two common types of interactions between the phosphate group and the rest of the protein. First, a phosphate group can engage in hydrogen bonds with the main-chain amide groups at the N-terminus of an α helix (**Figure 3.11a**). Such interactions help to neutralize the intrinsic dipole moment of the α helix (its propensity to carry a positive charge at the N-terminus and a negative charge at the C-terminus). The second major type of interaction is between the phosphate group and the positively charged guanidinium group of one or more arginine side chains (**Figure 3.11b**). Such electrostatic interactions can have the effect of inducing the formation or stabilization of local structure. In many cases, a new phosphoryl group will both disrupt preexisting interactions and induce new ones, leading to significantly different structures in the unphosphorylated and phosphorylated states. As detailed below, this type of phosphorylation-induced local ordering is commonly observed in the activation loop of protein kinases, where only the phosphorylated loop adopts a conformation that is compatible with substrate binding and catalysis.

Phosphorylation can also have longer-range effects on tertiary and quaternary structure (**Figure 3.12**). For example, phosphorylation can prevent the binding of a protein to a partner molecule or to another domain of the same protein. Addition of the negatively charged phosphate group to a binding surface may sterically and electrostatically block ligand or substrate binding, thus negatively regulating protein activity. Conversely, if the ligand is a negative regulator, then phosphorylation

 See Chapter 10 for more on modular interaction domains.

Figure 3.11

Common interactions made by phosphate groups in proteins. The doubly charged phosphate group is commonly observed to participate in several specific types of interactions. (a) It can interact with two successive free amide groups at the N-terminus of an α helix. The helix has a dipole moment oriented with positive charge at the N-terminal end; interaction with phosphate neutralizes this positive charge. (b) Phosphate can interact with two nitrogens in the guanidinium group of an arginine side chain.

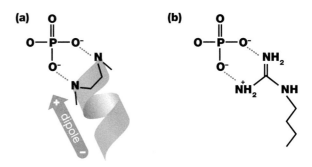

would result in an increase in protein activity. Phosphorylation can also promote new long-range intramolecular and intermolecular interactions. Signaling proteins often contain protein interaction domains, such as SH2 domains, that specifically recognize phosphorylated amino acid motifs. Interactions with these domains will only occur when the target protein is phosphorylated. This type of long-range, phosphorylation-dependent conformational change is a common method of functionally linking phosphorylation to enzyme activity, as detailed below for Src family kinases.

PROTEIN KINASES

Protein kinases are among the most numerous signaling enzymes in eukaryotes, playing a central role in many important signaling pathways. While protein kinase catalytic domains have a common overall structure and catalytic mechanism, these common elements are regulated in a variety of ways to allow kinases to adapt to different modes of control and to phosphorylate different classes of substrates. Since kinases are a canonical example of signaling enzymes, here we review the fundamental mechanisms of kinase catalysis and the ways in which this catalytic activity can be regulated by allosteric inputs.

The structure and catalytic mechanism of protein kinases are conserved

Both Ser/Thr kinases and Tyr kinases share a common folded structure for their catalytic domains. The fold of this bilobed structure, shown in **Figure 3.13**, is similar to those of distantly related metabolic kinases that phosphorylate small-molecule substrates. The active site of the enzyme, including binding sites for ATP and substrate peptide, lies at the groove between the two lobes of the domain.

The reaction that is catalyzed by kinases requires nucleophilic attack on the ATP γ phosphate by the hydroxyl group of a substrate amino acid (**Figure 3.14a**). Thus, the enzyme must be sterically compatible with both ATP and peptide substrates, and must provide properly positioned catalytic residues that increase the nucleophilicity of the attacking hydroxyl group, and stabilize the transition state and developing charge on the leaving group.

Given the precise requirements for this class of reactions, it is not surprising that protein kinases have a number of highly conserved catalytic residues, even across both Tyr kinases and Ser/Thr kinases. We will describe these key conserved catalytic residues using the residue numbering from cyclic AMP-activated *protein kinase A (PKA)*, a Ser/Thr kinase that was one of the first protein kinases to be structurally characterized (Figure 3.14a); other individual kinases have diverse residue numbering (and different insertions and deletions within the catalytic domain). The N-terminal lobe contains residues that are primarily involved in ATP coordination. This includes the phosphate-binding loop (P-loop), which is a flexible glycine-rich

long-range disruption

long-range ordering

Figure 3.12

Long-range effects of protein phosphorylation. (a) Phosphorylation at the interface of a tertiary or quaternary structural interaction can disrupt the interaction. (b) Phosphorylation that creates a docking site for a phospho-recognition domain can result in the formation of a new tertiary or quaternary interaction. The illustration depicts tertiary (intramolecular) rearrangements.

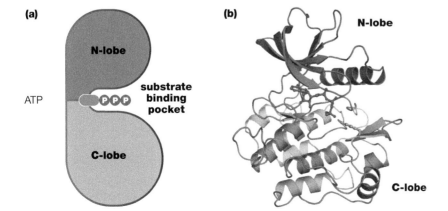

Figure 3.13

Structure of protein kinases.
(a) Schematic depiction of a protein kinase catalytic domain, showing the characteristic bilobed structure. The active site, which binds ATP and peptide substrates, lies between the N- and C-terminal lobes.
(b) X-ray crystal structure of the insulin receptor kinase domain in complex with ATP and substrate peptide (PDB 1IR3). Colors correspond with the diagram in part (a); peptide substrate is indicated in *orange*.

Figure 3.14

The active site of protein kinases. (a) Key catalytic residues are contributed by both the N- and C-lobes, including Lys72 and Asp184, which help to coordinate the ATP, and Asp166 on the catalytic loop *(orange)*, which acts as a general base to activate the substrate hydroxyl moiety. Proper positioning of these and other catalytic residues is dependent on tertiary interactions with key elements in the kinase structure, especially the C-helix *(purple)* in the N-lobe and the activation loop *(green)* in the C-lobe. In many kinases, the C-helix and activation loop are used as structural levers for controlling kinase activity. (b) Cyclin-dependent kinase (CDK) activation is regulated by proper positioning of the C-helix *(purple)* and the activation loop *(green)*. Dramatic repositioning of the C-helix occurs upon cyclin binding, while proper positioning of the activation loop occurs upon activating phosphorylation of Thr197. Together, these conformational changes lead to proper assembly of the active site and clearance of the peptide-binding site. Numbering of key residues is based on their positions in the protein kinase A (PKA) sequence. (b, Adapted from M. Huse and J. Kuriyan, *Cell* 109:275–282, 2002. With permission from Elsevier.)

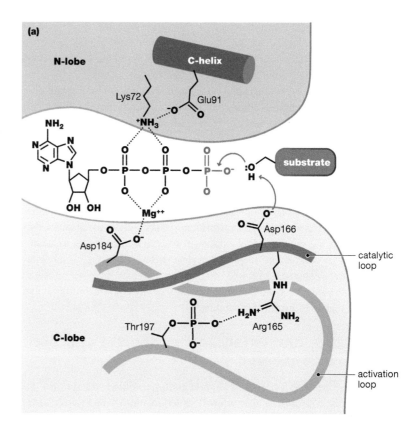

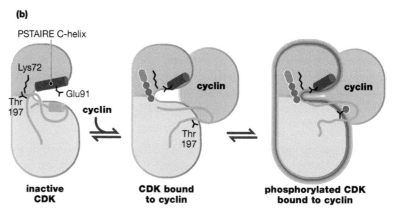

segment. In addition, Lys72 plays a critical role in coordinating the negatively charged phosphate moieties of ATP. Many of the remaining critical catalytic residues are contained in the slightly larger C-terminal lobe. Asp166 lies within a segment known as the catalytic loop; this residue serves as the general base that abstracts the proton from the attacking substrate peptide hydroxyl group. Two other C-lobe residues, Asp184 and Asn171, play a critical role in ATP binding.

The activation loop and C-helix are conserved molecular levers that conformationally control kinase activity

As signaling molecules, protein kinases typically function as switches that must exist in at least two conformations: an inactive conformation in which active-site residues are not aligned for catalysis; and an active conformation, stabilized by the proper upstream inputs, in which the active-site residues are optimally positioned for catalysis. There are several core mechanisms that are used by most protein kinases to regulate catalytic activity. Most of these mechanisms rely on a conserved set of structural elements—the activation loop and the C-helix, highlighted in Figure 3.14. Because of their central position and interactions with many catalytic residues, these two elements essentially act as regulatory levers that can control kinase activity. Diverse regulatory inputs that shift these levers result in the movement of catalytic residues in or out of proper position.

The **activation loop** is perhaps the most important regulatory segment in the protein kinase family. This loop is within the C-lobe and lies at the active site. The length, sequence, and conformation of the activation loop differ in individual kinases. However, in nearly all kinases, the activation loop is observed to undergo large conformational changes associated with distinct activity states (**Figure 3.14b**).

The activation loop controls kinase activity through two coordinated mechanisms. First, residues in the activation loop often participate in hydrogen bonds with residues immediately adjacent to catalytic residues. This structural coupling means that the position of the activation loop can alter catalytic efficiency. Second, the activation loop, in some cases, can directly occlude the binding site for the peptide substrate, thus blocking enzyme activity. In such cases, activating stimuli move the loop out of the way, allowing substrate access to the active site. In most protein kinases, activation requires phosphorylation of residues within the activation loop. These phosphorylation events usually alter the conformation of the loop, both exposing the peptide-binding site and properly positioning the catalytic residues.

The C-helix is another regulatory lever within the kinase domain. This helix is in the N-lobe and contains a conserved residue, Glu91, which forms a hydrogen-bond network with the conserved catalytic residue, Lys72. Precise positioning of Lys72 is required for kinase activity and its interaction with Glu91 ensures it is properly positioned for ATP binding. Thus, movement of the C-helix, and the subsequent movement of the Glu91-Lys72 pair, can dramatically alter the enzyme activity level. An example is provided by the **cyclin-dependent kinases** (**CDKs**), serine/ threonine kinases that control many important cellular events including the cell cycle. CDK regulation centers around the C-helix and activation loop. In the inactive CDK structure, the C-helix and the catalytic Lys72 residue are far out of optimal position. Activating inputs (particularly the binding of **cyclin**, a regulatory subunit) induce a large translation and 90° rotation of the C-helix, resulting in the proper positioning of the Glu91–Lys72 pair (Figure 3.14b).

Insulin receptor kinase activity is controlled via activation-loop phosphorylation

Control of kinase activity by phosphorylation of the activation loop is illustrated in the example of the insulin receptor tyrosine kinase (IRK) (**Figure 3.15**). The

 See Chapter 14 for more on the cell cycle and cyclin dependent kinases.

Figure 3.15

Activation-loop phosphorylation stabilizes the active conformation of the insulin receptor tyrosine kinase (IRK). (a) In the inactive, unphosphorylated state of the insulin receptor kinase, Tyr1162, located in the activation loop *(green)*, forms a hydrogen bond with the catalytic residue Asp1132 in the catalytic loop *(orange)*. This interaction occludes the active site. (b) In the active form of the kinase, three tyrosine residues (Tyr1158, Tyr1162, and Tyr1163) in the activation loop are phosphorylated (phosphates are highlighted in *red*). Phosphorylation of Tyr1162 (pTyr1162) disrupts the inhibitory interaction between it and the catalytic residue Asp1132, and pTyr1162 and pTyr1163 then form salt bridges with Arg residues elsewhere in the protein. The overall effect of these changes is a rightward shift of the activation loop *(green arrow)*, which opens up access to the active site for the peptide substrate *(blue)*. Note also that the N-terminal lobe of the kinase has rotated relative to the C-terminal lobe upon activation. ATP is not shown for clarity.

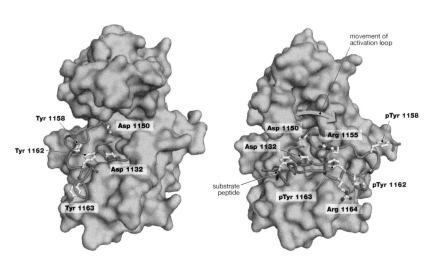

insulin receptor binds to insulin, the key hormone in glucose homeostasis, and mediates the physiological responses to it in cells throughout the body. The IRK exists as a heterotetramer that contains two identical catalytic domains linked by disulfide bonds. Activation of IRK requires phosphorylation on three residues in its activation loop: Tyr1158, Tyr1162, and Tyr1163. In the inactive, unphosphorylated state, Tyr1162 is hydrogen-bonded to the active-site Asp1132 residue located in the catalytic loop. This interaction positions the activation loop in a way that prevents both peptide and ATP binding. Phosphorylation of Tyr1162 destabilizes the autoinhibitory conformation of the activation loop, exposing the other tyrosines for phosphorylation. Phosphorylation of the loop then promotes new interactions between the phosphorylated tyrosines and Arg1155 and Arg1164, which configure the activation loop for catalysis and expose the catalytic cleft to substrates.

Phosphorylation mediates long-range conformational regulation of Src family kinases

The *Src family kinases,* a group of cytosolic tyrosine kinases that regulate processes such as cell adhesion and lymphocyte activation, illustrate how phosphorylation can induce long-range conformational changes that regulate activity (**Figure 3.16**). In particular, phosphorylation of a tyrosine side chain in the C-terminal tail of Src kinases (Tyr527) inhibits kinase activity through large-scale rearrangements involving two regulatory domains. In addition to the kinase domain, Src family members have two protein interaction modules: an SH2 and SH3 domain. The SH2 domain recognizes phosphotyrosine (pTyr) residues and can bind the pTyr on the C-terminal tail. This binding, in turn, promotes the interaction of the SH3 domain with another intramolecular recognition motif in the linker between the SH2 domain and the catalytic domain. These two intramolecular interactions together induce an inactive conformation of the kinase domain, essentially stabilizing a conformation that cannot efficiently phosphorylate substrates. Disruption of these autoinhibitory interactions, either by dephosphorylation of tyrosine in the tail or by binding of the SH2 or SH3 domains to competing external ligands, activates the enzyme.

This type of regulation depends on the dynamic nature of protein conformation. In order for the kinase to be activated in response to signals, the inactive conformation must be relatively unstable. For example, if we imagine that the SH2–pTyr527 interaction had very high affinity (was energetically very favorable), the two would almost never be dissociated. The pTyr527 site would not be exposed to phosphatases for dephosphorylation, and the SH2 domain would not be free to bind to other phosphotyrosine sites. Activation would be infrequent and would take a very long time. For this reason, intramolecular regulatory interactions usually have relatively low affinity, allowing for dynamic regulation.

Figure 3.16

Phosphorylation controls tertiary structure and activity of the Src family kinases. (a) Structure of the Src family tyrosine protein kinase Hck in its inactive form. The catalytic domain is linked to two small protein-binding domains *(orange)*—an SH3 domain and an SH2 domain—that make intramolecular interactions that clamp the catalytic domain in its inactive conformation. These intramolecular interactions are disrupted upon activation, allowing the SH2 and SH3 domains to interact with other proteins. The C-helix is colored *purple,* activation loop *green,* and catalytic loop *orange.* Side chains of Tyr416 and pTyr527 are shown in stick format. (b) Schematic representation of the inactive and active conformations of a generic Src family kinase. Phosphorylation of the C-terminal tyrosine (Tyr527) stabilizes the inactive conformation through interaction with the SH2 domain, while the active conformation is stabilized by phosphorylation of Tyr416 in the activation loop. Note that the active conformation can also be stabilized by binding of the SH2 or SH3 domains to ligands in *trans,* even if the C-terminal site is phosphorylated.

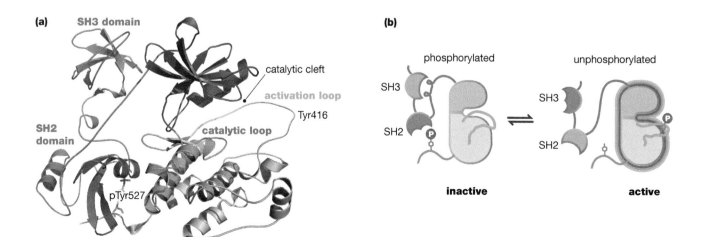

(a) SH3 domain

SH2 domain

catalytic cleft

activation loop

Tyr416

catalytic loop

pTyr527

(b)

SH3

SH2

phosphorylated

P

inactive

SH3

SH2

unphosphorylated

P

active

Full activation of the Src kinase also requires the subsequent autophosphorylation of Tyr416 within the activation loop, similar to other kinases such as IRK discussed above. Phosphorylation of Tyr416 is generally performed by another Src family kinase in *trans*, for example, after kinase aggregation or co-localization in response to signals. Thus, Src is actually controlled by phosphorylation in two ways: a negative regulatory phosphorylation at the C-terminal tail is responsible for tertiary interactions that lock the kinase in an inactive conformation, while a positive regulatory phosphorylation in the activation loop is required to locally order the active site for optimal activity.

Multiple binding interactions regulate protein kinase substrate specificity

Eukaryotic cells contain many different protein kinases (~500 distinct genes in humans; ~100 in yeast), raising the question of what determines which specific proteins will serve as substrates for each kinase. The **substrate specificity** of kinases is determined by a combination of factors (**Figure 3.17**). First, as for most enzymes, the active sites of kinases help determine specificity. For example, the substrate-binding pocket of Tyr kinases is considerably deeper than that of Ser/Thr kinases, which allows them to specifically accommodate the much larger tyrosine side chain. In addition, the active site often has adjacent pockets that show preferences for particular side chains at residues flanking the residue to be phosphorylated. CDKs, for example, almost exclusively phosphorylate serine or threonine residues that are immediately followed by a proline. However, unlike most canonical enzymes, kinases also use interactions that lie far outside of their active sites to determine substrate specificity.

Many Ser/Thr kinases have **docking sites** on the kinase domain that are distant from the active site and which recognize specific peptide motifs. Kinase substrates often have a combinatorial structure with a docking motif and a substrate motif separated by a linker element; the presence of two specificity-determining sequences greatly increases the ability of the kinase to discriminate among different substrates. In some cases, docking sites can play a regulatory role. For example, several members of the AGC Ser/Thr kinase family have a docking site on the N-terminal lobe referred to as the PIF pocket or hydrophobic motif pocket. Peptide docking motifs that bind in the PIF pocket not only aid in determining substrate specificity, but they can also allosterically activate the catalytic domain of the kinase.

Other domains that lie outside of the kinase domain can also help determine substrate specificity; these are known as accessory domains. For example, the SH2 and SH3 domains of Src family kinases are important both in regulation of kinase activity and in substrate recognition. Proteins that contain SH2- and SH3-binding motifs can serve as optimal substrates because these interactions preferentially target the kinase to these substrates. And because the SH2 and SH3 domains normally inhibit kinase activity through intramolecular interactions, as discussed above, substrate binding also serves to activate the kinase. In this way, maximal activity is attained only when the kinase is bound to an appropriate substrate, decreasing the likelihood of off-target activity that could have deleterious effects.

In some cases, accessory subunits that noncovalently associate with a kinase domain can help determine substrate specificity. For example, the cyclins that associate with CDKs have two major functions: as described earlier (see Figure 3.14b), they are allosteric activators that cause the kinase to adopt an activated conformation; but they also contain a docking site, so that proteins containing a complementary docking motif are preferentially phosphorylated by CDKs (typically, conformational changes that activate the kinase lead to an increased k_{cat}, while additional binding interactions lead to lower values of K_m). CDKs can associate with different cyclin subunits at different stages of the cell cycle. Since each of these cyclin

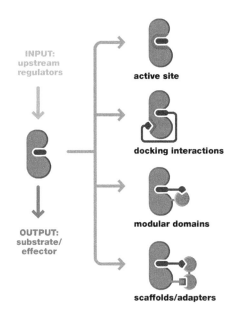

INPUT: upstream regulators

OUTPUT: substrate/ effector

active site

docking interactions

modular domains

scaffolds/adapters

Figure 3.17

Mechanisms of substrate specificity in protein kinases. Multiple mechanisms that contribute to substrate specificity are depicted, with specificity-determining regions highlighted in *pink* and substrates in *dark brown*. The active site of a kinase can recognize specific sequence motifs flanking the residue to be phosphorylated. Some kinases, particularly Ser/Thr kinases, have docking sites away from the active site that can recognize docking-motif peptides within substrates. Kinases may also contain modular binding domains that recruit target substrates. Finally, kinases can use third-party accessory proteins, such as adaptors or scaffolds. These proteins bind the kinase, but also provide interaction sites for specific substrates. When bound to different scaffolds or adaptors, a kinase might display quite different substrate specificities, thereby dramatically increasing the functional flexibility of one kinase. (Adapted from R.P. Bhattacharyya et al., *Annu. Rev. Biochem.* 75:655–680, 2006. With permission of Annual Reviews.)

subunits recognizes distinct docking motifs, this provides a mechanism for CDKs to phosphorylate different sets of substrates at different times. Often the broad class of accessory proteins that help direct kinases and other signaling enzymes toward specific substrates are known as scaffold proteins, which are discussed in detail at the end of this chapter.

As can be seen by these examples, accessory domains and subunits often play a role in regulating kinase activity and determining its resulting substrate specificity. It is worth noting that tyrosine kinases generally use diversification of accessory domains to achieve distinct specificity—these kinases tend to be modular, multi-domain proteins. In contrast, serine/ threonine kinases tend to be smaller proteins that achieve distinct substrate specificity by associating with different accessory subunits.

Histidine kinases and prokaryotic two component systems are described in Chapter 4.

Protein kinases can be divided into nine families

The canonical protein kinases, for the most part, appear to have evolved in eukaryotes. Prokaryotes do have a distinct type of kinase, histidine kinase, which is not found in most eukaryotes. Genomic sequencing studies have revealed that prokaryotes also have a number of Protein Kinase-Like (PKL) genes, which appear to comprise an ancient, distantly related family (which also includes some atypical eukaryotic kinases). The functions of prokaryotic kinase-like genes have not been well characterized.

Within eukaryotes, the protein kinase family has expanded dramatically. In the human genome, there are ~500 putative protein kinases, making them one of the largest families of enzymes in the genome. Of these, Ser/Thr kinases far outnumber the Tyr kinases: there are ~400 Ser/Thr kinases and ~90 Tyr kinases. Because of their greater number and their distribution among diverse eukaryotic species, Ser/Thr kinases are thought to have evolved considerably earlier than Tyr kinases.

Tyrosine kinases are found almost exclusively in metazoans (multicellular animals). The exception to this rule is that they are also found in choanoflagellates, the closest unicellular organisms to metazoans. This observation has led to the proposal that tyrosine kinase signaling first emerged about one billion years ago, near the point at which multicellular organisms evolved. The added signaling capacity provided by this novel system may have played a key role in facilitating the emergence of multicellular animals, with their increased needs for cell–cell signaling.

The evolution of signaling machinery during the emergence of metazoans is discussed in Chapter 13.

In humans, the protein kinase family can be divided into nine distinct subfamilies based on sequence homology. In addition to the tyrosine kinase (TK) family, there are eight families of Ser/Thr kinases (**Figure 3.18**). Presumably, the utility of the protein kinase as a fundamental regulatory device led to its expansion and diversification into this large number of distinct subfamilies.

PROTEIN PHOSPHATASES

Protein phosphatases catalyze the removal of covalent phosphate modifications on protein side chains. As mentioned above, this reaction functionally opposes the phosphorylation reaction catalyzed by kinases, although it is not the chemical reverse reaction. Rather, the phosphatase reaction is a distinct reaction that cleaves a phosphorylated protein into inorganic phosphate and the free protein. Together, the kinase and phosphatase reactions form a unidirectional cycle in which both reactions are under kinetic control.

As with kinases, protein phosphatases are classified based on the phosphorylated amino acid substrates on which they act. Thus, there are two classes: the serine/ threonine (Ser/Thr) phosphatases and tyrosine (Tyr) phosphatases.

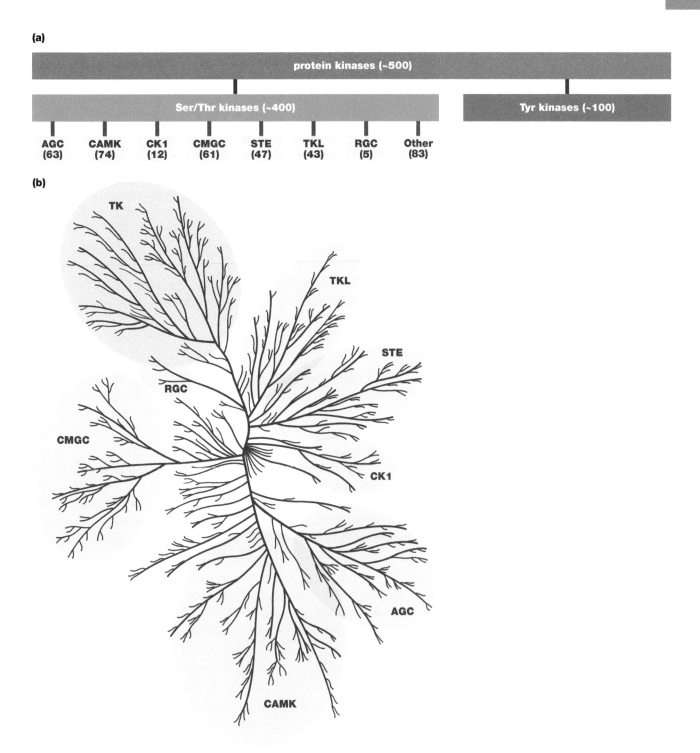

Figure 3.18

Subfamilies of protein kinases. (a) Classes of Ser/Thr kinases and Tyr kinases, with the approximate numbers in each class within the human genome indicated. Canonical members of each class include protein kinase A (AGC), calmodulin-dependent protein kinase (CAMK), casein kinase 1 (CK1), cyclin-dependent kinase (CMGC), MAP/Erk kinase (STE), Raf (TKL), receptor guanylyl cyclase A (RGC), and Src (tyrosine kinase). Kinases classified in "other" include well-defined families, such as the Polo family, that cannot be grouped with one of the major classes. (b) Phylogenetic tree showing the relationship between human kinases. (b, Adapted from G. Manning et al., *Science* 298:1912–1934, 2002. With permission from AAAS.)

Serine/threonine phosphatases are metalloenzymes

There are three major families of Ser/Thr phosphatases: the PPP, PPM, and FCP families (**Figure 3.19**). The PPP phosphatases share a core catalytic domain of 280 residues, and are found in all eukaryotes ranging from yeast to man. They are also found in bacteria and archaebacteria, indicating an ancient origin from a common pre-eukaryotic ancestor. There are 13 PPP family members in humans and 12 in yeast. PPP phosphatases are metalloenzymes that bind a pair of metal ions (typically Fe^{3+} and either Zn^{2+} or Mn^{2+}). This bimetal center contributes to catalysis in several ways: the metals coordinate and orient the substrate, they stabilize charge building in the transition state, and they activate a water molecule such that it can perform a nucleophilic attack on the substrate phosphate (**Figure 3.20**).

PPP phosphatases show specificity for phosphoserine and phosphothreonine, but in some cases are capable of acting on phosphotyrosine. Typically, these enzymes show only modest sequence specificity for the protein in which the phospho-modification occurs. In most cases, PPP phosphatases are found in complexes with other proteins that determine which specific substrates are targeted. Each of these phosphatases can participate in several distinct complexes, allowing their localization to diverse areas in the cell and participation in a broad range of distinct pathways (**Figure 3.21**). A few PPP members such as PP1 and PP2B (also known as calcineurin) also have docking grooves on their catalytic subunits, which contribute to substrate specificity by recognizing specific peptide motifs in substrates.

The second family of Ser/Thr phosphatases, the PPM phosphatases, includes PP2C and pyruvate dehydrogenase phosphatase. This class of enzymes is also present in eukaryotes, bacteria, and archaebacteria, suggesting a pre-eukaryotic origin. There are 10 PPM proteins in humans. Like the PPP family, the PPM phosphatases are metalloenzymes, but differ from PPP family members in that they require Mg^{2+}. The active site of the PPM phosphatases is structurally similar to that of PPP phosphatases but there is no sequence similarity, suggesting that these two families of phosphatases evolved independently (a case of evolutionary convergence). As with the PPP phosphatases, catalysis by PPM phosphatases occurs via activation of a water molecule that directly hydrolyzes the substrate phosphoester bond. Most PPM family members are monomeric, and thus their substrate preferences appear to be determined largely by accessory domains that occur within the same polypeptide chain as the catalytic domain.

Consistent with their similar mechanisms of action, both the PPP and PPM classes of Ser/Thr phosphatases are sensitive to the inhibitor okadaic acid, which has been an important tool in analysis of signaling pathways involving Ser/Thr phosphorylation.

The third family of Ser/Thr phosphatases, the FCP phosphatases, was discovered relatively recently. These enzymes are dependent on Mg^{2+} and two conserved Asp residues for catalysis, which involves a phosphoryl-Asp reaction intermediate. They dephosphorylate the C-terminal domain of RNA polymerase, which plays an important role in transcriptional regulation. In addition, emerging evidence suggests that these phosphatases can regulate mitogen signaling pathways.

Most tyrosine phosphatases utilize a catalytic cysteine residue

Protein tyrosine phosphatases (PTPs) can be divided into several different families (see Figure 3.19). There are two major classes based on the identity of the key catalytic residue. The cysteine-based phosphatases are far more common than the aspartate-based phosphatases; for example, the human genome encodes 103 Cys-based phosphatases and just 4 Asp-based phosphatases. The Cys-based

(a)

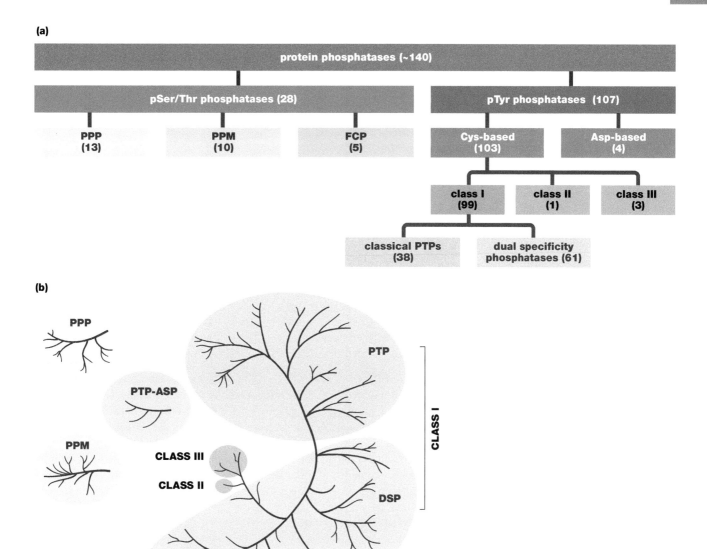

(b)

Figure 3.19

Subfamilies of protein phosphatases. (a) Classes of protein phosphatases, with the approximate numbers in each class within the human genome indicated. Phosphoserine/threonine (pSer/Thr) phosphatases can be divided into three classes: PPP, PPM, and FCP. Canonical members of each class include PP1 and calcineurin (PPP), PP2C (PPM), and FCP1 (FCP). Several phosphotyrosine (pTyr) phosphatases use an active-site aspartate (for example, Eya protein), but the majority use an active-site cysteine and can be divided into three classes. Class I Cys-based phosphatases can be further subdivided into the classical protein tyrosine phosphatases (PTPs), which only act on phosphotyrosine, and the dual-specificity phosphatases (DSPs), which can act on both phosphotyrosine and phosphoserine/threonine. Canonical class I PTPs include PTP1B (a classical PTP) and the MAP kinase phosphatases, which are DSPs. Some members of the DSP family (for example, PTEN) can act on nonprotein substrates such as phospholipids. The only class II PTP in humans is LMPTP, and the three class III PTPs are Cdc25A, Cdc25B, and Cdc25C. (b) Phylogenetic trees showing the relationship between human phosphatases. The large dendrogram shows the large family of class I Cys-based phosphatases, which encompasses both the protein tyrosine and dual-specificity phosphatases. The other unrooted trees show the other more distantly related families of phosphatases. (b, Adapted from S.C. Almo et al., *J. Struct. Funct. Genomics* 8:121–140, 2007. With permission from Springer Science and Business Media.)

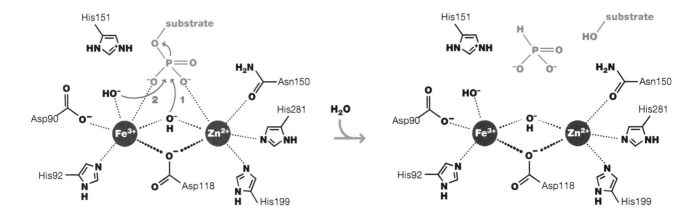

Figure 3.20

Reaction mechanism of serine/ threonine phosphatases. Ser/Thr phosphatases have a bimetal active site, with the metal ions coordinating the phosphorylated substrate. The phosphoryl group is transferred to water in a single step with no covalent intermediate. Two alternative mechanisms have been proposed involving either (1) attack by a bridging hydroxide, or (2) attack by a terminal, metal-bound hydroxide. In either case, a His residue (His151) has been suggested to act as a general acid. Residue numbering is for human calcineurin (PP2B).

phosphatases can be further subdivided into three classes. By far the largest class of Tyr phosphatases in eukaryotes is the class I Cys-based PTPs (99 in humans), which includes the classical PTPs and the dual-specificity phosphatases (DSPs, also known as DUSPs or VH1-like phosphatases). Class II Cys-based PTPs (of which there is only one example in humans) are of low molecular weight and related to bacterial arsenate reductase. Class III Cys-based PTPs include the cell-cycle regulatory phosphatase Cdc25 (there are three Cdc25 enzymes in humans).

The three classes of Cys-based tyrosine phosphatases are fairly diverse, but they share several common features (**Figure 3.22**). All have a catalytic motif comprising cysteine and arginine residues separated by five other amino acids (Cx_5R). The conserved cysteine acts as a nucleophile that attacks the substrate phosphotyrosine, forming a covalent phosphocysteine intermediate and releasing the substrate tyrosine. In the second step of the reaction, a water molecule then hydrolyzes the phosphocysteine intermediate to yield the free phosphate group. The conserved arginine residue in the catalytic motif functions to stabilize the transition state, by coordinating with the phosphoryl group throughout the reaction, while another conserved residue—an aspartic acid—acts as a general acid, donating a proton to the hydroxyl of the substrate tyrosine (the leaving group). The aspartate then acts as a general base to activate water for nucleophilic attack in the second half-reaction.

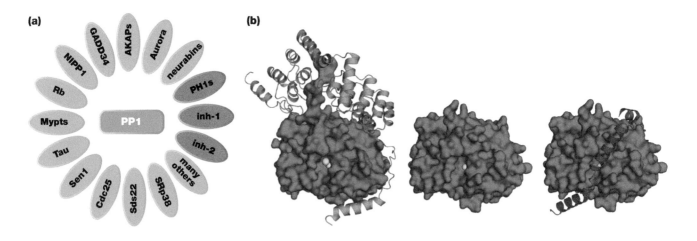

Figure 3.21

Serine/threonine phosphatases can form different holoenzyme complexes. (a) Regulatory proteins, including modulators (green) and inhibitors (pink), control Protein Phosphatase 1 (PP1) function. (b) Structures of PP1 holoenzyme complexes. PP1 (blue) is shown in surface rendering, with catalytic metal ions colored yellow; modulatory or inhibitory subunits are depicted as ribbons. Left: structure of PP1β in complex with the modulator MYPT1 (green). MYPT1 provides additional surfaces for substrate recognition and thus controls PP1 substrate specificity. Center and right: PP1γ in absence (center) and presence (right) of inhibitor-2 (pink). Inhibitor-2 blocks access to the active site. (Adapted from D.M. Virshup and S. Shenolikar, Mol. Cell 33:537–545, 2009. With permission from Elsevier.)

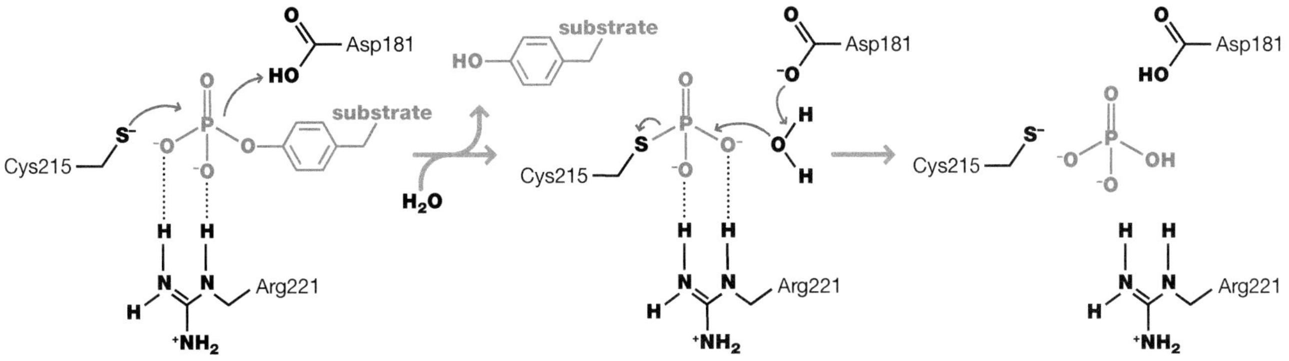

Figure 3.22

Reaction mechanism of tyrosine phosphatases. Class I Cys-based Tyr phosphatases, such as PTP1B or Shp2, utilize an active-site cysteine residue. The cysteine side chain acts as the initial nucleophile, resulting in release of the protein portion of the substrate and formation of a covalent, thiophosphate intermediate. In the second half-reaction, the phosphate is released after nucleophilic attack by water. In some PTPs, activity of the enzyme is regulated by reversible oxidation of the catalytic cysteine (in PTP1B, the catalytic cysteine and a neighboring serine can reversibly form a sulfenyl amide ring).

Other residues that are not absolutely conserved assist in the cysteine nucleophilic attack by lowering its pK_a, thus favoring the more nucleophilic deprotonated form.

Unlike Ser/Thr phosphatases, the Cys-dependent Tyr phosphatases are metal-independent and insensitive to the inhibitor okadaic acid. They are, however, sensitive to the inhibitor vanadate, which stereochemically mimics the pentacoordinate phosphate transition state. Phosphate normally has four coordinated oxygen atoms, but in the nucleophilic displacement of the dephosphorylation reaction, it must pass through a pentacoordinate transition state in which the attacking nucleophile and the leaving group are both present (**Figure 3.23**).

Because these PTPs use a catalytic cysteine residue, they are, in some cases, subject to oxidative regulation. For example, when the phosphatase PTP1B is oxidized, the catalytic cysteine and a neighboring serine residue form a sulfenyl amide ring, which inactivates the enzyme. However, this moiety can be reversibly reduced to restore enzyme activity. This unusual mode of regulation is thought to allow reactive oxygen species and nitrogen oxides to modulate phosphotyrosine signaling.

(a)   pentavalent transition state

(b)   vanadate

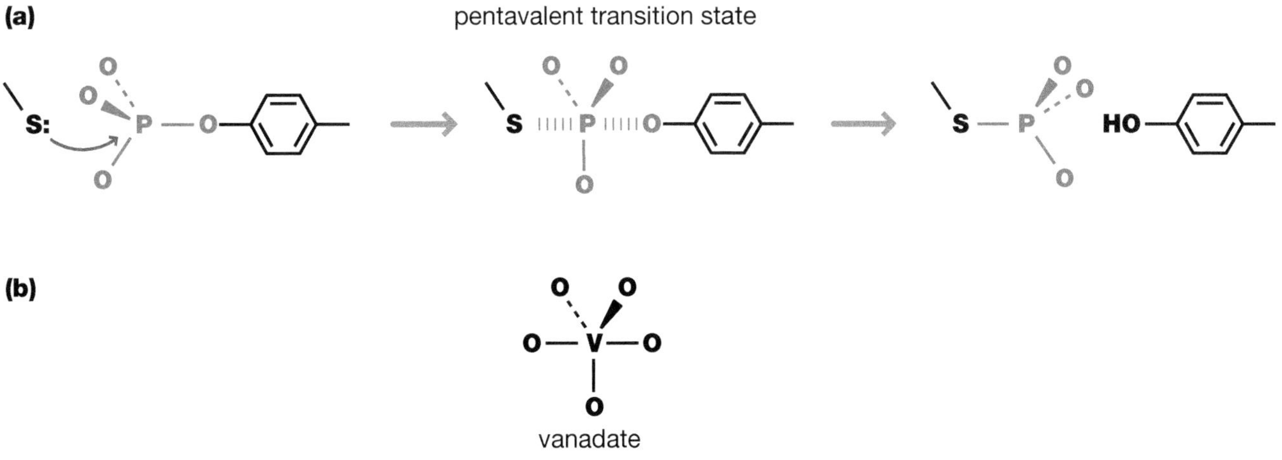

Figure 3.23

Vanadate mimics the transition state of the dephosphorylation reaction. (a) In the catalytic reaction of Cys-based tyrosine phosphatases, nucleophilic attack on the phosphorus atom by the active site cysteine proceeds through a pentacoordinate transition state. (b) This transition state is mimicked by the pentacoordinate structure of orthovanadate, which is a potent inhibitor of tyrosine phosphatases. Note that the dephosphorylation reaction also involves a second pentacoordinate intermediate, when water attacks the covalent phosphoenzyme intermediate to release inorganic phosphate (see Figure 3.22).

As mentioned above, the class I Cys-based PTPs can be further subdivided into two distinct subclasses—the classical PTPs, which only act on phosphotyrosine, and the dual-specificity phosphatases (DSPs). In humans, there are 38 classical PTPs and 61 DSPs. In many cases, the term PTP is used to refer only to the classical PTPs. The classical PTPs are highly selective for phosphotyrosine, most likely because of their deep active-site pocket, which only the longer phosphotyrosine residue can reach into. By contrast, dual-specificity phosphatases can, in some cases, hydrolyze phosphoserine and phosphothreonine. In fact, some members of the DSP family can also act on nonprotein substrates including phospholipids and RNA.

Tyrosine phosphatases are regulated by modular domains while serine/threonine phosphatases often associate with regulatory accessory subunits

Tyrosine phosphatases differ significantly from Ser/Thr phosphatases in their numbers and their protein architectures. There are around 100 Tyr phosphatases and a similar number of Tyr kinases in humans. This contrasts with the Ser/Thr phosphatases, which are vastly outnumbered by Ser/Thr kinases (~30 Ser/Thr phosphatases versus ~400 Ser/Thr kinases).

See Chapter 10 for more on the modular architecture of signaling proteins

Diversification of Tyr phosphatases occurs mostly with respect to protein architecture—most Tyr phosphatases are complex modular proteins in which the catalytic phosphatase domain is one of several modular domains within the same polypeptide chain (**Figure 3.24**). Highly diverse combinations of accessory domains are linked to both classical PTP and DSP domains. There are even PTP domains that are found in transmembrane proteins, in a class of proteins referred to as receptor PTPs. These diverse accessory domains are thought to determine the targeting of these proteins to particular subcellular sites and to specific substrates. The accessory domains can also participate in allosteric autoinhibitory interactions, thus regulating the phosphatase activity much as accessory domains regulate tyrosine kinase activity. This contrasts with the Ser/Thr phosphatases, where individual phosphatase polypeptides generally participate in a number of alternative holoenzyme complexes. In this case, various accessory subunits in the complexes are thought to play key roles in functional diversification: they target the enzyme to specific sites, determine substrate preferences, and mediate regulation.

G PROTEIN SIGNALING

G proteins store and transmit information based on their conformational state, and thus provide another canonical example of the tight coupling between protein activity and conformation. Serving as the basis for one of the most widely used and important signaling mechanisms in eukaryotes, G proteins play crucial roles in diverse signaling pathways including hormone signaling, cytoskeletal regulation,

Figure 3.24

Modular domain structure of tyrosine phosphatases. (a) Domain structures of three representative PTPs, showing multiple accessory domains in addition to the catalytic domain *(blue)*. TM, transmembrane segment. (b) X-ray crystal structure of Shp2 in the inactive state. Two SH2 domains *(yellow and green)* play a dual role in regulating Shp2: they allosterically regulate the phosphatase domain (by occluding the active site in the inactive state), and they target the phosphatase to specific locations and substrates. The active-site cysteine is shown in *pink*.

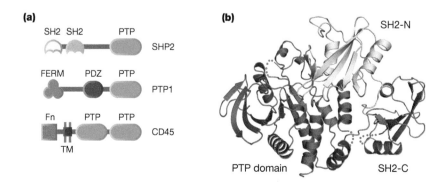

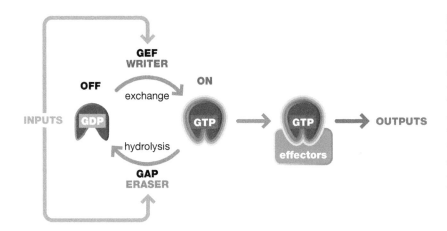

Figure 3.25

G proteins are conformational switches whose state is controlled by two opposing enzymes. The GDP-bound conformation of the G protein is inactive, while the GTP-bound conformation is active and able to bind downstream effectors. Both nucleotide exchange and GTP hydrolysis are extremely slow in G proteins on their own. Activation (exchange of GDP for GTP) is speeded up by guanine nucleotide exchange factor (GEF) enzymes, while inactivation is promoted by GTPase-activator protein (GAP) enzymes. The activity of GEFs and GAPs is regulated by signaling inputs.

vesicular trafficking, and nuclear import and export. G proteins are named for their ability to bind guanine nucleotides—both guanosine triphosphate (GTP) and guanosine diphosphate (GDP). The primary function of G proteins is to serve as conformational switches: they adopt significantly different conformations depending on whether GTP or GDP is bound (**Figure 3.25**). In general, the GTP-bound state is the "active" conformation, which is capable of binding and modulating the activity of downstream effectors, while the GDP-bound state is the "inactive" conformation, with much lower affinity for these effectors.

G proteins are conformational switches controlled by two opposing enzymes

In the absence of regulatory inputs, cycling between the active and inactive states of G proteins is very slow. G proteins can hydrolyze bound GTP to GDP (thereby switching themselves from the active to inactive conformation), but they are very poor enzymes. In fact, the reaction halftime for GTP hydrolysis (the time at which there is a 50% chance that GTP will have been cleaved to GDP and P_i) ranges from several minutes to more than an hour for different G proteins. In addition, the release of bound GDP after hydrolysis is also very slow—both GDP and GTP bind G proteins with high affinity (K_d in the nM to pM range), and thus their dissociation rates (k_{off}) are necessarily low.

This system is very useful for signaling, however, because regulatory enzymes can overcome these kinetic barriers. The nucleotide-binding state of G proteins is controlled by two opposing enzymes: **guanine nucleotide exchange factors** (**GEFs**), which activate G proteins, and **GTPase-activator proteins** (**GAPs**), which inactivate them. GEFs activate G proteins by catalyzing the release of GDP and the subsequent binding of GTP. GAPs inactivate G proteins by catalyzing the hydrolysis of bound GTP to GDP. Thus, these opposing enzymes form a kinetically controlled "writer/eraser" system analogous to kinase/phosphatase systems described earlier in this chapter.

G protein regulation is also similar to phosphorylation because the cycle of regulatory reactions, while under tight kinetic control, is thermodynamically favorable. Inactivation of a G protein is favorable because it is coupled to hydrolysis of the high-energy phosphodiester bond of GTP. Activation of a G protein, via rebinding of GTP, is favorable because the cell provides a constant excess of GTP over GDP (approximately tenfold excess), while the affinities and association rates for GTP or GDP are similar. Thus, the cell's production of GTP provides the energy to drive G protein signaling, while GEFs and GAPs provide the kinetic controls that harness this energy for regulatory control.

The presence of the GTP γ-phosphate determines the structure of G protein switch I and II regions

The nucleotides GTP and GDP differ only by the presence of a terminal phosphate on GTP (the three phosphates on GTP are designated as α, β, and γ, starting from the nucleotide ring). Despite its small size, the highly charged γ-phosphate moiety plays a critical role in controlling the conformation of G proteins (**Figure 3.26**; see also Figure 15.8). The guanine nucleotide binds in a conserved pocket. The γ-phosphate, when present, interacts with two loops known as **switch I and switch II**, forming hydrogen bonds with the main-chain atoms of two invariant Gly and Thr residues. These interactions cause significant structural rearrangements in switch I and II. Because these regions form a critical part of the binding site for downstream effectors, these conformational changes dramatically affect the ability of the G protein to bind such effectors. Thus, in a sense, the G protein functions to convert a single phosphate difference into a large conformational change.

In its GTP-bound conformation, a particular G protein often is capable of binding to many different downstream effectors. Thus, its activation can induce a variety of downstream effects through its interaction with different effectors. Although the switch I and II regions are critical for binding to all effectors, the other residues on the surface of the G protein that contribute to the binding site may vary from effector to effector. For this reason, it has been possible to construct specific point mutants of G proteins that bind to some effectors, but not to others. Such mutants can be helpful in teasing out which effectors are important for a particular downstream effect of G protein activation.

There are two major classes of signaling G proteins

A typical eukaryotic cell contains well over 150 different G proteins, involved in diverse signaling pathways. These G proteins can be divided into several distinct superfamilies, including two families that play key roles in cell signaling. The first family is the **small G proteins**; these monomeric G proteins are often referred to as small GTPases. The second family is the **heterotrimeric G proteins**, which have α, β, and γ subunits. A third family of G proteins, which we will not discuss here, plays a central role in translation (this family includes the elongation factor EF-Tu).

All of the G protein superfamilies have at their core a 20 kD **G domain**, which binds the guanine nucleotide and can adopt alternative conformations depending on whether GDP or GTP is bound. The small G proteins essentially consist of a single G domain, while the heterotrimeric G proteins contain a G domain in their Gα subunits (**Figure 3.27**). Below, we describe the mechanisms of regulation of both classes of G proteins.

Figure 3.26

The molecular basis of the G protein conformational change. (a) The GDP-bound conformation of a G protein is depicted schematically. (b) In the GTP-bound form, the γ-phosphate group on the bound nucleotide hydrogen-bonds with the main-chain atoms of conserved Thr and Gly residues, leading to conformational rearrangements of the switch I and II regions of the protein. These rearrangements create an effector-binding site. The amino acid numbering corresponds to the small GTPase Ras. See Figure 15.8 for a comparison of the X-ray crystal structures of the GDP- and GTP-bound forms of Ras. (Adapted from G. Petsko and D. Ringe, *Protein Structure and Function*. Oxford: Oxford University Press, 2004.)

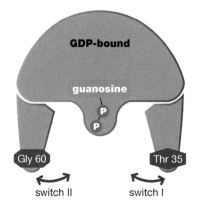

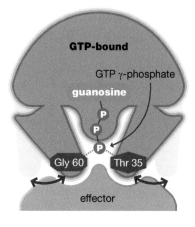

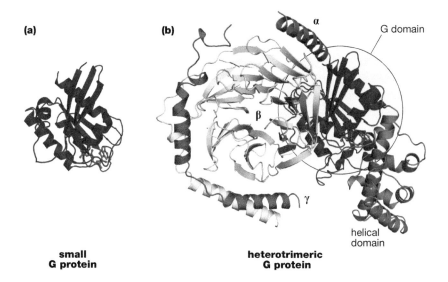

(a)

**small
G protein**

(b)

α

β

γ

G domain

helical
domain

**heterotrimeric
G protein**

Figure 3.27

Structure of G proteins. X-ray crystal structures of (a) Ras, a small G protein, and (b) a heterotrimeric G protein complex. Bound GDP *(orange)* is shown in ball-and-stick format. The α subunit of the heterotrimeric G protein consists of a G domain *(purple)* homologous to small G proteins, and an additional helical domain *(blue)*. The β *(yellow)* and γ *(pink)* subunits are tightly associated via a coiled-coil interaction.

Subfamilies of small G proteins regulate diverse biological functions

The small G proteins consist of a single 20–25 kD domain. The founding member of this family is **Ras**, a central regulator of cell proliferation and differentiation. Ras was first identified as an *oncogene* (a gene whose disregulated activity can lead to the uncontrolled growth characteristic of cancer). There are approximately 150 small G proteins in humans, and these can be further subdivided into at least five major subfamilies: the Ras, Rho, Rab, Arf, and Ran families (Table 3.1). Each of these subfamilies is, in general, involved in regulating distinct cellular functions: Ras proteins regulate pathways involved in cell differentiation and proliferation; Rho proteins regulate the cytoskeleton and control cell shape and movement; Rab and Arf proteins regulate membrane vesicle-associated processes including vesicle formation, trafficking, and secretion; Ran proteins control nuclear export and import, formation of the nuclear envelope, and mitotic spindle formation.

Not only do G protein families differ in the particular set of downstream effector functions that they control, but even within these families there are further functional subdivisions. For example, the Rho family G proteins, in their active state, interact uniquely with key cytoskeletal regulatory proteins. The approximately 25 human Rho subfamily members can be further divided, however, into the RhoA, Rac1, and Cdc42 subclasses. Each subclass is associated with distinct functions in cytoskeletal regulation. For example, Rac1-related G proteins are associated with formation of actin-based protrusive structures such as lamellipodia, while RhoA-related G proteins are associated with formation of actin–myosin contractile structures (Figure 3.28).

Table 3.1

Small G protein subfamilies: functions, downstream effectors, and upstream GEF and GAP domain

Subfamily	Canonical members	Function	GEF domains	GAP domains
Ras	K-Ras, Rap 1A	Cell proliferation and differentiation	RasGEF (cdc25 homology domain)[a]	RasGAP, RapGAP
Rho	RhoA, Cdc42, Rac1	Cell shape and movement	RhoGEF (Dbl homology domain), DOCK domain[b]	RhoGAP
Rab	Rab23, Rab4A	Vesicular trafficking and secretion	Mss4 domain, Sec2 domain, VSP9 domain	RabGAP
Arf	Arf6, Arl4	Vesicular trafficking and secretion	ArfGEF (Sec7 domain)	ArfGAP
Ran	Ran	Nuclear import	RanGEF	RanGAP

[a] RasGEF (Cdc25 homology) domains are often associated with an N-terminal Ras exchanger motif (REM) domain.

[b] RhoGEF (Dbl homology) domains are often associated with a C-terminal pleckstrin homology (PH) domain.

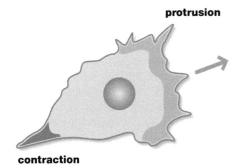

protrusion

contraction

Figure 3.28

Differential activation of the G proteins Rac and Rho contributes to directed cell motility. Rac activity (green) is concentrated at the leading edge of the cell, where it promotes actin-mediated protrusion. Rho activity *(orange)* is concentrated at the back of the cell, where it promotes actin–myosin-mediated contraction.

 See Chapter 8 for more on GPCRs

Many upstream receptors feed into a small set of common heterotrimeric G proteins

Heterotrimeric G proteins contain three subunits—α, β, and γ (see Figure 3.27). The 50 kD Gα subunit contains the conserved 20–25 kD G domain that is homologous to the small G proteins. This domain binds GTP or GDP and regulates the conformational change of the protein. The Gα subunit also contains an unrelated helical domain. The Gβ and Gγ subunits do not themselves have any enzymatic activity.

When the Gα subunit is bound to GDP, it also associates with the Gβ and Gγ subunits to form the heterotrimer. However, when the Gα subunit is in the GTP-bound, or "active," state, it dissociates from the Gβ and Gγ subunits (which stay tightly associated to each other) (**Figure 3.29**). Dissociation of the subunits occurs because the switch I and II regions in the Gα subunit—the regions that undergo the largest nucleotide-dependent conformational shifts—form part of the heterotrimer binding interface. When dissociated, both the Gα and Gβγ subunits can bind to various downstream effectors (channels and enzymes like adenylyl cyclase), changing their activity and thus leading to biological effects. Heterotrimeric G proteins are activated by **G-protein-coupled receptors** (**GPCRs**). In humans, there are many hundreds of distinct G-protein-coupled receptors, making them the most highly represented class of signaling proteins.

There are only 16 genes for Gα subunits in humans, which can be divided into four main families: $G\alpha_s$, $G\alpha_i$, $G\alpha_{q/11}$, and $G\alpha_{12/13}$. Although these Gα subunits have a similar mechanism of activation, they have different effector-binding properties (**Table 3.2**). For example, the $G\alpha_s$ isoform is primarily responsible for activation of adenylyl cyclase and production of the signaling mediator cAMP. Thus, a large number of diverse upstream receptors feed into a small set of common G proteins.

The mechanism by which downstream signaling is directed to specific functional outputs using this limited set of common G proteins is still unclear. However, there is growing evidence that mechanisms such as cell-specific expression of receptors and restricted subcellular localization mediated by scaffold proteins can help to limit and direct the downstream effectors that are targeted by specific receptor–G protein complexes.

Figure 3.29

The activity cycle of heterotrimeric G proteins. In the basal state, the Gα subunit is bound to GDP and in complex with the Gβγ subunits. Activation of the G-protein-coupled receptor (GPCR) by ligand binding leads to conformational changes that induce its guanine nucleotide exchange factor (GEF) activity, resulting in exchange of GDP for GTP in the Gα subunit. The resulting conformational changes in the Gα subunit lead to its dissociation from the receptor and from the Gβγ subunits. The Gα and Gβγ subunits then are competent to bind downstream effectors. Reversion to the GDP-bound state leads to reassociation of the heterotrimeric complex. GAP, GTPase-activator protein; RGS, regulators of G protein signaling.

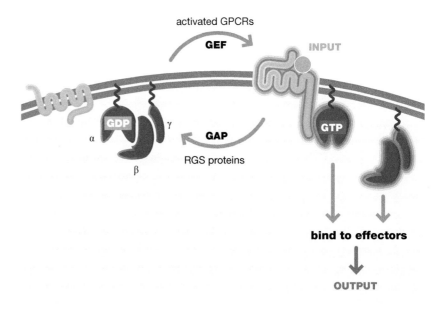

activated GPCRs

GEF

INPUT

GDP

γ

α

GAP

RGS proteins

β

GTP

bind to effectors

OUTPUT

Family	Subtype	Effectors
$G\alpha_s$	$G\alpha_{s(S)}$	Adenylyl cyclases ↑
	$G\alpha_{s(L)}$	Maxi K channel ↑
	$G\alpha_{s(XL)}$	Src tyrosine kinases (Src, Hck) ↑
	$G\alpha_{olf}$	GTPase of tubulin ↑
$G\alpha_{i/o}$	$G\alpha_{o1}$	adenylyl cyclase ↓ $G\alpha_{i,o,z}$
	$G\alpha_{o2}$	Rap1 GAPII-dependent
	$G\alpha_i$	Erk/MAP kinase activation ↑
	$G\alpha_z$	Ca^{2+} channels ↓ $G\alpha_{i,o,z}$
	$G\alpha_t$	K^+ channels ↑ $G\alpha_{i,o,z}$
	$G\alpha_{gust}$	GTPase of tubulin ↑ $G\alpha_i$
		src tyrosine kinases (c-Src, Hck) ↑ ($G\alpha_i$)
		Rap1 GAP ↑ ($G\alpha_z$)
		GRIN1-mediated activation of cdc42 ↑ ($G\alpha_{i,o,z}$)
		cGMP-PDE ↑ ($G\alpha_t$)
$G\alpha_{q/11}$	$G\alpha_q$	Phospholipase Cβ isoforms ↑
	$G\alpha_{11}$	p63-RhoGEF ↑ ($G\alpha_{q/11}$)
	$G\alpha_{14}$	Bruton's tyrosine kinase ↑ ($G\alpha_q$)
	$G\alpha_{15}$	K^+ channels ↑ ($G\alpha_q$)
$G\alpha_{12/13}$	$G\alpha_{12}$	Phospholipase D ↑
	$G\alpha_{13}$	Phospholipase Cε ↑
		NHE-1 ↑
		iNOS ↑
		E-cadherin-mediated cell adhesion ↑
		p115RhoGEF ↑
		PDZ-RhoGEF ↑
		Leukemia-associated RhoGEF (LARG) ↑
		Radixin ↑
		Protein phosphatase 5 (PP5) ↑
		AKAP110-mediated activation of PKA ↑
		hsp90 ↑

Table 3.2

Families of G_α subunits and their effectors

Table modified from G. Milligan and E. Kostenis, *Br. J. Pharmacol.* 147(1): S46–S55, 2006.

Up arrow indicates activation; down arrow indicates repression.

REGULATORY ENZYMES FOR G PROTEIN SIGNALING

As was mentioned above, for nearly all G proteins the intrinsic rates of the nucleotide exchange reaction and the nucleotide hydrolysis reaction are extremely slow. In the cell, the actual rates of these reactions (which change in response to signals within seconds), are kinetically controlled by two opposing classes of enzymes: the guanine nucleotide exchange factors (GEFs) that accelerate the rate of nucleotide exchange, and the GTPase-activator proteins (GAPs) that accelerate the rate of GTP → GDP hydrolysis (see Figure 3.25). Thus, GEFs kinetically control the activation of the G protein, while GAPs kinetically control its deactivation. Upstream signaling inputs can therefore control the amount of active G protein by coordinately regulating the activity of either the GEFs or GAPs.

There are diverse GEFs and GAPs, larger in number than the G proteins themselves. GEFs and GAPs are themselves highly regulated, and the diversity of these proteins provides a variety of ways to link diverse upstream signaling

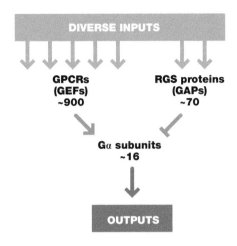

Figure 3.30

Signal processing by heterotrimeric G proteins. Signal inputs are received by an enormous number of G-protein-coupled receptors (GPCRs) that act as guanine nucleotide exchange factors (GEFs), and by regulators of G protein signaling (RGS) proteins that act as GTPase-activator proteins (GAPs). These inputs are funneled through a relatively small number of Gα subunits. The total number of each class of protein in humans is provided.

 GPCRs are described in greater detail in Chapter 8; the visual signaling system is discussed in Chapter 14

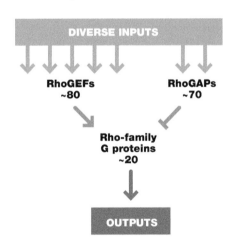

Figure 3.31

Signal processing by Rho family small G proteins. The activities of a relatively large number of RhoGEFs and RhoGAPs are regulated by signal inputs. These inputs modulate the activity of a somewhat smaller number of Rho family small G proteins. The total number of each class of protein in humans is provided. GEF, guanine nucleotide exchange factor; GAP, GTPase-activator protein.

inputs to the control of a common set of G proteins. We shall see in the following sections the mechanisms by which GEFs and GAPs interact with G proteins and modulate their activity.

G-protein-coupled receptors act as GEFs for heterotrimeric G proteins

The activation of heterotrimeric G proteins is controlled by G-protein-coupled receptors (GPCRs). GPCRs have a common overall structure with seven membrane-spanning segments, but different GPCRs respond to a very wide range of extracellular inputs. Typically, when a GPCR detects its ligand, it undergoes a conformational change that allows the receptor to act as a GEF and catalyze the exchange of GDP for GTP on the Gα subunit (see Figure 3.29). Interestingly, the exchange mechanism requires the Gβ and Gγ subunits. This requirement maximizes the efficiency of G protein activation by preventing the receptor from targeting already-activated (GTP-bound) Gα subunits.

Signaling mediated by heterotrimeric G proteins is extremely widespread. GPCRs are found in eukaryotes ranging from yeast to human and are the most numerous type of receptor in the human genome (~900; nearly 5% of the total number of genes) (**Figure 3.30**). They transduce diverse signals, including those induced by light, odorants, hormones, lipids, and proteins. Rhodopsin, which is found in the rod and cone cells of the retina, is an example of a GPCR that responds to light. As discussed below, regulators of G protein signaling (RGS) proteins serve as GAPs to counteract the effects of GPCRs.

Distinct GEF and GAP domains regulate specific small G protein families

The activity of small G proteins is linked to diverse upstream signals by GEF and GAP proteins that act in a coordinate manner to control the levels of the active, GTP-bound G protein. The number of GEF and GAP proteins exceeds the number of downstream G proteins. For example, there are ~20 Rho family G proteins but ~80 GEF and ~70 GAP proteins that act on them. Presumably, the large number of GEF and GAP proteins functions as an adaptor layer that allows diverse upstream inputs to plug into common G-protein-mediated output responses (**Figure 3.31**).

Like many signaling proteins that are involved in transmitting and processing information, these GEFs and GAPs are modular, multidomain proteins. Many of these domains, which differ widely in the individual examples, function as protein or lipid interaction domains that regulate or localize enzymatic function. However, all GEFs and GAPs contain one or more dedicated catalytic domains with intrinsic GEF or GAP activity. There are a relatively small number of types of GEF and GAP domains, with usually one type serving to regulate one subfamily of G proteins. For example, Rho family G proteins generally are activated by GEF proteins that have a Dbl homology (DH) catalytic domain or an unrelated DOCK domain. In contrast, Ras family G proteins are generally activated by GEF proteins containing a Cdc25 homology catalytic domain. Thus, GEFs and GAPs have distinct functional regions: a catalytic region that determines its specific output, and multiple regulatory domains that determine how and by what it is regulated (**Figure 3.32**). A summary of types of GEF and GAP domains, and what G proteins they act on, is shown in Table 3.1.

GEF and GAP proteins show several forms of regulation (**Figure 3.33**). First, the protein or lipid interaction domains in the protein can mediate regulated changes in localization. Second, many of these proteins are autoinhibited by intramolecular interactions, in which other domains in the protein interact with the catalytic domain to inhibit its intrinsic activity. This type of modular regulation can

Rho GEFs

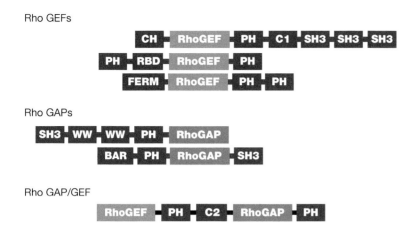

Rho GAPs

Rho GAP/GEF

Figure 3.32

Representative domain architectures of RhoGEFs and RhoGAPs. The core catalytic domains (*blue* for GEF, *orange* for GAP) are combined with diverse regulatory domains that determine input control and localization *(brown boxes).* Some proteins contain both GEF and GAP activities, which may control signaling dynamics by coordinating both activation and inactivation of the G protein. GEF, guanine nucleotide exchange factor; GAP, GTPase-activator protein.

(a)

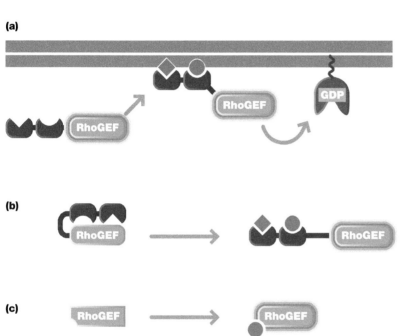

(b)

(c)

Figure 3.33

Three major mechanisms of GEF and GAP regulation. (a) GEF and GAP catalytic domains can be recruited to the sites of their target G protein on the membrane, (b) they can be activated by relief of autoinhibition, or (c) they can be allosterically activated. Many GEF and GAP proteins exhibit multiple mechanisms of regulation. Activating ligands are indicated as *orange* circles and diamonds.

lead to activation when the autoinhibitory interactions are disrupted by upstream inputs such as ligand binding or phosphorylation. Third, certain interactions with the catalytic domains can also allosterically increase the intrinsic GEF or GAP activity.

These forms of regulation are observed for the GEF Sos, which, as part of its function, can activate the G protein Ras via its Cdc25 GEF domain (Sos also has a Dbl domain and can act as a RhoGEF as well). Sos can be activated by recruitment to the membrane where it can act on Ras (which is exclusively localized on the membrane). Recruitment is mediated by the adaptor protein Grb2, which binds to tyrosine-phosphorylated proteins on the membrane via its SH2 domain. Domains N- and C-terminal to the Cdc25 domain also act to autoinhibit Sos exchange activity, and these autoinhibitory interactions may be relieved by interaction with upstream ligands like Grb2. Finally, Sos exchange activity is also further stimulated by the active state of Ras, via an allosteric binding interaction. This allosteric activation is postulated to contribute to positive feedback in Ras activation. Another example of autoinhibitory regulation is the GEF Epac, which acts on the small G proteins Rap1 and Rap2. Epac is activated by cAMP, which relieves autoinhibition of the catalytic domain mediated by a cyclic nucleotide binding (CNB) domain.

 See Chapter 10 for a discussion of modular domains and the architecture of signaling proteins

 Membrane recruitment as an activation mechanism is discussed in more detail in Chapter 5

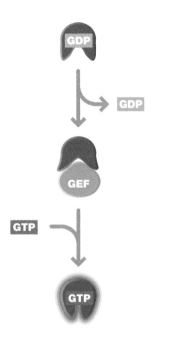

Figure 3.34

General mechanism of GEF action.
Guanine nucleotide exchange factor (GEFs)
act to pry apart the nucleotide-binding
regions of the G protein, promoting the
release of GDP. The GEF binds most tightly
to the G protein when it is not bound to
nucleotide (essentially the transition state for
the nucleotide exchange reaction).

GEFs catalyze GDP/GTP exchange by deforming the nucleotide-binding pocket

As mentioned above, most G proteins bind both GTP and GDP with high affinity and have half-times for dissociation in the range of minutes to hours. GEFs, in general, act by modifying the structure of the nucleotide- binding pocket, such that the affinity for the nucleotide is significantly decreased (**Figure 3.34**). It is important to note that the GEF does not favor dissociation of GDP over GTP, nor does it favor rebinding of GTP over GDP. Rather, the unidirectional nature of the G protein cycle (preferential binding of GTP) is driven by the much higher concentration of GTP over GDP maintained in the cell.

Several structures of small G proteins bound to GEFs have been solved; these show how various structurally unrelated catalytic GEF domains interact with G proteins (**Figure 3.35**). Despite their sequence and structural divergence, all the GEF domains bind near or at the nucleotide-binding pocket of their substrate G proteins. In most cases, some element of the GEF domain acts to pry apart the switch I and II regions of the G protein from the rest of the structure, thus severely deforming the nucleotide-binding pocket. In these structures, the GEF sterically occludes the Mg^{2+}-binding sites that are required for phosphate-group binding, as well as key residues in the switch I or II regions involved in nucleotide binding. Thus, these diverse catalytic domains utilize a convergent molecular mechanism.

GAPs order the catalytic machinery for hydrolysis

As mentioned above, GAP activity can accelerate the hydrolysis of GTP by G proteins by several orders of magnitude. An efficient hydrolysis reaction, in principle, requires a water molecule that is properly oriented, polarized, and occluded from bulk solvent for optimal nucleophilic attack. It also requires stabilization of the negatively charged γ-phosphate in the transition state. In most G proteins, the catalytic residues that perform these functions are either improperly oriented or missing. In the small G protein Ras, the Gln61 residue that interacts with the nucleophilic water is well ordered in the Ras–RasGAP complex, but is disordered in free Ras (**Figure 3.36**). In addition, a key arginine in RasGAP, referred to as the

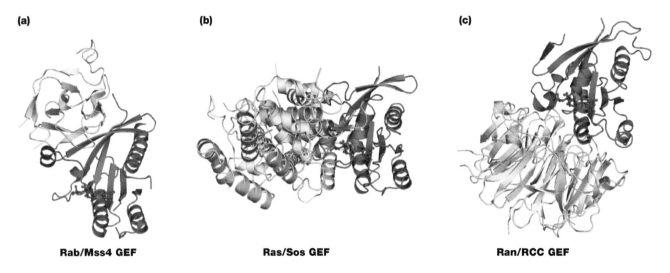

(a) Rab/Mss4 GEF **(b)** Ras/Sos GEF **(c)** Ran/RCC GEF

Figure 3.35

Structures of small G proteins bound to their GEFs. X-ray crystal structures of three G protein–GEF complexes illustrate the variety of structures and mechanisms used by different GEFs to promote release of GDP. In all structures, the G protein *(blue)* is shown in the same orientation; GEFs are colored *yellow*. Regions of the G protein that undergo significant conformational change upon GEF binding are colored in *pink*. GDP *(orange* ball-and-stick structure) is shown in the position it would occupy in the native G protein (uncomplexed with GEF). (Adapted from J.L. Bos, H. Rehmann and A. Wittinghofer, *Cell* 129:865–877, 2007. With permission from Elsevier.)

arginine finger, is inserted into the active site, and appears to stabilize the transition state by neutralizing the γ-phosphate. This catalytic function is missing in isolated Ras, and is provided in *trans* by RasGAP. Interestingly, mutations at the Gln61 position of Ras are frequently found in human tumors. Such mutants are constitutively active because they are unable to efficiently hydrolyze GTP, even in the presence of GAPs.

Although different GAPs are very diverse and have unrelated structures (**Figure 3.37**), they all use similar mechanisms of providing or ordering missing or disordered catalytic elements required for nucleotide hydrolysis. In some cases, the residues that order the nucleophilic water are, in turn, ordered by the GAP interaction; in other cases, such residues are provided by the GAP itself. In nearly all cases, the arginine finger that stabilizes the transition state is provided in *trans* by the GAP.

Regulators of G protein signaling (RGS) proteins act as GAPs for heterotrimeric G proteins

Like small G proteins, Gα subunits have intrinsic but extremely slow GTPase activity. Thus, rapid deactivation of responses requires factors that can catalyze GTP hydrolysis. **Regulators of G protein signaling (RGS) proteins** function as GAPs, accelerating the rate of GTP hydrolysis of the activated Gα subunits. More than 20 different RGS proteins have been identified. In several cases, it appears that heterotrimeric G proteins operate within multiprotein complexes containing specific GPCRs, RGS proteins, and downstream effectors, leading to tightly controlled activation, downstream signaling, and deactivation.

RGS proteins use a slightly different mechanism to that described above for small G protein GAPs. In this case, the arginine finger exists in the Gα subunits, provided by the helical domain that is linked to the Ras-like G domain. Thus, the heterotrimeric G proteins have a "built-in" arginine finger and RGS proteins are thought to play a role primarily in properly ordering the catalytic residues in the Gα subunit to optimize catalysis.

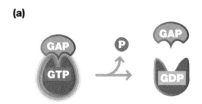

(a)

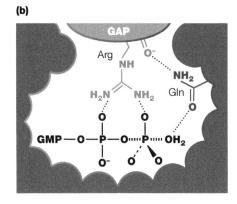

(b)

Figure 3.36

General mechanism of GAP action. GAPs bind above the nucleotide-binding pocket of the active G protein and promote GTP hydrolysis. (b) Many GAPs, including those acting on Ras and Rho, orient a catalytic residue (Gln) in the G protein to polarize the water molecule that attacks the γ-phosphate of GTP. They may also insert an "arginine finger" (Arg) into the active site, which plays a key role in stabilizing the transition state of the substrate. (b, Adapted from J.L. Bos, H. Rehmann & A. Wittinghofer, *Cell* 129:865–877, 2007. With permission from Elsevier.)

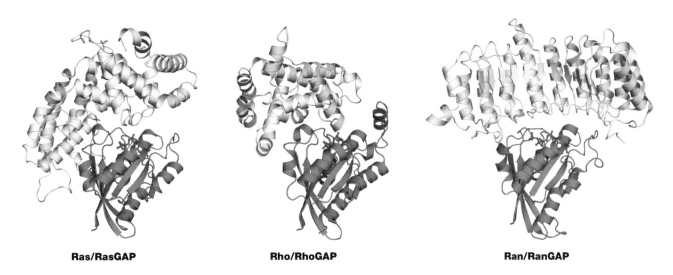

Ras/RasGAP	**Rho/RhoGAP**	**Ran/RanGAP**

Figure 3.37

Structures of small G proteins bound to their GAPs. X-ray crystal structures of three small G protein–GAP complexes illustrate the structural diversity of GAPs. In all structures, the G protein *(blue)* is shown in the same orientation; GAPs are colored *yellow* and GTP is *orange*. (Adapted from J.L. Bos, H. Rehmann & A. Wittinghofer, *Cell* 129:865–877, 2007. With permission from Elsevier.)

Lipid modification and the role of GDIs and GDFs are discussed in more detail in Chapter 5

Additional mechanisms are used to fine-tune the activity of G proteins

Many G proteins have lipid groups attached at their C-termini that are required for membrane localization and function. This modification is used to provide an additional mechanism for controlling G protein activity, especially for small G proteins of the Rho and Rab families. **Guanine nucleotide dissociation inhibitors** (**GDIs**) are proteins that can bind specific G proteins and shield their prenyl groups, thus maintaining the G protein in the cytoplasm. Thus, GDIs lock G proteins in their GDP-bound state and prevent their localization to the membrane. **GDI displacement factors** (**GDFs**) are enzymes that can promote the release of GDIs from G proteins, thus allowing their localization to the membrane and subsequent activation by GEFs. Thus, GDIs and GDFs provide an additional layer of regulation for prenylated G proteins.

The activity of heterotrimeric G proteins is also modulated by proteins containing GoLoco motifs (GoLoco domains). These 19-residue motifs bind to specific Gα subunits, but only in the inactive GDP-bound state. Thus, proteins containing the GoLoco motif act as GDIs that can, in principle, inhibit G protein activation. Nonetheless, in some cases, GoLoco domains can play positive roles. For example, binding of GoLoco domains to inactivated Gα subunits can prolong activity of dissociated Gβγ subunits, because they competitively block reassociation of the heterotrimer. The GoLoco domain in the protein PINS (partner of inscrutable) has been found to modulate heterotrimeric G protein activity in a GPCR-independent manner as part of the machinery that positions the mitotic spindle and attaches microtubules to the cell cortex during cell division.

SIGNALING ENZYME CASCADES

So far we have examined in detail the properties of individual signaling enzymes, such as kinases, phosphatases, GEFs, and GAPs. But in the cell, signaling enzymes do not function individually, but rather are embedded within pathways and networks in which they function as one relay node within a much larger network. These higher-order signal transduction networks are constructed by functionally connecting the individual nodes. This often means that individual signaling enzymes are linked in series into **cascades**, where the output of one enzyme directly or indirectly regulates the activity of the next enzyme. Below, we discuss conserved kinase and G protein cascades and how they are used in signaling.

The three-tiered MAP kinase cascade forms a signaling module in all eukaryotes

The role of kinase aggregation in signaling across the plasma membrane is discussed in Chapter 8

As discussed earlier, many protein kinases require phosphorylation on their activation loops to become active. Thus, it is inherently easy for kinases to be regulated in series, as their activation usually requires the action of some upstream kinase. As a result, it is fairly common to find conserved pathways with kinases acting in series in a cascade. (A number of other kinases are activated by phosphorylation by a different molecule of the same type; for example, after kinase homodimerization or aggregation.)

Among the most prevalent kinase cascades in all eukaryotes are the **mitogen-activated protein (MAP) kinase** cascades. This is a pathway module composed of three kinases that act in series, and it is utilized in a remarkably wide variety of cellular responses (**Figure 3.38a**). The most upstream member is referred to as a MAP kinase kinase kinase (MAPKKK). Activation of a MAPKKK by upstream inputs allows it to phosphorylate and, in turn, activate a MAP kinase kinase (MAPKK) on two Ser/Thr residues in its activation loop. The MAPKK, in turn, activates the most downstream member of the module, **MAP kinase (MAPK)**, by

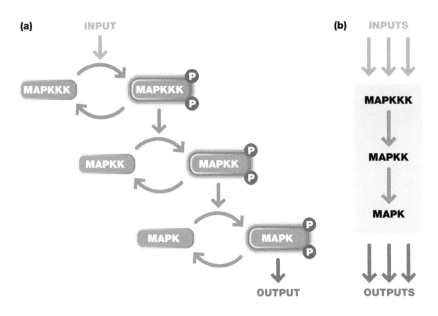

(a) INPUT

(b) INPUTS

MAPKKK

MAPKK

MAPK

OUTPUT   OUTPUTS

Figure 3.38

Figure 3.38

MAP kinase (MAPK) cascades. (a) MAPK cascades consist of three kinases that successively phosphorylate and activate one another. The MAP kinase kinase kinase (MAPKKK), when activated by inputs, can in turn phosphorylate and activate the MAP kinase kinase (MAPKK). The activated MAPKK can phosphorylate the MAPK on two positions in its activation loop, leading to activation of the MAPK. (b) MAPK modules are often activated by a large number of different inputs, and can result in activation of specific programs that involve a large and diverse set of MAPK substrates.

phosphorylating it on its activation loop on specific Ser/Thr and Tyr residues (note that MAPKKs are one of the few members of the Ser/Thr kinase family that can catalyze a Tyr phosphorylation reaction). When activated, MAPKs phosphorylate a large number of downstream targets. In many cases, this involves translocation to the nucleus, where phosphorylation of specific transcription factors results in changes in gene expression patterns.

These three kinases form what is essentially an obligate cascade, because in most cases the only significant substrates for MAPKKKs are MAPKKs, and the only significant substrates for MAPKKs are MAPKs. Different MAPKKKs can be regulated by a variety of specific inputs, and different MAPKs can phosphorylate many different targets (**Figure 3.38b**). Thus, because both the input and output of the module are flexible, these three-tiered cascades can be used to mediate a host of different response behaviors. Cells often contain a number of individual proteins that belong to each of the three kinase families, thus further flexibility results from using different combinations of individual MAPK cascade components. In principle, the phosphatases that dephosphorylate MAPKKKs, MAPKKs, and MAPKs could be important points of regulation. Relatively little is known, however, about these reverse reactions and which specific enzymes are involved.

In mammalian cells, the canonical MAPK cascade involves the MAPKKK Raf (which is recruited to the membrane and activated by the small G protein Ras), the MAPKK MEK (MAPK/Erk kinase), and the MAPK Erk (**Figure 3.39**). Some of the most important substrates of Erk are mitogenic transcription factors, such as Fos and Jun. The Raf–MEK–Erk module is used to mediate cell responses to various mitogenic signals (explaining the origin of the name "MAP kinase"). This module is also used in immune-cell activation, in developmental pathways, and in many other contexts. However, there are a number of other related MAPK modules in mammalian cells, which involve distinct kinase family members and which are used to mediate other types of behaviors including responses to stress or cytokines. These include pathways that activate the JNK (c-Jun N-terminal kinase) and p38 families of MAP kinases (also referred to as stress-activated protein kinases or SAPKs).

The number of MAPK cascade components found in different eukaryotic organisms is shown in **Figure 3.40a**. Even a simple eukaryote such as budding yeast contains several distinct MAPK cascades, all sharing the same three-tiered structure, but varying with respect to the specific kinases involved, and the physiological inputs and outputs that are linked by the cascade (**Figure 3.40b**). There is a particularly large expansion of MAPK components in plants.

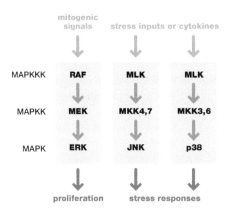

mitogenic signals   stress inputs or cytokines

	MAPKKK	RAF	MLK	MLK
MAPKK	MEK	MKK4,7	MKK3,6
MAPK	ERK	JNK	p38

proliferation   stress responses

Figure 3.39

Examples of different mammalian MAPK modules. The three major families of MAPK cascades in mammals are illustrated. The Erk MAPK family controls cell proliferation programs, while the JNK and p38 families control stress responses (and are often referred to as stress-activated protein kinases or SAPKs).

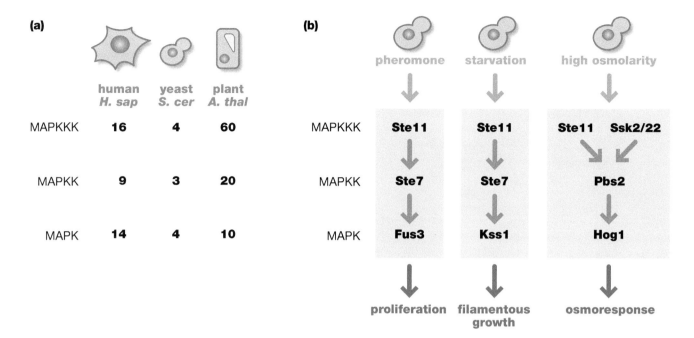

Figure 3.40

Alternative linkages in MAPK cascades.
(a) Different species vary in the number of individual MAPK cascade components. In principle, there are many ways in which these components can be linked. *H. sap, Homo sapiens; S. cer, Saccharomyces cerevisiae; A. thal, Arabidopsis thaliana.*
(b) In budding yeast, there are at least three physiologically distinct MAPK cascades, which respond to mating pheromone (mating pathway), nitrogen starvation, and high osmolarity stress. Activation of each cascade results in a different output. However, the three cascades utilize shared components, including the MAPKKK Ste11 and the MAPKK Ste7. In some cases, scaffold proteins organize specific components to provide specificity (see Figure 3.41).

Scaffold proteins often organize MAPK cascades

The large number of related MAPK cascade components, and the potential cross-talk that could occur between these components, raises the issue of how specific signaling responses can be generated in the cell. The very same kinase can play a role in two distinct pathways, in which case the problem of pathway specificity is particularly acute.

In many cases, such specificity is accomplished through the use of **scaffold proteins**—proteins that interact with multiple proteins within a pathway, organizing them into a single complex. Not surprisingly, MAPK cascades were one of the first pathways shown to be organized by scaffolds. Some of the scaffold proteins that are found in yeast and mammalian MAPK pathways are shown in **Figure 3.41**.

Scaffolds are thought to modulate MAPK signaling in multiple ways (**Figure 3.42**). Most simply, scaffold proteins can promote efficient signaling between component proteins in a pathway by increasing their proximity to each other. They are also thought to insulate the components from cross-talk with incorrect but potentially competing partner proteins, by sequestering them into distinct complexes that may be localized in distinct regions of the cell. For example, in yeast, the MAPK pathway that responds to mating pheromone is organized by the scaffold protein Ste5, while the MAPK pathway that responds to high salt (osmotic stress) is organized by the protein Pbs2, which serves as both the scaffold and the MAPKK in this pathway (see **Figure 3.41**). These scaffolds are thought to control information

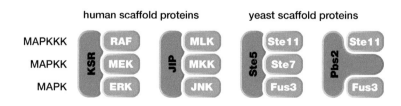

Figure 3.41

Scaffold proteins physically organize MAPK cascades. The KSR (kinase suppressor of Ras) and JIP (JNK interacting protein) proteins organize the mammalian Erk (proliferation) and JNK (stress response) pathways, respectively. The Ste5 and Pbs2 proteins organize the yeast Fus3 (mating) and Hog1 (osmotic stress) pathways.

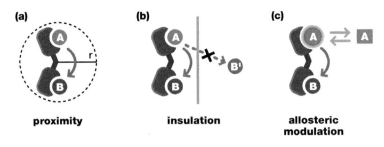

(a) proximity **(b)** insulation **(c)** allosteric modulation

Figure 3.42

Mechanisms by which scaffold proteins control information flow. (a) Scaffolds can increase the effective concentration of two interacting enzymes for one another through tethering and enhanced proximity. (b) Scaffolds can also serve to insulate a protein from alternative partners and substrates. (c) Finally, scaffolds can serve as allosteric regulators for pathway members. For example, a pathway protein may be inactive (or inactivatable) unless it is associated with the scaffold, thus preventing its ability to signal when associated with alternative partners.

input and output of the common pathway component, the MAPKKK Ste11. The population of Ste11 that is associated with the Ste5 scaffold is thought to play a role only in mating pheromone signaling, while the population of Ste11 that is associated with the Pbs2 scaffold is thought to play a role only in osmolarity signaling. Thus, Ste11 that is activated by one pathway will not cross-activate the other pathway, which is crucial for proper behavior and survival of the cell.

There is also evidence that scaffold proteins can control kinase activity allosterically. For example, the mammalian scaffold protein KSR appears to allosterically activate the MAPKKK Raf. Thus, Raf associated with the scaffold is more active than Raf that is not associated with the scaffold. The yeast mating scaffold protein, Ste5, has been shown to allosterically modulate the mating MAPK Fus3 so that only when associated with the scaffold does it become a good substrate for a co-associated Ste7 MAPKK enzyme. In essence, Fus3 has evolved to be a "locked," inactivatable kinase, but the Ste5 scaffold acts as a specific "key" allowing it to be activated by Ste7. This behavior is thought to be important for signaling specificity, because the yeast starvation response is mediated by "free" Ste7 that is not associated with a scaffold. This lock-and-key control of the mating MAPK Fus3 prevents its misactivation by Ste7 in response to starvation.

G protein activity can also be regulated by signaling cascades

Although enzymatic cascades are often associated with kinases, cascades composed of other classes of signaling enzymes can also be found. Cascades involving G protein regulatory enzymes have been found that control cell morphology and intracellular trafficking. For example, endocytic trafficking pathways are characterized by sequential progression of an endosome through different states, each associated with a distinct Rab small G protein. In some cases, an upstream-activated Rab G protein is found to recruit and/or activate a GEF that then activates a downstream Rab, essentially forming a G-protein-mediated cascade (**Figure 3.43a**). In some cases, an activated Rab G protein will also recruit a GAP that will inactivate the G protein upstream of it in the cascade. This forward GEF cascade interwoven with a backward GAP cascade is thought to restrict the overlap between the two Rab proteins, leading to a sharp and irreversible transition from one step to the next along the trafficking pathway.

Scaffold proteins have also been identified that mediate small G protein signaling. For example, the yeast protein Bem1 is thought to act as a scaffold that binds the GEF enzyme Cdc24, its target (the G protein Cdc42), and the protein kinase Ste20, a downstream effector of activated Cdc42 (**Figure 3.43b**).

Figure 3.43

G protein signaling cascades. (a) During vesicle trafficking, activation of one G protein (Rab1) leads to the activation of a second G protein (Rab2). In this case, activated Rab1 binds to and activates the Rab2 GEF in a cascade-like manner. In addition, activated Rab1 can also lead to a negative feedback loop in which it also recruits and activates the GAP for Rab1. This leads to the inactivation of Rab1. Together, these linkages lead to the temporal sequence of events in which a Rab1-marked vesicle is converted in an all-or-none fashion into a Rab2-marked vesicle. (b) G protein signaling can also be controlled by scaffold proteins. The yeast protein Bem1 organizes a G protein cascade that controls cell polarity. Bem1 can bind the GEF Cdc24, its substrate (the G protein Cdc42), and the Cdc42 effector Ste20 (PAK kinase), thus leading to efficient signaling through the pathway. Moreover, binding of Bem1 to sites of activated Cdc42 is thought to provide positive feedback by localizing Cdc24, and thereby activating more Cdc42 at these sites.

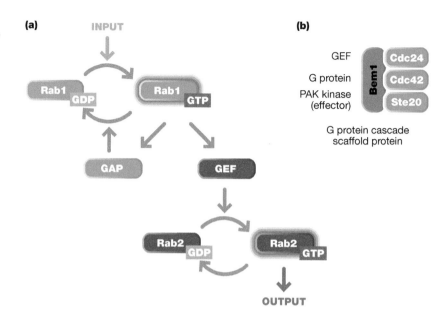

SUMMARY

Enzymes have a remarkable ability to catalyze specific biological reactions. They greatly increase the rate of thermodynamically favorable reactions, and can promote thermodynamically unfavorable reactions by coupling them to more favorable ones. The activity of signaling enzymes is frequently regulated by allosteric changes in conformation induced by upstream signaling inputs, which allows these enzymes to function as relay nodes to transmit information. The most common inputs that induce conformational change are ligand binding and post-translational modification. Canonical signaling enzymes, such as kinases, are optimized to allow coupling between conformational changes and catalytic function.

Many cell signaling processes are regulated by opposing enzyme "writers" and "erasers" that switch substrates between active and inactive states. For example, kinases and phosphatases control the addition and removal of phosphate modifications. Phosphorylation is a particularly useful class of information-carrying mark because both the phosphorylation and dephosphorylation reactions are thermodynamically favorable, but kinetically slow. Thus, the distribution between phosphorylated and dephosphorylated states can be kinetically controlled by the appropriate enzymes. Another important class of enzymatic "writers" and "erasers" in cell signaling are the GEFs and GAPs, which activate and deactivate G proteins. G proteins can bind either GTP or GDP, but are only active in the GTP-bound state. GEF enzymes activate G proteins by catalyzing the exchange of bound GDP for GTP, while GAP proteins inactivate G proteins by accelerating their intrinsically slow GTPase activity.

Signaling enzymes are often organized into cascades in which an upstream enzyme activates a downstream enzyme in series. Such cascades often form evolutionarily conserved modules, such as the three-tiered MAPK cascades used by all eukaryotes to transmit signals. These cascades are often organized by scaffold proteins, and they can be flexibly configured to control the flow of cellular signals in different ways.

QUESTIONS

1. What aspects of enzyme catalysis are described by the kinetic parameters k_{cat}, K_m, and k_{cat}/K_m? The isolated catalytic domains of many signaling enzymes often have K_m values that are high relative to the substrate concentrations at

which they operate in the cell. Why might this be, and how might this lend itself to particular mechanisms for regulating overall catalytic activity?

2. Compare and contrast some of the selective pressures on signaling enzymes with those acting on enzymes involved in metabolism, in particular enzymes involved in high-flux metabolic reactions.

3. If you were to design a new post-translational modification system for signal transmission, what key features would you include?

4. Treatment of a cell with mitogen X leads to tyrosine phosphorylation of a number of target proteins, consistent with a model in which a tyrosine kinase is activated by mitogen stimulation. Interestingly, treatment of the cell with vanadate (an inhibitor of phosphotyrosine phosphatases) also leads to the accumulation of the same set of tyrosine-phosphorylated species. Can you explain why this increase in target protein phosphorylation is observed with vanadate? What are the implications of your model for how kinases and phosphatases operate as a regulatory system, and the speed at which signals will be turned on and off by this system?

5. You are given a purified preparation of protein kinase and test its ability to phosphorylate a peptide substrate at various concentrations of enzyme. You notice that at low concentrations of enzyme, the rate of substrate phosphorylation is initially low but increases substantially over time. However, if you use a higher concentration of kinase, this apparent lag is not seen. How can you explain these results, and how would you test your hypothesis?

6. In mammalian cells, the hydroxyl amino acids (serine, threonine, and tyrosine) are the major sites of regulatory phosphorylation. In bacteria, histidine phosphorylation predominates in signaling. The free energy of hydrolysis (ΔG) of phosphohistidine is larger than for the phosphorylated hydroxyl amino acids, and the rate of spontaneous hydrolysis is much faster. What differences between eukaryotic and bacterial cells could account for the selective pressure for eukaryotic cells to use phosphorylation of the hydroxyl amino acids instead of histidine for their signaling pathways?

7. Summarize the general mechanisms by which diverse regulatory inputs that act on kinase domains can increase the enzyme's ability to phosphorylate substrates.

8. Summarize how the gain of a single phosphate group in the GTP-bound state of a G protein (versus the GDP-bound state) can lead to changes in the interactions of the G protein.

9. G proteins are almost always active (and transmit a downstream signal) in their GTP-bound forms and inactive when bound to GDP. In principle, however, it would be possible for the GDP-bound form to be active and the GTP-bound form to be inactive. How might you design a signaling system where the GDP-bound form is active in transmitting a signal? Explain how receptor binding woutld lead to activation of the G protein, and conversely how the G protein would eventually be down-regulated. In this system, what kind of G protein mutants would be constitutively active (that is, active in the absence of receptor stimulation)?

10. Some of the most common mutations associated with cancer are found in the small G protein Ras. Oncogenic mutants of Ras lead to constitutive signaling. Suggest several possible mechanisms for why these forms of Ras show constitutive signaling. Conversely, mutations of Ras have been isolated that have dominant negative activity (that is, they block signaling by the normal, endogenous G protein). Suggest possible mechanisms for how these mutants block signaling.

BIBLIOGRAPHY

PRINCIPLES OF ENZYME CATALYSIS

Fersht A (1999) *Structure and Mechanism in Protein Science.* New York: WH Freeman and Company.

Hunter T (2012) Why nature chose phosphate to modify proteins. *Philos. Trans. R. Soc. Lond. B Biol. Sci.* 367, 2513–2516.

Knowles JR (1980) Enzyme-catalyzed phosphoryl transfer reactions. *Annu. Rev. Biochem.* 49, 877–919.

Lassila JK, Zalatan JG & Herschlag D (2011) Biological phosphoryl-transfer reactions: Understanding mechanism and catalysis. *Annu. Rev. Biochem.* 80, 669–702.

Walsh CT (1978) *Enzymatic Reaction Mechanisms.* New York: WH Freeman and Company.

Westheimer FH (1987) Why nature chose phosphates. *Science* 235, 1173–1178.

Wolfenden R (2006) Degrees of difficulty of water-consuming reactions in the absence of enzymes. *Chem. Rev.* 106, 3379–3396.

ALLOSTERIC CONFORMATIONAL CHANGES

Cui Q & Karplus M (2008) Allostery and cooperativity revisited. *Protein Sci.* 17, 1295–1307.

Holehouse AS & Kragelund BB (2023) The molecular basis for cellular function of intrinsically disordered protein regions. *Nat. Rev. Mol. Cell Biol.* doi: 10.1038/s41580-023-00673-0.

Kern D & Zuiderweg ER (2003) The role of dynamics in allosteric regulation. *Curr. Opin. Struct. Biol.* 13, 748–757.

Lim WA (2002) The modular logic of signaling proteins: Building allosteric switches from simple binding domains. *Curr. Opin. Struct. Biol.* 12, 61–68.

PROTEIN PHOSPHORYLATION AS A REGULATORY MECHANISM

Johnson LN & Lewis RJ (2001) Structural basis for control by phosphorylation. *Chem. Rev.* 101, 2209–2242.

Westheimer FH (1987) Why nature chose phosphates. *Science* 235, 1173–1178.

PROTEIN KINASES

Hubbard SR (1997) Crystal structure of the activated insulin receptor tyrosine kinase in complex with peptide substrate and ATP analog. *EMBO J.* 16, 5572–5581.

Huse M & Kuriyan J (2002) The conformational plasticity of protein kinases. *Cell* 109, 275–282.

Johnson JL, Yaron TM, Huntsman EM, et al. (2023) An atlas of substrate specificities for the human serine/threonine kinome. *Nature* 613, 759–766. doi: 10.1038/s41586-022-05575-3.

Johnson LN, Noble ME & Owen DJ (1996) Active and inactive protein kinases: Structural basis for regulation. *Cell* 85, 149–158.

Kannan N, Taylor S, Zhai Y, et al. (2007) Structure and functional diversity of the microbial kinome. *PLoS Biol.* 5, e17.

Manning G, Whyte DB, Martinez R, et al. (2002) The protein kinase complement of the human genome. *Science* 298, 1912–1934.

Meharena HS, Chang P, Keshwani MM, et al. (2013) Deciphering the structural basis of eukaryotic protein kinase regulation. *PLoS Biol.* 11, e1001680.

Reményi A, Good MC & Lim WA (2006) Docking interactions in protein kinase and phosphatase networks. *Curr. Opin. Struct. Biol.* 16, 676–685.

Taylor SS, Zhang P, Steichen JM, et al. (2013) PKA: Lessons learned after twenty years. *Biochim. Biophys. Acta* 1834, 1271–1278.

PROTEIN PHOSPHATASES

Almo SC, Bonanno JB, Sauder JM, et al. (2007) Structural genomics of protein phosphatases. *J. Struct. Funct. Genomics* 8, 121–140.

Alonso A, Sasin J, Bottini N, et al. (2004) Protein tyrosine phosphatases in the human genome. *Cell* 117, 699–711.

Barford D (1995) Protein phosphatases. *Curr. Opin. Struct. Biol.* 5, 728–734.

Chen MJ, Dixon JE & Manning G (2017) Genomics and evolution of protein phosphatases. *Sci Signal.* 10(474), eaag1796. doi: 10.1126/scisignal.aag1796.

Cohen PT (2004) Overview of protein serine/threonine phosphatases. In *Protein Phosphatases* (Arino J., ed.), pp. 1–20. Berlin: Springer.

Fauman EB & Saper MA (1996) Structure and function of the protein tyrosine phosphatases. *Trends Biochem. Sci.* 21, 413–417.

Guan KL & Dixon JE (1991) Evidence for protein-tyrosine-phosphatase catalysis proceeding via a cysteine-phosphate intermediate. *J. Biol. Chem.* 266, 17026–17030.

Karisch R, Fernandez M, Taylor P, et al. (2011) Global proteomic assessment of the classical protein-tyrosine phosphatome and "Redoxome". *Cell* 146, 826–840.

Nguyen H & Kettenbach AN (2023) Substrate and phosphorylation site selection by phosphoprotein phosphatases. *Trends Biochem Sci.* 48(8), 713–725. doi: 10.1016/j.tibs.2023.04.004.

Tonks NK (2006) Protein tyrosine phosphatases: From genes, to function, to disease. *Nat. Rev. Mol. Cell Biol.* 7, 833–846.

G PROTEIN SIGNALING

Hamm HE & Gilchrist A (1996) Heterotrimeric G proteins. *Curr. Opin. Cell Biol.* 8, 189–196.

Heo WD & Meyer T (2003) Switch-of-function mutants based on morphology classification of Ras superfamily small GTPases. *Cell* 113, 315–328.

Milligan G & Kostenis E (2006) Heterotrimeric G-proteins: A short history. *Br. J. Pharmacol.* 147(1), S46–S55.

Sprang SR (1997) G protein mechanisms: Insights from structural analysis. *Annu. Rev. Biochem.* 66, 639–678.

Venkatakrishnan AJ, Deupi X, Lebon G, et al. (2013) Molecular signatures of G-protein-coupled receptors. *Nature* 494, 185–194.

Vetter IR & Wittinghofer A (2001) The guanine nucleotide-binding switch in three dimensions. *Science* 294, 1299–1304.

REGULATORY ENZYMES FOR G PROTEIN SIGNALING

Bos JL, Rehmann H & Wittinghofer A (2007) GEFs and GAPs: Critical elements in the control of small G proteins. *Cell* 129, 865–877.

Cherfils J & Zeghouf M (2013) Regulation of small GTPases by GEFs, GAPs, and GDIs. *Physiol Rev.* 93(1), 269–309. doi: 10.1152/physrev.00003.2012.

Ross EM & Wilkie TM (2000) GTPase-activating proteins for heterotrimeric G proteins: Regulators of G protein signaling (RGS) and RGS-like proteins. *Annu. Rev. Biochem.* 69, 795–827.

Rossman KL, Der CJ & Sondek J (2005) GEF means go: Turning on RHO GTPases with guanine nucleotide- exchange factors. *Nat. Rev. Mol. Cell Biol.* 6, 167–180.

Schmidt A & Hall A (2002) Guanine nucleotide exchange factors for Rho GTPases: Turning on the switch. *Genes Dev.* 16, 1587–1609.

SIGNALING ENZYME CASCADES

Bhattacharyya RP, Reményi A, Yeh BJ & Lim WA (2006) Domains, motifs, and scaffolds: The role of modular interactions in the evolution and wiring of cell signaling circuits. *Annu. Rev. Biochem.* 75, 655–680.

Bose I, Irazoqui JE, Moskow JJ, et al. (2001) Assembly of scaffold-mediated complexes containing Cdc42p, the exchange factor Cdc24p, and the effector Cla4p required for cell cycle-regulated phosphorylation of Cdc24p. *J. Biol. Chem.* 276, 7176–7186.

Ferrell Jr. JE (1996) Tripping the switch fantastic: How a protein kinase cascade can convert graded inputs into switch-like outputs. *Trends Biochem. Sci.* 21, 460–466.

Good M, Zalatan JG & Lim WA (2011) Scaffold proteins: Hubs for controlling the flow of cellular information. *Science* 332, 680–686.

Johnson GL & Lapadat R (2002) Mitogen-activated protein kinase pathways mediated by ERK, JNK, and p38 protein kinases. *Science* 298, 1911–1912.

Rivera-Molina FE & Novick PJ (2009) A Rab GAP cascade defines the boundary between two Rab GTPases on the secretory pathway. *Proc. Natl. Acad. Sci. USA* 106, 14408–14413.

Role of Post-Translational Modifications in Signaling

4

Signal transduction requires change in some component of the cell in response to an incoming signal. One of the most commonly used mechanisms for changing the properties of proteins is the covalent modification of their structure. Such changes, collectively termed post-translational modifications (PTMs), range from the addition of small chemical groups as in phosphorylation and methylation, to more substantial additions such as lipid groups or entire polypeptides, or even cleavage of the peptide backbone through proteolysis. These PTMs are catalyzed by specific enzymes and, in many cases, are reversed through the action of other enzymes, so change can be rapid, specific, and tightly regulated. The functional consequence of post-translational modification is to change the activity of the modified protein—its binding or enzymatic activity, localization, or conformation. The variety of PTMs possible, and the large number of different sites on a protein that can be modified, expand enormously the possible states for each protein. Given the rapidity, variety, and regulatory potential of PTMs, it is no surprise that they are centrally important for virtually all signal transduction mechanisms.

THE LOGIC OF POST-TRANSLATIONAL REGULATION

PTMs enable the regulation of cellular processes by providing a way to change protein properties rapidly. In the absence of PTMs, any major changes in the protein complement of the cell would require new protein synthesis, which is energetically costly to the cell. Moreover, the synthesis of a new protein is a relatively slow process. Furthermore, without post-translational modification, the structural diversity of proteins would be limited to what could be built from the 20 standard amino acids. Thus, post-translational modification greatly expands the diversity and dynamic

 Methods to detect and characterize PTMs are discussed in Chapter 15.

DOI: 10.1201/9780429298844-4

behavior of the proteins encoded by the genome (~20,000 protein-encoding genes in humans). Below, we briefly introduce the most common PTMs used in signaling.

Proteins can be covalently modified by the addition of simple functional groups

A variety of small chemical groups can be transferred enzymatically to the side chains of proteins (**Figure 4.1**). Such groups can change the surface charge distribution, hydrophobicity, hydrogen-bonding capacity, or conformation of the proteins they modify. One of the most prevalent of these modifications is **phosphorylation**, the transfer of the terminal phosphate group from ATP to proteins, most commonly to the hydroxyl groups of serine, threonine, or tyrosine side chains. The enzymes that perform the transfer are *protein kinases*, and the enzymes that reverse it by removing phosphate are *protein phosphatases*.

***N*-acetylation** involves the transfer of the acetyl group from acetyl CoA to the terminal ε-amino group of lysine side chains of target proteins. This modification is often seen on the N-terminal tails of histones, the major protein component of the nucleosomes that package genomic DNA into chromatin; thus, acetylation can regulate chromatin structure and gene expression. The reaction is catalyzed by **histone acetyl transferase (HAT)** and can be reversed by the action of **histone deacetylase (HDAC)**. This nomenclature reflects the prominence of histones as acetylation targets, but given the importance of this modification to a variety of other proteins, the terms **lysine acetyl transferase (KAT)** and **lysine deacetylase (KDAC)** are more accurate.

Figure 4.1

Modification of proteins with small functional groups. Structures of commonly modified amino acid side chains and their modified forms are depicted, along with the enzymes that add and remove each modification. Modifications are highlighted in *pink*.

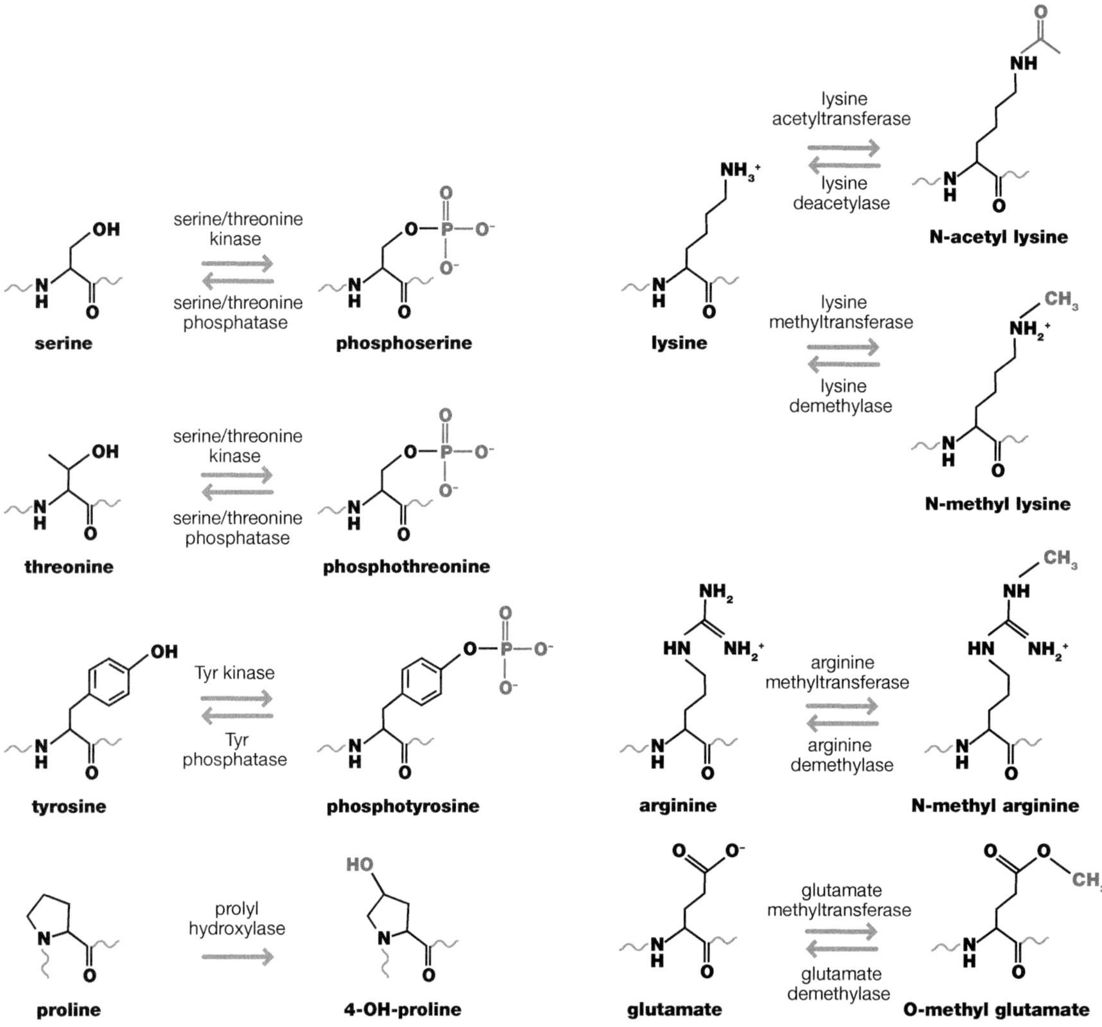

The simple single-carbon methyl group can also be transferred to proteins. The targets of *N*-**methylation** are the amino groups of lysine and arginine side chains, and like acetylation, this is a common modification of histone tails. N-methyl transferases transfer the methyl group from S-adenosyl methionine (SAM) to the protein; these transferases can be divided into **lysine methyl transferases (KMTs)** and **protein arginine methyl transferases (PRMTs)**. The sequential addition of multiple methyl groups to the same lysine nitrogen can generate di- and trimethylation. Unlike many other PTMs, N-methylation is very stable, and relatively few **lysine demethylases (KDMs)** and **arginine demethylases (RDMs)** catalyzing the reverse reactions are known. In prokaryotes, *O*-**methylation** of glutamate side-chain oxygens is important for transmembrane signaling, for example, in bacterial *chemotaxis*. Finally, the hydroxyl group can be transferred to several amino acid side chains; **proline hydroxylation** is important for signaling in a number of contexts, such as the cellular response to hypoxia.

Proteins can also be covalently modified by the addition of sugars, lipids, and even proteins

Larger organic compounds (carbohydrates and lipids) can also be transferred to proteins. Virtually all proteins exposed to the extracellular environment—secreted proteins and the extracellular portions of transmembrane proteins, as well as those residing in the lumen of the endoplasmic reticulum (ER) or Golgi apparatus—are modified by the addition of complex, branched carbohydrate chains. This process, termed **glycosylation**, can occur either on the hydroxyl groups of serine or threonine (*O*-glycosylation) or the amino group of asparagine (*N*-glycosylation). The enzymes that sequentially add sugar groups and trim the resulting polysaccharides to their final form are confined to the lumenal face of the ER and Golgi. Although glycosylation can profoundly affect the folding, trafficking, and ultimately the function of proteins, there are few examples in which changes in glycosylation store or transmit a signal, so these modifications will not be further discussed here. However, it has lately been appreciated that a very different type of glycosylation can play an active role in signaling: the addition of single N-acetyl glucosamine (GlcNAc) sugar groups to serine and threonine residues of cytosolic and nuclear proteins (sometimes termed GlcNAcylation) (**Figure 4.2a**).

Lipids are another important class of molecules that can be added to proteins, particularly simple fatty acids such as myristate and palmitate, as well as more complex prenyl groups such as farnesyl and geranylgeranyl. Given the very hydrophobic nature of these lipids, their addition often restricts a target protein to cellular membranes, though there are some examples of lipid-modified proteins that reside in the cytosol (for example, when the lipid group can be sheltered in a hydrophobic

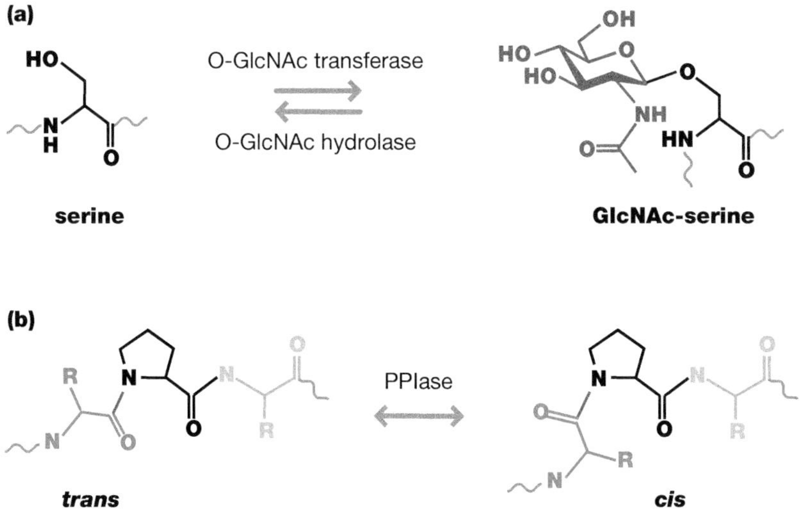

(a)

serine   →   O-GlcNAc transferase / O-GlcNAc hydrolase   →   GlcNAc-serine

(b)

trans   ←PPIase→   cis

Figure 4.2

Glycosylation and proline isomerization. (a) Addition of a single N-acetyl glucosamine group (GlcNAc, highlighted in *pink*) to a serine residue. (b) Proline cis-trans isomerization. Peptidyl prolyl cis-trans isomerase (PPiase) catalyzes rotation about the peptide bond (*pink*) N-terminal to proline. The amino acids N-terminal and C-terminal to proline are depicted in *blue* and *green*, respectively.

Lipid modifications and their roles in protein localization are discussed in greater detail in Chapter 5.

Figure 4.3

ADP-ribosylation. (a) ADP-ribosyltransferases use NAD⁺ as a substrate to transfer the ADP-ribosyl group to nucleophilic groups in various amino acid side chains; in this example, ADP-ribose is transferred to a glutamate residue. The ADP-ribose group is hydrolyzed by ADP-ribosylglycohydrolases. (b) ADP-ribose can also be added to other ADP-ribose groups, generating long poly-ADP-ribose (PAR) chains. Often different enzymes catalyze addition and removal of mono-ADP-ribose groups versus PAR chains.

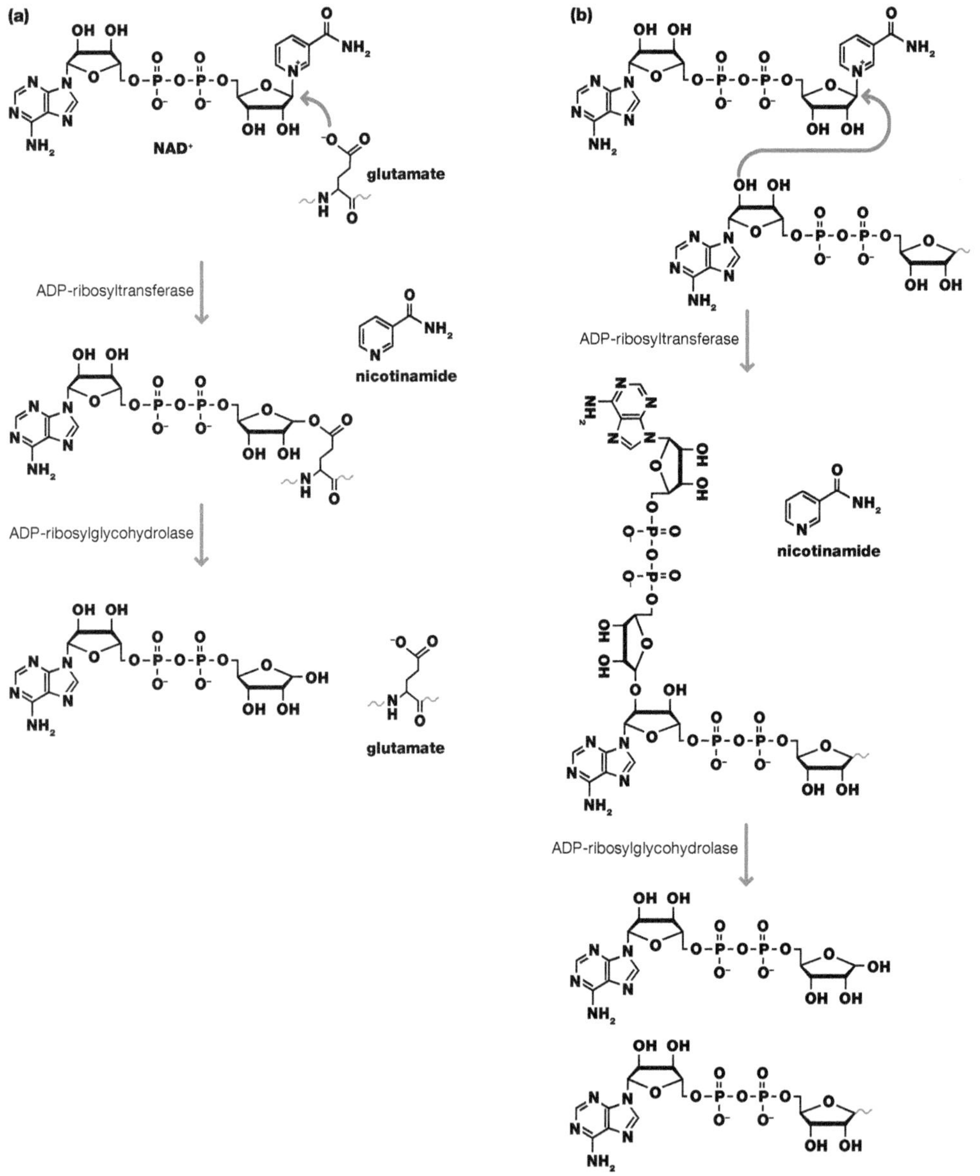

pocket on the protein surface). In many cases, lipids are added co-translationally (during translation or very soon thereafter) and are stable, and therefore are not well suited for transmitting signals. Some lipid modifications, however, particularly **S-palmitoylation** (the addition of the palmitic acid group to the sulfur of cysteine side chains), are more dynamic and may play a more active role in signaling.

Another relatively common protein PTM is **ADP-ribosylation**, which is the enzymatic addition of an ADP-ribose group to a variety of amino acid side chains including Glu, Asp, and Lys (**Figure 4.3**). The addition reaction uses nicotine adenine dinucleotide (NAD⁺) as a substrate, and is catalyzed by ADP-ribosyltransferases (ARTs); ADP-ribose groups are removed by ADP-ribose glycohydrolases (ARHs). Some ARTs have the ability to add ADP-ribose to existing ADP-ribose groups on proteins, leading to the formation of long poly-ADP-ribose (PAR) chains, which can be linear or branched. ADP-ribosylation plays a key role in a number of cellular activities; one of the best characterized is the DNA damage response, where the ART PARP1 is recruited rapidly to sites of DNA damage. The PAR chains that it generates then help recruit and assemble the DNA repair machinery at the site. Because of the importance of PARP to DNA damage repair, PARP inhibitors are now sometimes used in cancer therapy to enhance the killing of tumor cells with DNA damage. Another well-characterized role of ADP-ribosylation is in microbial pathogenesis. Some microbial pathogens, including those that cause cholera and diphtheria, secrete toxins into cells that ADP-ribosylate key signaling proteins, and thereby disrupt their function.

Entire protein chains can also be covalently linked to another protein. **Ubiquitin** is a 76-residue protein that is enzymatically added via its C-terminus to lysine side chains in a target protein, generating an isopeptide bond linking the two. In some cases, long chains of ubiquitin can be generated by sequential addition. Ubiquitylation is an important and widely used signal, often used to target proteins for proteolytic degradation. Several other ubiquitin-like proteins (UBLs), such as SUMO and Nedd8, are also transferred to target proteins by similar mechanisms (termed sumoylation and neddylation). This type of modification can be considered the post-translational addition of a modular protein domain to an existing protein.

The cleavage of the protein backbone itself, termed *proteolysis*, can be considered the most drastic of all PTMs. Proteolytic cleavage of proteins can destroy their activity or can generate biologically active molecules from larger, inert precursors. The enzymes that mediate proteolysis are termed proteases. The unique role of proteolysis in cellular signaling is the subject of Chapter 9.

On the other extreme is the relatively subtle conformational modification of proline residues in protein, termed **prolyl *cis-trans* isomerization**. This is not strictly speaking a post-translational modification, as there is no change in covalent structure of the protein; instead, it is a switch in the conformation of the proline residue (rotation around the peptide bond) from the *cis* to the *trans* conformation (**Figure 4.2b**). Spontaneously, this reaction proceeds very slowly, but it can be speeded up greatly through the action of peptidyl prolyl *cis-trans* isomerase (PPIase). This conformational switch plays a role in some signaling mechanisms.

Post-translational modifications can alter protein structure, localization, and stability

Of course, for PTMs to be able to transmit information, they must in some way impact the biological activity of their host proteins. They do this by changing the physical structure of the protein (generally the shape, charge, and hydrophobicity of the surface) which, in turn, affects how that protein behaves and interacts with other molecules in the cell. For example, phosphorylation of a hydroxyl amino acid such as serine converts a relatively small, uncharged, moderately hydrophilic side chain into a much bulkier one carrying a strong negative charge. Adding the acetyl group to the amino nitrogen of lysine side chains also increases bulk, but also partially quenches the strong positive charge of the unmodified amino group (**Figure 4.4**). Such changes at the local level can have profound effects on the more global level of protein function. Some of these effects are introduced below.

PTMs are often tightly coupled to changes in protein conformation (see, for example, **Figure 4.5**). Such changes may be in the conformation of the polypeptide backbone itself, particularly when the modified residues exist in relatively unstructured loops. Alternatively, post-translational modification may alter the intramolecular arrangement of the different folded domains of the protein. The resulting conformational changes again can have diverse and profound secondary effects on protein activity, such as stimulating or inhibiting catalytic activity, altering the ability to bind to other proteins, or enabling further PTMs by unmasking or burying potential modification sites.

Another important consequence of post-translational modification is to alter protein–protein interactions. Modification can either increase or decrease the affinity of a protein for one or more binding partners (that is, can create or destroy a binding site) by virtue of its effects on the charge distribution, hydrogen-bonding possibilities, and shape of the binding surface (**Figure 4.5b** and **4.5c**). Indeed, as discussed in the next section, the coupling of post-translational modification to changes in protein–protein interaction is a very common theme in signaling. In

Figure 4.4

Chemical effects of protein modification.
(a) Serine and phosphoserine and
(b) lysine and N-acetyl lysine are depicted
in stick format, with the molecular surface
superimposed. Both the phosphate and
acetyl groups greatly increase the bulk of
the side chain. Electrostatic potential is also
indicated on the surface (red = negative,
blue = positive potential). Phosphorylation
introduces strong negative charge, while
acetylation decreases positive charge
compared to the unmodified side chains.

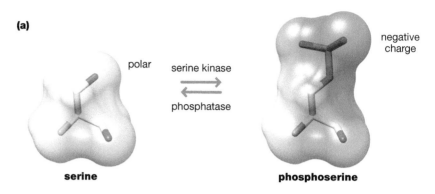

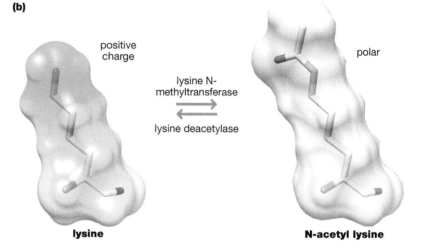

(a) change conformation, activity

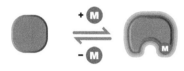

(b) promote protein binding

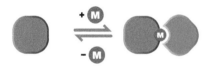

(c) prevent protein binding

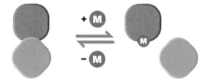

(d) change subcellular localization

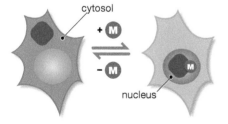

(e) change proteolytic stability

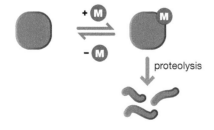

this way, changes in the activity of enzymes that mediate PTMs are converted into changes in the physical interactions between proteins. Post-translational modification can also lead to changes in the interaction of proteins with other cellular components, for example, nucleic acids or membrane lipids. Changes in protein interaction often lead, in turn, to secondary effects, such as changes in subcellular localization (for example, if the protein is cytosolic and its binding partner is an integral membrane protein) or further changes in post-translational modification (as when a binding partner is itself an enzyme that can modify the first protein).

Changes in subcellular localization can also be a direct consequence of PTMs. For example, lipid modification directs many proteins to stably interact with cellular membranes. There are also several examples where transmembrane proteins are proteolytically processed in response to signals, releasing from the membrane an active fragment that can diffuse away and exert its effects elsewhere in the cell, as in the case of *Notch* signaling. (Notch is a transmembrane receptor that, upon binding to an extracellular ligand, is cleaved to release an intracellular fragment that translocates to the nucleus and regulates gene transcription.) Furthermore, post-translational modification can alter the dynamics of the shuttling of proteins between different subcellular compartments (termed **protein trafficking**). There are many cases where the phosphorylation of nuclear localization or nuclear export

Figure 4.5

Diverse effects of protein modification. (a) Post-translational modification (*pink* circle) of a substrate protein (*light brown*) results in conformational changes and activation of the substrate. (b) Modification creates a binding site for another protein (*orange*). (c) Modification of the substrate prevents the binding of a protein (*blue*) that binds to the unmodified substrate. (d) Modification results in a change in subcellular localization of the substrate from the cytosol to the nucleus. (e) Modification leads to proteolytic degradation of the substrate.

sequences of a protein either enhances or blocks its interaction with the nuclear import or export machinery, thus altering the distribution of the protein between the nucleus and cytosol. In the same vein, ubiquitylation of plasma membrane proteins often serves as a signal for their internalization (and possible degradation) through the process of *endocytosis*.

Finally, PTMs can alter the proteolytic stability, and therefore the expression level, of a protein. Some phosphorylation events can target a protein for proteolytic degradation, whereas other modifications can specifically stabilize a protein. As we will learn later in this chapter, phosphorylation is often coupled to a second modification, ubiquitylation, which then leads to proteolytic degradation of the ubiquitylated target.

Notch signaling is described in more detail in Chapter 5.

Post-translational control machinery often works as part of "writer/eraser/reader" systems

A common theme among these diverse PTMs is that they are often controlled by "writer/eraser/reader" systems. In Chapters 1 and 3, we discussed how enzymes that can mark proteins by PTMs can be termed "writers", while enzymes that can remove these post-translational marks can be termed "erasers". In some cases, the PTMs directly alter protein structure and activity. However, in many cases, the physical changes in the marked proteins are interpreted indirectly by cytosolic "readers"—proteins that contain modular domains that bind only to proteins containing the appropriate post-translational mark. Examples of such systems are shown in **Figure 4.6**. It is important to note that in such systems, it is changes in the overall level of post-translational marking that lead to downstream signaling. We are naturally biased to focus on the affirmative act of protein marking by writers, but the removal of preexisting marks by erasers may be at least as important in some contexts.

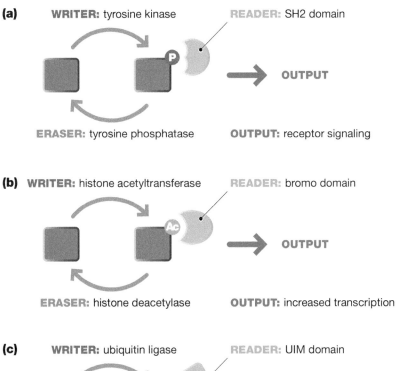
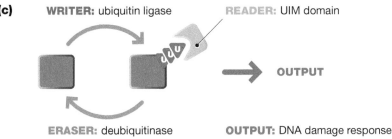

(a) WRITER: tyrosine kinase   READER: SH2 domain

OUTPUT

ERASER: tyrosine phosphatase   OUTPUT: receptor signaling

(b) WRITER: histone acetyltransferase   READER: bromo domain

OUTPUT

ERASER: histone deacetylase   OUTPUT: increased transcription

(c) WRITER: ubiquitin ligase   READER: UIM domain

OUTPUT

ERASER: deubiquitinase   OUTPUT: DNA damage response

Figure 4.6

Three writer/eraser/reader systems.
(a) In signaling by tyrosine kinase growth factor receptors, tyrosine phosphorylation of the activated receptor creates binding sites for effector proteins containing SH2 domains. (b) Acetylation of histones by histone acetyl transferase creates binding sites for chromatin modifying factors with bromo domains. (c) Ubiquitylation of proteins associated with damaged DNA creates binding sites for signaling and repair complexes containing UIM domains.

Such modular writer/eraser/reader systems lend themselves to the rapid diversification and adaptation of the signaling machinery in the course of evolution. This is because they provide a ready means to link together two previously unlinked proteins into a new signaling pathway, simply by generating a new site of post-translational modification by point mutation, or by adding a modular binding domain through recombination.

All of the PTMs described in this chapter are catalyzed by specific enzymes (some covalent modifications, such as the oxidation of certain amino acid side chains, can occur spontaneously, but these are generally not exploited in signal transduction mechanisms). Thus, the state of post-translational modification of a protein necessarily depends on the activity and abundance of the enzymes that perform the modification and those that can remove it. Whether or not a particular site is modified depends on the local concentration of the modifying enzymes, their overall state of activation, and their activity toward that specific site.

For a particular reaction to occur (a writer adding a mark, or an eraser removing one), all three requirements must be satisfied: little if any modification will occur if there is no enzyme in the vicinity, if the enzyme is not catalytically active, or if the active enzyme cannot efficiently modify the particular site. Because each of these parameters may be modified by upstream signals, however, PTMs can respond to environmental changes and thus transmit information.

Post-translational modifications allow very rapid signaling and transmission of spatial information

Because post-translational modification is mediated by enzymes, it can be very rapid and lead to massive amplification of a signal. Enzymes can be extremely efficient catalysts, with a single enzyme molecule being capable of modifying many substrate molecules per second. Thus, only a few activated enzyme molecules can exert an enormous effect in a very short period of time. These properties are useful for signaling mechanisms that need to operate on relatively short time scales (seconds to minutes). By contrast, changes in transcription leading to synthesis of new proteins typically require much longer times to exert their effects (minutes to hours). The only signaling mechanisms with the potential for more rapid response are those driven by changes in membrane potential and ion flow, used extensively in the nervous system where speed is essential.

The enzymes responsible for post-translational modification are often activated in specific places in the cell, and they and their products have finite diffusion rates, and so such systems can be used to transmit spatial information. These systems can be used to detect where an input signal originates, and can control resulting spatial responses, including complex changes in cellular organization or morphology. This type of information cannot be propagated by transcriptional regulatory systems.

INTERPLAY BETWEEN POST-TRANSLATIONAL MODIFICATIONS

Many proteins can be modified at many different sites and in a variety of ways. An important consequence of this diversity of modification is to greatly increase the number of potential molecular states of a protein (**Figure 4.7**). To take a simple example, if a protein has ten possible phosphorylation sites and each site can be phosphorylated or dephosphorylated independently, 2^{10} (1024) distinct phosphorylated states are possible. When other potential modifications are considered, the number of possible states for a single protein can become virtually infinite. In principle, each of these different states can have a distinct biological activity. Thus, the genomic coding potential is expanded enormously by the ability of each protein to be modified in various ways after translation. This **combinatorial complexity**

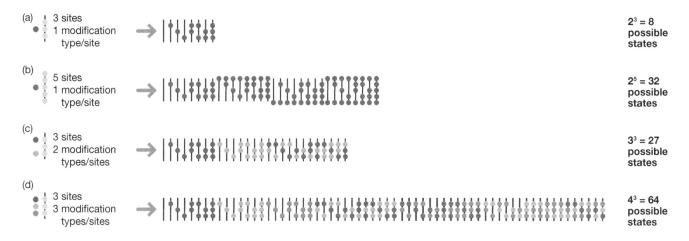

also creates problems in characterizing the properties of proteins, because in most cases, it is not easy to physically separate each of the modified states for analysis, so one can only assay the average properties of the ensemble of modified states.

In this section, we will briefly consider some of the ways that individual PTMs can interact with each other to regulate protein activities. We will then delve more deeply into a specific example, p53, which uses a host of PTMs to integrate signals from throughout the cell and thereby regulate cell-cycle progression and programmed cell death.

A post-translational modification can promote or antagonize other modifications

There are a number of examples where one PTM can directly affect either the likelihood or nature of additional modifications. Such interrelationships can be important in allowing proteins to integrate and process multiple upstream signals and perform logical operations.

Since many PTMs target the same amino acid residues, a single residue can be modified in different, mutually exclusive ways. Competition for sites by different modifying enzymes can result in a switch between two (or more) different states of the modified protein. Lysine residues are commonly used in this way; the terminal amino group can be subject to acetylation, methylation, ubiquitylation, and modification by other ubiquitin-like groups (**Figure 4.8**). Similarly, the same serine and threonine residues can be subject either to phosphorylation or to *N*-GlcNAcylation. An example of the latter is provided by the transcription factor Myc, an important mediator of cell growth and proliferation. The balance between phosphorylation and *N*-GlcNAcylation of certain serine residues can be regulated by upstream signals, and the different modified forms have very different activities and different functional consequences.

Another way in which modifications can affect each other is when one type of modification is required for subsequent modifications of another type. An excellent example of this is provided by *cyclin-dependent kinases (CDKs)*, which regulate cell-cycle transitions. CDKs phosphorylate specific protein targets on serine and threonine residues. These phosphorylated sites then serve as binding sites for a large ubiquitin ligase complex which polyubiquitylates the phosphorylated protein, thereby targeting it for destruction by the proteasome (**Figure 4.9**).

In this pathway, three distinct PTMs (phosphorylation, ubiquitylation, proteolysis) are functionally linked—one modification promotes the next by making the modified protein a better substrate for the next modifying enzyme. In effect, multiple writer/eraser/reader systems are linked together. The ultimate result, in the case of CDKs, is to convert the reversible activation of a kinase into the essentially irreversible destruction of the substrates of that kinase.

Figure 4.7

Multiplicity of modification sites and modification types greatly increases the number of possible states for a protein. A number of different scenarios are presented; proteins are depicted as vertical lines and modifications as colored circles. (a) A single modification is possible (two possible states: modified and unmodified) at three possible sites. (b) The number of possible sites is increased to five. (c) Two different modifications are possible at the same site (for example, lysine acetylation and lysine methylation), generating three possible states per site. (d) Three different modifications are possible at each site (for example, acetylation, methylation, and ubiquitylation), generating four possible states per site.

 Cyclin-dependent kinases and cell-cycle control are discussed in more detail in Chapter 14.

Figure 4.8

Switching between distinct states of post-translational modification. (a) Lysine residues can be acetylated, methylated, or ubiquitylated. Switching between each state requires removal of the first modification before addition of the second. (b) Serine residues can be either phosphorylated [by a serine/threonine (S/T) kinase] or GlcNAcylated.

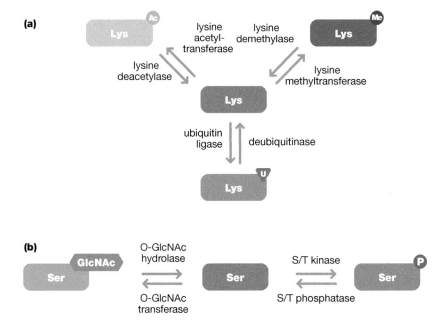

Apoptosis is discussed in more detail in Chapter 9.

Figure 4.9

Linked writer/eraser/reader systems. During the cell cycle, proteins phosphorylated by cyclin-dependent kinases are targeted for destruction in the proteasome. (a) The SCF ubiquitin ligase complex acts as both a reader (of phosphorylation) and a writer (of ubiquitylation). (b) Phosphorylation of a substrate by cyclin-dependent kinase leads to binding of SCF via its F-box domain. SCF polyubiquitylates the substrate, which is then recognized by ubiquitin-binding domains (UBDs) of the proteasome. Proteasome binding results in the proteolytic destruction of the protein.

Different modifications can also be mutually antagonistic—the presence of one modification can prevent other modifications in the same protein or associated proteins. One type of modification may block interaction with other modifiers by decreasing the affinity of their interaction, or modification may make a protein a poorer substrate for the second enzyme. This is seen in the case of modification of histone tails, which regulate chromatin structure and thus the ability of DNA to be transcribed (discussed in more detail below). For example, a threonine residue in histone H3 is phosphorylated by a protein kinase (specifically, protein kinase C-β_1) after androgen hormone stimulation. This PTM blocks the ability of a lysine demethylase, LSD1, to remove methyl groups from a nearby lysine residue. The resulting increase in histone methylation at this position is important for the induction of transcription by the hormone (**Figure 4.10**).

p53 is tightly regulated by a wide variety of post-translational modifications

A particularly well-characterized illustration of the diversity of post-translational modifications is provided by **p53**, a master regulator of cellular responses to a wide range of environmental stresses such as DNA damage. Depending on the specific stress and cellular context, p53 can induce momentary cell-cycle arrest, while the cell attempts to repair any damage, or permanent cell-cycle arrest and *apoptosis* (a form of programmed cell death). Thus, p53 prevents cells from replicating inappropriately and passing on damaged genomic DNA, earning it the nickname "guardian of the genome." It is the most commonly mutated gene in human cancers, underscoring its central importance as a check

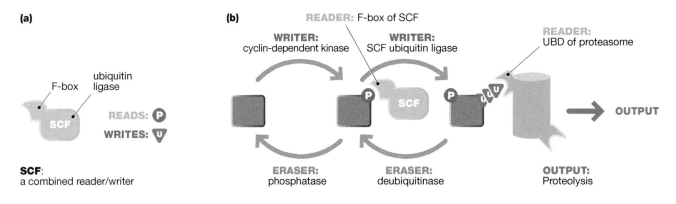

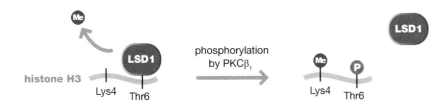

on inappropriate cell proliferation (such genes, that antagonize cell proliferation and survival pathways and promote tumorigenesis when mutated, are termed **tumor suppressors**).

Not surprisingly, given the importance of its task and the potentially fateful consequences of its activation, p53 is very tightly regulated by a wide variety of inputs. The primary way in which environmental conditions communicate to p53 is through a vast array of PTMs, which include phosphorylation, acetylation, methylation, ubiquitylation, sumoylation, and neddylation (**Figure 4.11a**). Roughly 10% of the amino acids of p53 have been shown to be subject to at least one form of PTM, and many can be modified in more than one way. Further regulation of p53 occurs through its interactions with over a hundred binding partners and by its subcellular localization but, as we shall see, these too are primarily controlled by PTMs.

The level and activity of p53 are regulated by ubiquitylation and acetylation

When active, p53 binds to specific DNA sites and induces the transcription of genes that regulate the cell cycle, DNA repair, apoptosis, and other activities. In unstressed cells, however, the levels of p53 are very low. This is mostly due to the polyubiquitylation of p53 on a number of C-terminal lysine residues by a ubiquitin ligase, Mdm2 (**Figure 4.11b**). As we will discuss below and in Chapter 9, polyubiquitylated proteins are rapidly degraded by a specialized proteolytic structure, the *proteasome*. Also associated with p53 are *deubiquitinases* like HAUSP, which can remove ubiquitin chains and thereby stabilize p53. In addition to polyubiquitylation, p53 can be monoubiquitylated at low levels of Mdm2. This modification leads to the export of p53 to the cytosol, where it can regulate apoptosis and another form of cellular self-destruction, autophagy.

Figure 4.10

Regulation of histone methylation by phosphorylation. Under basal conditions, the lysine demethylase LSD1 binds to the N-terminal tail of histone H3 and removes methyl groups from Lys4. Upon androgen stimulation, protein kinase C-β_1 (PKCβ_1) phosphorylates Thr6, preventing the binding of LSD1 and leading to increased methylation of Lys4.

Figure 4.11

Regulation of p53 by post-translational modification. (a) The domain structure of p53 is depicted to scale, along with known sites of post-translational modification. For p53 domains: TA, transcriptional transactivation; Pro, proline-rich domain; NLS, nuclear localization signal; Tet, tetramerization domain; NES, nuclear export signal; Reg, regulatory domain. (b) Mechanism of activation of p53. Top: in its basal, unactivated state, p53 is bound to Mdm2, which polyubiquitylates it and targets it for destruction by the proteasome. Cell stresses such as DNA damage lead to phosphorylation of p53, loss of Mdm2 binding, and recruitment of acetyl transferases such as p300/CBP. The resulting acetylation leads to recruitment of transcriptional activators. p53 is stabilized, tetramerizes, and binds specific DNA sequences to induce transcription of p53-responsive genes. (a, adapted from K.A. Boehme & C. Blattner, *Crit. Rev. Biochem. Mol. Biol.* 44:367–392, 2009. With permission from Informa Healthcare.)

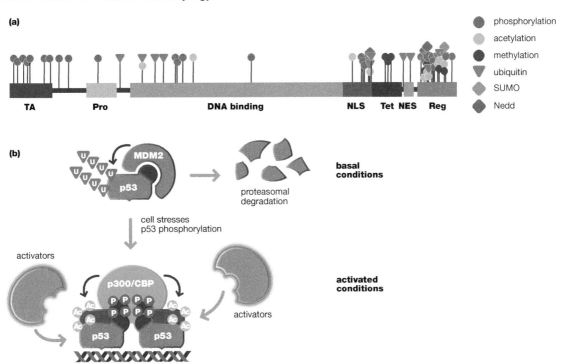

Clearly, the extent of ubiquitylation is critical for regulating the stability of p53 and thus determining its overall concentration in the cell. The ubiquitylation of p53 is, in turn, regulated by phosphorylation. A variety of protein kinases, such as ATM, ATR, and Chk1/2 (which are activated by DNA damage), can phosphorylate p53 and render it less susceptible to polyubiquitylation by Mdm2 and other ubiquitin ligases. This is an example of one type of PTM (phosphorylation) negatively regulating another modification (ubiquitylation).

The acetylation of a number of lysine residues on p53 dramatically enhances the recruitment of transcriptional coactivators, thereby stimulating transcription of specific p53-dependent promoters. The extent of acetylation is thought to determine, at least in part, which specific promoters are activated by p53 (that is, those that promote cell-cycle arrest or those that promote apoptosis). The primary acetyl transferase for p53 is termed the p300/CBP complex, though others exist. Importantly, the most prominent sites of acetylation are the same lysine residues that are polyubiquitylated by Mdm2, so their modification (ubiquitylation versus acetylation) serves as a switch to toggle between two distinct states: unstable and transcriptionally inactive versus stable and transcriptionally active. This is an example of two alternative modifications of the same residues resulting in very different functional outputs.

Additional modifications further fine-tune p53 activity

In addition to phosphorylation, ubiquitylation, and acetylation, other PTMs contribute to p53 regulation (see Figure 4.11a). Many of these target the same lysine residues that are subject to ubiquitylation and acetylation, thus adding to the rich complexity of p53 regulation. For example, methylation of lysine and arginine residues of p53 can either enhance or suppress transcriptional activities on particular promoters, most likely through regulating interactions with other specific transcriptional coactivators. Transfer of the ubiquitin-like proteins SUMO and Nedd8 to lysine residues can also regulate the transcriptional activity of p53, its subcellular localization, or both. Clearly, the number of lysines on p53 subject to modification, and the diverse modifications that are possible at each, have the potential to generate an enormous number of different activity states for this protein.

An additional layer of complexity is provided by a dense network of regulatory relationships connecting various p53-modifying enzymes. For example, activation of p53 leads to increased expression of Mdm2 which, in turn, polyubiquitylates p53 and down-regulates its activity (an example of a *negative feedback loop*). In a second example, ubiquitin ligases and acetyl transferases not only compete to modify the same lysine residues on p53, but can also modify each other and thereby either inhibit or stimulate their respective activities.

The rich diversity of PTMs of p53 greatly expands the ways in which its abundance, activity, and localization can be regulated. This not only enables p53 to respond to and integrate diverse cellular inputs, but also allows it to have diverse output activities that can be finely tuned to the needs of the system.

PROTEIN PHOSPHORYLATION

In eukaryotic cells, protein phosphorylation is a very common modification. It is estimated that more than a third of all human gene products are phosphorylated, and this fraction is almost certain to grow as more sensitive methods are used to detect phosphorylation. The hydroxyl amino acids (serine, threonine, and tyrosine) are by far the most common targets of phosphorylation in eukaryotes. In Chapter 3, we discussed some of the properties of the phosphate group that make it particularly useful for signaling: in short, phosphorylation allows the cell to use a readily available raw material (ATP) to induce stable and significant alterations in protein structure and function. Although the phosphoester linkage is relatively stable to spontaneous hydrolysis, in the cell its reactions can be rapidly catalyzed through the opposing actions of protein phosphatases and protein kinases.

The importance of phosphorylation in regulating signaling pathways is suggested by the increasing importance of kinase inhibitors in the treatment of human disease. Since all protein kinases use ATP as a substrate of the phosphoryl transfer reaction, ATP analogs (small molecules that mimic ATP, but cannot be used for phosphate transfer) are useful as protein kinase inhibitors. Many of these compounds are now used clinically as drugs, for example, to inhibit kinases that cause cancer when inappropriately active.

Phosphorylation is often coupled with protein interactions

In addition to its direct effects on protein structure, phosphorylation has another important role in signaling: it can dramatically affect the interaction of a protein with other proteins in the cell. As already noted, both serine/threonine and tyrosine phosphorylation lie at the heart of writer/eraser/reader systems, in which proteins containing modular phosphorylation-specific binding domains "read" changes in phosphorylation by binding specifically to certain proteins only after they are phosphorylated.

 How phosphorylation can allosterically alter protein structure is discussed in Chapter 3.

This type of system was first appreciated in signaling by receptors with tyrosine kinase activity. When cells were stimulated with ligands for such receptors, in many cases the most abundantly phosphorylated substrate was found to be the receptor itself. This rather puzzling observation raised the question of how the signal was transmitted, in the absence of significant phosphorylation of downstream substrates. The discovery of a modular domain that binds specifically to peptides in the tyrosine-phosphorylated state (the SH2 domain) provided a solution to the puzzle: autophosphorylation of the receptor led to the recruitment of SH2-containing proteins from the cytosol to the receptor on the membrane. This change in localization brought SH2-containing enzymes into close proximity with their substrates on the membrane, thereby increasing their activity (**Figure 4.12a**).

 Signaling by receptor tyrosine kinases is discussed in more detail in Chapter 8.

The SH2 domain is the founding member of the family of phosphospecific binding modules, of which more than ten are now known. The SH2 domain is the major binding partner for tyrosine-phosphorylated sites, though a few other phosphotyrosine-binding domains have been identified. In the case of phosphoserine and phosphothreonine, a larger number of modular binding domains are known, likely reflecting the earlier evolutionary origin and higher overall prevalence of serine/threonine phosphorylation compared to tyrosine phosphorylation.

Phosphospecific binding domains have a positively charged pocket that interacts directly with the phosphate group, and adjacent surfaces that interact with amino acids surrounding the phosphorylated site. For the SH2 domain for example, roughly half of the binding energy is provided by the phosphate group, and half by the surrounding residues. In this way, at normal physiological concentrations of binding domain and substrate, only those substrates that are phosphorylated and which also contain favorable residues in the vicinity can bind to a significant extent; the affinity of the domain for the unphosphorylated site, or to phosphate group alone, is too weak to support binding.

In the case of tyrosine phosphorylation, the receptor itself is often phosphorylated on many different sites, each of which is recognized by a subset of SH2-domain-containing proteins, providing the potential for assembly of a large, multicomponent signaling structure nucleated by the phosphorylated receptor (see Figure 4.12a). In other cases, the receptor phosphorylates an intracellular scaffold protein on many sites, which can then recruit many distinct effectors (**Figure 4.12b**). For example,

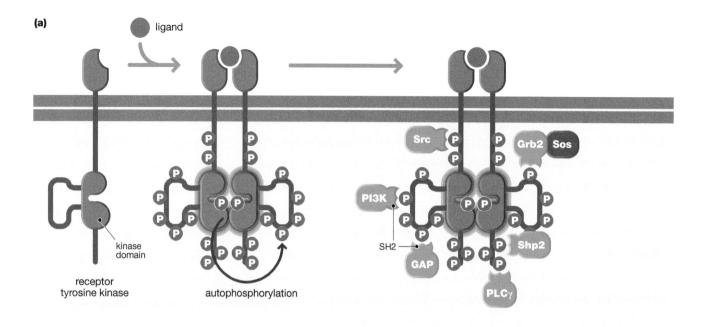

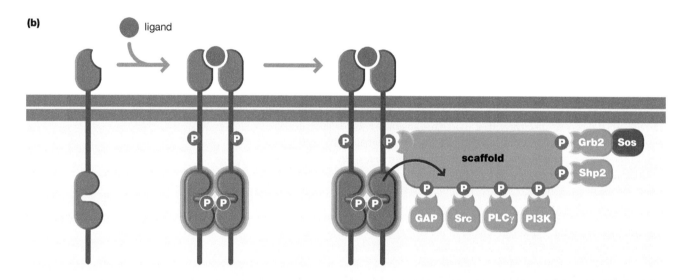

Figure 4.12

Receptor tyrosine kinases create phospho-binding sites. (a) Binding of ligand induces dimerization of the receptor, activation of the catalytic domain, and tyrosine autophosphorylation on a number of sites. These sites serve to recruit effector proteins with phosphotyrosine-binding domains such as SH2 domains. The identity of the effectors bound depends on the sites that are phosphorylated and the effectors expressed in the cell. (b) Some receptor tyrosine kinases do not autophosphorylate extensively, but instead phosphorylate scaffold proteins, which then serve to recruit downstream effectors. A number of common effectors are depicted: Src, Src family nonreceptor tyrosine kinase; PI3K, phosphatidylinositol 3-kinase; PLCγ, phospholipase Cγ; Shp2, tyrosine phosphatase Shp2. GAP is a Ras GTPase activator protein, Grb2 is an adaptor protein, and Sos is a Ras guanine nucleotide exchange factor.

in signaling by the insulin receptor, the scaffold IRS1 (insulin receptor substrate 1) is phosphorylated by the activated receptor on around ten different tyrosine residues, and serves as a platform for assembly of a downstream signaling complex consisting of proteins with phosphotyrosine-binding domains. IRS1 itself contains a PTB domain, another phosphotyrosine-binding domain, which presumably helps recruit the scaffold to the activated (autophosphorylated) receptor.

The domains that bind phosphoserine/phosphothreonine, and the consequences of this binding, are much more varied than in the case of phosphotyrosine.

A number of domain families including FHA, WW, BRCT, Polo-box, MH2, and WD40 have members that can bind phosphothreonine- or phosphoserine-containing motifs. There is also a family of small proteins, termed the *14-3-3 proteins*, which specifically bind to proteins phosphorylated on serine or threonine. Unlike other modular phosphoprotein-binding domains, 14-3-3 proteins do not contain any other functional domains, though they do form homo- or heterodimers. 14-3-3 dimers interact either with multiple phosphorylated sites on the same phosphoprotein or with two different phosphoproteins. They regulate activity upon binding either by sterically blocking access to other binding partners or by inducing conformational changes.

Kinases and phosphatases vary in their substrate specificity

In every cell, there are many hundreds of thousands of different potential phosphorylation sites (serine, threonine, or tyrosine residues) located on the surface of proteins. This poses a problem, as the specificity of signal output may depend on the phosphorylation of one or a few sites within this vast excess of irrelevant sites. As we have seen, kinases and phosphatases select the substrates they act upon at a number of levels (see Figure 3.17). For example, the catalytic cleft itself can discriminate between different potential phosphorylation sites depending on the amino acids flanking the potential phosphorylation site. Kinases vary considerably in their intrinsic specificity, however, and some (for example, most tyrosine kinases) are quite promiscuous in the variety of peptides that can be efficiently phosphorylated.

A second level of specificity is provided by contacts with the substrate outside of the catalytic cleft, including other regions of the kinase itself, or proteins associated with the kinase. Substrate specificity can be further enhanced by association with scaffold proteins. Scaffolds bind a kinase (or other enzyme) and its potential substrates, aligned like peas in a pod; thus, substrates are presented at high concentration and at optimal orientation for phosphorylation by the kinase. Scaffolds can also anchor proteins to specific subcellular locations, thereby altering the potential substrates that might be encountered by the kinase.

Phosphospecific binding modules such as 14-3-3 proteins and SH2 domains can also contribute, albeit indirectly, to specificity in the cell. Not only do such proteins play a critical role as readers of the presence or absence of phosphorylation, but they also have the potential to protect sites from dephosphorylation. The constitutive rates of protein dephosphorylation in the cell can be very high, so that any sites that are not protected in some way will rapidly be dephosphorylated. In such a setting, a relatively nonspecific protein kinase might phosphorylate many sites, most of which are irrelevant, and only those sites that successfully bind to reader proteins will be protected and thus persist (**Figure 4.13**). Thus, the functional relationship between readers, writers, and erasers may allow some of the burden for substrate specificity to be borne by the binding proteins.

 See Chapter 3 for a more extensive discussion of scaffolds.

Figure 4.13

Protein-binding domains can contribute to apparent kinase specificity.
(a) A variety of proteins phosphorylated on a variety of different sites is depicted. Under conditions of high nonspecific phosphatase activity, most sites will be rapidly dephosphorylated. Sites that are bound by specific phosphoprotein-binding domains, however, may be protected from phosphatase activity and thus come to predominate in the population. (b) The likelihood that a site will be bound by a phosphoprotein-binding domain (and thus be protected from dephosphorylation) depends both on kinase specificity (the likelihood that the site will be phosphorylated) and on the specificity of the phosphoprotein-binding domain (the likelihood that a site, if phosphorylated, will bind).

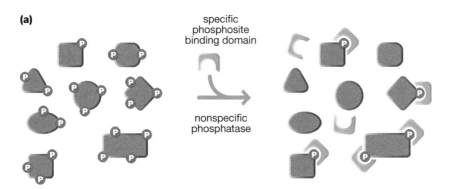

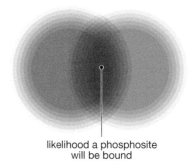

Multiple phosphorylation of proteins can arise by different mechanisms

When more than one site is phosphorylated on the same protein, there are a number of ways in which these phosphorylation events can depend on each other. For instance, phosphorylation by one kinase can make a protein a better (or worse) substrate for a second kinase that phosphorylates different sites. This is seen in glycogen synthase kinase 3 (GSK-3), which is a key serine/threonine kinase in a number of signaling pathways. Typically, GSK-3 only recognizes and efficiently phosphorylates substrates after they have first been **primed** by phosphorylation by another kinase, such as CK1 or CK2 (**Figure 4.14a**). Crystal structures revealed that the substrate-binding cleft of GSK-3 contains a positively charged pocket that recognizes the phosphate group of the primed substrate, promoting substrate binding and subsequent phosphorylation of the second site.

When a substrate can be phosphorylated at multiple sites by the same kinase, this can occur either by a **distributive** or **processive** mechanism (**Figure 4.14b** and **4.14c**). In distributive phosphorylation, each site is phosphorylated independently—for each site, the kinase binds, transfers phosphate to the substrate, and then dissociates. By contrast, in processive phosphorylation, the kinase remains associated with the substrate and phosphorylates multiple sites sequentially. In this latter case, the unphosphorylated and highly phosphorylated states are much more prevalent than states of phosphorylation at just one or a few sites.

Processive phosphorylation can be mediated by domains or binding sites on the kinase itself, which interact with phosphorylated sites on the substrate after the initial phosphorylation. For example, mitotic CDKs have multiple sites that bind to phosphorylated substrates (one on the cyclin, and one on a CDK-associated

Figure 4.14

Diverse modes of multiple phosphorylation. (a) Priming: a kinase phosphorylates a substrate efficiently only after the substrate has been previously phosphorylated on a different site, often by a different kinase, creating a structure that fits a binding pocket on the kinase. (b) Processive phosphorylation: a kinase binds to a substrate and phosphorylates multiple sites on that substrate before dissociating. (c) Distributive phosphorylation: a kinase phosphorylates only one site before dissociating from the substrate. Multiple phosphorylation requires multiple rounds of binding, catalysis, and dissociation.

(a) priming

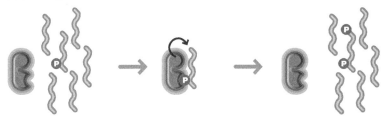

(b) processive phosphorylation

(c) distributive phosphorylation

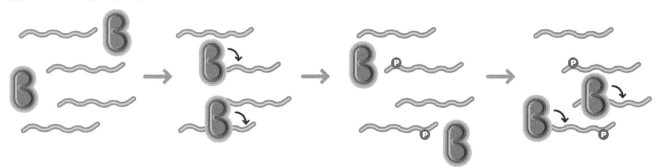

adaptor protein called Cks1). Another example is provided by nonreceptor tyrosine kinases, which often contain an SH2 domain that associates with tyrosine-phosphorylated substrate proteins and facilitates their processive phosphorylation.

Priming, distributive phosphorylation, and processive phosphorylation all can lead to phosphorylation of multiple sites, but differ significantly in how the final state of phosphorylation depends on the concentration and activity of the kinases responsible. These mechanisms allow the generation of complex output behaviors in systems regulated by protein phosphorylation. For example, requiring multiple distributive modifications for an enzyme to be active is one way to generate switchlike activation.

Switchlike activation and ultrasensitivity are described in Chapter 11.

Histidine and other amino acids can be phosphorylated, especially in prokaryotes

In multicellular animals, phosphorylation of serine, threonine, and tyrosine accounts for almost all protein phosphorylation. However, other amino acid side chains can also be phosphorylated, including histidine, arginine, and aspartate, and such modifications are important for signaling in prokaryotes and some eukaryotes.

Both histidine and aspartate phosphorylations are key elements of a common and highly conserved prokaryotic signaling mechanism—the **two-component system**. In bacteria, two-component systems are the most common means of transducing information from outside the cell to the interior. In their most typical form, they consist of two proteins: a **histidine kinase** and a **response regulator (RR)**. The histidine kinase transfers the γ-phosphate of ATP to one of its own histidine residues (an autophosphorylation). This phosphate is then rapidly transferred to the carboxyl group of an aspartate side chain on the RR. Phosphorylation induces conformational changes in the RR domain that lead to downstream effects (**Figure 4.15**). In many bacterial two-component systems, the phosphorylated RR is a transcriptional activator that binds DNA and regulates gene transcription.

A particularly well-studied two-component system regulates bacterial **chemotaxis** (the ability of bacteria to swim toward or away from environmental cues). In this case, chemotactic receptors on the cell surface are associated with a histidine kinase (CheA). When the receptors bind a noxious chemical (chemorepellant), CheA transfers its phosphate to a RR (CheY). Phosphorylated CheY binds to the flagellar motor and regulates its activity by changing the direction of flagellar

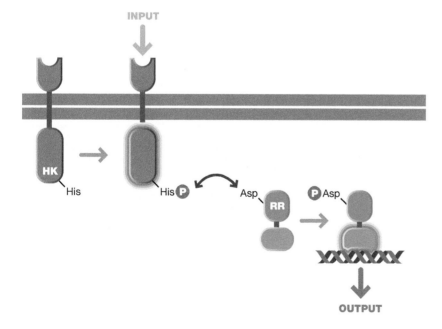

Figure 4.15

Two-component signaling systems. A histidine kinase (HK) responds to input signals by becoming activated and autophosphorylating on a histidine residue. This phosphate is then transferred to an aspartate on a response regulator (RR) protein, inducing conformational changes leading to signal output. Many RR proteins bind to genomic DNA and regulate transcription upon phosphorylation.

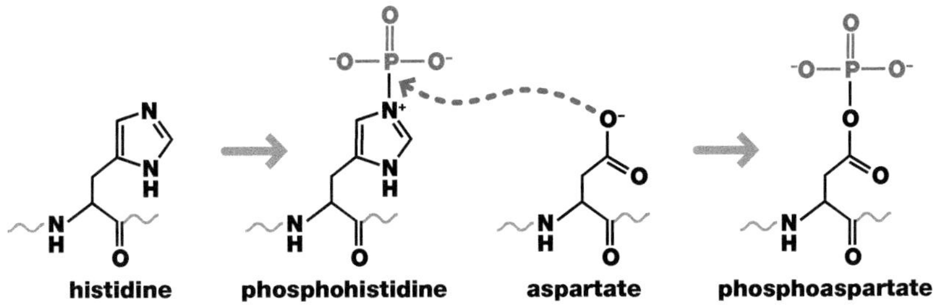

Figure 4.16

Phosphohistidine and phosphoaspartate. Structures of histidine, aspartate, and their phosphorylated derivatives are shown. Attack of phosphohistidine by aspartate during the phosphotransfer reaction is indicated by the gray dotted arrow. Phosphate groups are highlighted in *pink*.

histidine   phosphohistidine   aspartate   phosphoaspartate

rotation which, in turn, changes the direction of swimming by inducing tumbling (see Chapter 11; Figure 11.23).

Several aspects of the two-component system make it distinct from the phosphorylation of hydroxyl amino acids more common in eukaryotes. The biochemical nature of the phosphate linkages is different from that of the phosphoester linkage in hydroxyl amino acids—a phosphoramidate in the case of histidine phosphorylation, and an acyl phosphate in the case of aspartate phosphorylation (**Figure 4.16**). These two linkages are very high-energy bonds, which are kinetically much more susceptible to hydrolysis than the phosphoesters under physiological conditions. In practice, this means that phosphohistidine and phosphoaspartate are much more transient, with half-lives of minutes in most cases. In addition to the rapid spontaneous hydrolysis, both the histidine kinase and RR domains have intrinsic phosphatase activity for the RR domain, ensuring that any signal is relatively transient.

Most bacterial cells contain a relatively large set of histidine kinases linked to receptors, and RRs linked to effector domains. This raises the question of whether histidine kinase and RR domains couple in exclusive pairs, or whether they have overlapping sets of interactions. A study of the bacterium *Caulobacter crescentus*, which encodes 62 histidine kinases and 44 RRs, found that most histidine kinases were quite specific in transferring phosphate to only one or a few RRs at physiological concentrations and time scales. Greater promiscuity was seen in *in vitro* experiments done at high concentrations or for an extended time. Thus, specificity is ensured and unwanted cross-talk is avoided at the kinetic level; only a small fraction of possible interactions occur rapidly enough at the relatively low kinase and RR concentrations seen in the cell for phosphotransfer to occur before dephosphorylation of the histidine kinase.

Two-component systems and histidine phosphorylation are also present in eukaryotes

Two-component systems are also common in plants, slime molds, and fungi, with the RRs feeding into typical eukaryotic signaling pathways such as those involving MAP kinases or cAMP. However, two-component systems are not found in multicellular animals (metazoans). The enzymes responsible for histidine autophosphorylation, phosphotransfer to the RR, and dephosphorylation bear no resemblance to the kinases and phosphatases that modify hydroxyl amino acids in eukaryotes, so this mode of signaling is apparently evolutionarily distinct. One hypothesis for why two-component systems were lost in metazoans (and replaced with kinases and phosphatases that modify hydroxyl amino acids) is that the lability of phosphohistidine and phosphoaspartate made it difficult to transmit signals reliably in larger cells. In a very small bacterial cell, the distance between the cell membrane and the target (for example, chromosomal DNA) is short, so spontaneous dephosphorylation is unlikely before the signal can be transmitted. These distances become considerably larger in most eukaryotic cells, giving more time for the signal to be lost by nonenzymatic dephosphorylation. The more stable phosphorylation of hydroxyl amino acids allows signals to be more precisely controlled over longer time scales and distances.

Despite this, histidine phosphorylation has been observed in mass spectrometric analysis of metazoan proteins, and may be regulated in some specialized instances. For example, it has been shown that the KCa3.1 K$^+$ channel is activated by histidine phosphorylation of its C-terminus, and that this modification is important for T cell activation. Phosphorylation is mediated by NDPK-B, a member of the nucleotide diphosphate kinase family, and phosphate removal is mediated by a protein histidine phosphatase (PHPT-1) (**Figure 4.17**). It is not clear whether this is merely the first example of a more extensive class of physiologically important and regulated histidine phosphorylation events, or whether it is an interesting one-off solution to a very specific biochemical problem. In general, the study of histidine phosphorylation and other transient phosphorylated species is hampered by their lability under typical conditions of cell lysis and analysis.

ADDITION OF UBIQUITIN AND RELATED PROTEINS

Entire globular proteins can also be covalently added to proteins, resulting in major changes to the structure and activity of the modified target. The addition of the small, 76-residue protein ubiquitin or its relatives to target proteins has profound effects on the biological activity of the modified protein. For example, the addition of long chains of ubiquitin (polyubiquitylation) often tags proteins for destruction by a specialized protein-degrading complex, the proteasome. By contrast, addition of single ubiquitin units (monoubiquitylation) or shorter polyubiquitin chains is used to target proteins for endocytosis, or to mediate specific protein–protein interactions. Ubiquitin and polyubiquitin chains are recognized by **ubiquitin-binding domains (UBDs)** on other proteins.

Specialized enzymes mediate the addition and removal of ubiquitin

The addition of ubiquitin to a substrate involves three distinct proteins working in series (**Figure 4.18**). First, an **E1 ubiquitin activating enzyme** uses the energy of ATP hydrolysis to covalently attach the C-terminus of ubiquitin to a cysteine residue on the E1 protein. Second, the activated ubiquitin is transferred to a cysteine on an **E2 ubiquitin-conjugating enzyme**. Finally, the ubiquitin is transferred to an amino group (generally a lysine side chain) on the substrate protein with the help of an **E3 ubiquitin ligase**. In vertebrates, there are two E1 enzymes, ~50 E2 enzymes, and many hundred E3 ligases. It is the E3 ligase, and to a lesser extent the E2 enzyme, that determines which substrates will be ubiquitylated and also the nature of the linkage of the resulting ubiquitin chains. A similar set of enzymes is used to transfer other ubiquitin-like (UBL) peptides—such as SUMO, Nedd8, and ISG15—to their substrates. The UBL peptides have the same overall structure and fold as ubiquitin, though they direct interaction with distinct binding partners and thus confer distinct biological activities to the proteins they modify.

There is considerable variety in how the ubiquitin subunits are linked together in polyubiquitin chains (**Figure 4.19**). For proteins targeted for degradation, the C-terminal glycine of each ubiquitin is generally coupled to Lys48 of the preceding

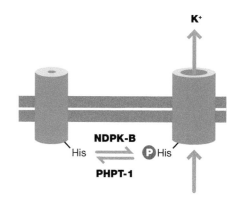

Figure 4.17

Regulation of a mammalian K$^+$ channel by histidine phosphorylation. The potassium (K$^+$) channel KCa3.1 is closed in the basal state. Upon histidine phosphorylation by the kinase NDPK-B, the channel opens, allowing K$^+$ to exit the cell. Dephosphorylation and channel closing is mediated by the phosphatase PHPT-1.

Figure 4.18

The ubiquitylation machinery. The successive action of E1, E2, and E3 enzymes results in the transfer of ubiquitin (orange triangle) to substrate protein (light brown).

(a)

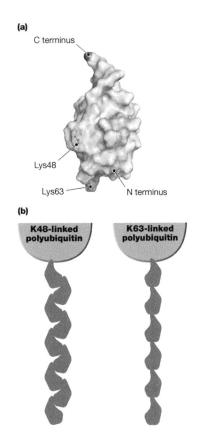

(b)

Figure 4.19

Ubiquitin structure. (a) Three-dimensional structure of human ubiquitin, with the most common sites of linkage (C-terminus, N-terminus, Lys48, Lys63) indicated. (b) Major forms of polyubiquitin. Linkage of the C-terminus to Lys48 (K48) or to Lys63 (K63) generates polyubiquitin chains with different structures (*orange*). The polyubiquitylated protein is indicated in *light brown*.

Aggregation-induced receptor activation is discussed in Chapter 8.

E3 ligases that regulate the cell cycle are discussed in Chapter 9.

subunit, whereas linkage via Lys63 is usually used to tag proteins for other fates such as endocytosis. Linkage to the N-terminus (generating head-to-tail or linear chains) or to the other five lysines of ubiquitin also occurs. Generally, polyubiquitylation results in long, unbranched chains connected by the same linkage, but mixed linkages and branched structures are also possible given the number of possible attachment points. Ubiquitin is thus rather unusual among PTMs used for signaling in the structural diversity of possible modifications that can be generated from a single molecular building block.

Deubiquitinases (DUBs) represent the flip side of the ubiquitin-conjugating machinery—proteins that can erase the post-translational marks, potentially saving the marked proteins from destruction or reversing other ubiquitin-mediated activities. Approximately 80 DUBs are known in the human genome. These enzymes are, in fact, a specialized type of protease, catalyzing cleavage of the isopeptide bond between a lysine amino group and the C-terminus of ubiquitin. Different DUBs can be specific for different substrates and different types of linkages in polyubiquitin chains. Furthermore, they are subject to regulation by post-translational modification and regulated protein–protein interactions.

E3 ubiquitin ligases determine which proteins will be ubiquitylated

Consistent with their pivotal role in determining what substrates will be modified, the E3 ligases exhibit a wide variety of structures and binding specificities. These proteins essentially function as adaptors, bringing the E2 and its activated ubiquitin into close proximity to the lysine to be modified on the target protein. Most E3 ligases fall into two major classes: the RING group and those containing HECT domains. The RING E3 ligases are by far the largest group, with more than 600 examples in humans. These proteins interact with the E2–ubiquitin complex via the zinc-coordinating RING finger motif, and facilitate ubiquitin transfer to substrate proteins. Generally, it is the E2 enzymes and not the RING E3 ligases that specify the type of linkage (for example, Lys48 versus Lys63). Thus, different combinations of E2 enzymes and RING E3 ligases can mediate transfer of polyubiquitin chains of different types to a huge variety of distinct substrates (**Figure 4.20**). The HECT domain class of E3 ligases is defined by the relatively large HECT domain, which recognizes the E2–ubiquitin complex and catalyzes ubiquitin transfer first to itself, and then to substrate proteins.

With respect to signaling, the critical issues for all E3 ligases are how they choose substrates and how their activity is regulated. Many E3 ligases contain well-characterized modular protein-binding domains such as WW, WD40, and SH2 domains, which mediate interaction with specific substrate proteins. Other E3 ligases interact indirectly with substrates through additional E3-associated adaptors. In some cases, one of the primary substrates of ubiquitylation is the E3 ligase itself; autoubiquitylation can be promoted by E3 dimerization or oligomerization. This is very analogous to the autophosphorylation of receptors with kinase activity upon ligand-induced dimerization or clustering.

In many cases, either the catalytic or binding activity can be regulated by other PTMs, either of the substrate or the E3 ligase itself. Phosphorylation of a potential substrate often increases its affinity for the E3 ligase; for example, the E3 ligase Cbl, which contains an SH2 domain, interacts specifically with substrates when they are tyrosine phosphorylated. In another example, the large E3 ligase complexes that regulate the cell cycle bind specifically to targets phosphorylated by cyclin-dependent kinases. In this way, a relatively transient modification (phosphorylation) can be converted to a signal that may doom a protein to degradation. The importance of E3 ligases in signaling processes is indicated by examples in which their mutation is implicated in diseases such as cancer and neurological disorders.

Ubiquitin-binding domains read ubiquitin-mediated signals in diverse cellular activities

For the most part, the functional consequences of ubiquitylation are mediated by proteins that bind directly to the ubiquitin units of the modified protein. Ubiquitin binds specifically to a number of structurally distinct families of binding domains, collectively termed UBDs. Most UBDs bind to the same small hydrophobic patch on the surface of ubiquitin surrounding Ile41, though different UBDs engage other residues surrounding this patch as well (see Figure 10.14a in Chapter 10). In many cases, multiple UBDs, either on the same protein or on associated proteins, engage multiple ubiquitin units on a modified protein, thereby increasing the affinity and specificity of the interaction through avidity and/or cooperative binding. Furthermore, some UBDs recognize specific ubiquitin–ubiquitin linkages, either by recognizing the interface between the two ubiquitin units, or through geometric constraints. In this way, proteins containing UBDs can discriminate between monoubiquitylated and polyubiquitylated proteins, and polyubiquitylated proteins with different linkages (for example, Lys48 versus Lys63). Below, we provide a few examples of how different ubiquitin modifications are sensed in a variety of signaling contexts.

Proteins on the cell surface that are to be degraded are internalized in vesicles (*endocytosis*), and then sent to the **lysosome**, a membrane-enclosed intracellular compartment where the protein and lipid components of the vesicle are broken down by digestive enzymes. Membrane proteins are tagged for destruction either by multiple monoubiquitylation or by Lys63-linked polyubiquitylation. Lysosomal targeting is mediated by a series of multiprotein complexes termed the ESCRT machinery (see Figure 8.28 in Chapter 8). Each *ESCRT complex* has distinct UBDs that interact with ubiquitylated cargo proteins. In particular, the ESCRT-0 complex, which first binds to internalized ubiquitylated targets, plays a critical role in identifying and concentrating those proteins that will be sent to the lysosome. This complex contains multiple UBDs, and in some cases (for example, the UIM domain of Hrs), a single ubiquitin-binding domain can bind two ubiquitin subunits simultaneously (**Figure 4.21a**). The multiplicity of ubiquitin-binding sites in the

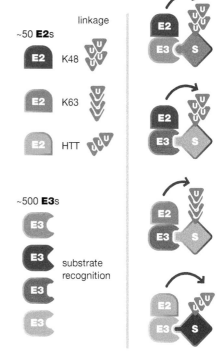

Figure 4.20

Combinations of E2 and E3 enzymes can modify many different substrates with different polyubiquitin linkages.
E2 conjugating enzymes (50 genes in humans) generally determine the linkage of polyubiquitin chains; for example, Lys48 (K48), Lys63 (K63), or head-to-tail (HTT). E3 ligases generally specify the substrates to be modified. Combinations of different E2 and E3 enzymes generate a wide diversity of possible substrates and linkages.

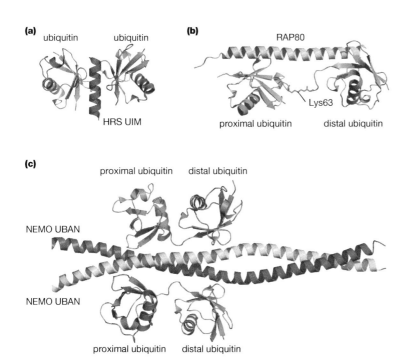

Figure 4.21

Recognition of different polyubiquitin linkages by ubiquitin-binding domains. (a) The UIM motif of Hrs can bind simultaneously to two monoubiquitin molecules. The two ubiquitin molecules bind on opposite sides of the UIM and with a similar binding mode. (b) Structure of Lys63-linked diubiquitin bound to receptor-associated protein Rap80 (coordinates provided by S. Fukai, University of Tokyo, Japan). (c) The UBAN domain in NEMO binds two linear (head-to-tail) diubiquitins. (Adapted from I. Dikic, S. Wakatsuki & K.J. Walters, *Nat. Rev. Mol. Cell. Biol.* 10:659–671, 2009. With permission from Springer Nature.)

ESCRT-0 complex, each of which alone has rather low affinity, promotes the cooperative binding of targets that contain multiple ubiquitins.

Another area where ubiquitylation plays a key role is in the cellular response to DNA damage. In the case of double-strand breaks—a particularly troublesome form of DNA damage because it can lead to the loss of chromosome fragments if not repaired before cell division—a key early event is the recruitment of an E2/E3 complex (Ubc13/RNF8) that adds Lys63-linked polyubiquitin chains to histones in the vicinity of the break. This signal is then amplified by the recruitment of a large effector complex that includes the tumor suppressor protein BRCA1, itself a ubiquitin E3 ligase. Recruitment of this complex is mediated by the UBDs of an adaptor protein, Rap80. Rap80 contains two ubiquitin-binding UIMs that work together to recognize Lys63-linked diubiquitin. The linker region between the two UIMs positions them in such a way that they can productively interact only when the linkage is via Lys63—the distance between two ubiquitins is too short if the linkage is via Lys48 (**Figure 4.21b**).

Head-to-tail (linear) polyubiquitin chains play an important role in the *NF-κB* signaling pathway. NF-κB is a transcription factor whose activation is critical for innate and adaptive immune responses. A key step in this pathway is the activation of IKK, a multisubunit serine/threonine kinase. IKK activation is controlled by a regulatory subunit, NEMO, which contains a UBD (the UBAN domain) that is highly specific for linear polyubiquitin chains. Upstream signals lead to the addition of linear polyubiquitin chains to NEMO by LUBAC (linear ubiquitin chain assembly complex); this is likely to lead to conformational changes in the IKK complex, as the UBAN domain of one NEMO molecule interacts *in trans* with the linear polyubiquitin chains of another. The UBAN domain contacts both ubiquitins in a linear diubiquitin chain, but productive contacts are only possible when the two are connected by a head-to-tail linkage (**Figure 4.21c**). The NF-κB pathway is interesting in that it exploits three distinct types of polyubiquitin chains: Lys63 for activation of upstream kinases, Lys48 for targeting inhibitory subunits for degradation, and head-to-tail for activating IKK.

HISTONE ACETYLATION AND METHYLATION

Chromosomal DNA is packaged into chromatin through its association with **histone** proteins. The overall structure of chromatin is regulated by the post-translational modification of histones and other proteins that interact with them. Chromatin structure has a direct impact on virtually all activities of the genome, including transcription, replication, repair, genomic imprinting, and chromosome segregation. Many cell signaling events directly or indirectly affect chromatin by changing its post-translational marks which, in many cases, provide binding sites for modification-specific readers. In this section, we will focus on the remodeling of chromatin by PTMs such as methylation and acetylation, particularly in transcriptional regulation.

Chromatin structure is regulated by post-translational modification of histones and associated proteins

The basic unit of chromatin is the **nucleosome**, consisting of eight histone subunits arranged in a disclike structure, around which is wrapped ~147 base pairs of DNA. A typical nucleosome contains two molecules each of histone 2A (H2A), histone 2B (H2B), histone 3 (H3), and histone 4 (H4) (**Figure 4.22a**). The simplest higher-order chromatin structure is an extended conformation where individual nucleosomes are arranged like beads on a string; a linker histone, histone 1 (H1), associates with adjacent nucleosomes and with the DNA that connects them (**Figure 4.22b**). Higher-order conformations result from interactions between nucleosomes, leading to packaging of the chromatin into more dense and compact

The NF-κB pathway is described in Chapter 9.

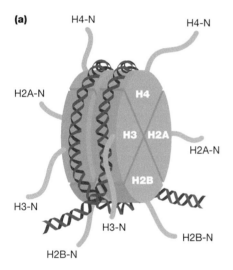

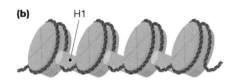

Figure 4.22

Nucleosome structure. (a) Arrangement of histone subunits in a standard nucleosome is depicted schematically. There are two subunits each of histones H2A, H2B, H3, and H4, around which ~147 nucleotides of genomic DNA are wound (brown). The N-terminal tails of the histones protrude from the nucleosome and are accessible for post-translational modification. (b) Multiple nucleosomes are arrayed along the DNA in chromatin fibers. Histone H1 (green) bridges two adjacent nucleosomes.

fibers. In general, the chromatin of genes that are actively transcribed or accessible for transcriptional activation is in a relatively loosely packed form termed euchromatin, while less transcriptionally active regions are in a more tightly packed and inert form termed heterochromatin. During mitosis and meiosis, when chromosomes need to be bundled into an easily handled form for nuclear division, the chromatin condenses dramatically compared to its state in interphase cells.

Activities that use the genomic DNA as a template, such as transcription, require at least temporary loosening of the higher-order interactions of chromatin; without transient histone dissociation, transcription factors and RNA polymerase would be unable to access the DNA for binding. The primary means of modulating chromatin structure is by post-translational modification of the histones. The bulk of each histone is a globular domain that contacts DNA and other histones, and is mostly inaccessible for modification. However, each histone has a fairly long, positively charged N-terminal tail that projects out from the nucleosome core and thus is accessible for modification by a variety of enzymes. In particular, a number of lysine residues in the histone tails can be either *N*-methylated, *N*-acetylated, ubiquitylated, or sumoylated. Arginine residues are subject to *N*-methylation and deimination, serine and threonine residues may be phosphorylated, and prolines are subject to *cis-trans* isomerization. A number of histone tail side chains are also ADP-ribosylated.

The discovery of these modifications, along with proteins that seemed to bind specifically to different modified sites on histones, led to the proposal that specific patterns of modifications determine the activity state of the chromatin. While this concept has been useful in focusing research and thought, the situation is much more complicated than a simple "one modification, one output" model. As discussed above for p53, it is now clear that the post-translational modification of histones is highly complex, with many marks being present simultaneously and in a host of different possible combinations. Furthermore, some histone-binding proteins are themselves modified, providing additional variation. Histone modifications are also highly dynamic: many of the marks turn over very rapidly, and thus are in dynamic equilibrium and subject to change over short time scales.

Chromatin modification is intimately associated with gene transcription. Transcription is the ultimate end point of many signal transduction pathways that exert their long-term effects through modulating the expression of specific gene products. For transcription to occur, RNA polymerase (RNA Pol II in the case of protein-coding genes) must bind to the promoter region of the template DNA. This is generally facilitated by a variety of **transcription factors**, which bind DNA and facilitate binding of the polymerase. Histones must dissociate from the site of transcription initiation to allow polymerase binding. Once polymerization of the RNA message has initiated, histones must be transiently stripped from the template to allow passage of the polymerase; generally, histones rapidly rebind after polymerase transit to prevent adventitious initiation of transcription at internal sites. Thus, transcription requires an elaborate and highly dynamic set of changes to chromatin.

Two writer/eraser/reader systems are based on protein methylation and acetylation

The lysine acetylation of histone tails is generally associated with actively transcribed chromatin. Some of the effects of histone acetylation are intrinsic; that is, they directly regulate interactions among nucleosomes. For example, acetylation of Lys16 of histone H4 (abbreviated as H4K16Ac) prevents the compaction of nucleosome arrays by preventing contacts between H4 subunits on one nucleosome and H2B subunits on an adjacent nucleosome, most likely because of the loss of positive charge on lysine when it is acetylated. However, most effects of acetylation are extrinsic, due to the specific binding to lysine-acetylated sites of "reader" proteins with small modular protein domains such as bromo domains.

Consistent with the connection between transcription and acetylation, a number of transcriptional activators and proteins that associate with transcriptional start sites have histone acetyl transferase (HAT) activity. Conversely, the activity that removes acetyl groups, histone deacetylase (HDAC), is frequently found in transcriptional co-repressor complexes. Both HAT and HDAC activities are most often found in large multifunctional proteins or protein complexes that also contain one or more bromo domains. In this way, both the addition and removal of acetyl groups to histone tails is processive: the complex remains stably associated with acetylated chromatin, allowing multiple sites in the vicinity to be modified. The rate of turnover of acetyl groups on histones is quite high, with half-lives of several minutes, though turnover is slower for a minority of sites. This likely reflects a relatively low basal level of acetylation in actively transcribed regions, with transient increases associated with the loading and passage of RNA polymerase.

The other major class of histone modification is *N*-methylation of lysine and arginine residues, by lysine methyl transferases (KMTs) and protein arginine methyl transferases (PRMTs), respectively. Either one, two, or three methyl groups can be added to the ε-nitrogen of lysine side chains; in the case of arginine, one or two methyl groups can be added, with the two either symmetric (one methyl group per amino group) or asymmetric (two methyl groups on one amino group) (**Figure 4.23**). Protein methylation was once thought to be essentially irreversible, but a number of lysine demethylases (KDMs) have now been identified that readily remove methyl groups from lysine. Only a few arginine demethylases have been found, exemplified by a subset of the JmjC class of KDMs that also can act on

Figure 4.23

Diverse modes of protein methylation.
Modifications are highlighted in *pink*.
(a) Mono-, di-, and trimethylated lysine.
(b) Monomethylated and cis and trans dimethylated arginine. (c) The deimination of monomethyl arginine by peptidyl arginine deiminase yields N-methyl citrulline. Unmethylated arginine can similarly be deiminated to generate citrulline, which cannot be methylated by arginine methyl transferases.

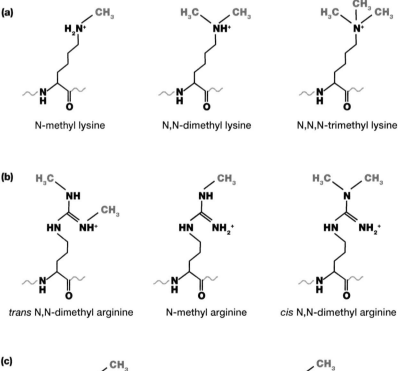

arginine. Arginine methylation can also be counteracted in the cell by deimination of the side chain, converting arginine to citrulline (see Figure 4.23c).

The readers for protein methylation marks represent a rather diverse group of binding domains with varying degrees of specificity and affinity. Domains that bind specifically to methylated lysines include the "royal family" domains (tudor, chromo, and MBT), WD40 domains, and the PHD finger (see Figure 10.13 in Chapter 10). Domains that have been shown to bind specifically to methylated arginines include a subset of tudor and BRCT domains. As is the case for enzymes that modify protein acetylation, the enzymes that add or remove methyl groups from histones are often part of larger complexes that contain one or more reader domains as well, leading to a complex interplay between existing marks and the addition or removal of new marks.

Lysine methylation of histones is associated both with activation and repression of transcription. For example, trimethylation of histone H3 at Lys4 (H3K4me3) is enriched on actively transcribed genes, while the trimethylation at Lys9 of the same protein (H3K9me3) is enriched on transcriptionally silenced genes. Even the very same mark can have widely divergent effects, depending on the specific context of associated proteins and other marks. In contrast, arginine methylation is most always associated with transcriptional activation.

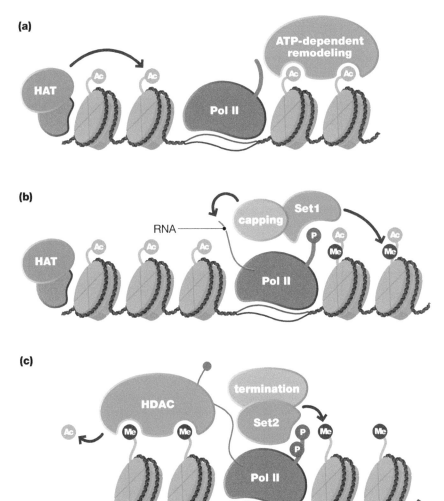

Figure 4.24

Dynamic chromatin modifications during transcription. (a) Transcriptional activators recruit histone acetyl transferases (HAT), leading to acetylation of histones in the vicinity of the site of transcription initiation. In turn, acetylation recruits ATP-dependent chromatin remodeling complexes and promotes binding of RNA polymerase II (Pol II). (b) As transcription is initiated, the C-terminal tail repeats of Pol II are phosphorylated on Ser5, recruiting RNA capping factors and the Set1 methyl transferase, which adds methyl groups to Lys4 of histone H3. Nascent RNA transcript is depicted in *orange*. (c) As transcription proceeds, the Pol II C-terminal repeats are phosphorylated on Ser2, leading to recruitment of termination factors and the Set2 methyl transferase, which adds methyl groups to Lys36 on histone H3. This recruits histone deacetylases (HDAC), which remove acetyl groups from nearby histones.

Chromatin modification in transcription is dynamic and leads to highly cooperative interactions

In order for a gene to be transcribed, signals must lead to the recruitment of RNA polymerase to the promoter of the gene. Most often this is accomplished by increased binding of transcriptional activators near the promoter of the gene. This may be due to increased transport of the transcriptional activators to the nucleus, increased affinity for their binding sites on DNA, increased binding to transcriptional coactivators, or increased synthesis, but in some way, their activity and/or concentration in the nucleus must be increased as a result of the signal. These transcriptional activators, in turn, recruit a number of other proteins, including chromatin modifying complexes and RNA Pol II itself. Alternatively, some signals lead to the inactivation of transcriptional repressors already bound to the DNA, thereby relieving repression and allowing the recruitment of factors that promote transcription.

Transcriptional activators such as p300/CBP or PCAF/Gcn5 generally associate with or contain HAT activities that acetylate chromatin in the promoter region. Acetylated histones, in turn, recruit a variety of factors that help promote transcription, including general transcription factors and chromatin remodeling factors such as the SWI/SNF complexes, which use the energy of ATP to loosen the association of histones with DNA near the site of transcription initiation. Transcriptional activators also recruit the RNA Pol II holoenzyme, which then binds to the template DNA and is poised to initiate transcription (**Figure 4.24a**).

RNA polymerase itself becomes phosphorylated in the course of transcription, and this PTM leads to important changes in protein interactions—yet another writer/eraser/reader system that intersects with the systems based on histone methylation and acetylation. RNA Pol II contains a long C-terminal domain (CTD) composed of a number of heptameric repeats, with the consensus sequence YSPTSPS. The serine, threonine, and tyrosine residues are subject to phosphorylation, and the prolines to *cis-trans* isomerization. One of the key events of transcription initiation is the phosphorylation of the CTD on the fifth residue of the heptameric repeat (Ser5) by a cyclin-dependent kinase, Cdk7, which is a component of the general transcription factor TFIIH.

Ser5 phosphorylation of the CTD repeats has at least two important consequences. First, it decreases the affinity of the polymerase for general factors bound to the promoter, such as the Mediator complex, allowing the polymerase to escape the promoter and begin transcript elongation. Second, it provides specific binding sites for a variety of factors, including RNA capping enzymes that process the 5' end of the RNA transcript. The phosphorylated CTD also binds the Set1 lysine methyl transferase, which leads to increased histone H3K4 methylation. H3K4 methylated sites, in turn, recruit more complexes that promote open and active chromatin structure (HATs, KMTs, and ATP-dependent remodeling complexes) (**Figure 4.24b**).

Once the polymerase clears the promoter and is actively transcribing the gene, another cyclin-dependent kinase (a component of the P-TEFb transcription elongation complex) phosphorylates the CTD repeats on Ser2. This, in turn, recruits a second KMT (Set2), which methylates another site on the tail of histone 3, Lys36 (H3K36Me). One consequence of H3K36Me is the recruitment of HDACs, which remove acetyl groups from histones in the protein-coding region of the gene. This is thought to be important to reset the histones within the coding region to a basal, closed state after the passage of the polymerase, helping prevent inappropriate polymerase recruitment and transcription initiation from internal sites within the gene. Finally, the Ser2-phosphorylated RNA Pol II CTD helps recruit termination and polyadenylation factors needed to terminate the new transcript and process its 3' end (**Figure 4.24c**). Specific phosphatases are also recruited that target the CTD and reset the polymerase back to its basal, unphosphorylated state.

The above description is highly simplified and is only meant to provide an overview of the complex and highly coordinated events of transcription. For example, the effects of DNA methylation and the phosphorylation, ubiquitylation, sumoylation, ADP-ribosylation, and proline isomerization of histones and associated proteins are not considered. But even at this relatively simple level of description, it is apparent that writer/eraser/reader systems provide a flexible way to control complex and dynamic cell behaviors. A sequential, linked series of PTMs, each creating binding sites for new factors that generate additional modifications (or remove previous ones), is a powerful and recurrent theme in signaling.

SUMMARY

Post-translational modification provides a rapid and efficient way to change the activity of proteins. The addition and removal of these modifications are catalyzed by the activity of specific enzymes. The large number of possible modification sites, along with the diversity of possible modifications, enormously expands the number of potential protein states beyond what can be encoded by the genome. Modifications can directly affect protein activity, but in many cases, the modifications are "read" by proteins that specifically bind to the modified sites. Phosphorylation of the hydroxyl amino acids (serine, threonine, tyrosine) is the most widespread PTM in signaling in metazoans. The addition of ubiquitin and its relatives is also widely used, and affords considerable diversity in the length of the chains that are added and how the subunits are linked together. The structure and activity of chromatin are dynamically modulated by the post-translational modification of histones and histone-associated proteins, especially by acetylation and methylation.

QUESTIONS

Answers to these questions can be found online at www.routledge.com/9780367279370.

1. In analyzing how a cell responds to stress, you discover several induced phosphorylation events on a membrane-associated protein, X. Phosphorylation of protein X on Thr122 induces its interaction with protein Y. Describe possible alternative mechanisms by which this modification might induce binding to protein Y. Propose experiments to distinguish between these possibilities.

2. Phosphorylation of the protein X from Question 1 on Ser54 disrupts its interaction with protein Z. Describe possible mechanisms by which this modification might disrupt binding to protein Z. How might you distinguish between these possibilities?

3. Sometimes, phosphorylation of a protein on a specific residue leads to the degradation of that protein. Describe the general mechanisms by which degradation can be mediated by phosphorylation, and the intervening steps between phosphorylation and degradation.

4. **Figure 4.14** shows several different mechanisms by which a protein could be phosphorylated on multiple distinct sites. One mechanism is *priming*, where one kinase phosphorylates the target, creating a binding site for a second kinase. Recruitment of the second kinase subsequently leads to phosphorylation on a second (or additional) site. What kinds of kinetic responses might result from a protein that is controlled by a priming mechanism? Describe additional features of the signaling system that would be necessary in your model.

5. How might regulation of the activity of enzymes that mediate protein PTMs (writers and erasers) generate signals that are highly localized and which transmit spatial information? What elements might limit the ability of enzymes to

generate localized signals? What elements might enhance their ability to generate localized signals?

6. Efforts are underway to map the full complement of protein PTMs in various cells under different conditions. Is it realistic to expect that such a task will ever be completed? What are the major challenges in such an effort?

7. Is has been suggested that some experimentally detected PTMs of specific sites might have no function, or may even be detrimental to the cell. Under what conditions might this be the case?

8. A number of enzymes that modify proteins contain a reader domain that binds to the modified substrate. For example, nonreceptor tyrosine kinases have SH2 domains that bind to tyrosine-phosphorylated peptides; and histone acetyl transferases have bromo domains that bind acetylated lysines. What effect will these reader domains have on modification by such enzymes, and under what conditions would such an arrangement be useful?

9. How might you engineer a system, based on a ubiquitin E3 ligase, to degrade a particular protein of choice in the cell?

BIBLIOGRAPHY

THE LOGIC OF POST-TRANSLATIONAL MODIFICATION

Lüscher B, Bütepage M, Eckei L, et al. (2018) ADP-ribosylation, a multifaceted posttranslational modification involved in the control of cell physiology in health and disease. *Chem. Rev.* 118(3), 1092–1136. doi: 10.1021/acs.chemrev.7b00122.

Macek B, Forchhammer K, Hardouin J, et al. (2019) Protein post-translational modifications in bacteria. *Nat Rev Microbiol.* 17(11), 651–664. doi: 10.1038/s41579-019-0243-0.

Pandey N & Black BE (2021) Rapid detection and signaling of DNA damage by PARP-1. *Trends Biochem. Sci.* 46(9), 744–757. doi: 10.1016/j.tibs.2021.01.014.

Seet BT, Dikic I, Zhou MM & Pawson T (2006) Reading protein modifications with interaction domains. *Nat. Rev. Mol. Cell Biol.* 7, 473–483.

Walsh CT (2006) *Posttranslational Modification of Proteins: Expanding Nature's Inventory.* Englewood, CO: Roberts and Co. Publishers.

Wang ZA & Cole PA (2020) The chemical biology of reversible lysine post-translational modifications. *Cell Chem. Biol.* 27(8), 953–969. doi: 10.1016/j.chembiol.2020.07.002.

INTERPLAY BETWEEN POST-TRANSLATIONAL MODIFICATIONS

Boehme KA & Blattner C (2009) Regulation of p53- insights into a complex process. *Crit. Rev. Biochem. Mol. Biol.* 44, 367–392.

Butkinaree C, Park K & Hart GW (2010) O-linked beta-N-acetylglucosamine (O-GlcNAc): Extensive crosstalk with phosphorylation to regulate signaling and transcription in response to nutrients and stress. *Biochim. Biophys. Acta* 1800, 96–106.

Liu Y, Tavana O & Gu W (2019) p53 modifications: Exquisite decorations of the powerful guardian. *J. Mol. Cell. Biol.* 11(7), 564–577. doi: 10.1093/jmcb/mjz060.

Lothrop AP, Torres MP & Fuchs SM (2013) Deciphering post-translational modification codes. *FEBS Lett.* 587, 1247–1257.

Meek DW & Anderson CW (2009) Posttranslational modification of p53: Cooperative integrators of function. *Cold Spring Harb. Perspect. Biol.* 1(6), a000950. doi: 10.1101/cshperspect. a000950.

Metzger E, Imhof A, Patel D, et al. (2010) Phosphorylation of histone H3T6 by PKCbeta(I) controls demethylation at histone H3K4. *Nature* 464, 792–796.

Prabakaran S, Lippens G, Steen H & Gunawardena J (2012) Post-translational modification: Nature's escape from genetic imprisonment and the basis for dynamic information encoding. *Wiley Interdiscip. Rev. Syst. Biol. Med.* 4, 565–583.

Yang XJ (2005) Multisite protein modification and intramolecular signaling. *Oncogene* 24, 1653–1662.

PROTEIN PHOSPHORYLATION

Asfaha JB, Örd M, Carlson CR, et al. (2022) Multisite phosphorylation by Cdk1 initiates delayed negative feedback to control mitotic transcription. *Curr. Biol.* 32(1), 256–263. doi: 10.1016/j. cub.2021.11.001.

Gao R & Stock AM (2009) Biological insights from structures of two-component proteins. *Annu. Rev. Microbiol.* 63, 133–154.

Jadwin JA, Curran TG, Lafontaine AT, et al. (2018) Src homology 2 domains enhance tyrosine phosphorylation in vivo by protecting binding sites in their target proteins from dephosphorylation. *J. Biol. Chem.* 293(2), 623–637. doi: 10.1074/jbc.M117.794412.

Jin J & Pawson T (2012) Modular evolution of phosphorylation-based signalling systems. *Philos. Trans. R. Soc. Lond. B Biol. Sci.* 367, 2540–2555.

Johnson LN & Lewis RJ (2001) Structural basis for control by phosphorylation. *Chem. Rev.* 101, 2209–2242.

Podgornaia AI & Laub MT (2013) Determinants of specificity in two-component signal transduction. *Curr. Opin. Microbiol.* 16, 156–162.

Srivastava S, Zhdanova O, Di L, et al. (2008) Protein histidine phosphatase 1 negatively regulates CD4 T cells by inhibiting the K+ channel KCa3.1. *Proc. Natl Acad. Sci. USA.* 105, 14442–14446.

Ubersax JA & Ferrell Jr. JE (2007) Mechanisms of specificity in protein phosphorylation. *Nat. Rev. Mol. Cell Biol.* 8, 530–541.

ADDITION OF UBIQUITIN AND RELATED PROTEINS

Dikic I, Wakatsuki S & Walters KJ (2009) Ubiquitin-binding domains - from structures to functions. *Nat. Rev. Mol. Cell Biol.* 10, 659–671.

Harper JW & Schulman BA (2021) Cullin-RING ubiquitin ligase regulatory circuits: A quarter century beyond the F-box hypothesis. *Annu. Rev. Biochem.* 90, 403–429. doi: 10.1146/annurev-biochem-090120-013613.

Kerscher O, Felberbaum R & Hochstrasser M (2006) Modification of proteins by ubiquitin and ubiquitin-like proteins. *Annu. Rev. Cell Dev. Biol.* 22, 159–180.

Komander D (2009) The emerging complexity of protein ubiquitination. *Biochem. Soc. Trans.* 37, 937–953.

Metzger MB, Hristova VA & Weissman AM (2012) HECT and RING finger families of E3 ubiquitin ligases at a glance. *J. Cell Sci.* 125, 531–537.

Searle MS, Garner TP, Strachan J, et al. (2012) Structural insights into specificity and diversity in mechanisms of ubiquitin recognition by ubiquitin-binding domains. *Biochem. Soc. Trans.* 40, 404–408.

HISTONE ACETYLATION AND METHYLATION

Barth TK & Imhof A (2010) Fast signals and slow marks: The dynamics of histone modifications. *Trends Biochem. Sci.* 35, 618–626.

Berger SL (2007) The complex language of chromatin regulation during transcription. *Nature* 447, 407–412.

Campos EI & Reinberg D (2009) Histones: Annotating chromatin. *Annu. Rev. Genet.* 43, 559–599.

Kouzarides T (2007) Chromatin modifications and their function. *Cell* 128, 693–705.

Morgan MAJ & Shilatifard A (2020) Reevaluating the roles of histone-modifying enzymes and their associated chromatin modifications in transcriptional regulation. *Nat. Genet.* 52(12), 1271–1281. doi: 10.1038/s41588-020-00736-4.

Zentner GE & Henikoff S (2013) Regulation of nucleosome dynamics by histone modifications. *Nat. Struct. Mol. Biol.* 20, 259–266.

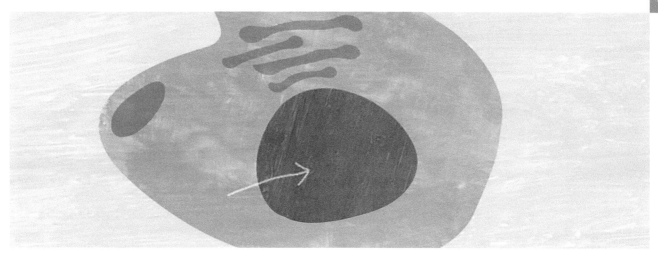

Subcellular Localization of Signaling Molecules

5

One of the defining properties of a cell is the nonuniform distribution of its constituents. The most basic division is between the aqueous cytosol and the water-insoluble lipid environment of the plasma membrane that surrounds it. The plasma membrane serves a dual role both as a physical barrier separating the cell's contents from the environment, and as a dynamic interface mediating the selective passage of information and material to and from the interior of the cell. In eukaryotes, internal membranes serve to further compartmentalize key cellular functions such as transcription, replication, and energy production.

Even within the same intracellular compartment, many components are unevenly distributed, and this lack of homogeneity is essential for many of the cell's activities. Nearly all cells display some functionally important asymmetry, or **polarity**. For example, a migrating cell must protrude at the front and de-adhere at the back. In another example, epithelial cells have clearly distinct apical and basal faces. These polarity differences are due to an asymmetric distribution of specific molecular components. Here we will address the more general questions of how cell signaling mechanisms take advantage of the nonuniform distribution of cell components to transmit information, and how signals can modulate the distribution of those components.

LOCALIZATION AS A SIGNALING CURRENCY

One way of describing the uneven distribution of cell components is in terms of differences in their local concentration across the cell. **Local concentration** is the concentration of a component at a specific location, irrespective of the overall amount present elsewhere in the cell. Although there may be only a few molecules of some component in the entire cell, if they are all co-localized within a very small volume, then their local concentration at that spot will be very high. Local concentration is particularly important for cell physiology because most macromolecules

DOI: 10.1201/9780429298844-5

cannot freely diffuse throughout the cell; their movements in the crowded intracellular environment are highly constrained by physical barriers and by their interactions with other cellular components. The cell is not an idealized, well-mixed solution in equilibrium, and so very different things will occur at different sites due to differences in the local concentrations of components that are found there.

Subcellular localization plays a critical role in signaling because it directly affects what reactions can occur. For both enzymatic and binding reactions, the actual rate of reaction is proportional to the concentration of each of the reactants. Obviously, if an enzyme is localized in the interior matrix of the mitochondrion and a potential substrate is found only in the nucleus, the two proteins will never encounter each other and the enzyme cannot act upon that substrate. In effect, the concentration of these two proteins toward each other is zero, no matter what their overall (average) concentrations might be per cell. On the other hand, the high local concentrations that result when two components are co-localized can drive reactions very efficiently, both thermodynamically and kinetically.

Changes in subcellular localization can transmit information

Signaling mechanisms exploit the translocation of molecules among a number of distinct cellular compartments. As shown in **Figure 5.1**, molecules can be partitioned to the cytosol, nucleus, or membrane. Even within the membrane, molecules can have functionally distinct sublocalizations—bulk plasma membrane is distinct from membrane domains enriched in particular lipid or protein components, or from the membranes associated with physical structures such as the primary cilium, cell–cell junctions, or cell poles, for example. Furthermore, proteins and nucleic acids that interact with each other via multiple, relatively low-affinity interactions in some cases form dense aggregates or *biomolecular condensates*, which can have the properties of membraneless organelles. In this chapter, we will focus on two of the most common cellular translocations involved in signaling—the movement of proteins to and from the nucleus, or to and from cell membranes, and then go on to discuss the role of biomolecular condensates in signaling.

The rationale behind regulated nuclear localization is obvious: genomic DNA and other chromatin constituents such as histones are found exclusively in the nucleus. Access to this compartment is therefore absolutely essential for the activity of proteins that act on chromatin, such as transcription factors. As we will see below, the cell has elaborate mechanisms that regulate nuclear import and export, providing ample opportunity for regulation in signaling.

The rationale for changes in membrane localization is similar, as many signaling proteins and their substrates are exclusively found on membranes. These include transmembrane proteins and proteins containing covalently attached lipid groups, as well as lipids themselves, which are frequent targets of modification in signaling. Although the plasma membrane surrounds the cytosol and thus is available to interact with cytosolic components, a molecule is much more likely to encounter a membrane-bound partner if it is localized to the membrane itself than if it is free to diffuse in the cytosol. This is because the effective volume within which the molecule can roam is much smaller when it is confined to the membrane, so its local concentration with respect to membrane components is correspondingly higher.

The magnitude of this effect can be illustrated by a hypothetical example. If we assume a spherical cell 20 μm in diameter, from simple geometry we can calculate its volume and surface area (roughly 4000 and 1200 μm², respectively). But what is the effective volume for a protein that is confined to the plasma membrane? We can estimate this by calculating the volume of cytosol within 5 nm of

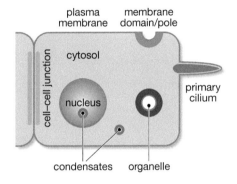

Figure 5.1

Subcellular locations important in signaling. The biological activity of a signaling protein can be very different depending on whether it is localized to the cytosol, nucleus, or membranes. Membranes can be further divided into different functional domains, including bulk plasma membrane, specialized regions of the plasma membrane, or the membranes surrounding organelles or intracellular vesicles. Signaling proteins can also be concentrated within or excluded from various biomolecular condensates. Trafficking of proteins or other components between all of these different compartments can transmit signaling information.

the membrane (this distance is roughly the diameter of an average protein domain). In this example, the volume of cytosol inhabited by a membrane-bound protein is then $(1200 \, \mu m^2) \times (0.005 \, \mu m) = 6 \, \mu m^3$ (**Figure 5.2**). Therefore, a protein would be concentrated by a factor of almost 700-fold (from 4000 to $6 \, \mu m^3$) when it is confined to the membrane, compared to when it is uniformly distributed throughout the cell. If the protein is an enzyme whose substrate is localized to the membrane, it is therefore much more likely to encounter and interact productively with that substrate.

Although the spherical cell used as an example is obviously idealized, comparable effects are seen when actual values for cell volume and surface area are used. For example, in one experiment, mouse fibroblasts spreading on a tissue-culture dish were found to have an average cytosolic volume of $16,000 \, \mu m^3$ and a surface area of $8400 \, \mu m^2$, and relocation of a cytosolic protein to the membrane in this case would lead to an almost 400-fold increase in local concentration.

Subcellular localization can be regulated by a variety of mechanisms

The mechanisms that control the constitutive targeting of proteins to specific subcellular compartments are a major focus of research in cell biology, and will not be discussed in great detail here. Broadly speaking, however, short peptide sequences termed sorting signals or targeting signals are often sufficient to target a protein bearing that sequence to a particular compartment or organelle. Such targeting sequences often work by promoting association with specialized proteins, termed trafficking proteins, that deliver them to their proper cellular address.

Of more interest for cell signaling, however, are the ways in which the localization of proteins and other signaling molecules can be dynamically regulated in response to signals. As discussed in Chapter 4, post-translational modification is one way that signaling can directly or indirectly affect protein localization. For example, phosphorylation can destroy or create binding sites for partners involved in the transport or tethering of a protein to a specific site. Post-translational modifications can also have more direct effects, as in the addition of hydrophobic lipid groups that promote membrane association, or in proteolytic cleavage that physically releases one part of a protein from another that tethers it to a specific subcellular localization. Another widely used mechanism involves small modular domains that bind to specific membrane lipids such as **phosphoinositides** (phosphatidylinositol-derived lipids which may be further phosphorylated at specific positions on the inositol head group) whose levels are regulated during signaling. The membrane localization of a protein bearing such a domain can be regulated by localized changes in the concentration of the cognate lipid.

CONTROL OF NUCLEAR LOCALIZATION

The movement of macromolecules such as proteins between the nucleus and cytosol is tightly controlled. The two compartments are separated from each other by the double lipid bilayer of the nuclear envelope, which is perforated by a relatively small number of aqueous pores. The **nuclear pore complex** is a large, multiprotein complex that serves as a selective gateway regulating the passage of macromolecules into and out of the nucleus. Relatively small macromolecules, up to a size of 40–60 kD, can passively diffuse through the nuclear pore and thus rapidly equilibrate between the two compartments. Any macromolecule larger than this must be actively ferried through the nuclear pore by specialized transport proteins in a process that requires energy. Because this process is highly regulated, it provides ample scope for modulation during signaling.

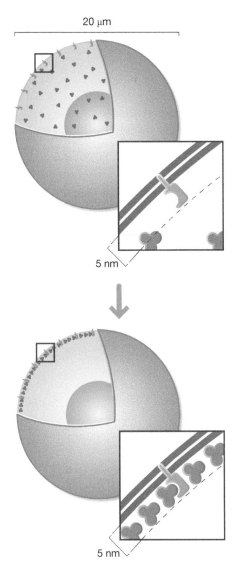

20 µm

5 nm

5 nm

Figure 5.2

Effect of membrane localization on local concentration. (a) A protein (*pink*) is uniformly distributed throughout the cell. Its local concentration in the vicinity of a membrane-associated binding partner (*blue*) is relatively low. (b) The protein is localized to the plasma membrane. This results in an increase in local concentration in the vicinity of its binding partner of nearly 700-fold (see the text for details).

Short, modular peptide motifs direct nuclear import and export

The efficient nuclear import or export of proteins is mediated by short (often less than ten residues) amino acid sequences in the protein itself. A **nuclear localization signal** (**NLS**) is sufficient to direct a protein containing it to enter the nucleus, while a **nuclear export signal** (**NES**) is sufficient to direct nuclear export. As will be discussed below, these sequences bind to specific transporter proteins, which negotiate passage of the protein through the nuclear pore complex. These targeting sequences are modular, in the sense that they can be moved to different sites on the protein, or even to different proteins, without losing their ability to direct import or export. Although some NLS and NES motifs can be recognized from their primary sequence (for example, one class of NLS is lysine-rich and highly basic, while an NES often contains a characteristic leucine-rich motif), the precise sequence and/or conformational requirements are not fully understood.

All soluble proteins are synthesized in the cytosol, so the existence of nuclear export signals implies that some proteins can be both imported into the nucleus and exported back out into the cytosol. Indeed, many proteins can be shown to contain multiple functional NLS and NES motifs. The localization of such proteins is thus in dynamic equilibrium as they shuttle between the cytosol and nucleus. This has important implications for signal transduction. First, changes in the rate of import and/or export can rapidly affect the overall distribution of a protein between the nucleus and the cytosol. Second, the constant cycling between the two compartments allows such a protein to continuously sample conditions in both the cytosol and the nucleus, and respond rapidly to changes in conditions in both compartments.

Nuclear transport is controlled by shuttle proteins and the G protein Ran

The binding of specific transport proteins to the NLS or NES of a protein mediates its nuclear import or export. The most prominent class of such transport proteins is termed the **karyopherins**, which can be divided into the **importins** and **exportins** depending on whether they function primarily to import cargo to the nucleus or export it out to the cytosol. In mammalian cells, many different exportins and importins are expressed, each recognizing different classes of cargo proteins.

G proteins and their regulation are described in Chapter 3.

The vectorial transport of proteins bound to karyopherins is controlled by a nuclear–cytoplasmic gradient of the GTP-binding protein Ran (**Figure 5.3**). Like all G proteins, Ran can exist in GTP-bound and GDP-bound states, which differ in their binding partners. Rcc1 is a guanine nucleotide exchange factor (GEF) that promotes release of GDP and binding of GTP to Ran. Rcc1 is found exclusively in the nucleus. By contrast, GTPase-activator proteins (GAPs) for Ran, which promote cleavage of GTP bound to Ran into GDP plus phosphate, are exclusively cytosolic. Because of this asymmetric distribution of Ran regulators, there is a much higher concentration of Ran-GTP in the nucleus than in the cytosol. The binding of exportins to their cargo is stimulated by the binding of Ran-GTP. Trimolecular exportin/cargo/Ran-GTP complexes pass through the nuclear pore complex, whereupon cargo is released in the cytosol once the GTP bound to Ran is hydrolyzed. Conversely, importins bind to their cargo in the cytosol in the absence of Ran-GTP, and release it in the nucleus upon Ran-GTP binding. Note that each import–export cycle results in the cleavage of one molecule of GTP. It is the energy provided by GTP hydrolysis that drives the transport of proteins against their concentration gradient, which is thermodynamically unfavorable. Several examples of how nuclear import can be regulated to mediate diverse cell signaling responses are described below.

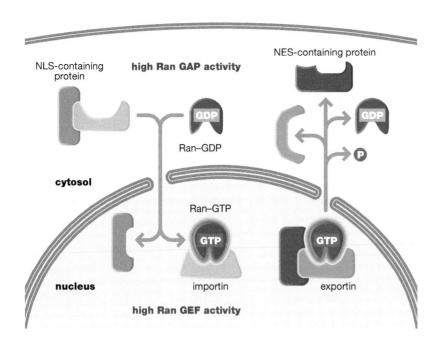

Figure 5.3

Nuclear import and export are regulated by karyopherins and Ran. The nucleus contains high levels of Ran-GTP, while Ran-GDP predominates in the cytosol. Proteins containing nuclear localization signal (NLS) sequences bind importins (*green*) in the cytosol, pass through the nuclear pore, and are released in the nucleus by the binding of Ran-GTP to the importin. Proteins containing a nuclear export signal (NES) bind exportins (*blue*) and Ran-GTP in the nucleus are exported and are released in the cytosol when GTP bound to Ran is hydrolyzed.

Phosphorylation of transcription factor Pho4 regulates nuclear import and export

An elegant example of how the balance of nuclear import and export can be exploited in signaling is provided by the yeast transcription factor Pho4. Pho4 regulates the transcription of genes needed for growth under conditions of low phosphate. Therefore, when cells experience phosphate starvation, Pho4 must localize to the nucleus so it can bind to its DNA targets. When phosphate is abundant, however, Pho4 activity is not needed and the protein is found predominantly in the cytosol. This regulation of subcellular localization of Pho4 by phosphate has been shown to depend on the activity of Pho80/85, a nuclear cyclin–dependent protein kinase. Pho80/85 is active under normal conditions when phosphate is abundant, and inactivated upon phosphate starvation.

How does Pho80/85 regulate the subcellular localization of Pho4? It turns out that the phosphorylation of Pho4 directly affects both its import and export (**Figure 5.4**). First, phosphorylation of two sites within an NES motif promotes its binding to an exportin, leading to much more efficient nuclear export of Pho4 when the kinase is active. Second, phosphorylation of a site within the NLS abolishes the binding of an importin, thus preventing import of Pho4 into the nucleus. Finally, phosphorylation of a third region of Pho4 eliminates binding to another transcription factor, Pho2; in the absence of Pho2 binding, Pho4 cannot bind strongly to its specific binding sites on the promoters of phosphate-responsive genes and thus activate their transcription. Upon inhibition of Pho80/85 under conditions of phosphate

Figure 5.4

Regulation of Pho4 by phosphorylation. (a) The subcellular localization of the transcription factor Pho4 (indicated in *pink*) changes from cytosolic under normal conditions to nuclear when cells are starved of phosphate. (b) Pho4 contains a nuclear export signal (NES), a nuclear localization signal (NLS), and a region that binds the transcriptional coactivator Pho2. Under normal conditions (top), Pho4 is phosphorylated by the Pho80/85 kinase at multiple sites. As a result, it binds exportin but not importin or Pho2. Pho4 is exported to the cytosol and cannot activate transcription. Under phosphate starvation (bottom), Pho80/85 is inactivated, and unphosphorylated Pho4 binds to importin and to Pho2, but not to exportin. Pho4 is now localized to the nucleus and can activate transcription.

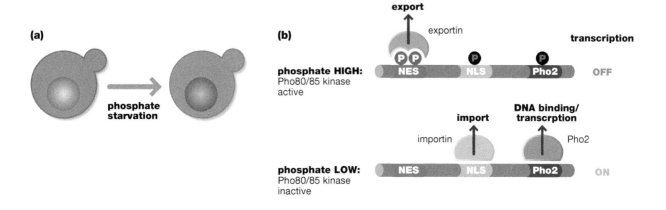

starvation, a combination of dephosphorylation and new synthesis leads to a rapid increase in the pool of unphosphorylated, nuclear Pho4 and thus induction of the transcription of Pho4-dependent genes.

Phosphorylation of Pho4 dramatically alters the nuclear/cytoplasmic equilibrium in three distinct ways: increasing nuclear export, decreasing nuclear import, and decreasing binding to transcriptional cofactors and DNA in the nucleus. A very important feature of this mechanism is that Pho4 activity is cooperatively regulated at multiple steps, making the response sharper and more switchlike than it would be if only one step was regulated.

The use of phosphorylation to regulate the partitioning of proteins between nucleus and cytoplasm is quite common in signaling, and additional examples will be discussed later. One frequently used mechanism involves the phosphorylation of serine and threonine residues that create binding sites for *14-3-3 proteins*. These small proteins bind specifically to phosphorylated sites on their partners. In some cases, 14-3-3 binding serves to sterically block the binding of importins or of DNA, thereby excluding the bound protein from the nucleus or inhibiting its activity. For example, this mechanism is used to regulate the activity of the FOXO family of transcription factors, and also the Cdc25 family of protein phosphatases that regulate cell-cycle progression (see Figure 10.12).

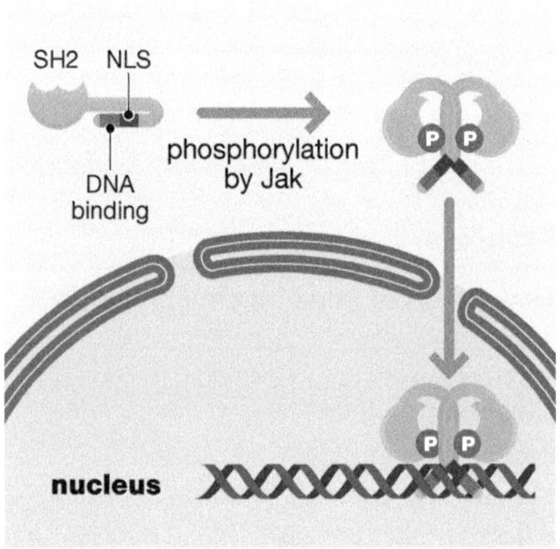

Nuclear import of STATs is regulated by phosphorylation and conformational change

In Chapter 3, we discussed a number of examples where conformational change was functionally linked to changes in enzymatic or binding activity. Changes in subcellular localization can also be coupled to conformational changes, as illustrated by the **STAT** (**signal transducer and activator of transcription**) proteins. These transcription factors function in a common signaling module termed the **JAK–STAT pathway**, which relays signals from a variety of cytokines and hormones such as growth hormone, erythropoietin, and interferon.

The seven closely related members of the STAT family each contain a DNA-binding domain, an SH2 domain that recognizes phosphotyrosine, and a conserved tyrosine phosphorylation site. In unstimulated cells, the STAT proteins reside in the cytosol in a latent, inactive state (**Figure 5.5**). Stimulation of cell-surface cytokine receptors by their ligands leads to activation of JAK-family tyrosine kinases, which then phosphorylate STAT proteins on specific tyrosine residues. Phosphorylation of STATs causes a conformational change that facilitates formation of active homodimers or heterodimers with other phosphorylated STAT molecules. This dimerization is mediated by mutual interaction of the SH2 domain and tyrosine-phosphorylated site on one partner with the corresponding tyrosine-phosphorylated site and SH2 domain of the other. This bivalent interaction is particularly stable due to the *avidity* effect.

The active STAT dimer translocates to the nucleus where it binds specific DNA sequences to stimulate gene transcription. In this case, both the NLS (which interacts with importins) and the DNA-binding domain are not functional in the unphosphorylated form of STAT, and are only fully assembled and active upon conformational change and dimerization. This dual control of activity, at the levels of rate of nuclear import and binding to specific DNA targets, again makes the response to activation more switchlike.

Localization of MAP kinases is regulated by association with nuclear and cytosolic binding partners

Many extracellular signals lead to the activation of serine/threonine kinases of the MAP kinase family. These kinases are activated by phosphorylation by upstream

Figure 5.5

Nuclear localization of activated STAT. In unstimulated cells, STAT is located in the cytosol and its nuclear localization signal (NLS) and DNA-binding domains are not functional. Upon phosphorylation by JAK-family kinases, STATs dimerize through interactions between SH2 domains and phosphotyrosine. Conformational changes expose NLS and DNA-binding domains that mediate nuclear localization and binding to target sequences in DNA, respectively.

 Avidity is described in Chapter 2.

 MAP kinase cascades are discussed in Chapter 3.

kinases (MAP kinase kinases, or MAPKKs). Substrates for MAP kinases are found in various subcellular localizations, but perhaps the best characterized of these substrates are nuclear transcription factors. To phosphorylate these nuclear substrates, however, the activated MAP kinase must itself be localized to the nucleus.

When the subcellular localization of a MAP kinase such as Erk2 is tracked following activation, often a dramatic relocation from the cytosol to the nucleus is seen, corresponding with the appearance of the phosphorylated, active kinase (see, for example, Figure 15.16). The precise mechanism of this translocation is not yet completely resolved. Erk2 itself has no apparent NLS or NES sequences, and a number of experimental approaches have indicated that translocation does not, in most cases, require energy. Indeed, MAP kinases are small enough (~40 kD) to enter the nucleus through passive diffusion, and it is likely that this mechanism is sufficient to explain the observed rates of translocation. What then is responsible for the change in distribution upon activation? This is thought to be largely the result of differences in the binding partners for the unphosphorylated (inactive) and phosphorylated (active) forms of the MAP kinase. In the case of Erk2, the inactive form binds tightly to Mek, its upstream MAPKK, which is localized almost entirely in the cytosol; furthermore, the complex is too large to diffuse passively into the nucleus. Phosphorylated Erk2, by contrast, does not bind strongly to Mek, and thus is free to diffuse into the nucleus, where it presumably preferentially binds to nuclear proteins (**Figure 5.6**).

This relatively straightforward picture is almost certainly an oversimplification of the actual mechanisms controlling MAP kinase subcellular localization. For example, in some situations, activated Erk can be found in the cytosol where a number of important substrates are localized. There is also some evidence for energy-dependent import and export of MAP kinases, perhaps in complex with proteins that contain bona fide NLS or NES motifs. It is also worth noting that once again, activation is intimately coupled to change in subcellular localization, thus increasing the apparent difference between the inactive and active states for nuclear substrates such as transcription factors.

Notch nuclear localization is regulated by proteolytic cleavage

The **Notch** signaling pathway is commonly used during development to specify cell fate and to regulate morphogenesis. It provides a mechanism for clusters of adjacent cells to communicate with each other and coordinate their activities. The receptors of the Notch family are transmembrane proteins containing an intracellular domain linked to an extracellular domain that interacts with specific ligands, such as Delta, on adjacent cells. Ligand engagement leads to a series of proteolytic cleavages of Notch, the most important consequence of which is the release of the intracellular domain from the membrane. This domain (termed the Notch intracellular domain, or NICD) then translocates to the nucleus where it interacts with a DNA-binding protein and other cofactors to regulate gene transcription (**Figure 5.7**). The NICD contains NLS motifs that presumably interact with importins and mediate nuclear import once the domain is freed from the plasma membrane.

The fact that nuclear localization is the result of proteolytic processing confers several interesting properties to this signaling pathway. Most obviously, the release of the active NICD is irreversible; thus, this system is not well suited to respond to rapid, repeated changes in stimulus levels over time. Also, signal amplification is not possible, as one activated receptor generates just a single activated transcription factor. This pathway is also remarkable for its directness, as it does not involve any intervening molecules between receptor engagement and nuclear activity; thus, Notch signaling is less likely than other signaling pathways to respond to diverse upstream stimuli or lead to many different downstream effects.

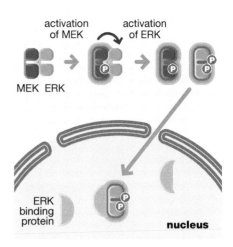

Figure 5.6

Dynamic localization of MAP kinases. The MAP kinase Erk2 binds to different partners in its inactive and activated states. Inactive Erk binds tightly to its upstream activator Mek, which is predominantly found in the cytosol. Upon activation by Mek, Erk dissociates from Mek. Erk is then free to diffuse passively into the nucleus, where it associates with nuclear proteins.

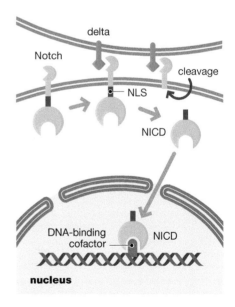

Figure 5.7

Localization of Notch is regulated by proteolysis. Notch is a transmembrane receptor. When engaged by its ligand Delta on an adjacent cell, Notch is cleaved within the juxtamembrane region, releasing the Notch intracellular domain (NICD). A nuclear localization signal (NLS) within the NICD facilitates its import to the nucleus, where it associates with cofactors and DNA to regulate transcription.

General properties of membranes and their role in signaling are described in Chapter 7.

CONTROL OF MEMBRANE LOCALIZATION

Cellular membranes, and in particular the plasma membrane, are the site of many signaling reactions. In the case of the plasma membrane, this is in large part due to the presence of transmembrane receptors conveying information from the outside environment to the cytosol. Membranes also offer a variety of lipids that can attract protein-binding partners and provide substrates for lipid-modifying enzymes. Furthermore, the two-dimensional surface of the membrane serves to restrict diffusion and increase the efficiency of interactions between membrane-associated proteins compared to those in the cytosol.

Signaling can involve both proteins that are constitutively localized to the membrane, such as most receptors, and proteins that are conditionally localized to the membrane. Below, we discuss both classes of protein and the diverse mechanisms that are used to target such proteins.

Proteins can span the membrane or be associated with it peripherally

The different ways in which proteins can be associated with membranes are illustrated in **Figure 5.8.** Some proteins exist as integral membrane proteins that actually span the lipid bilayer, while peripheral membrane proteins can associate with membranes by virtue of covalent lipid modification, through specialized domains that bind directly to membranes, or by indirect binding to other membrane-bound proteins.

Integral membrane proteins such as transmembrane receptors are inserted into the membrane during the process of translation on the endoplasmic reticulum, and are ultimately transported to their target membranes by vesicular transport. Different portions of such proteins face the extracellular (or lumenal) and cytosolic faces of the membrane. Because of the enormous energetic cost of dragging bulky hydrophilic protein domains through the lipid membrane after their initial synthesis, integral membrane proteins typically cannot be liberated from the membrane other than through proteolytic cleavage (as in the case of Notch, as discussed above).

Proteins can be covalently modified with lipids after translation

Some proteins are anchored in the membrane through the covalent addition of bulky hydrophobic lipid groups. Such post-translational lipid modifications provide more scope for regulation than is possible for integral membrane proteins, because the lipid addition and removal steps can potentially be controlled by signals. Furthermore, the energetic cost of extracting the lipid group from the membrane is not prohibitive, so lipid-modified proteins can, in some cases, shuttle between the membrane and cytosolic compartments. Several common lipid modifications will be discussed below, and their properties are summarized in **Table 5.1.**

The **glycosylphosphatidylinositol (GPI) anchor** involves the addition of a modified phosphatidylinositol lipid group to the C-terminus of a protein within the lumen of the endoplasmic reticulum (**Figure 5.9**). GPI-modified proteins then proceed through the normal secretory pathway to the outer leaflet of the plasma membrane. Because they are anchored to the membrane via the phospholipid head

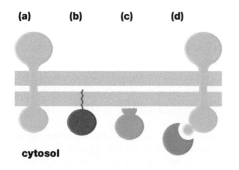

(a) **(b)** **(c)** **(d)**

cytosol

Figure 5.8

Mechanisms for localizing proteins to membranes. Mechanisms include: (a) an integral membrane protein, (b) covalent lipid modification, (c) association via a membrane-binding domain, and (d) association by binding to another membrane protein.

Table 5.1

Properties of common lipid modifications of proteins

Modification	Site of attachment	Strength of membrane association	Reversible?
Glycosylphosphatidylinositol (GPI) anchor	C-terminus	Strong	No
N-Myristoylation	N-terminus	Weak	No
S-Acylation	Internal cysteine	Weak	Yes
Prenylation	C-terminus	Weak	No

group, they can be released from the cell surface through the action of extracellular phospholipases. This cleavage is, however, irreversible.

***N*-Myristoylation** involves the addition of the myristoyl group, a 14-carbon fatty acid, to glycine at the N-terminus of a cytosolic protein just after translation, following removal of the initiator methionine (see Figure 5.9). This modification is thought to be essentially irreversible. A large number of signaling proteins are *N*-myristoylated, including heterotrimeric G protein α subunits and other G proteins, nonreceptor tyrosine kinases, and the catalytic subunit of protein kinase A (PKA). The membrane association mediated by myristoylation is relatively weak, and thus some myristoylated proteins can partition between the membrane and cytosol depending on protein conformation, interactions with other proteins, and other post-translational modifications that are present. Thus, myristoylation can serve as one component of a regulatable membrane switching mechanism. Indeed, in some proteins, such as the catalytic subunit of PKA, the myristoyl group is buried in a hydrophobic groove on the protein surface and thus does not lead to membrane association under most conditions. However, in many other cases, myristoylation is sufficient to target a protein to the membrane. Thus, engineered myristoylation sites are often used experimentally to test the consequences of forced membrane localization of a protein of interest.

***S*-Acylation** involves the addition of an acyl group, most often the 16-carbon fatty acid palmitate, to cysteine residues of a protein (see Figure 5.9). Unlike most other lipid modifications, *S*-acylation is readily reversible in the cell, and thus, proteins can cycle to and from the membrane upon addition or removal of lipid. As in the case of myristoylation, *S*-acylation leads to relatively weak membrane association, and thus, it often cooperates with other mechanisms of membrane association to mediate tighter membrane binding. For example, some Src family nonreceptor tyrosine kinases are thought to be transiently directed to the plasma membrane by *N*-myristoylation, and then to become more stably associated after addition of palmitate groups by membrane-associated protein *S*-acyl transferases.

The addition of relatively bulky isoprenoid lipids, such as farnesyl or geranylgeranyl groups, to the C-terminus of a protein is known as **prenylation** (see Figure 5.9). The lipid is added to the cysteine within a so-called **CAAX box** motif (where A is an aliphatic amino acid and X is any amino acid), and this is followed by proteolytic removal of the AAX and methylation of the terminal carboxyl group. Like other lipid modifications, prenylation can be sufficient to direct membrane association, but stable binding often requires additional interactions. One of the most important classes of prenylated proteins includes various G proteins such as the Ras family of oncogenic small GTPases, whose function requires their membrane association. Small molecules that block prenylation, such as the statin class of anticholesterol drugs (which inhibit the rate-limiting step in isoprenoid biosynthesis), were tested as potential anti-cancer drugs for their ability to inhibit the function of prenylated proteins such as Ras family G proteins.

Modular lipid-binding domains are important for regulated association of proteins with membranes

Many signaling proteins can associate with membranes via small, modular lipid-binding domains. These domains include such examples as the PH, FYVE, and PX domains. In many cases, a lipid-binding domain is highly specific for binding to a particular lipid head group. For example, different examples of the *PH domain* bind to specific phosphoinositides and not to other closely related molecules. Membrane association by such domains is therefore controlled by the amount and local concentration of these lipids which, in turn, are controlled by lipid kinases, lipid phosphatases, and phospholipases whose activity can be regulated in signaling. The affinity of such domains for their lipid targets is often relatively

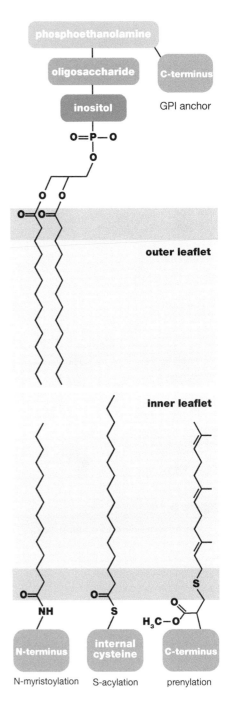

Figure 5.9

Protein lipid modifications. The structures and linkages of common lipid modifications are depicted. The outer (extracellular) leaflet of the plasma membrane is oriented toward the top of the figure. Site of attachment to protein (*green*) is also indicated for each type of modification. GPI, glycosylphosphatidylinositol.

(a)

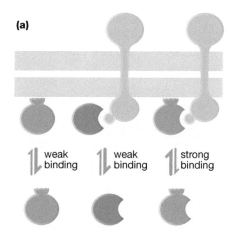

(b)

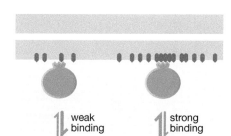

Figure 5.10

Cooperative binding to membranes.
(a) Cooperative binding via multiple independent interactions. Binding via a membrane-binding domain (left) or via association with another membrane protein (middle) is relatively weak in both cases, and most of the protein will therefore remain in the cytosol. When both interactions are present, however, binding is much stronger and most of the protein is associated with the membrane. (b) This binding domain has multiple sites for binding a specific membrane phospholipid (*pink*). When the local density of the lipid is low (left), binding is weak and most of the protein is cytosolic. When local density of the lipid is high (right), cooperative binding leads to very strong binding to the membrane.

 Lipid kinases and lipid phosphatases are discussed in Chapter 7.

low, so other factors (for example, interaction with other membrane-associated proteins) may be necessary for stable binding (**Figure 5.10a**). In this way, stable membrane association of a protein can be made to depend cooperatively on multiple inputs, allowing for integration of multiple signals and the generation of more switchlike, all-or-none changes in localization.

A second class of membrane-binding motifs includes various positively charged protein surfaces that interact electrostatically with membranes, which generally carry a net negative charge due to the phosphate groups of phospholipids. These protein surfaces include positively charged amphipathic helical segments. Other types of positively charged motifs are somewhat specific for particular highly phosphorylated lipids, such as **phosphatidylinositol 4,5-bisphosphate [PI(4,5) P$_2$]**. PI(4,5)P$_2$ is a relatively rare membrane lipid that plays a disproportionately large role in signaling through its ability to bind cytosolic proteins, and its ability to generate soluble second messengers after cleavage.

Such positively charged motifs have interesting potential regulatory properties. For example, they have the potential to be highly multivalent (to bind to multiple phosphate groups on the membrane); because of this, binding to the membrane surface can be highly cooperative and thus vary dramatically with relatively small differences in the local concentration of the lipid-binding partners, such as phosphoinositides (**Figure 5.10b**). Furthermore, because the interaction of such motifs with membranes is dependent on their positive charge, their activity can be abolished by post-translational modifications, such as phosphorylation, that reduce their positive charge. This mechanism is used, for example, to regulate the membrane localization of the MAP kinase scaffold protein Ste5 during the cell cycle in yeast. In this case, phosphorylation of a positively charged membrane-localization motif by a cyclin-dependent kinase, which is activated in the G$_1$ phase of the cell cycle, abolishes Ste5 membrane localization.

Some lipid-modified proteins can reversibly associate with membranes

If the membrane binding of proteins containing bulky lipid groups is to be regulated in signaling, there must be a mechanism to shield the hydrophobic lipid group from the hydrophilic environment of the cytosol when it is not embedded in the membrane. One such mechanism involves allosteric changes in the three-dimensional structure of the protein itself. For example, some isoforms of the Abl nonreceptor tyrosine kinase are modified by *N*-myristoylation. In the inactive conformation of the kinase, the myristoyl group folds up into a hydrophobic groove on the surface of the protein and helps maintain and stabilize the inactive conformation. Activation of Abl is associated with a concerted conformational change that eliminates the myristoyl binding pocket and thus exposes the free myristoyl group, which can then insert into the membrane. By this means, the kinase activity of Abl can be functionally coupled to membrane association.

Lipid groups can also be shielded by association with specific binding proteins. This is best understood in the case of prenylated small G proteins, for example, the Rho and Rab families. In the case of the Rab family, a protein called RabGDI (Rab guanine nucleotide dissociation inhibitor) binds to membrane-associated Rab proteins and can extract them from the membrane by shielding their prenyl groups in a hydrophobic pocket (**Figures 5.11** and **5.12**). The RabGDI–Rab complex is soluble and can then be transported to new sites in the cell, where interaction with target membranes leads to dissociation of the GDI and insertion of the prenyl group of the G protein into the lipid bilayer. This process may be facilitated by so-called *GDI displacement factors* (*GDFs*) on the target membrane, which interact with the GDI and stimulate release of the bound G protein. In some cases, guanine nucleotide exchange factors (GEFs) may also coordinate with the GDI in coupling

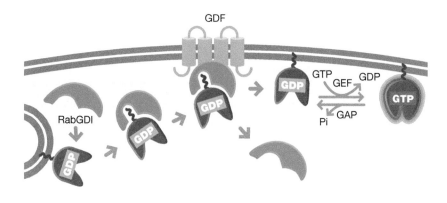

Figure 5.11

The RabGDI cycle. Rab guanine nucleotide dissociation inhibitor (RabGDI) binds to GDP-bound Rab (*purple*) and extracts it from the membrane. The soluble RabGDI–Rab complex can then move through the cytosol to a new target membrane. Dissociation of RabGDI, which can be facilitated by GDI displacement factors (GDFs), allows insertion of Rab into the target membrane. Guanine nucleotide exchange factors (GEFs) on the target membrane can activate Rab by inducing release of GDP and binding of GTP. GAP, GTPase-activator protein; P$_i$, inorganic phosphate.

membrane binding to the activation state of the G protein. RhoGDI (Rho guanine nucleotide dissociation inhibitor) plays a similar role in regulating the membrane association and guanine nucleotide binding of Rho family G proteins such as Rac and Cdc42.

Coupling effector protein activation to membrane recruitment is a common theme in signaling

Activation of plasma membrane receptors often leads to the membrane recruitment of effector molecules which, in turn, become activated and transmit downstream signals. Below, we discuss the example of the Akt kinase, just one of many protein kinases activated as a direct consequence of regulated membrane recruitment. In such cases, the enzyme is activated by the presence of activating enzymes or cofactors that are localized on the membrane.

In other cases, membrane localization of an enzyme may not increase its state of activity, but its output is increased (that is, the number of reactions it can catalyze per unit time) simply by bringing it into close proximity with its substrates. The most striking example of this is activation of the small G protein Ras by receptor tyrosine kinases. This is a crucial step in signaling pathways that control proliferation and differentiation. The enzyme that actually activates Ras is a GEF called Sos. While Ras is tightly associated with membranes by virtue of covalent lipid modifications, its activator Sos is cytosolic in unstimulated cells, and therefore only rarely encounters its membrane-bound substrate. Upon receptor activation, however, Sos is recruited to the membrane where Ras is found, thus greatly increasing the efficiency of Ras activation by Sos (**Figure 5.13**).

Akt kinase is regulated by membrane recruitment and phosphorylation

The **Akt** family of serine/threonine kinases (also called the PKB family) provides a good example of activation by membrane recruitment. Akt plays a central role in transmitting signals that regulate cell growth, proliferation, and survival. Its activation depends on phosphoinositides. Specifically, Akt is dependent on **phosphatidylinositol 3,4,5-trisphosphate [PI(3,4,5)P$_3$]** which is generated by a lipid kinase, **phosphatidylinositol 3-kinase (PI3K)**. PI3Ks can be activated by upstream signals by several distinct mechanisms, which themselves involve recruitment to the membrane through protein–protein interactions, coupled with conformational changes. As we have seen before, membrane recruitment increases the availability of substrate for an enzyme, such as PI3K, that targets membrane components. The net result of PI3K activation is a high local concentration of its product PI(3,4,5)P$_3$.

It is this increased density of PI(3,4,5)P$_3$ that triggers activation of Akt. Like many kinases, Akt is activated by phosphorylation on several regulatory sites. At least one of these critical activating phosphorylations is performed by a second protein

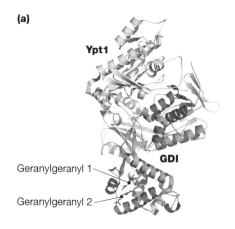

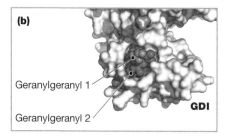

Figure 5.12

Structure of Rab–RabGDI complex. (a) Ribbon diagram of the X-ray crystal structure of the yeast Rab family G protein Ypt1 (*yellow*) complexed with RabGDI (*blue*). Ypt1 is modified by addition of two geranylgeranyl lipid groups, indicated in *pink* and *orange*. Note the C-terminal 10 residues of Ypt1 are not resolved in the crystal structure. (b) Space-filling model showing the two Ypt1 geranylgeranyl groups (*pink* and *orange*) fitting into hydrophobic grooves on the surface of GDI (*white*). (Adapted from O. Pylypenko et al., *EMBO J.* 25:13–23, 2006. With permission from Springer Nature.)

Phosphoinositide kinases and phosphatases will be discussed in more detail in Chapter 7.

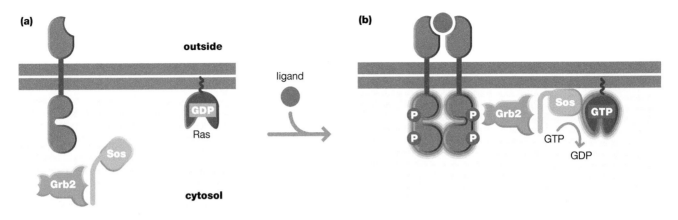

Figure 5.13

Activation of Ras. (a) in unstimulated cells, the Ras activator Sos is associated with the adaptor protein Grb2 in the cytosol; its concentration at the membrane is relatively low. (b) A receptor tyrosine kinase is activated by binding its ligand and autophosphorylates, creating binding sites for the SH2 domain of Grb2. Thus, Grb2 and Sos are recruited to the membrane. The concentration of Sos at the membrane is now high, and Ras is activated.

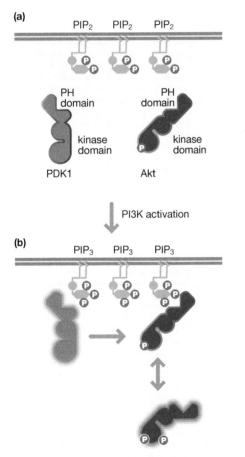

Figure 5.14

Activation of Akt. (a) In unstimulated cells, both PDK1 and Akt are cytosolic and inactive. PIP$_2$, phosphatidylinositol 4,5-bisphosphate. (b) Activation of phosphatidylinositol 3-kinase (P13K) by upstream signals leads to high local concentrations of the lipid phosphatidylinositol 3,4,5-trisphosphate (PIP$_3$) on the membrane. Both PDK1 and Akt bind to PIP$_3$ via their PH domains. Once recruited to the membrane, PDK1 phosphorylates Akt, promoting its activation. Activated Akt can dissociate from the membrane to phosphorylate substrates in other subcellular locations. Phosphorylation at a second site by a different kinase (the mTorc2 complex) is also required for full activation of Akt.

kinase, PDK1. Both Akt and PDK1 contain a PH domain (a modular lipid-binding domain) that binds specifically to PI(3,4,5)P$_3$. Thus, the activating enzyme (PDK1) and its substrate (Akt) are both recruited to the same patch of membrane, where PDK1 then phosphorylates and activates Akt (**Figure 5.14**). Conformational changes upon membrane binding may also facilitate phosphorylation of Akt by PDK1. Akt phosphorylation induces further conformational changes that decrease its affinity for PDK1 and membranes, freeing Akt to diffuse throughout the cytosol and nucleus to phosphorylate targets in these compartments. Thus, the membrane plays a transient, albeit critical, role in Akt activation.

MODULATION OF SIGNALING BY MEMBRANE TRAFFICKING

Cellular membranes are highly dynamic. Vesicles containing lipids and membrane proteins continuously shuttle from the endoplasmic reticulum and Golgi apparatus toward the cell surface. At the same time, plasma membrane components are internalized and sorted for different fates, such as destruction in lysosomes or recycling back to the surface. Thus, the many signaling proteins that are intimately associated with membranes can also be vectorially transported through these mechanisms. The resulting movement of signaling proteins between membrane compartments with distinct subcellular locations, physical properties, and constituents, can profoundly influence their behavior, either positively or negatively. In this section, we will consider a few specific examples of where the transport of membrane proteins is important for signal output.

Proteins can be internalized by a variety of mechanisms

Cell-surface proteins such as receptors are internalized by **endocytosis**, a process whereby a portion of the membrane invaginates and ultimately pinches off into the cytosol, forming a vesicle termed an **endosome** that is no longer physically connected to the outside environment.

Receptors and other membrane components can be internalized by a variety of specific mechanisms (**Figure 5.15**). Perhaps the best characterized of these endocytic pathways is clathrin-mediated endocytosis. This involves the cooperative assembly of the protein **clathrin** into a hollow framework or coat around a patch of membrane, which ultimately pinches off. In the case of receptors, this process is often promoted by their ubiquitylation which, in turn, is mediated by ubiquitin ligases that are recruited upon receptor activation. Once free of the plasma membrane, the vesicle sheds its clathrin coat and the vesicle (now termed an early endosome) and its contents are targeted to their next destination. By contrast, caveolin-mediated endocytosis occurs through the invagination of patches of membrane, termed **caveolae**, enriched in the lipid cholesterol and the protein caveolin. Finally, macropinocytosis involves large-scale rearrangements of the actin cytoskeleton (often termed circular dorsal ruffles), leading to engulfment of a large area of membrane, its associated proteins, and extracellular fluid. For a given protein or lipid molecule, the kinetics of its internalization and its ultimate fate (that is, recycling to the cell surface versus destruction in the lysosome) are highly dependent on the specific mechanism of internalization.

Internalization of receptors can modulate signal transduction

 Receptor down-regulation is discussed in Chapter 8.

The most straightforward way in which membrane transport affects signaling is through the internalization of cell-surface receptors. Once internalized, receptors no longer have access to extracellular ligands, and may be targeted for destruction in the lysosome, and so this is often used as a mechanism to down-regulate receptor signaling after stimulation. There are also examples where receptors are stored in intracellular vesicles and transported to the cell surface in response to signaling inputs. This mechanism is used to up-regulate the number of receptors for the neurotransmitter AMPA in the dendritic spines of neurons when they are exposed to certain types of stimuli, which is thought to be important for learning and memory.

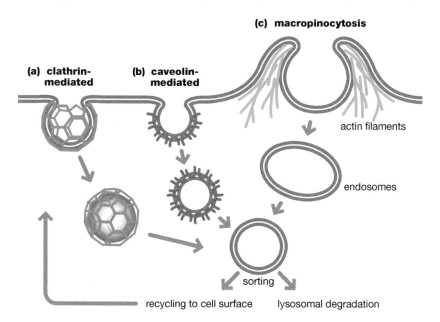

Figure 5.15

Mechanisms of endocytosis.
(a) Clathrin-mediated endocytosis.
(b) Caveolin-mediated endocytosis.
(c) Macropinocytosis. After endocytosis, vesicles are sorted and either recycled to the cell surface or directed to lysosomes for degradation.

Any ligand already bound to its receptor is likely to remain bound for some time after internalization, due to the generally slow off-rate for receptor–ligand interactions and the small volume of the lumen of the vesicle, which ensures that the local concentration of ligand remains quite high even if it dissociates from the receptor. However, the pH in the lumen of an endocytic vesicle generally decreases over time en route to the lysosome, and some ligand–receptor interactions are sensitive to this acidification. For example, among the closely related ligands and receptors of the epidermal growth factor (EGF)/EGF receptor family, differences in the acid stability of binding likely lead to differences in the time that internalized receptor–ligand complexes remain active. This, in turn, affects the fraction of receptors that are fated for degradation versus recycling back to the surface.

There is considerable evidence that activated receptors, such as the EGF receptor, continue to signal after internalization. This makes sense, as the intracellular portion of the receptor that transmits downstream signals remains exposed to the cytosol, whether the receptor is on the cell surface or an endosome. In some cases, however, the vesicle environment restricts the output signal compared to the plasma membrane. One way in which this can occur is in the availability of substrates. To provide a specific example, tyrosine kinase receptors such as the EGF receptor transmit signals in part by recruiting enzymes such as *phospholipase C (PLC)* and PI3K that target membrane lipids. The substrate for both of these enzymes is PI(4,5)P$_2$. It is easy to see how, in a small vesicle, receptor-associated PLC or PI3K activity would rapidly deplete any available PI(4,5)P$_2$. By contrast, on the cell surface, a much larger pool of substrate is available, and thus the potential magnitude of the downstream signal is much greater.

 TGFβ signaling is discussed in greater detail in Chapter 8.

TGFβ signaling output depends on the mechanism of receptor internalization

The transforming growth factor β (TGFβ) receptor provides an excellent example of how the mode of endocytic transport can profoundly affect signal output. TGFβ family receptors possess intrinsic serine/threonine kinase activity. Ligand binding induces the heterodimerization and activation of the receptors, which then phosphorylate and activate the downstream effectors SMAD2 or SMAD3. This process is facilitated by the scaffold protein SARA. Phosphorylated SMAD2 and SMAD3 then dissociate from the receptor and bind another subunit, SMAD4, and this complex translocates to the nucleus where it induces the transcription of specific target genes.

TGFβ receptors are constitutively internalized by two distinct mechanisms, one of which promotes and one of which inhibits signal output (**Figure 5.16**). The early endosomes resulting from clathrin-mediated endocytosis promote activation.

Figure 5.16

Two pathways for internalization of TGFβ receptors. Transforming growth factor β (TGFβ) receptors are internalized via clathrin-mediated endocytosis or caveolin-mediated endocytosis. In the clathrin-mediated pathway (left), downstream signaling is promoted by recruitment of SARA and SMAD2 to phosphatidylinositol 3-phosphate [PI (3)P]-rich early endosomes. In the caveolin pathway (right), signaling is suppressed by recruitment of SMAD7 and the SMURF ubiquitin ligase, leading to receptor ubiquitylation and degradation.

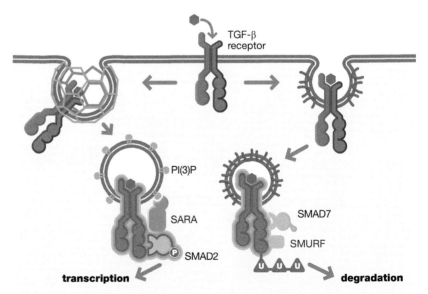

This is because these vesicles are highly enriched in the SARA protein, which promotes SMAD2 and SMAD3 phosphorylation and downstream signaling. This enrichment of SARA is likely due to high levels of the lipid **phosphatidylinositol 3-phosphate [PI(3)P]** in these vesicles, and the fact that SARA has a FYVE domain that specifically binds PI(3)P. Thus, clathrin-mediated endocytosis promotes TGFβ signaling, and blocking this pathway inhibits signal output. On the other hand, caveolin-dependent endocytosis of TGFβ receptors inhibits signaling. The vesicles resulting from this process are enriched for the "inhibitory SMAD" SMAD7, which recruits the ubiquitin ligases SMURF1 and SMURF2. SMURF-mediated ubiquitylation of the receptor targets it for degradation. Thus, the dynamic balance between the clathrin-dependent and caveolin-dependent endocytic pathways determines the ultimate strength of the downstream signal.

Retrograde signaling allows effects distant from the site of ligand binding

While it is clear that internalized receptors have the potential to generate signals, in the case of neurons, there is good evidence that receptor internalization is actually required for what is termed retrograde signaling. Neurotrophins are a family of ligands that engage receptor tyrosine kinases, and they transmit cell growth and survival signals in neurons. Because nerve processes can be very long (more than a meter long in the case of humans), transmission to the cell body of a signal generated by engagement of neurotrophin receptors at the end of an axon would be extremely slow and uncertain if it relied solely on the passive diffusion of protein intermediates or other signaling molecules. To circumvent this problem, it is thought that internalized vesicles containing the activated receptors, along with associated downstream effector proteins, are actively transported along microtubules to the cell body by the molecular motor dynein (**Figure 5.17**). Such vesicles have been termed "signaling endosomes."

Ras isoforms in distinct subcellular locations have different signaling outputs

Ras provides another example where the specific membrane localization of a signaling protein appears to affect its signal output. As we have already seen, Ras family small GTPases act as switches to regulate processes such as cell proliferation and differentiation. There are several isoforms of Ras in mammals that differ in their

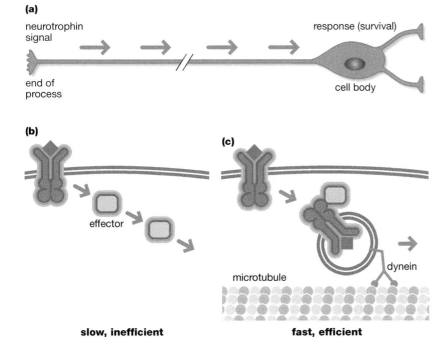

(a)

neurotrophin signal

response (survival)

end of process

cell body

(b)

effector

slow, inefficient

(c)

microtubule

dynein

fast, efficient

Figure 5.17

Retrograde signaling by neurotrophins.
(a) Neurotrophins activate their receptors at the end of neuronal processes, which can be a great distance from the cell body that must respond to the signal. (b) Transmission of the signal would be slow and inefficient if effectors (*light blue*) activated by the receptor had to travel to the cell body via passive diffusion. (c) Endocytosed receptor and associated effectors are actively transported to the cell body by the motor protein dynein.

	Ras localization	Mating	Morphology
wild type	PM + ER	+	+
no Ras	–	–	–
Ras mutant 1	PM	+	–
Ras mutant 2	ER	–	+

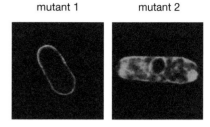

mutant 1   mutant 2

Figure 5.18

Correlation of subcellular localization of Ras with signaling output. In wild-type fission yeast cells, Ras localizes to both the plasma membrane (PM) and endomembranes (ER). The ability of Ras mutants with more restricted localization to rescue defects in mating and cell morphology is indicated in the table on the left by plus and minus symbols. On the right are photomicrographs of cells expressing the engineered Ras mutants (fused to cyan fluorescent protein). Mutant 1 localizes exclusively to the plasma membrane, while mutant 2 localizes exclusively to endomembranes. (Adapted from B. Onken, H. Wiener, M.R. Philips & E.C. Chang, *Proc. Natl Acad. Sci. USA* 103:9045–9050, 2006. With permission from National Academy of Sciences, USA.)

membrane-localization signals. While all forms are farnesylated at the C-terminus (see above), this lipid modification is not sufficient for stable membrane association, which requires a second signal. In the case of the H-Ras and N-Ras isoforms, this is provided by S-acylation with one or two palmitate groups, respectively, while the major K-Ras isoform contains a stretch of basic amino acid residues that promotes plasma membrane localization through interaction with negatively charged membrane lipids. Palmitoylation of Ras occurs on the Golgi apparatus, and is followed by vesicular transport to the plasma membrane. Thus, N-Ras and H-Ras are thought to cycle between the Golgi and plasma membranes, with the rate of cycling and relative distribution among them determined by the rates of palmitoylation and depalmitoylation. K-Ras does not require palmitoylation and thus is highly enriched in plasma membranes, though this can be regulated by the phosphorylation of residues flanking the polybasic site.

Studies have shown that the plasma membrane and Golgi pools of Ras are activated and deactivated with different kinetics following stimulation, most likely because the specific GEFs and GAPs that can activate and inactivate Ras are differentially distributed in these compartments. Experiments in which activated Ras isoforms are artificially targeted to either the Golgi or endoplasmic reticulum showed that the downstream signaling outputs are distinct, presumably because of the availability of different Ras effectors in these compartments.

In the fission yeast *Schizosaccharomyces pombe*, there is a single Ras protein that normally localizes to both the endoplasmic reticulum and plasma membrane. In this organism, Ras regulates both cell morphology and the mating response, and genetic studies have shown that these two pathways employ different upstream GEFs and downstream effectors. In the absence of wild-type Ras, expression of a Ras mutant localized exclusively to the endoplasmic reticulum rescues the morphology defects but not the mating defect (**Figure 5.18**). Conversely, a Ras mutant exclusively localized to the plasma membrane rescues the mating defect but not the morphology defect. These experiments clearly highlight the dependence of signal output on subcellular localization.

BIOMOLECULAR CONDENSATES

We have already noted that the cell consists of a highly concentrated solution of proteins, nucleic acids, and other biomolecules, and that these components are neither well mixed nor homogeneously distributed. In recent years, it has become appreciated that the physical properties of biomolecules can lead to their segregation into distinct phases within the cell, in which some components are highly enriched while others are largely excluded. These **biomolecular condensates** are now recognized to be important in promoting specific reactions by colocalizing reactants or segregating molecules apart, and they are implicated in many important cell behaviors including signaling. On the other hand, aberrant condensates are thought to contribute to a number of disease states. Here we will briefly consider the properties of biomolecular condensates and how they assemble, and some of their proposed roles in signaling.

Condensate formation is driven by multivalent interactions

Cell biologists have long been aware of subcellular structures that are visible by light microscopy, yet are not bounded by lipid membranes. The nucleolus, for example, is a relatively large, spherical structure found in the nuclei of eukaryotic cells. Nucleoli are essentially ribosome factories, where ribosomal RNAs (rRNAs) are synthesized, processed, and assembled with ribosomal proteins before export of the finished ribosomal subunits to the cytosol. Other examples include stress granules, in which mRNAs and incompletely folded proteins are co-localized in cells subjected to stresses such as high temperature, and ribonucleoprotein (RNP) granules of various types, which consist of mRNAs and RNA binding proteins. Recent work has begun to define more clearly the properties of such membraneless organelles, and the physical principles that underlie their behavior.

The formation of biomolecular condensates is driven by biophysical processes in which the energy of the system is minimized by a **phase transition**. We are very familiar with phase transitions in nonbiological systems, such as the crystallization of water into ice as the temperature drops, or the precipitation of salt crystals as water evaporates from a saturated solution. We are also familiar with **phase separation**, such as when oil and water separate into distinct layers after mixing. Under some conditions, biomolecules similarly will undergo phase transitions to find lower energy states, spontaneously segregating into distinct phases within the cell.

Biomolecular condensates often have the properties of liquid droplets (think of droplets of oil suspended in a polar solvent like water). The transition from a homogeneous solution to one in which some components segregate into a distinct, more concentrated liquid phase is termed **liquid–liquid phase separation (LLPS)**. Early experiments showed that condensates in the cell could fuse together into larger droplets, could be deformed by shear forces, and exhibited surface tension, just like droplets of oil in water. Furthermore, while the overall structure of the condensate is maintained over time, the individual molecules that comprise it are highly dynamic. For example, fluorescence recovery after photobleaching (FRAP) experiments show that individual molecules diffuse quite rapidly throughout the condensate, and readily exchange between the solution and the condensate. In some cases, however, the condensed phases are metastable and may eventually form insoluble protein aggregates (essentially another phase transition to a solid state). The abnormal accumulation of such protein aggregates is thought to play a role in a number of neurodegenerative diseases in humans, such as Alzheimer's Disease, Amyotrophic Lateral Sclerosis (ALS), and Huntington's Disease.

In simple single-component systems, as the concentration is increased, at some point a critical concentration will be reached and droplets of condensate begin to form. As the concentration is increased further, more and more of the volume of the solution consists of the dense (condensate) phase, though the concentration of the component in the condensate and in the solvent phases remains the same. This is analogous to adding increasing amounts of a salt into water—eventually (at the solubility limit), the salt will no longer dissolve, and there will be two phases (liquid and solid in this case). As even more salt is added, the overall amount of the solid increases, but the concentration of salt in each of the two phases (the liquid and in the solid) remains the same. While this simple example helps us understand the formation of biomolecular condensates, the behavior of actual condensates in the cell is typically more complex, given the rich variety of components in cellular fluids and in the condensates themselves. Theoretical frameworks and computational models that accurately describe the behavior of complex, multi-component condensates are still areas of very active research.

As noted in Chapter 2, formation of biomolecular condensates is often driven by weak, multivalent interactions. Intuitively, it is easy to see how long, flexible molecules with many low-affinity binding sites might aggregate together into clusters; the low affinity of individual interactions means that individual bonds will rapidly break and re-form, allowing a molecule to diffuse within a cluster, or between the cluster and the surrounding solution.

Some of the first condensates to be studied *in vitro* consisted of solutions of two classes of molecules: one with multiple SH3 domains, and the other with multiple proline-rich motifs (PRMs) to which the SH3 domains bound (**Figure 5.19**).

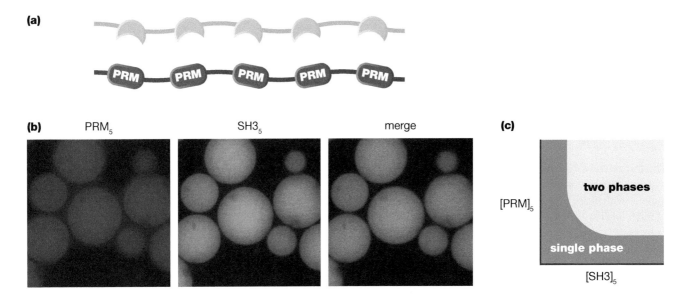

(b) PRM$_5$ SH3$_5$ merge

(c) [PRM]$_5$ two phases   single phase   [SH3]$_5$

Figure 5.19

Biomolecular condensates. Flexible molecules that interact via multiple relatively low-affinity sites can generate condensed phases. (a) In this example, one molecule contains five SH3 domains, and a second molecule contains five proline-rich motifs (PRMs) that bind to the SH3 domains. The two molecules are each fluorescently labeled so they can be visualized by fluorescence microscopy (*green* and *magenta*).
(b) Fluorescence micrographs of mixtures of the two proteins, illustrating the formation of condensed-phase droplets containing both molecules. Scale bar = 20 μm. (c) Phase diagram illustrates that two phases form only when the concentrations of both molecules exceed a threshold. (b, courtesy of Zeynep Baltaci and Jon Ditlev.)

Another example is provided by the combination of RNA molecules and proteins containing multiple RNA binding domains. A common thread is that the interacting molecules are typically highly flexible (lack a rigid structure), which makes interaction with many different partners possible. Many proteins found in condensates contain **intrinsically disordered regions (IDRs)**, which do not fold into stable structures, but instead are flexible and sample many different conformational states. Often, such IDRs contain interspersed charged residues and planar aromatic residues, which can mediate low-affinity interactions mediated either by electrostatic (charge-charge) interactions or via interaction of aromatic residues either with other aromatic residues (π-π stacking) or aromatic residues with amino groups (cation-π interactions).

In addition to the flexible, multivalent "scaffold" proteins that initiate condensate formation, other proteins may be enriched or excluded from such condensates, depending on their ability to interact with the scaffold proteins. Proteins that are enriched in a particular class of condensate are sometimes called "clients." Although they may not bind to scaffold proteins through multivalent interactions, they are enriched because of the high local concentration of binding sites within the scaffold, making it difficult for them to escape into the solvent phase.

Biomolecular condensates can make reactions more efficient and specific

A variety of functions have been ascribed to biomolecular condensates, including sequestering cell components from each other, buffering the concentrations of proteins in the cytosol, and serving as organizational hubs for complex biochemical activities. Probably the most compelling role, and the one that is most relevant to cell signaling, is in increasing the rate and specificity of biochemical reactions, primarily by concentrating the reactants together.

The law of mass action tells us that the rate of a reaction is proportional to the concentrations of the individual reactants. Since the concentrations of biomolecules in condensates can be much higher than their bulk concentrations in solution (typically more than ten-fold, and up to 100-fold higher in some cases), reaction rates may be increased dramatically in condensates. For example, one might expect that if two reactants are each concentrated by 20-fold in the condensate, that would result in a 400-fold increase in reaction rate relative to bulk solution. In terms of specificity, condensates could function similarly to molecular scaffolds in promoting a specific (desired) reaction compared to undesired side reactions, especially if competing reactants are excluded from the condensate. The nucleolus, for example, is likely to make ribosome assembly both more rapid and more efficient by concentrating in a relatively small volume the rRNA genes and the specialized RNA polymerases needed for their transcription, the enzymes needed for the processing of rRNA, and ribosomal proteins. In fact, the nucleolus is thought to have a particularly complex structure consisting of several distinct condensed phases, each of which performs a separate step in the ribosome synthesis and assembly process. Given the frequent involvement of condensates in RNA-mediated reactions, some have suggested that such condensates might have been critical for the first emergence of living organisms, by providing a "reaction crucible" in which the complex biochemical reactions needed for life could evolve in the absence of lipid membranes.

Biomolecular condensates have been firmly implicated in several cell signaling systems, including T cell receptor and receptor tyrosine kinase signaling. In these cases, the condensate forms on the plasma membrane, and thus is largely two-dimensional (in contrast to the more spherical, three-dimensional condensates discussed up to this point). In each case, signaling is initiated by binding of a ligand to a receptor, which leads to tyrosine phosphorylation of multiple sites on the receptor, recruitment of multivalent adaptor and scaffold proteins via SH2-pTyr interactions, and recruitment of downstream effectors via additional multivalent interactions. In the case of T cell receptor signaling, virtually the entire signaling machinery has been reconstituted *in vitro* on supported lipid bilayers using purified proteins, and the formation of phase-separated condensates and downstream biochemical activities observed in real time by fluorescence microscopy (**Figure 5.20**).

Several general themes emerge from such studies on the role of condensates in signal transduction. First, these systems often depend on regulatory inputs, such as phosphorylation, to initiate formation of the condensate. As with any signaling system, changes in the properties of the system in response to input are required for it to transmit information. The ability to convert one type of signal (such as phosphorylation of a receptor) into a change in the physical state of the system (from isolated receptors to a condensed state in which many receptors, scaffold proteins, and effectors are highly concentrated together) is another example of the interlocking input/output behaviors seen in all signaling systems. Second, condensate formation may help explain why in almost all instances of signaling from membrane receptors, the receptors are observed to aggregate or cluster together upon activation.

Finally, one proposed role of condensates in such systems is to increase the membrane **dwell time**, of signaling enzymes (the time they spend associated with the membrane). This provides the enzyme additional time to perform complex, multi-step reactions before it dissociates from the membrane. For any reaction that requires multiple steps (and this is probably true of most signaling reactions),

 The role of molecular scaffolds in reaction specificity is discussed in Chapter 3.

 Signaling by the T cell receptor and by the RTK PDGF are discussed in detail in Chapter 14.

Figure 5.20

Condensates in T cell receptor signaling reconstituted *in vitro*. In this experiment, a supported lipid bilayer (SLB) was reconstituted with purified signaling proteins, and the formation of condensates and polymerized actin detected by fluorescence microscopy. (a) Components are depicted schematically. The SLB contains purified T cell receptor ζ chain (TCR), tyrosine kinase Lck, and scaffold protein LAT. Other purified proteins provided in solution include tyrosine kinase ZAP70; adaptors Grb2, Gads, SLP76, and Nck; signaling enzymes Sos1, PLCγ1, and NWASP; and the Arp2/3 complex and monomeric actin. LAT, ZAP70, and actin are fluorescently labeled (*blue*, *green*, and *red*, respectively). (c) Fluorescence micrographs of time course after addition of ATP, which initiates substrate phosphorylation by Lck and ZAP70. At early time points, formation of condensates containing LAT and ZAP70 on the SLB is visible; at later time points, actin polymerization mediated by NWASP and ARP2/3 is seen. Actin polymerization was shown to be dependent on formation of condensates. (For details, see Su, et al., *Science* 352(6285):595–599, 2016; Panel b courtesy of Jon Ditlev.)

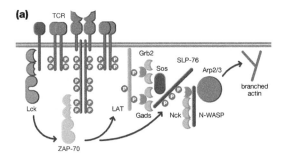

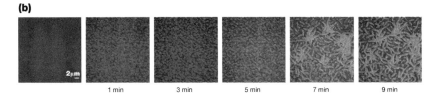

1 min   3 min   5 min   7 min   9 min

Figure 5.21

Kinetic proofreading is promoted by condensates. For complex biochemical reactions such as activation of a signaling enzyme, it takes time to reach maximal reaction probability. Here the distribution of reaction probability is plotted vs. time, illustrating that transient binding is unlikely to result in reaction, while sustained binding via condensate formation or other mechanisms greatly increases the likelihood of reaction. Mean reaction time (half-time for the reaction) indicated by dotted line. (Adapted from Huang, et al., *Proc. Natl. Acad. Sci. USA*. 113(29):8218–8223, 2016.)

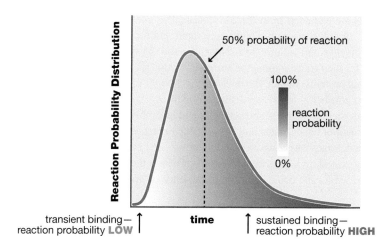

it takes time to reach the maximum probability of reaction; if the two reactants diffuse away from each other too rapidly, the likelihood of reaction will be low. This type of behavior is often called a **kinetic proofreading** mechanism, because chance encounters are unlikely to lead to reaction, as opposed to the more sustained interactions mediated by bona fide signaling complexes (**Figure 5.21**). This has been shown, for example, in the case of activation of the small G protein Ras by Sos, a key step in TCR signaling. Multivalent interactions between phosphorylated receptor, the adaptor/scaffold LAT, the adaptor protein Grb2, and the Ras GEF Sos increase the dwell time of Sos on the membrane, and thus provide sufficient time for Sos to become activated, and in turn to activate Ras. By contrast, a Sos molecule that associates by chance to the membrane would quickly dissociate, and thus would be unlikely to activate Ras.

SUMMARY

The activity of a signaling protein is highly dependent on its location within the cell. Many signaling proteins undergo regulated changes in localization as a means to transmit information. Signaling proteins that control transcription often undergo regulated changes in nuclear localization. Another key location for coordination of signaling is at the cell membrane. Many signaling proteins are localized to the plasma membrane, either constitutively or in a signal-regulated manner. Coordinated localization of proteins at the membrane often controls when and if signaling partner proteins can effectively communicate with one another. Membranes are dynamically redistributed throughout the cell by vesicular transport. Signaling proteins associated with the membrane, such as receptors, can be internalized into endosomes and other organelles, which can be used to further modulate signal output. Not all subcellular locations are defined by lipid membranes, however. Biomolecular condensates are phase-separated regions that can concentrate or exclude certain cell components. Condensates typically form via multivalent, relatively low-affinity interactions between flexible molecules, and can make signaling reactions within them both more efficient and more specific.

Answers to these questions can be found online at www.routledge. com/9780367279370

QUESTIONS

1. Explain the basic principles by which changes in the subcellular localization of signaling proteins can be used as a mechanism to transmit information.

2. In your studies of how a cell responds to stimulation by a specific extracellular growth factor, you discover that a downstream protein translocates from the nucleus to the cytoplasm upon stimulation. How is it possible for a protein

localized mostly in the nucleus to respond to signaling changes in the cytosol or plasma membrane?

3. What are the different mechanisms by which phosphorylation can regulate the nuclear–cytoplasmic localization of a protein?

4. In what ways can internalization of a membrane-bound receptor affect signaling pathways involving that receptor, and the dynamics of the response?

5. How could the signaling properties of a protein differ depending on its subcellular localization?

6. Small G proteins play a central role in determining subcellular localization, not only for nuclear import/export (Ran), but also in vesicle transport and trafficking (Rab family). Which aspects of G proteins and their regulation make them well suited to this role?

BIBLIOGRAPHY

CONTROL OF NUCLEAR LOCALIZATION

Aaronson DS & Horvath CM (2002) A road map for those who don't know JAK-STAT. *Science* 296, 1653–1655.

Bridges D & Moorhead GBG (2005) 14-3-3 proteins: A number of functions for a numbered protein. *Sci. STKE* 2005(296), re10.

Burack WR & Shaw AS (2005) Live cell imaging of ERK and MEK. *J. Biol. Chem.* 280, 3832–3837.

Chatterjee-Kishore M, van den Akker F & Stark GR (2000) Association of STATs with relatives and friends. *Trends Cell Biol.* 10, 106–111.

Ebisuya M, Kondoh K & Nishida E (2005) The duration, magnitude and compartmentalization of ERK MAP kinase activity: Mechanisms for providing signaling specificity. *J. Cell Sci.* 118, 2997–3002.

Fortini ME (2009) Notch signaling: The core pathway and its posttranslational regulation. *Dev. Cell* 16, 633–647.

Komeili A & O'Shea EK (1999) Roles of phosphorylation sites in regulating activity of the transcription factor Pho4. *Science* 284, 977–980.

Ranganathan A, Yazicioglu MN & Cobb MH (2006) The nuclear localization of ERK2 occurs by mechanisms both independent of and dependent on energy. *J. Biol. Chem.* 281, 15645–15652.

Riddick G & Macara IG (2005) A systems analysis of importin-α-β mediated nuclear protein import. *J. Cell Biol.* 168, 1027–1038.

Stark GR & Darnell Jr. JE (2012) The JAK-STAT pathway at twenty. *Immunity* 36, 503–514.

Stewart M (2007) Molecular mechanism of the nuclear protein import cycle. *Nat. Rev. Mol. Cell Biol.* 8, 195–208.

Weis K (2003) Regulating access to the genome: Nucleocytoplasmic transport throughout the cell cycle. *Cell* 112, 441–451.

CONTROL OF MEMBRANE LOCALIZATION

Behnia R & Munro S (2005) Organelle identity and the signposts for membrane traffic. *Nature* 438, 597–604.

Calleja V, Alcor D, Laguerre M, et al. (2007) Intramolecular and intermolecular interactions of protein kinase B define its activation in vivo. *PLoS Biol.* 5, e95.

Carlton JG & Cullen PJ (2005) Coincidence detection in phosphoinositide signaling. *Trends Cell Biol.* 15, 540–547.

Cho W (2006) Building signaling complexes at the membrane. *Sci. STKE* 321, pe7.

Dransart E, Olofsson B & Cherfils J (2005) RhoGDIs revisited: Novel roles in Rho regulation. *Traffic* 6, 957–966.

Fivaz M & Meyer T (2003) Specific localization and timing in neuronal signal transduction mediated by protein-lipid interactions. *Neuron* 40, 319–330.

Hantschel O, Nagar B, Guettler S, et al. (2003) A myristoyl/phosphotyrosine switch regulates c-Abl. *Cell* 112, 845–857.

Hentschel A, Zahedi RP & Ahrends R (2016) Protein lipid modifications—More than just a greasy ballast. *Proteomics* 16(5), 759–782. doi: 10.1002/pmic.201500353.

Manning BD & Toker A (2017) AKT/PKB signaling: Navigating the network. *Cell* 169(3), 381–405. doi: 10.1016/j.cell.2017.04.001.

Pylypenko O, Rak A, Durek T, et al. (2006) Structure of doubly prenylated Ypt1: GDI complex and the mechanism of GDI-mediated Rab recycling. *EMBO J.* 25, 13–23.

Strickfaden SC, Winters MJ, Ben-Ari G, et al. (2007) A mechanism for cell-cycle regulation of MAP kinase signaling in a yeast differentiation pathway. *Cell* 128, 519–531.

MODULATION OF SIGNALING BY MEMBRANE TRAFFICKING

Di Guglielmo GM, Le Roy C, Goodfellow AF & Wrana JL (2003) Distinct endocytic pathways regulate TGF-β receptor signalling and turnover. *Nat. Cell Biol.* 5, 410–421.

Eisenberg S & Henis YI (2008) Interactions of Ras proteins with the plasma membrane and their roles in signaling. *Cell. Signal.* 20, 31–39.

Haugh JM (2002) Localization of receptor-mediated signal transduction pathways: The inside story. *Mol. Interv.* 2, 292–307.

Le Roy C & Wrana JL (2005) Clathrin- and non-clathrin-mediated endocytic regulation of cell signalling. *Nat. Rev. Mol. Cell Biol.* 6, 112–126.

Onken B, Wiener H, Philips MR & Chang EC (2006) Compartmentalized signaling of Ras in fission yeast. *Proc. Natl Acad. Sci. USA.* 103, 9045–9050.

Rocks O, Peyker A & Bastiaens PI (2006) Spatio-temporal segregation of Ras signals: One ship, three anchors, many harbors. *Curr. Opin. Cell Biol.* 18, 351–357.

Zweifel LS, Kuruvilla R & Ginty DD (2005) Functions and mechanisms of retrograde neurotrophin signalling. *Nat. Rev. Neurosci.* 6, 615–625.

BIOMOLECULAR CONDENSATES

Brangwynne CP, Eckmann CR, Courson DS, et al. (2009) Germline P granules are liquid droplets that localize by controlled dissolution/condensation. *Science* 324(5935), 1729–1732. doi: 10.1126/science.1172046.

Case LB, Ditlev JA & Rosen MK (2019) Regulation of transmembrane signaling by phase separation. *Annu. Rev. Biophys.* 48, 465–494. doi: 10.1146/annurev-biophys-052118-115534.

Huang WY, Yan Q, Lin WC, et al. (2016) Phosphotyrosine-mediated LAT assembly on membranes drives kinetic bifurcation in recruitment dynamics of the Ras activator SOS. *Proc. Natl. Acad. Sci. USA* 113(29), 8218–8223. doi: 10.1073/pnas.1602602113.

Jaqaman K & Ditlev JA (2021) Biomolecular condensates in membrane receptor signaling. *Curr. Opin. Cell Biol.* 69, 48–54. doi: 10.1016/j.ceb.2020.12.006.

Li P, Banjade S, Cheng HC, et al. (2012) Phase transitions in the assembly of multivalent signalling proteins. *Nature* 483(7389), 336–340. doi: 10.1038/nature10879.

Lyon AS, Peeples WB & Rosen MK (2021) A framework for understanding the functions of biomolecular condensates across scales. *Nat. Rev. Mol. Cell Biol.* 22(3), 215–235. doi: 10.1038/s41580-020-00303-z.

Shin Y & Brangwynne CP (2017) Liquid phase condensation in cell physiology and disease. *Science* 357(6357), eaaf4382. doi: 10.1126/science.aaf4382.

Su X, Ditlev JA, Hui E, et al. (2016) Phase separation of signaling molecules promotes T cell receptor signal transduction. *Science* 352(6285), 595–599. doi: 10.1126/science.aad9964.

van der Lee R, Buljan M, Lang B, et al. (2014) Classification of intrinsically disordered regions and proteins. *Chem. Rev.* 114(13), 6589–6631. doi: 10.1021/cr400525m.

Second Messengers: Small Signaling Mediators

6

Much of the information transmitted within cells is carried by large macromolecules, such as proteins or nucleic acids. As we have seen in previous chapters, incoming signals can alter proteins in a number of ways—they can induce conformational changes, changes in complex formation, or post-translational modifications. However, information can also be carried by much smaller and simpler molecules; these include Ca^{2+}, various lipid-derived mediators, and the cyclic nucleotides cAMP and cGMP. In this chapter, we discuss these **small signaling mediators** and their special properties.

PROPERTIES OF SMALL SIGNALING MEDIATORS

When signaling inputs cause a change in the concentration of a small signaling mediator, that change is detected by downstream effector proteins that bind the mediator. Mediator binding leads to conformational changes in the effectors and ultimately to changes in their activity. Thus, information is conveyed by the concentration and distribution of these mediators, and how they change over time. These small signaling mediators are often referred to as **second messengers**, a term that derives largely from their historical discovery as signals that were produced downstream of hormone stimulation, where the hormone itself was considered the first messenger.

Signaling by small-molecule mediators differs in several respects from more typical signaling mechanisms based only on changes in proteins. Changes in the concentration of the mediators can be quite rapid and can result in an enormous amplification of an input signal. And since most small signaling mediators are highly diffusible, their effects can spread rapidly throughout the cell.

DOI: 10.1201/9780429298844-6

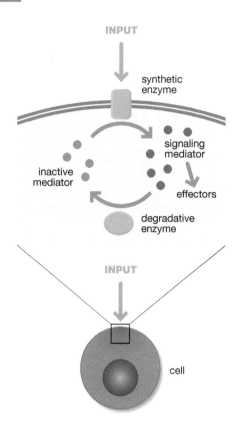

Figure 6.1

Small signaling mediators are controlled by a balance of synthesis and degradation. The concentration and distribution of signaling mediators are coordinately controlled by opposing activities – the enzyme that synthesizes the mediator, and the enzyme that degrades or removes it. In the case of Ca^{2+}, levels are controlled in a similar fashion by the opposing activities of channels and pumps. Small signaling mediators, because they are highly diffusible, can act at greater distances than macromolecular signaling molecules.

Small signaling mediators are controlled by an interplay of their production and elimination

For a small signaling mediator to be an effective information-carrying molecule, its concentration must differ significantly under unstimulated and stimulated conditions. The steady-state concentration of these mediators is determined by a balance between the production of the mediator and its elimination. Usually, stimulation of the system by an input leads to an increase in the rate of synthesis of the mediator (for example, the allosteric activation of the enzyme that synthesizes the mediator), but stimulation can also involve a decrease in the breakdown or elimination of the mediator (**Figure 6.1**). This situation is analogous to the opposing activities of "writers" and "erasers" in the case of post-translational modifications of proteins. To provide a specific example, the mediator cAMP is produced from ATP by the enzyme adenylyl cyclase, and is eliminated by the enzyme cAMP phosphodiesterase. Many signaling pathways involve stimulation of adenylyl cyclase by heterotrimeric G proteins, leading to a large, transient increase in cAMP concentration in the cell.

Unlike other mediators, the concentration of Ca^{2+} is not regulated by its synthesis and degradation. Instead, the concentration of Ca^{2+} in the cytoplasm is regulated by the opposing activities of pumps that remove it, and channels that allow it to flow back in. In the basal state, Ca^{2+} concentrations in the cytoplasm are kept low by Ca^{2+} pumps. These use the energy of ATP to actively transport Ca^{2+} against its concentration gradient either out of the cell or into the endoplasmic reticulum, which serves as an intracellular reservoir of Ca^{2+}. Upon signaling, Ca^{2+} channels open and the ions rapidly enter the cell, flowing down the concentration gradient, leading to a rapid increase in Ca^{2+} concentration in the cytoplasm. Upon cessation of the signal, channels close and pumps once again restore the basal concentration gradient. Thus, the opposing actions of channels and pumps play a role equivalent to the synthesis and degradation of other mediators by enzymes.

In all of these cases, the proteins that increase and decrease mediator concentrations function in a coordinated fashion in many different signaling contexts. A variety of different upstream inputs can be linked to production of a particular mediator. Similarly, as discussed below, a single mediator can activate a wide range of downstream effectors.

Small signaling mediators exert their effects by binding downstream effectors

Small signaling mediators such as cAMP or Ca^{2+} exert their biological effects by binding to a wide range of downstream enzymes and channels that, in response, are either allosterically activated or repressed. In some cases, the mediator exerts its effects indirectly by binding to regulatory subunits, as in the case of protein kinase A (cAMP binds to an inhibitory subunit that releases the active kinase subunit). Lipid mediators can also bind and regulate effector proteins by recruiting them to the membrane. Often, proteins that respond to the same mediator will contain evolutionarily related modular domains or motifs that transduce signals from that mediator, such as cAMP- or Ca^{2+}-binding modules. These binding modules are functionally analogous to "reader" domains that recognize post-translational modifications, as discussed in Chapter 4.

The fact that the small signaling mediators exert their effects solely through binding to downstream effector proteins distinguishes them from other ions and small molecules that participate in signaling. For example, in excitable cells such as neurons, the regulated opening and closing of channels for Na^+, K^+, and other ions lead to very rapid changes in membrane potential that can be amplified and propagated. These changes in voltage across the membrane provide the basis of neuronal signaling, but are not considered further here. Note that in this case, the ions primarily exert their influence by carrying electric charge, not by binding and allosterically changing the activity of effector proteins.

For effector proteins to properly read and respond to the mediator signal, it is critical that their dissociation constants for binding (and activation/repression) be properly tuned so that binding is low at the basal steady-state concentration of mediator, but high under activated steady-state concentrations. If, for example, the dissociation constant of an effector was lower than the basal steady-state concentration of a mediator (affinity was too high), then most of the effector would already be bound to mediator before stimulation; a further increase in mediator concentration upon signaling would have little or no impact.

The tuning of affinity to the physiological concentration range of ligands is discussed in Chapter 2.

Small signaling mediators can lead to fast, distant, and amplified signal transmission

The rate of diffusion in solution is inversely proportional to the radius of the diffusing molecule. Because signaling mediators are relatively small, they can, in principle, rapidly diffuse throughout a cell to transmit information. Signaling through such mediators can occur on the millisecond time scale and can rapidly cover distances of up to an entire cell. **Figure 6.2** shows the relative sizes of a macromolecular signaling protein (a protein kinase catalytic domain; radius ~50Å) and the small signaling mediators cAMP (radius ~5Å) and Ca^{2+} (radius ~1Å). Also shown are the relative concentration profiles for each molecule at a given time after diffusion from an equal concentration point source. Molecules in the smaller size range have the potential to spread farther and faster than macromolecules. Some mediators, such as the gas nitric oxide, can even diffuse through membranes and mediate communication from cell to cell. The actual distance that small-mediator signals propagate, however, is also determined by the lifetime of the molecule. For example, nitric oxide is very short-lived, and therefore it only propagates within a few cells. Similarly, in the case of Ca^{2+}, the high ambient concentration of Ca^{2+}-binding proteins results in the rapid sequestration of Ca^{2+} in the cytosol. But as discussed in the next section, this interplay between production, diffusion, and destruction/sequestration can lead to signaling outputs with complex spatial and temporal behaviors.

Small signaling mediators also have the potential to greatly amplify signals. In the case of mediators that are generated by enzymes, the activation of only a small number of synthetic enzymes can result in the production of an enormous number of mediator molecules. Similarly, even the momentary opening of a small number of membrane channels will allow the passage of many Ca^{2+}

Figure 6.2

Relative sizes and diffusion of protein and small-molecule signaling mediators. (a) Relative sizes of a typical signaling protein [the kinase domain of protein kinase A (PKA)] and two small signaling mediators, cAMP and Ca^{2+}. (b) Theoretical distribution of molecules from a point source after a fixed time of diffusion. Diffusion rates are calculated using the radii given in part (a) using the Stokes-Einstein equation, $D = kT/(6\pi\eta r)$, where k is Boltzmann's constant, T is the absolute temperature, η is solvent viscosity, and r is the radius. As can be seen, small-molecule mediators can diffuse further and faster than macromolecular ones.

(a)

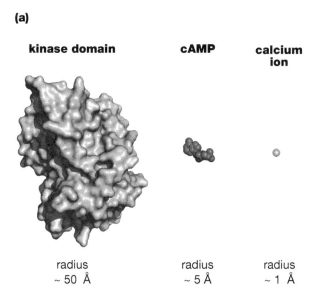

kinase domain	cAMP	calcium ion
radius ~ 50 Å	radius ~ 5Å	radius ~ 1Å

(b)

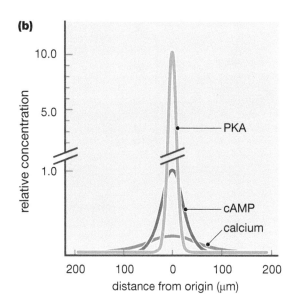

ions. In either case, if the initial concentration of the mediator is low, it will rapidly and massively increase. It is the potential for enormous amplification combined with the potential for rapid spread of the resulting signal throughout the cell that sets signaling by small-molecule mediators apart from other signaling mechanisms.

Small signaling mediators can generate complex temporal and spatial patterns

Because the levels of small signaling mediators are controlled by opposing production and elimination activities, coordinated regulation of both can yield complex and diverse signaling dynamics. **Figure 6.3a** compares signaling mediator levels to the fill level of a vessel with water flowing in and out. The steady-state level of water can be changed either by tuning the IN flow rate or the OUT flow rate (or both). Similarly, change in the levels of second messengers can be coordinately controlled both by their production and their destruction.

Figure 6.3b shows a schematic time profile of change in mediator concentration in response to a transient stimulus that increases the synthesis rate. Efficient restoration of basal mediator levels, after the input stimulus ends, is dependent on a sufficiently rapid basal rate of mediator degradation. Transient increases in mediator level could also be induced by distinct complementary mechanisms: an input stimulus could decrease the degradation rate (**Figure 6.3c**), or coordinately increase mediator synthesis rate and decrease degradation rate.

Small signaling mediators can also be used to generate more complex spatiotemporal patterns. For example, highly local activation of synthetic enzymes, but widespread distribution of degradation enzymes (or sequestrating binding proteins, in the case of Ca²⁺), can lead to sharp gradients of mediator. Temporal patterns, such as rapid adaptation or oscillation, can also be produced by an interplay between the synthesis and elimination enzymes. An example is provided later

Figure 6.3

The rise and fall of small signaling mediator concentration in response to the beginning and end of an upstream stimulus. (a) Control over the steady-state level of a small-molecule signaling mediator is analogous to control over the steady-state level of water flowing into and out of a bucket. An increase in steady-state level can be achieved by either increasing the flow rate of water into the bucket or by decreasing the flow rate out of the bucket (or by a combination of both mechanisms). (b) A transient rise in mediator concentration can be induced by an increase in synthesis. Restoration of mediator concentration to a low, basal level after removal of the stimulus is dependent on a high basal degradation rate. (c) A similar increase in mediator concentration can also be induced by a transient decrease in degradation rate.

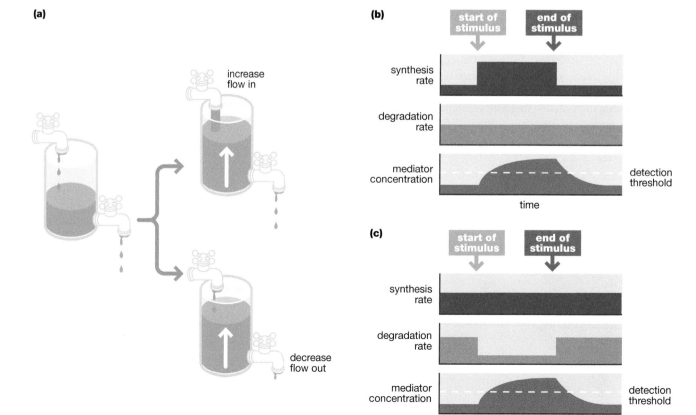

in this chapter of highly complex wave patterns that can be generated in the case of Ca^{2+} signaling.

CLASSES OF SMALL SIGNALING MEDIATORS

The various classes of small signaling mediators are listed in **Table 6.1**, along with the enzymes or proteins involved in their production and elimination, and regulatory inputs and targets. In this section, we will focus in detail on the mechanism of action and regulation of a few prominent mediators: cyclic nucleotides and the lipid-derived signaling mediators inositol trisphosphate (IP_3) and diacylglycerol (DAG). The following section will focus on the special properties of signaling mediated by Ca^{2+}.

Small signaling mediators have a wide range of physical properties

While small signaling mediators share the defining property of rapid and potentially widespread effects, their actual physical properties are quite variable. Typical mediators such as cyclic nucleotides and Ca^{2+} are strictly water-soluble, and thus cannot pass through the lipid bilayer. By contrast, some lipid-derived mediators are quite hydrophobic and therefore remain confined to membranes. These can only diffuse laterally in two dimensions within the plane of the membrane. Some other lipid-derived mediators are somewhat water-soluble, and they can thus partition both to the membrane and to the cytosol or extracellular space, allowing them to transit from one cell to another.

Table 6.1

Classes of commonly used small-molecule signaling mediatorsa

Small signaling mediator	Synthesis/ production enzyme	Degradation/ elimination enzyme	Location (potential range)	Inputs	Output targets
Cyclic nucleotides					
cAMP	Adenylyl cyclase	cAMP phosphodies-terase	Intracellular (long range)	GPCRs	Kinases, channels, GEFs
cGMP	Guanylyl cyclase	cGMP phosphodies-terase	Intracellular (long range)	Nitric oxide, rhodopsin	Kinases, channels
Lipid-derived mediators					
Inositol trisphos-phate (IPs)	Phospholipase C (PLC)	Dephosphorylation	Intracellular (long range)	RTKs, GPCRs	Ca^{2+} channels
Diacylglycerol (DAG)	Phospholipase C (PLC)	Phosphorylation	Membrane restricted	RTKs, GPCRs	Kinases
Phosphoinositides (e.g., PIP_3)	PI kinases	PI phosphatases	Membrane restricted	RTKs, GPCRs	Kinases, many PH domains
Eicosanoids (e.g., prostaglandins)	Cyclooxygenase, pros-taglandin synthases	Dehydrogenase, reductase	Extracellular	PLA_2	GPCRs
Ceramide	sphingomyelinase	Ceramidase	Membrane restricted	TNF and IL-2 receptors	Kinases/phos-phatases
Sphingosine phos-phate	Ceramidase/sphingo-sine kinase	Sphingosine phos-phatase	Extracellular	TNF and IL-2 receptors	GPCRs
Other					
Ca^{2+}	Ca^{2+} channels (release from ER stores)	Ca^{2+}-ATPase pumps, Ca^{2+}-sequestering proteins	Intracellular (0.1–1 μm)	Channels, IP_3	Channels, kinases, enzymes, cytoskeletal proteins (calmodulin)
Nitric oxide (NO)	Nitric oxide synthase (NOS)	Reaction with O_2, heme binding	Extracellular	Ca^{2+}/calmodulin, phosphorylation	cGMP (guanylyl cyclase)

GPCRs, G-protein-coupled receptors; GEFs, guanine nucleotide exchange factors; RTKs, receptor tyrosine kinases; PIP_3, phosphatidylinositol 3,4,5-trisphosphate; PI, phosphatidylinositol; PLA_2, phospholipase A_2; TNF, tumor necrosis factor; IL-2, interleukin-2; ER, endoplasmic reticulum.[a]This table shows selected examples from the classes, and is not an exhaustive list.

NO signaling will be discussed in more detail in Chapter 8.

The small free-radical gas molecule nitric oxide (NO) is an interesting special case that easily traverses cell membranes by virtue of its small size and uncharged nature. NO is produced by the enzyme nitric oxide synthase (NOS), which converts L-arginine into citrulline and NO. NOS is activated by Ca^{2+}/calmodulin. Once produced, NO is able to diffuse across the cell membrane to adjacent cells. Its effects are highly transient and local because NO is intrinsically very unstable and reacts spontaneously with oxygen or heme. The major target of NO is the enzyme guanylyl cyclase, which produces cGMP (itself another signaling mediator). NO plays an important role in smooth muscle relaxation and vessel dilation, thereby regulating blood flow and pressure.

The cyclic nucleotides cAMP and cGMP are produced by cyclase enzymes and destroyed by phosphodiesterases

The **cyclic nucleotides cAMP** and **cGMP** are used as signaling mediators by a wide variety of organisms ranging from bacteria to vertebrates. Both cAMP and cGMP are synthesized—from the high-energy precursors ATP and GTP, respectively—by cyclase enzymes (**Figure 6.4**). Vertebrate adenylyl cyclases are large transmembrane proteins that contain evolutionarily conserved cytoplasmic cyclase catalytic domains. These enzymes are allosterically regulated by heterotrimeric G proteins (**Figure 6.5**). Certain Gα subunits activate adenylyl cyclase, whereas others can inhibit it. Vertebrates have two classes of guanylyl cyclases, including a soluble form that is activated by the upstream signaling mediator nitric oxide (NO), and transmembrane (receptor) forms that can be regulated by various upstream ligands.

cAMP and cGMP are degraded by phosphodiesterase enzymes (PDEs), which convert the active molecules to the inactive forms 5'-AMP and 5'-GMP, respectively. Although most signaling pathways raise the steady-state level of cyclic nucleotides by activating their production (via cyclases), the phosphodiesterases can also be regulated. Ca^{2+}/calmodulin can activate cAMP phosphodiesterase, while $Gα_i$ can activate cGMP phosphodiesterase, leading in both cases to a decrease in cyclic nucleotide levels. Phosphodiesterases can also be an important drug target (**Figure 6.6**). The erectile dysfunction drug sildenafil (Viagra®) is an inhibitor of cGMP phosphodiesterase. In vascular smooth muscle cells, blocking

Figure 6.4

Synthesis and degradation of cyclic nucleotides. The structures of the signaling mediators cAMP and cGMP are shown, along with their precursors and inactive products.

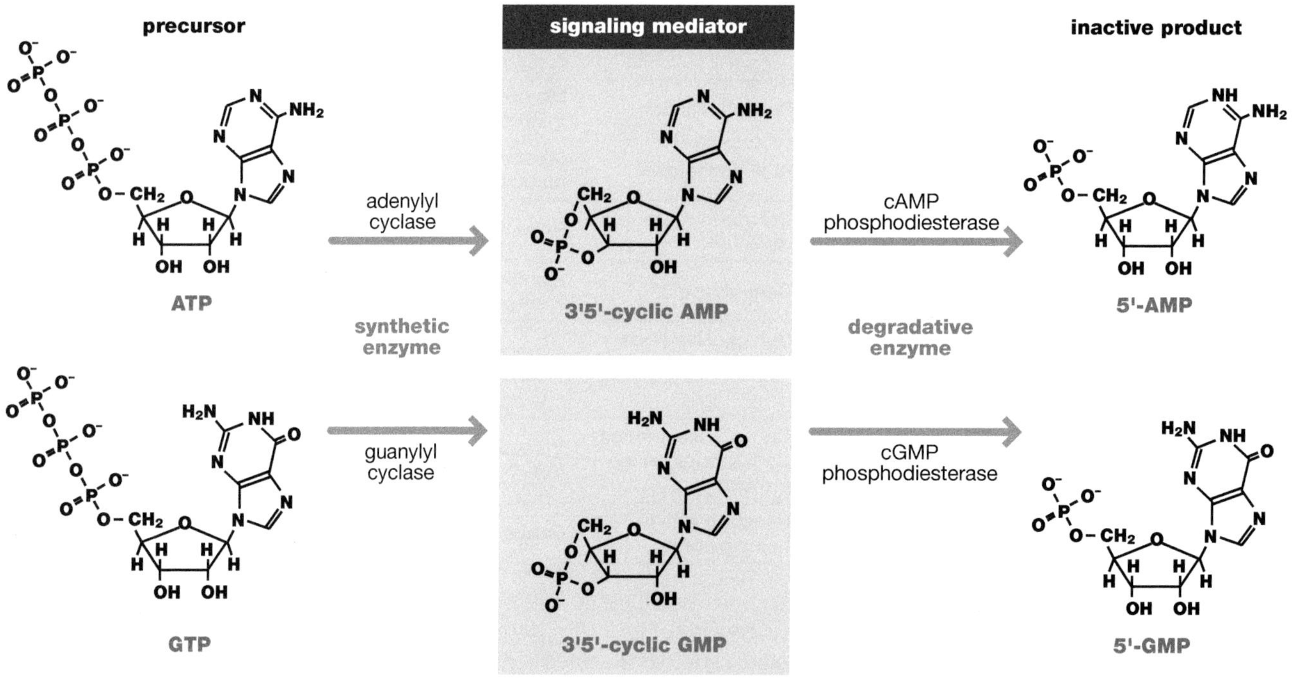

precursor	signaling mediator	inactive product

ATP → (adenylyl cyclase / synthetic enzyme) → 3'5'-cyclic AMP → (cAMP phosphodiesterase / degradative enzyme) → 5'-AMP

GTP → (guanylyl cyclase) → 3'5'-cyclic GMP → (cGMP phosphodiesterase) → 5'-GMP

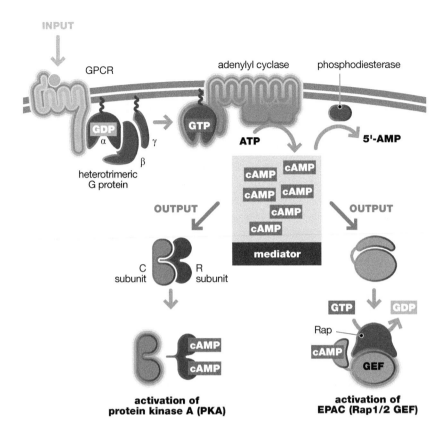

Figure 6.5

Signaling via the small signaling mediator cAMP. The enzyme that synthesizes cAMP, adenylyl cyclase, is activated by heterotrimeric G proteins. cAMP is degraded by the enzyme phosphodiesterase. Most of the effects of cAMP are mediated through regulation of protein kinase A (PKA), and EPAC, a guanine nucleotide exchange factor (GEF) for the small G proteins Rap1 and Rap2. Binding of cAMP to the regulatory (R) subunit of PKA causes dissociation of the catalytic (C) subunit. Binding of cAMP to EPAC relieves autoinhibition of the GEF activity of this protein. GPCR, G-protein-coupled receptor.

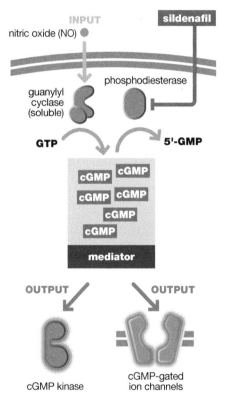

the destruction of cGMP (combined with low basal cyclase activity) leads to a buildup of high steady-state levels of cGMP. This causes smooth muscle relaxation and blood vessel dilation, resulting in increased blood flow to the penis.

Cyclic nucleotides regulate diverse cellular activities

The primary targets of cAMP in vertebrates are *protein kinase A* (*PKA*) and EPAC, a regulator of small G proteins (see Figure 6.5). Both share related cAMP-binding modules, also known as cyclic nucleotide-binding domains, which are found in species ranging from bacteria to humans. The details of activation of PKA by cAMP are discussed in the next section. EPAC is a guanine nucleotide exchange factor (GEF) that activates Rap1 and Rap2, two small G proteins that play a key role in regulating cellular adhesion interactions with the extracellular matrix. EPAC has a catalytic GEF domain that is regulated by several other domains, including cyclic nucleotide-binding domains that are related to those found in the PKA regulatory subunit (EPAC1 has one such domain, EPAC2 has two regulatory domains). These domains are involved in autoinhibition of the catalytic GEF domain, which is relieved by cAMP binding.

cGMP has several downstream targets, including the family of cGMP-dependent protein kinases (cGK, also known as protein kinase G or PKG) and cGMP-regulated ion channels (see Figure 6.6). In addition, this mediator appears to regulate some cGMP phosphodiesterases, the enzymes that break down cGMP. This represents a form of feedback regulation that may be important for controlling signaling dynamics. In general, the effectors of cGMP are involved in inducing vascular smooth muscle relaxation and increased blood flow, and they also play a key role in vertebrate vision.

Figure 6.6

Signaling via the small signaling mediator cGMP. One of the enzymes that synthesizes cGMP, soluble guanylyl cyclase, is activated by nitric oxide (NO). cGMP is degraded by the enzyme phosphodiesterase. cGMP regulates various effectors, including kinases and ion channels, stimulating vascular smooth muscle relaxation. The drug sildenafil (Viagra®) functions by inhibiting the cGMP phosphodiesterase, leading to a buildup of cGMP.

 The visual signal transduction system is described in Chapter 14.

The regulatory (R) subunit of protein kinase A is a conformational sensor of cAMP binding

Protein kinase A (**PKA**), or cyclic AMP-dependent protein kinase, is one of the best-understood examples of an enzyme whose activity is directly regulated by a small signaling mediator. As its name suggests, the ability of PKA to phosphorylate substrates is entirely dependent on the presence of cAMP. Thus, the role of PKA is to convert changes in intracellular cAMP concentration into changes in the phosphorylation of serine and threonine residues in substrate proteins. PKA has two subunits: a catalytic (C) subunit and a regulatory (R) subunit that suppresses its activity when bound (**Figure 6.7a**). The kinase activity of the catalytic subunit is tightly controlled by its association with the regulatory subunit, which acts as a conformational switch controlled by cAMP binding. The ability of the regulatory unit to inhibit catalytic activity is largely due to its pseudosubstrate-like inhibitory segment, which blocks the active site when it is tethered to the catalytic subunit (**Figure 6.7b**). The R subunit normally exists as a dimer, so the actual inactive complex is an R2C2 heterotetramer.

In the enzyme's inactive state (when levels of cAMP are low, typically below 10^{-8} M), the catalytic and regulatory subunits are tightly associated—the K_d is less than 10^{-9} M—such that essentially every catalytic subunit is complexed with a regulatory subunit. In this complex, the regulatory subunit sterically blocks substrates from binding to the enzyme active site of the catalytic subunit. Thus, when cAMP levels are low, the vast majority of PKA is in an inactive state, sequestered through stable interaction between the two subunits.

Cooperative binding is explained in Chapter 2.

Each regulatory subunit has two distinct binding sites for cAMP, which exhibit positive cooperativity—binding of cAMP to one site increases the binding affinity of the other site, resulting in a sigmoidal dependence of activity on the concentration of cAMP. Thus, as cAMP concentrations rise and approach the K_d for binding to the R subunit (10^{-8} to 10^{-7} M), relatively small differences in cAMP levels lead to large differences in binding. Once two molecules of cAMP are bound to each regulatory subunit, a conformational change alters its binding interface with the

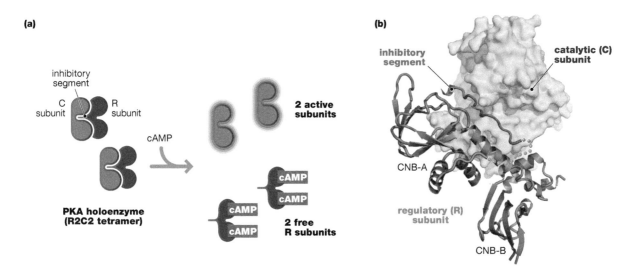

(a)

inhibitory
segment

C
subunit   R
subunit

cAMP

**PKA holoenzyme
(R2C2 tetramer)**

2 active
subunits

cAMP
cAMP
cAMP
cAMP
2 free
R subunits

(b)

inhibitory
segment

catalytic (C)
subunit

CNB-A

regulatory (R)
subunit

CNB-B

Figure 6.7

Regulation of PKA by cAMP binding. (a) In the absence of cAMP, PKA exists as a complex in which the catalytic (C) subunit is held in an inactive state by association with the regulatory (R) subunit. An inhibitory segment from the R subunit blocks the active site of the C subunit. When cAMP levels increase, the R subunit binds cAMP and dissociates from the C subunit, which is now active and can phosphorylate substrates. The inactive holoenzyme exists as a heterotetrameric complex containing two R and two C subunits. (b) Structure of a single regulatory and catalytic PKA heterodimer. CNB-A and CNB-B, cyclic nucleotide-binding domains A and B. (b, Adapted from P. Zhang, et al., *Science* 335:712–716, 2012. With permission from AAAS.)

catalytic subunit, decreasing its affinity and leading to dissociation of the catalytic subunit into its active form. The dissociated catalytic subunit can now phosphorylate substrates.

Although the regulation of PKA appears straightforward, in reality the on-off switch is part of a regulatory apparatus that is considerably more complex and subtle. For example, there are actually several distinct R subunits with slightly different biological properties. Furthermore, each of the R subunits interacts with specific scaffold proteins, termed *A-kinase anchoring proteins* (*AKAPs*), that tether the inactive complexes to specific subcellular localizations and are associated with distinct potential substrates (discussed below). Thus, the active catalytic subunit is only released in specific places in the cell where it is in close proximity to relevant substrates. We can see that even this relatively simple example actually involves a variety of coupled changes in protein conformation, protein–protein interactions, and subcellular localization, all induced by binding of a small signaling mediator.

Some small signaling mediators are derived from membrane lipids

Another common class of mediators is derived from the lipids that comprise cell membranes. The membrane lipid **phosphatidylinositol** and its derivatives give rise to three classes of intermediates. Cleavage of phosphatidylinositol 4,5-bisphosphate (PIP_2) by the enzyme phospholipase C yields a water-soluble head group, **inositol 1,4,5-trisphosphate** (**IP_3**), and **diacylglycerol** (**DAG**), which is highly hydrophobic and remains embedded in the membrane (**Figure 6.8**). Both of these products are active signaling mediators. Alternatively, modification of phosphatidylinositol by lipid kinases and phosphatases can give rise to a range of phosphoinositide species in which the head group is differentially phosphorylated. An important example is the mediator phosphatidylinositol 3,4,5-trisphosphate (PIP_3). These species can activate various effector enzymes, or can be recognized by lipid-recognition domains that result in conditional membrane recruitment.

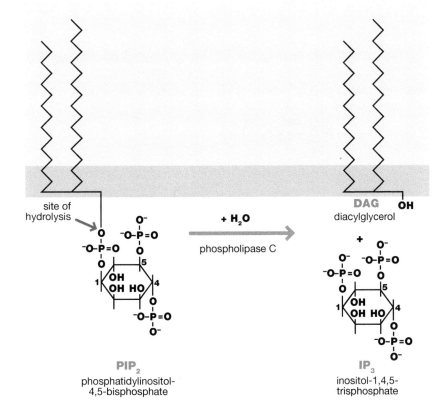

Figure 6.8

Hydrolysis of PIP_2 by phospholipase C yields the signaling mediators IP_3 and DAG. IP_3 is soluble and diffuses rapidly through the cytoplasm, whereas DAG remains in the membrane.

Eicosanoids, such as the prostaglandins, are critical inflammatory mediators that are synthesized from fatty acids such as arachidonic acid, which are derived from membrane lipids. These mediators, unlike most others, are sufficiently soluble in both lipids and water that they can leave the cell in which they are produced. They exert their biological effects by activating G-protein-coupled receptors (GPCRs) on neighboring cells, leading to a range of effects on inflammation, blood clotting, and vascular tone and premeability.

The third common class of lipid-derived mediators is those produced from *sphingomyelin*. Receptor-mediated activation of the lipase sphingomyelinase cleaves off the phosphocholine head group from sphingomyelin, producing the membrane-restricted signaling mediator ceramide, which can activate specific kinases and phosphatases. Alternative enzymatic cleavage and phosphorylation of sphingomyelin can produce the soluble mediators sphingosine and sphingosine 1-phosphate, which can also leave the cell of production.

 The roles of inositol lipids, eicosanoids, and sphingomyelin in signaling are discussed in greater detail in Chapter 7.

PLC generates two signaling mediators, IP$_3$ and DAG

As mentioned above, IP$_3$ and DAG are products of the hydrolysis of PIP$_2$ by phospholipase C (PLC). The β isoform of this enzyme (PLC-β) is activated by heterotrimeric G proteins, whereas the γ isoform (PLC-γ) is activated by tyrosine kinase signaling pathways. After cleavage, IP$_3$ can be inactivated by enzymatic dephosphorylation, while DAG can be further modified by phosphorylation to generate *phosphatidic acid*, which itself can mediate signals or can be further enzymatically processed into other bioactive lipids.

The major effect of IP$_3$ is to regulate intracellular Ca^{2+} levels. IP$_3$ can diffuse through the cytoplasm to the endoplasmic reticulum, where it binds and activates the endoplasmic reticulum–associated IP$_3$ receptor (also termed the IP$_3$-gated Ca^{2+} channel). Activation opens these channels, leading to the release of Ca^{2+} stores into the cytoplasm. As discussed below, this causes a rapid and sharp increase in intracellular calcium concentration that can be exploited for signaling.

Activation of protein kinase C is regulated by IP$_3$ and DAG

The **protein kinase C (PKC)** family of serine/threonine kinases illustrates how enzyme activity and membrane association can be tightly regulated in time and space by small signaling mediators. The PKC family consists of at least ten

Figure 6.9

Activation of protein kinase C (PKC) by IP$_3$ and DAG. Activation of heterotrimeric G proteins leads to activation of phospholipase C (PLC), which cleaves phosphatidylinositol 4,5-bisphosphate (PIP$_2$) to form diacylglycerol (DAG) and inositol trisphosphate (IP$_3$). IP$_3$ diffuses through the cytoplasm to activate IP$_3$-gated Ca^{2+} channels in the endoplasmic reticulum, releasing stored Ca^{2+} (*green circles*) into the cytoplasm. Ca^{2+} and DAG function cooperatively to activate PKC, which phosphorylates downstream targets. PLC can also be activated by tyrosine kinases. GPCR, G-protein-coupled receptor.

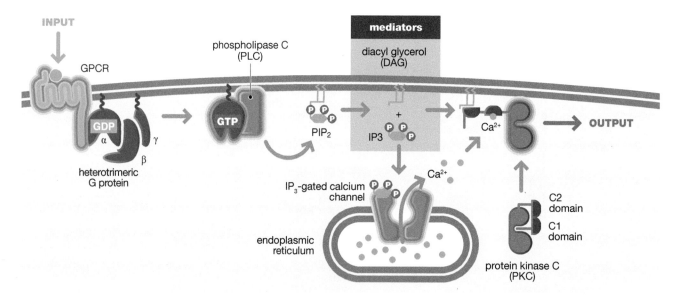

members in mammals, all of which contain a C-terminal kinase domain, which is normally inhibited by an N-terminal segment that binds to the catalytic site and blocks substrate binding. In addition, the different family members contain various regulatory domains, termed C1 and C2 domains, that allow them to respond to changes in the levels of intracellular Ca^{2+} and/or DAG. For typical PKC isoforms, the simultaneous binding of Ca^{2+} and DAG induces a conformational change that relieves inhibition and activates the catalytic domain.

In unstimulated cells, conventional PKC resides in the cytosol in its autoinhibited conformation. Upon activation of PLC, however, the resulting increased membrane density of DAG leads to recruitment of PKC to the membrane via its tandem C1 domains. Binding of Ca^{2+} to the C2 domain is also required for high-affinity binding to the membrane. Conformational changes induced by binding of DAG and Ca^{2+} are then sufficient to catalytically activate the enzyme (**Figure 6.9**). Thus, membrane recruitment of PKC is intimately linked to its catalytic activity. **Phorbol esters** are organic compounds that mimic the structure of DAG and thereby promote the activation of PKC *in vivo*. They were first isolated as the active ingredient in croton oil, which promotes tumorigenesis when applied to the skin of experimental animals. Phorbol esters are used experimentally to manipulate the activity of PKC in cells.

Since binding to neither DAG nor Ca^{2+} is sufficient to recruit and fully activate conventional PKC, this mechanism serves to integrate signals from DAG and Ca^{2+}, the local concentrations of which are likely to have very different spatial and temporal patterns. Other PKC isoforms differ in their dependence on small signaling mediators. So-called "novel" isoforms are dependent on DAG only, while "atypical" isoforms require neither calcium nor DAG, and are regulated by other mechanisms. Thus, different PKC isoforms represent a diverse palette of related kinases that can be activated in distinct kinetic and subcellular patterns.

CALCIUM SIGNALING

Calcium ions are among the most important intracellular signaling mediators. Calcium signaling depends on membrane Ca^{2+} pumps that generate and maintain a gradient of Ca^{2+} across cell membranes. Normally, the cytosol has very low concentrations of calcium ($\sim 10^{-7}$ M) compared to the concentration outside the cell, or in intracellular stores such as the endoplasmic reticulum, where it is approximately 10^{-3} M. Because of this 10,000-fold concentration gradient, the influx of Ca^{2+} ions from the environment or release from intracellular stores causes a very rapid and dramatic increase in cytoplasmic calcium concentration, which is used in various ways for signal transduction.

Activation of Ca^{2+} channels is a common means of regulation

Gated ion channels and their regulation are discussed in more detail in Chapter 8.

In signaling, changes in intracellular Ca^{2+} levels are initiated by channel opening (**Figure 6.10**). In the plasma membrane, there are diverse Ca^{2+} channels that allow Ca^{2+} to pass through the membrane into the cytosol. Most of these are "gated"; this means that they only open to allow passage of ions under certain conditions. Some Ca^{2+} channels are gated by the binding of specific ligands (**ligand-gated channels**), while others can be gated by environmental changes such as temperature, pH, or voltage across the membrane (for example, **voltage-gated channels**). These different channels allow diverse signals to be converted into the common currency of increased Ca^{2+} concentration. Small molecules that can block Ca^{2+} channels are an important class of drugs, and have also been useful experimental tools in studying Ca^{2+}-dependent signaling.

Another physiologically important site of Ca^{2+} storage is the endoplasmic reticulum (or the analogous sarcoplasmic reticulum in muscle cells). The endoplasmic

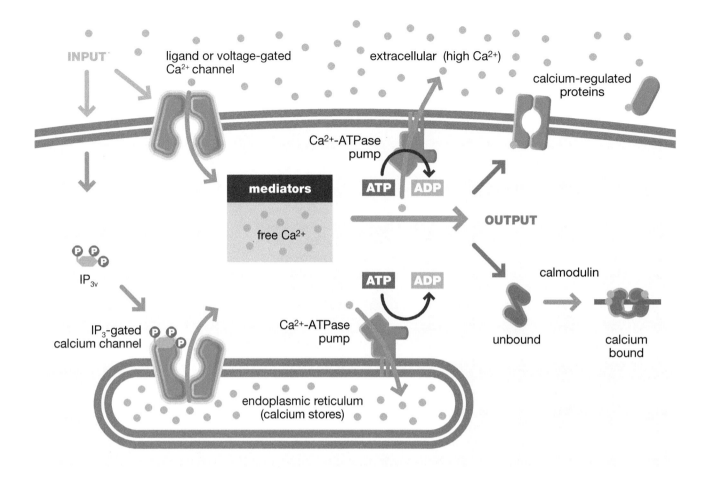

Figure 6.10

Use of Ca²⁺ as an intracellular signaling mediator. Cytosolic Ca^{2+} concentrations are kept low ($\sim10^{-7}$ M) by Ca^{2+}-ATPase pumps which drive Ca^{2+} into the extracellular environment and the endoplasmic reticulum. Cytosolic Ca^{2+} concentrations rise rapidly when Ca^{2+} channels open. These include ligand- or voltage-gated channels in the plasma membrane, or IP_3-gated channels in the membrane of the endoplasmic reticulum. The resulting increase in Ca^{2+} concentration directly activates some effectors, and activates others via the Ca^{2+}-binding protein calmodulin. Normal levels of Ca^{2+} are restored by the Ca^{2+}-ATPase pumps.

reticulum membranes are rich in IP_3-gated Ca^{2+} channels (IP_3 receptors). Local increases in the level of IP_3 (caused by PLC-mediated cleavage of PIP_2—see above) will activate these receptors, leading to Ca^{2+} release from the endoplasmic reticulum into the cytoplasm. The IP_3 receptors are also regulated by Ca^{2+} itself. Intermediate levels of Ca^{2+} can lead to further activation of the channels, leading to positive feedback. High levels of Ca^{2+}, however, inhibit the channel, leading to negative feedback. These unusual channel properties lead to unique spatiotemporal modes of Ca^{2+} signaling, such as Ca^{2+} waves, which are discussed below.

Ca²⁺ influx is rapid and local

Ca^{2+} can diffuse rapidly. However, the actions of intracellular Ca^{2+} tend to be highly transient and local. First, Ca^{2+} levels are rapidly restored by the action of Ca^{2+}-ATPase pumps. Second, because there is an extremely high concentration of Ca^{2+}-binding proteins in the cytoplasm, free Ca^{2+} has a very short half-life before it is bound to protein and thus, at least temporarily, sequestered. These Ca^{2+}-binding proteins act to buffer Ca^{2+} increases. These transient and spatially localized increases in Ca^{2+}, however, can be functionally important.

We know a great deal about Ca^{2+} dynamics in living cells because of biophysical tools that allow visualization of changes in intracellular Ca^{2+} concentrations.

One common approach takes advantage of specialized fluorescent dyes, such as Fura-2, whose spectral properties depend on Ca^{2+} concentration. More recently, genetically encoded *biosensors* for Ca^{2+} concentration have been developed. In one example, two fluorescent proteins were fused to calmodulin, so that Ca^{2+} binding changes the distance between the two fluorescent proteins, leading to changes in their spectral properties (*fluorescence resonance energy transfer*, or *FRET*). The use of genetically encoded biosensors avoids the need to introduce a chemical dye into the cell, facilitating the visualization of Ca^{2+} changes, for example, in individual neurons in the brain.

Intracellular Ca^{2+} exerts its effects by binding to downstream effector proteins. There is a large class of proteins that bind directly to Ca^{2+}. These include kinases (for example, PKC), Ca^{2+}-gated channels, cytoskeletal proteins, and synaptic-vesicle proteins (for example, synaptotagmin). In the synapses of neurons, Ca^{2+} binding to synaptotagmin plays a central role in stimulating the fusion of synaptic vesicles with the plasma membrane and the release of their contents. Many Ca^{2+}-binding proteins have a conserved structural motif, known as the **EF hand**, which has a conserved set of acidic residues positioned to chelate Ca^{2+}. Many downstream effectors, however, do not directly bind Ca^{2+}, but they instead are regulated by the common Ca^{2+} regulatory protein, calmodulin.

Calmodulin is a conformational sensor of intracellular calcium levels

The protein **calmodulin** (**CaM**) is an example of a calcium-binding sensor—upon binding calcium, it undergoes conformational changes that allow it to regulate the activity of a range of downstream effector proteins. Such a mechanism is useful for coupling a single input signal (change in Ca^{2+} concentration) to widespread and rapid changes in many different cellular activities, including enzyme action and the opening and closing of membrane channels. Calmodulin is a compact protein containing four EF hand calcium-binding sites. The affinity of calcium for the EF hand sites in calmodulin ranges from 5×10^{-7} to 5×10^{-6} M, ideally suited for sensing increases in calcium above the resting intracellular level (~10^{-7} M). The conformation of CaM changes dramatically upon calcium binding (**Figure 6.11**). These conformational changes expose a new peptide-binding surface, which allows CaM

Biosensors are described in more detail in Chapter 15.

(a) unbound
(apo-CaM)

(b) calcium bound
(Ca^{2+} CaM)

(c) Ca^{2+} CaM bound to a peptide

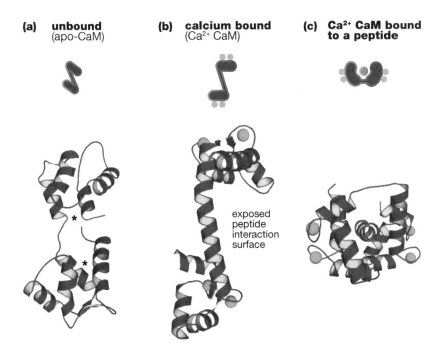

exposed peptide interaction surface

Figure 6.11

Calcium binding triggers changes in calmodulin (CaM) structure and binding. Atomic structure (bottom) and diagrammatic representation (top) of CaM in different states. (a) Apo-CaM (not bound to calcium). Asterisks indicate the positions of occluded peptide-interaction surfaces. (b) The Ca^{2+}-bound state, with four Ca^{2+} ions (*green circles*) bound to each CaM molecule. The Ca^{2+}-bound state has an altered conformation and the peptide-interaction surface is now exposed. (c) Ca^{2+}–CaM bound to a peptide target (*orange*). The altered surface of the Ca^{2+}-bound state allows CaM to bind with increased affinity to specific peptide targets on the surfaces of proteins. Note that the view in panel (c) is rotated relative to that in panels (a) and (b).

to interact favorably with a wide range of binding partners that include protein kinases and phosphatases, adenylyl cyclases, transcriptional regulators, and membrane channels and pumps. **Figure 6.11c** shows how Ca^{2+}-CaM can wrap around a recognized peptide, using the newly exposed binding surface. In the most familiar cases, unliganded CaM (termed apo-CaM) cannot bind the downstream target, but calcium-bound CaM (Ca^{2+}-CaM) binds with high affinity. This mechanism operates to regulate **Ca^{2+}/CaM-dependent protein kinases (CaMKs)**. Binding of Ca^{2+}-CaM to the CaMK induces activating conformational changes in its catalytic domain, allowing the enzyme to phosphorylate substrates on serine and threonine residues. The modes of CaM regulation can be different for different effectors. For example, in some cases, binding of CaM to the effector leads to its inhibition, not activation. Moreover, there are a few targets that bind constitutively to apo-CaM but are then released by calcium binding.

Signaling can lead to propagating Ca^{2+} waves

As described above, IP_3 receptors show an unusual Ca^{2+}-dependent function—they are activated by intermediate levels of intracellular Ca^{2+} but inhibited by higher levels of Ca^{2+}. As a result, upon local stimulation by a pulse of IP_3, the channels in one region of the cell will open and lead to a local influx of Ca^{2+}. Ca^{2+} will diffuse a short distance and will activate nearby IP_3 receptors, or the related ryanodine receptor, which is also a Ca^{2+}-gated Ca^{2+} channel. This positive feedback effect can lead to a propagating wave of increasing Ca^{2+} (**Figure 6.12**). These Ca^{2+} waves can

(a)

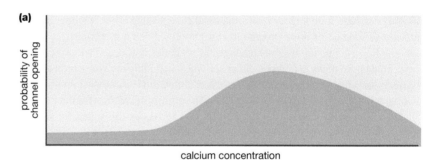

Figure 6.12

Intracellular Ca^{2+} signaling can show complex dynamics such as propagating waves. (a) The typical IP_3-gated Ca^{2+} channel responds to Ca^{2+} as well as IP_3. The channels show a bell-shaped dependence of activity on Ca^{2+} concentration. (b) When channels are first activated by IP_3, the initial rise in Ca^{2+} leads to a local positive feedback loop that spatially propagates channel opening and further Ca^{2+} influx. However, once Ca^{2+} concentration becomes further elevated, it has an inhibitory effect on the IP_3-gated channels, resulting in a slower negative feedback loop. This complex regulation can lead to propagating waves of Ca^{2+}. (c) A calcium wave in a rabbit urethral muscle cell, visualized by a calcium-sensitive dye (*yellow* and *red* colors correspond to high Ca^{2+} concentration). Images were taken at a rate of 30 frames per second. Some downstream signaling systems, such as CaM-Kinase II, can read the frequency of such waves as a measure of initial input. (c, Courtesy of Mark Hollywood.)

(b)

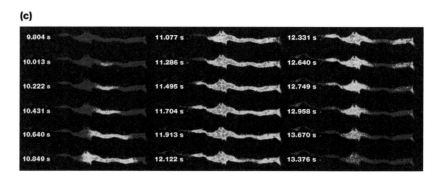

(c)

propagate further than individual Ca^{2+} ions could diffuse, given that the high ambient concentration of Ca^{2+}-binding proteins in the cell limits the diffusion of free ions. As the waves propagate, once local Ca^{2+} concentration increases beyond a certain point, the IP_3 receptors will be inhibited, limiting further Ca^{2+} influx. Thus, as these waves pass, the channels will shut, allowing pumps to restore low Ca^{2+} levels (creating troughs that follow the crests of the wave). Together, these phenomena can lead to traveling waves of Ca^{2+}, where the intensity of input is encoded in the frequency of the waves. Certain downstream effectors, such as CaMKs, are thought to be able to detect and respond to differences in Ca^{2+}-wave frequencies, providing a novel way of encoding signaling information. There is now evidence that other small signaling mediators such as cAMP may also exhibit waves and oscillatory behavior. The importance of such **dynamic encoding** of information, where the timing and/or duration of signals affects the output of the system, is becoming better appreciated in a number of signaling contexts.

SPECIFICITY AND REGULATION

A limited number of small signaling mediators are used to mediate many diverse biological responses. Thus, a key question is how signaling by these seemingly ubiquitous mediators can yield specific responses in cells. This is an important conceptual challenge, since a single mediator can be generated by many different upstream inputs, and potentially targets a diverse set of downstream effectors (**Figure 6.13**).

A further complication is that many of the small signaling mediators that we have discussed can directly or indirectly impact each other. These pathway interactions mean that it can be difficult to disentangle the effect of one signaling mediator from the effects of others. We have seen how IP_3, DAG, and Ca^{2+} all cooperate to regulate PKC activity. A number of other examples can be found, such as when elevated Ca^{2+} leads to NO synthesis in vascular endothelial cells, inducing cGMP synthesis in nearby vascular smooth muscle. In this section, we discuss how scaffold proteins provide one mechanism to impart spatial and temporal specificity to small-molecule signaling.

Scaffold proteins can increase input and output specificity of small-molecule signaling

Scaffold proteins help increase the specificity of small-mediator signaling by organizing upstream and downstream molecules (such as receptors and effectors, respectively) into localized complexes. For example, cAMP signaling specificity can be enhanced by a class of proteins known as **A-kinase anchoring proteins** (**AKAPs**) (**Figure 6.14a**). AKAPs contain multiple protein-binding sites, including one for protein kinase A (PKA), the major immediate downstream target for cAMP. AKAPs have an amphipathic α helix that binds selectively to a site on the regulatory (R) subunit of PKA. As discussed above, at basal levels of cAMP, the R subunit of PKA stably associates with the catalytic (C) subunit, which is thereby maintained in an inactive conformation. The affinity of AKAPs for the R subunit of PKA is very high (K_d of ~10^{-8} M), and thus, inactive PKA is effectively tethered to the AKAPs.

Individual AKAPs also possess targeting motifs that associate with specific subcellular compartments, such as the plasma membrane, the mitochondrion, or the centrosome. Thus, although PKA by itself lacks an intrinsic signal for subcellular localization, subsets of PKA are directed to specific sites within the cell by their association with AKAPs, and are held there in an inactive state in readiness for a cAMP signal. A rise in the local concentration of cAMP will trigger binding of the AKAP-associated R subunit and release of the catalytic subunit of PKA in an active conformation, which will preferentially phosphorylate substrates in the vicinity of

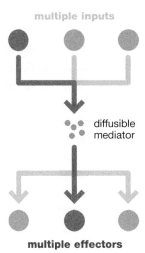

specific INPUT/OUTPUT

multiple inputs

diffusible mediator

multiple effectors

Figure 6.13

Small signaling mediators are generated by diverse signaling inputs and generate diverse signaling outputs. A key question is how these mediators can yield specific input/output linkages (an example is highlighted in *yellow*).

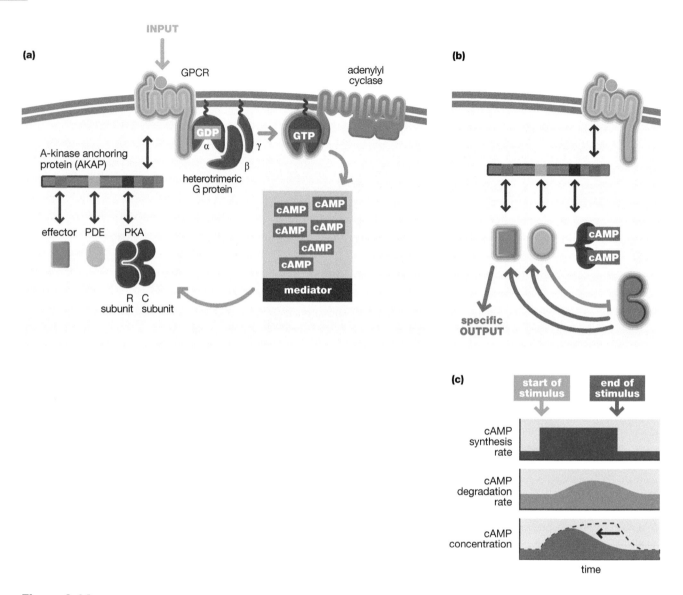

Figure 6.14

AKAP scaffolding proteins can enhance input/output specificity of cAMP-mediated signaling. (a) In the unstimulated condition, specific A-kinase anchoring proteins (AKAPs) are associated with specific G-protein-coupled receptor (GPCR) complexes. AKAPs contain targeting sites for specific upstream GPCRs (*pink*), binding sites for protein kinase A (PKA; *brown*), and binding sites for specific downstream effectors (PKA substrates; *orange*). In addition, some AKAPs contain binding sites for phosphodiesterase (PDE; *green*). For clarity, only one heterodimer of the R2C2 PKA tetramer is shown. (b) Activation of the receptor and adenylyl cyclase in the vicinity of the AKAP leads to activation of PKA and dissociation of the catalytic subunit, linking activation of specific receptors to phosphorylation of specific PKA outputs. In addition, phosphorylation and activation of PDE by PKA leads to degradation of cAMP. This generates a negative feedback loop that can control the amplitude and duration of cAMP-mediated signaling. (c) Schematic of the effect of AKAP on rates of cAMP synthesis and degradation. AKAP-mediated activation of PDE leads to more rapid signal termination than would be the case in the absence of AKAP (dotted line).

the AKAP scaffold (**Figure 6.14b**). Importantly, evidence suggests that the active PKA catalytic subunit usually does not stray far from its site of activation. Possible explanations for this relative lack of mobility include residual affinity for the regulatory subunit, membrane localization due to N-terminal myristoylation of the catalytic subunit, or inclusion in molecular condensates that maintain high local concentrations of components.

AKAPs have additional protein-binding sites (interaction motifs) that help refine the specificity with which cAMP production and PKA activity are coupled to specific upstream receptors and downstream targets. For example, some AKAPs contain targeting motifs that can localize the AKAP complex to the vicinity of a particular upstream GPCR. When this specific GPCR is activated and induces activation of nearby adenylyl cyclase molecules, the localized increase in cAMP will more rapidly activate the population of PKA associated with the AKAP scaffold. Thus, cAMP production and PKA activation in the complex are linked to only one of many possible upstream receptors.

AKAPs can also control pathway output downstream from PKA—they often have binding sites for specific substrates of PKA, such as ion channels. An AKAP scaffold can thereby link specific upstream receptors to specific downstream effectors.

AKAP scaffold proteins can also regulate dynamics of cAMP signaling

AKAP scaffolds can also be used to limit the timing of signaling. For example, certain AKAP proteins also contain binding sites for a phosphodiesterase (PDE), the enzyme that degrades cAMP and therefore shuts off signaling. This PDE can have two functions that are critical for PKA activity. In the absence of external signals, the PDE has a basal activity that suppresses the level of cAMP in the vicinity of the AKAP, and this, in turn, ensures that PKA is not spuriously activated in unstimulated cells. However, as the level of cAMP rises, it overwhelms the degradative capacity of the PDE, and PKA is activated to phosphorylate local substrates. In the absence of any further activity, the PDE will gradually degrade the cAMP, leading to inactivation of PKA.

One form of AKAP-associated PDE, however, is itself a PKA substrate that is stimulated upon phosphorylation. The activation of PKA induced by cAMP therefore stimulates the PDE to more efficiently degrade the local pool of cAMP. As a consequence of this negative feedback loop, there is a more rapid degradation of cAMP and inactivation of PKA (**Figure 6.14c**). The AKAP scaffold, therefore, tunes the kinetics of the response to a specific cAMP response signal by binding an inducible inhibitor.

Some AKAPs also bind to kinases other than PKA, such as protein kinase C (PKC), and to phosphatases such as calcineurin. It is easy to see how very complex response dynamics can result from the localized activation of multiple enzymes that can potentially reinforce or counteract each other's activities. Because these AKAP-mediated effects are largely restricted to the immediate vicinity of the scaffold, increasing the local concentration of cAMP can have very different outputs at different spots in the cell.

SUMMARY

A variety of small molecules serve as signaling intermediates. The levels of these small signaling mediators are determined by their relative rates of production and elimination. They exert their effects on downstream effector proteins by binding and inducing conformational changes that alter the activity of the effectors. Small signaling mediators enable relatively weak signaling inputs to induce rapid, widespread, and highly amplified effects in cells. Major classes of small signaling mediators include Ca^{2+}, the cyclic nucleotides cAMP and cGMP, and lipid-derived mediators such as DAG and IP_3.

Many different inputs regulate the same limited number of mediators, which in turn regulate a wide variety of outputs. However, scaffold proteins can provide spatial and temporal specificity by physically assembling particular components into localized complexes.

Answers to these questions can be found online at www.routledge.com/9780367279370

QUESTIONS

1. What physical and chemical properties distinguish small signaling mediators from larger signaling proteins? What kinds of functional differences might result from these properties?

2. In Chapters 3 and 4, we discuss the concept of enzymes that generate and remove post-translational modifications (writers and erasers). By controlling the extent of a modification, these enzymes coordinately control the flow of signaling information. For Ca^{2+} signaling—where the concentration of cytoplasmic Ca^{2+} is used to transmit information—which molecules play the roles of writers and erasers? What is the source of the energy driving Ca^{2+} signaling?

3. Under resting conditions, the basal rate of degradation of a small signaling mediator, such as cAMP, is relatively high. Why might this be important for cAMP-dependent signaling?

4. Describe two different ways that, in principle, an incoming signal could result in an increase in cAMP levels in a cell.

5. How are waves of Ca^{2+} generated in cells? How might oscillating waves provide an added dimension for encoding information?

6. Why is specificity a particularly acute problem for small signaling mediators such as cAMP? How do scaffold proteins mitigate this issue?

BIBLIOGRAPHY

CLASSES OF SMALL SIGNALING MEDIATORS

Bos JL (2006) Epac proteins: multi-purpose cAMP targets. *Trends Biochem. Sci.* 31, 680–686.

Kim C, Xuong NH & Taylor SS (2005) Crystal structure of a complex between the catalytic and regulatory (RIα) subunits of PKA. *Science* 307, 690–696.

Newton AC, Bootman MD & Scott JD (2016) Second messengers. *Cold Spring Harb. Perspect. Biol.* 8(8), a005926. doi: 10.1101/cshperspect.a005926.

Rehmann H, Prakash B, Wolf E, et al. (2003) Structure and regulation of the cAMP-binding domains of Epac2. *Nat. Struct. Biol.* 10, 26–32.

Taylor SS, Zhang P, Steichen JM, et al. (2013) PKA: Lessons learned after twenty years. *Biochim. Biophys. Acta* 1834, 1271–1278.

Zhang P, Smith-Nguyen EV, Keshwani MM, et al. (2012) Structure and allostery of the PKA RIIβ tetrameric holoenzyme. *Science* 335, 712–716.

CALCIUM SIGNALING

Bootman MD & Bultynck G (2020) Fundamentals of cellular calcium signaling: A primer. *Cold Spring Harb. Perspect. Biol.* 12(1), a038802. doi: 10.1101/cshperspect.a038802.

Chin D & Means AR (2000) Calmodulin: A prototypical calcium sensor. *Trends Cell Biol.* 10, 322–328.

Clapham DE (2007) Calcium signaling. *Cell* 131, 1047–1058.

SPECIFICITY AND REGULATION

Bucko PJ & Scott JD (2021) Drugs that regulate local cell signaling: AKAP targeting as a therapeutic option. *Annu. Rev. Pharmacol. Toxicol.* 61, 361–379. doi: 10.1146/annurev-pharmtox-022420-112134.

Greenwald EC & Saucerman JJ (2011) Bigger, better, faster: Principles and models of AKAP anchoring protein signaling. *J. Cardiovasc. Pharmacol.* 58, 462–469.

Welch EJ, Jones BW & Scott JD (2010) Networking with AKAPs: Context-dependent regulation of anchored enzymes. *Mol. Interv.* 10, 86–97.

Wong W & Scott JD (2004) AKAP signalling complexes: Focal points in space and time. *Nat. Rev. Mol. Cell Biol.* 5, 959–970.

Membranes, Lipids, and Enzymes That Modify Them

7

The water-impermeable lipid membrane that separates the cytosol from the outside environment is one of the defining features of all living cells. In eukaryotes, lipid membranes also partition the cytosol into different compartments and organelles, and membrane vesicles are used to shuttle proteins and other components between different sites in the cell. In addition to their important structural roles, however, membranes and their lipid building blocks also play a very active role in cell signaling. They provide the raw materials for generating intracellular signaling mediators such as diacylglycerol (DAG) and inositol trisphosphate (IP$_3$), and intercellular signaling molecules such as prostaglandins. They also play an important role in providing regulated binding sites for cytosolic proteins with lipid-binding domains. Because membrane lipids are arranged in two-dimensional sheets, the biochemical reactions in which they participate have different properties from more familiar reactions in aqueous solution, where all components can freely diffuse in three dimensions. In this chapter, we will examine the special properties of lipids and membranes, and the enzymes that modify them, in the context of their roles in signal transduction.

BIOLOGICAL MEMBRANES AND THEIR PROPERTIES

The properties that allow a biological membrane to serve as an effective barrier between the cell's contents and the environment are its impermeability to water and other hydrophilic compounds, and its ability to spontaneously organize into self-sealing sheets and vesicles. These special properties arise from the biophysical nature of its molecular components. The fundamental driving force is the high thermodynamic cost of placing hydrophobic chemical groups (such as aliphatic hydrocarbons, which lack more electronegative atoms such as oxygen or nitrogen) into a polar aqueous environment, and of placing polar or charged chemical groups

DOI: 10.1201/9780429298844-7

Figure 7.1

Organization and structure of the lipid bilayer. (a) Phospholipids spontaneously organize into bilayers, with the polar head groups of the lipids oriented toward the aqueous solution, and the fatty acid tails inside. In addition to large sheets (top), phospholipids can also organize into closed vesicles or liposomes (lower right). In some cases, particularly when the head group is bulky relative to the hydrophobic tail, amphipathic lipids can also organize into micelles (lower left). (b) A computer-generated dynamic simulation of the position of 100 phosphatidylcholine molecules arranged in a lipid bilayer in aqueous solution. In this representation, the lipid head groups are *red*, the fatty acid tails are *yellow*, and water molecules are *blue*. (Adapted from B. Alberts, et al., *Molecular Biology of the Cell*, 5th ed. Garland Science, 2008; and S.W. Chiu, et al., *Biophys. J.* 69:1230–1245, 1995. With permission from Elsevier.)

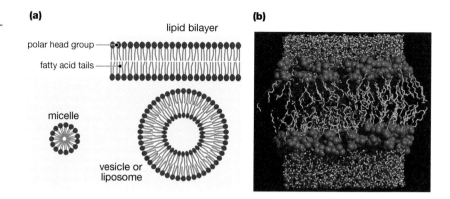

(a)

lipid bilayer

polar head group

fatty acid tails

micelle

vesicle or liposome

(b)

in a hydrophobic environment. Of course, this is why oil and vinegar separate in a bottle of salad dressing—the hydrophobic oil and the hydrophilic vinegar segregate away from each other into two distinct phases, minimizing the energetically unfavorable interface between the two.

Biological membranes consist of molecules in which relatively long hydrophobic tails are linked to polar head groups. Such molecules, containing both hydrophilic and hydrophobic portions, are said to be **amphipathic**. In aqueous environments, their lowest energy configuration is one in which the hydrophobic chains are arranged in the interior, shielded from the solvent, while the polar head groups are arrayed on the surface facing the solvent. While some amphipathic molecules readily form micelles—small, spherical structures with the hydrophobic portions sequestered in the interior—for the polar lipids that make up biological membranes, the most favorable arrangement is in a **lipid bilayer** (**Figure 7.1**). This consists of two layers of polar lipids, arranged in a sheet with the hydrophobic chains oriented inward and the polar head groups on the surface of the sheet facing the aqueous environment. As discussed below, lipid bilayers spontaneously seal into spherical vesicles or liposomes.

Biological membranes consist of a variety of polar lipids

The most abundant lipids in biological membranes are the phospholipids, or more properly the **glycerophospholipids** (**Figure 7.2**). All glycerophospholipids are built on a three-carbon glycerol backbone. Two of the hydroxyl groups of the glycerol are attached through an ester linkage to fatty acids, which consist of a carboxylic acid group linked to a long, hydrophobic aliphatic chain, generally 14–20 carbons in length. The third glycerol hydroxyl is linked to a phosphate group, which is highly polar and carries a negative charge at physiological pH. The phosphate itself is generally linked to other chemical groups, the most common being choline, serine, inositol, or ethanolamine; the resulting glycerophospholipids are termed phosphatidylcholine (PC), phosphatidylserine (PS), phosphatidylinositol (PI), and phosphatidylethanolamine (PE) (**Figure 7.3**). *Cis*-double bonds may be present in the aliphatic chains of the fatty acids (most typically in the fatty acid in the middle, or *sn*-2, position of the glycerol backbone), which generate kinks in the chain because of the planar nature of the double bonds (see Figure 7.2). As we will see below, such double bonds affect the biophysical properties of the membrane. Chains with double bonds are said to be unsaturated (that is, the carbons have fewer than the maximum number of hydrogen atoms associated with them). It is important to note that the specific fatty acids found on phospholipids can be quite variable, both in terms of the length of the chain and the position and number of double bonds, even when the head group (and thus the generic chemical name of the lipid) is the same.

Another major component of biological membranes is the sphingolipids. These resemble phospholipids in that they have similar polar head groups, but the

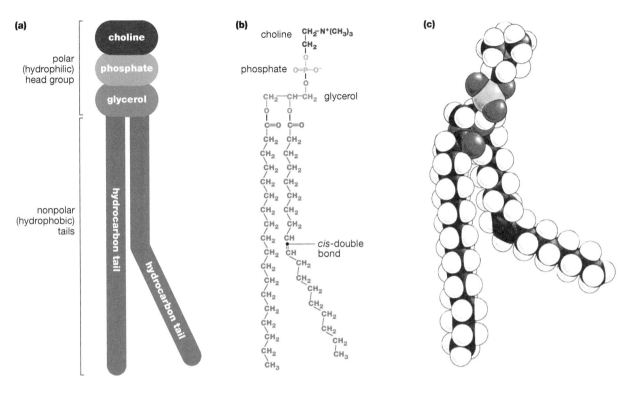

Figure 7.2

The structure of a glycerophospholipid. A glycerophospholipid is shown in (a) graphical, (b) chemical structure, and (c) space-filling representations. Note the kink in the fatty acid chain introduced by the planar *cis*-double bond. (Adapted from B. Alberts, et al., *Molecular Biology of the Cell*, 5th ed. Garland Science, 2008.)

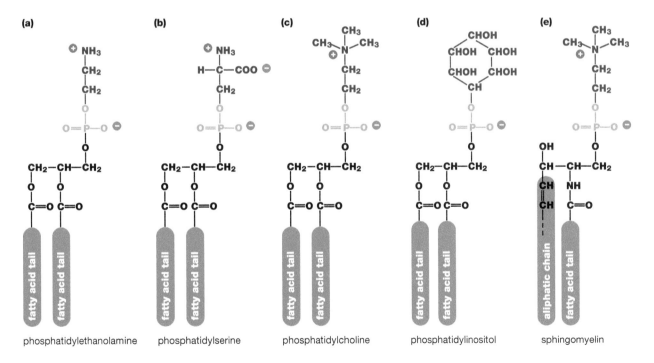

glycerol backbone and fatty acid tails are replaced by ceramide. Ceramide itself has two long aliphatic chains, derived by joining sphingosine (a sphingoid base) via an amide linkage to a fatty acid. For example, **sphingomyelin** (**Figure 7.3**) has a phosphorylcholine head group attached to ceramide. Since the aliphatic chains of sphingolipids tend to be long and saturated, these lipids are generally taller and narrower in shape than the glycerophospholipids, and thus can pack together more tightly.

Figure 7.3

The chemical structures of the most common membrane phospholipids.

(Adapted from B. Alberts, et al., *Molecular Biology of the Cell*, 5th ed. Garland Science, 2008.)

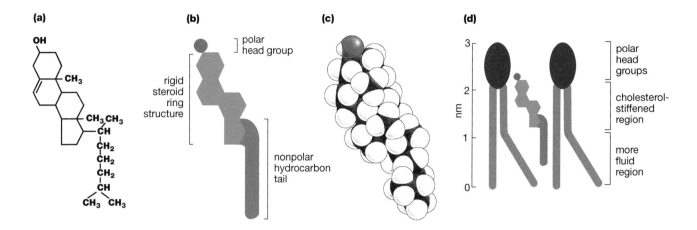

Figure 7.4

The structure of cholesterol. Cholesterol is shown in (a) chemical structure, (b) graphical, and (c) space-filling representations. (d) The interaction of cholesterol with phospholipids in a lipid bilayer, with the rigid ring structure aligned with the fatty acid chains of the phospholipids. (Adapted from B. Alberts, et al., *Molecular Biology of the Cell*, 5th ed. Garland Science, 2008.)

Apoptosis is discussed in Chapter 9.

The last major components of eukaryotic membranes are the sterols, predominantly **cholesterol** in mammals. Cholesterol is unlike the other major membrane lipids in that it is a rigid, planar, polycyclic compound that is relatively nonpolar (**Figure 7.4**). Through its interaction with the aliphatic chains of other membrane lipids, it exerts potent effects on the biophysical properties of the membrane, such as its fluidity (see below).

Structural properties of membrane lipids favor the formation of bilayers

Because of the amphipathic nature and roughly cylindrical shape of most membrane lipids, their most energetically favorable configuration in aqueous solutions is in a bilayer, with the polar head groups oriented outward and the fatty acid chains sandwiched between (see Figure 7.1). (By contrast, micelles are generally favored for amphipathic molecules with relatively large polar heads and small hydrophobic tails.) The lipid bilayer consists of two back-to-back layers, or leaflets. Because it is energetically very costly to expose the hydrophobic fatty acid chains to the aqueous environment, any rips or tears in the membrane tend to spontaneously reseal, a highly useful property for a barrier. Similarly, relatively small sheets will spontaneously form closed, spherical vesicles. Although individual lipids in a membrane experience considerable thermal motion, the hydrophobic core of the intact membrane presents a very effective barrier to water and to other hydrophilic compounds (see Figure 7.1b). However, a few signaling molecules—such as the small, uncharged gas nitric oxide, and hydrophobic organic molecules such as steroid hormones—are sufficiently small and lipid-soluble that they can pass through the hydrophobic core relatively freely.

The composition of the membrane determines its physical properties

The inner and outer leaflets of the cell membranes differ in their lipid composition. This is because most lipids cannot spontaneously "flip" between the two leaflets, because of the energetic cost of dragging the polar head group through the hydrophobic core of the membrane. For these lipids, transport between leaflets requires specialized enzymes and energy in the form of ATP. In general, the outer (extracellular) leaflet of the plasma membrane contains sphingomyelin and glycosphingolipids, and is enriched for PC, while the inner leaflet is enriched for PE. The phospholipids PS and PI, which carry a net negative charge, are exclusively found on the inner leaflet, helping maintain the slight negative charge on the inner face of the membrane (the resting potential). The exclusive presence of PS in the inner leaflet is actually used by cells as a signal: the presence of PS on the outside of cells undergoing *apoptosis*, or programmed cell death, is an "Eat Me" signal that induces surrounding cells to engulf the apoptotic cell fragments.

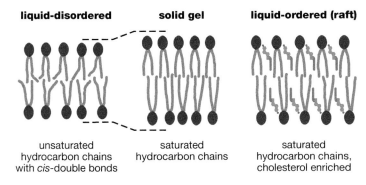

liquid-disordered	solid gel	liquid-ordered (raft)
unsaturated hydrocarbon chains with *cis*-double bonds	saturated hydrocarbon chains	saturated hydrocarbon chains, cholesterol enriched

Figure 7.5

Different organization states of a lipid bilayer. (a) When many of the fatty acid chains are unsaturated, the chains are not densely packed and the bilayer adopts a liquid-disordered phase, with high lateral mobility. (b) When most fatty acid chains are saturated, they align closely with each other and the membrane adopts a solid gel phase, with limited lateral mobility. (c) If cholesterol is present, a membrane can adopt a liquid-ordered phase that is both highly ordered yet relatively fluid. (Adapted from B. Alberts, et al., *Molecular Biology of the Cell*, 5th ed. Garland Science, 2008.)

The physical properties of the membrane are highly dependent on its component lipids and, in particular, the aliphatic chains of those lipids. Just as butter and other familiar fats undergo temperature-dependent phase transitions from a solid gel to a more highly disordered liquid, the aliphatic chains of membrane lipids can undergo similar phase transitions. In general, the more tightly packed and ordered the aliphatic chains, the less fluid the membrane will be (**Figure 7.5**). The presence of double bonds, which introduce kinks into the aliphatic chains, makes it more difficult for them to pack together tightly; thus, increasing the fraction of unsaturated fatty acids increases membrane fluidity (the industrial process of hydrogenation removes these double bonds in liquid vegetable oils to make shortening or margarine that gels at room temperature). In addition, the presence of relatively large amounts of membrane-associated protein (up to 50% by weight) and of rigid, nonpolar sterols can have dramatic effects on fluidity.

Membrane fluidity is relevant to signaling for several reasons. First, the amount of fluidity directly affects the behavior of proteins embedded in the membrane and of lipids that act as signaling intermediates (such as DAG) by affecting their rates of diffusion. Diffusion is much faster in a fluid membrane than in regions that are more ordered and gel-like. More importantly, since different phases can exist in the same membrane, this raises the possibility that different membrane domains will have different lipid and protein constituents as well as different physical properties. For example, the long, straight alkyl chains of sphingomyelin interact favorably with cholesterol, generating local lipid domains (**lipid rafts**) that are highly ordered, yet allow a high degree of lateral mobility (termed the liquid-ordered phase) (see Figure 7.5). Such structures have been difficult to visualize in living cells, implying that they are likely to be small and transient in nature. However, such lipid microdomains can potentially play an important role in signaling by concentrating key components and thereby increasing the efficiency of reactions, and by segregating components away from competing interactors or inhibitors. Such membrane domains are analogous to the protein- and nucleic acid-based *biomolecular condensates* that were previously discussed in Chapter 5.

There are fundamental differences between biochemistry in solution and on the membrane

The fact that membranes are two-dimensional has implications for the reactions that occur there. Intuitively, we can see that if two reactants are confined to the same membrane, they are more likely to interact than if they are free to wander about in solution. This is a simple consequence of less physical space to explore. However, the actual rate of diffusion of a molecule in the plane of the membrane is considerably slower than that of the same molecule in solution, so the absolute rate at which two molecules encounter each other in solution versus on the membrane may not be all that different. However, those interactions are likely to be much more efficient in two dimensions (fewer competing interactions to wade through, and in many instances, the interacting molecules are pre-oriented by their anchorage points to the membrane) (**Figure 7.6**).

Figure 7.6

Membrane association can make molecular interactions more efficient.

(a) Two interacting proteins (one *green*, one *blue*) bind productively only in a certain precise orientation (top). If molecules are free in solution, they can rotate around all three axes; only a small fraction of collisions will be correctly oriented to allow binding (middle). However, if the two molecules are confined to the membrane by an anchor (bottom), they can only rotate around one axis; thus a relatively large percentage of collisions will be oriented to allow binding. (b) For molecules in solution, a three-dimensional volume must be explored (top), while if molecules are confined to a membrane, only a two-dimensional plane must be explored, making interaction more likely (middle). In actual biological membranes, lateral diffusion is thought to be limited to relatively small areas separated by diffusion barriers (bottom), further increasing the likelihood of interaction between molecules.

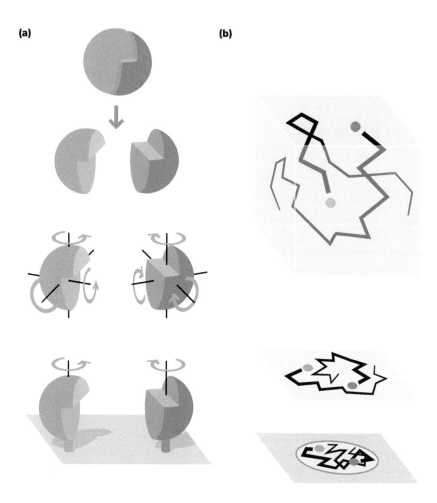

(a) (b)

Another consideration, touched on above, is the likelihood that different domains of the membrane have quite different compositions and physical properties, and that migration between these domains may be impeded to various extents. Experimental evidence from single-molecule tracking and fluorescence recovery after photobleaching (FRAP) suggests that membrane lipids and proteins that are embedded in the membrane diffuse significantly more slowly in actual biological membranes than in artificial membranes consisting of pure lipids. Furthermore, measurements of diffusion on very short time scales suggest that membrane components can rapidly diffuse within small domains (several microns in size), but move much more slowly between adjacent domains. These slower diffusion rates may be due to the physical constraints of a high density of membrane proteins. It also seems that some cytoskeletal structures may act to corral membrane proteins, impeding their ability to freely diffuse over the entire surface of the cell membrane. Distinct membrane domains that differ in fluidity could also contribute to this anomalous diffusion. Thus, signaling proteins that are strongly associated with membranes through transmembrane domains or via lipid anchors are likely to interact with their near neighbors in a much less transient way than if they were cytosolic.

Another unusual aspect of membrane signaling is the diversity of biophysical properties of the molecules involved. The signaling compounds that are generated by modification of membrane lipids can be either lipid-soluble, thus constrained to the membrane from whence they came, or water-soluble, and thus free to diffuse into the cytosol or extracellular fluid. In some cases, one reaction generates both types of molecules, as when phospholipase C cleaves phosphatidylinositol 4,5-bisphosphate [PI(4,5)P$_2$] to generate diacylglycerol (DAG), which is confined to the membrane, and inositol trisphosphate (IP$_3$), which is water-soluble. DAG does not contain a polar head group and so can flip relatively easily from one leaflet of the membrane to the other. Membrane-localized signaling mediators such as DAG work primarily by recruiting and, in some cases, activating cytosolic proteins. Other lipid signaling mediators such as sphingosine 1-phosphate (S1P) and lysophosphatidic acid (LPA) can partition into both the lipid and soluble phases. Thus, the downstream effects of signals involving membrane lipids may be exerted throughout the cell, or in specific membrane compartments, or even to distant cells, depending on the specific nature of the signaling intermediary that is generated.

A final distinctive feature of signaling that involves membrane lipids is the rapid metabolism of these compounds from one bioactive species to another. This interconversion can make it difficult to say with confidence which specific lipid species is responsible for a particular output. In other words, if the levels of a single bioactive lipid are increased experimentally, this increase will rapidly alter the levels of a slew of breakdown products and further metabolites that can be derived from it, each of which may have different (but possibly overlapping) biological activities. This presents considerable difficulties in attempting to tease out specific cause-and-effect relationships in lipid signaling pathways.

LIPID-MODIFYING ENZYMES USED IN SIGNALING

Signaling by membrane lipids involves their breakdown or chemical modification in response to input stimuli. These reactions are catalyzed by enzymes, most prominently phospholipases, lipid kinases, and lipid phosphatases. In this section, we will briefly outline the general properties of these enzymes. We will then explore their specific roles in a number of important signal transduction pathways in greater detail in the following section.

Cleavage of membrane lipids by phospholipases generates a variety of bioactive products

The enzymes that break down phospholipids are termed phospholipases. They are classified on the basis of the bonds that they cleave. The major phospholipases that participate in signaling are phospholipase A_2, phospholipase C, and phospholipase D (**Figure 7.7**).

Phospholipase A_2 (PLA$_2$) cleaves the fatty acid chain from the *sn*-2 (middle) position of the glycerol backbone of glycerophospholipids, generating a free fatty acid and a **lysophospholipid** containing a single fatty acid chain. Both of these products can generate bioactive components. In many cases, the fatty acid at the *sn*-2 position is arachidonic acid (AA). AA may itself activate some signaling proteins, and also serves as the basic building block of the **eicosanoids**, a large family of intercellular signaling lipids that includes the prostaglandins and leukotrienes, as discussed later in this chapter. Eicosanoids can leave the cell of origin and signal to nearby cells by binding to specific G-protein-coupled receptors (GPCRs). Lysophospholipid, the second product, is also sufficiently water-soluble that it can leave the membrane and signal to adjacent cells. Lysophosphatidic acid (LPA), the product of PLA$_2$ action on phosphatidic acid, can potently signal via a GPCR to regulate diverse activities such as proliferation, differentiation, and cell morphology. PLA$_2$ is a major component of many insect and snake venoms, presumably exerting its toxic effects by disrupting the biophysical integrity of membranes in the victim.

Phospholipase C (PLC) cleaves the phosphorylated head group from the phospholipid, leaving the uncharged DAG behind. In the case of PI-PLC, which is specific for phosphatidylinositol, the released phosphoinositol head group (such as IP$_3$) can diffuse rapidly in the cytosol and act as a soluble signaling intermediate,

Activation of PKC is discussed in Chapter 6.

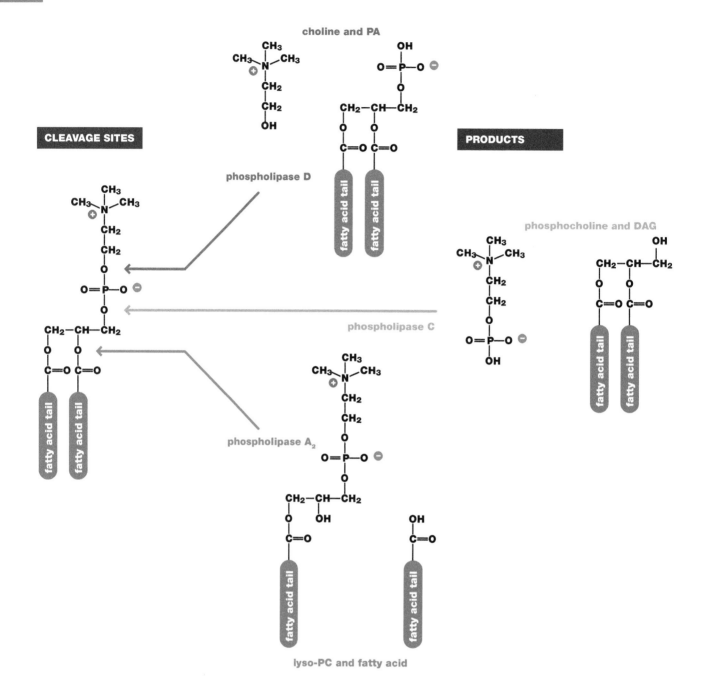

Figure 7.7

The sites of cleavage and the reaction products generated by phospholipase D (PLD), phospholipase C (PLC), and phospholipase A$_2$ (PLA$_2$). The substrate phospholipid in this example is phosphatidylcholine. PA, phosphatidic acid; DAG, diacylglycerol; lyso-PC, lysophosphatidylcholine.

while DAG, which remains in the membrane, acts as an activator of proteins such as protein kinase C (PKC).

Finally, **phospholipase D** (**PLD**) cleaves the unphosphorylated head group from the phospholipid, leaving behind **phosphatidic acid** (**PA**). PA not only serves as a substrate for PLA$_2$ to generate LPA, but itself plays a role in activation of the *mTOR* (*mechanistic target of rapamycin*) kinase complex, a key regulator of metabolic and stress signaling and cell growth, as described below.

A variety of lipid kinases and phosphatases are involved in signaling

The phosphorylation and dephosphorylation of membrane lipids, particularly those derived from PI (*phosphoinositides*), are also commonly regulated in signaling to generate a variety of bioactive compounds. As might be expected, lipid kinases (which add phosphate) and lipid phosphatases (which remove it) both play a role in changing

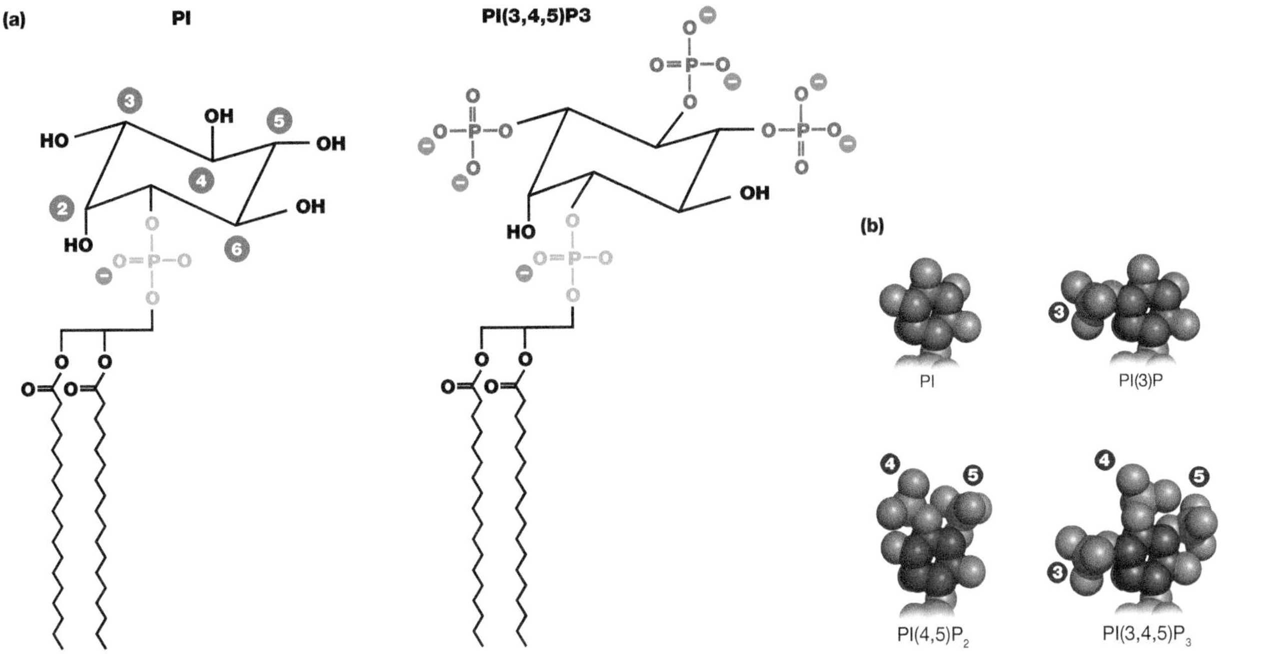

Figure 7.8

The structure of phosphoinositol lipids.
(a) Structure of phosphatidylinositol (PI) and phosphatidylinositol 3,4,5-trisphosphate [PI(3,4,5)P$_3$]. Numbering of carbons of the inositol ring is indicated in *orange*.
(b) Space-filling models of the head groups of PI and the major phosphoinositides recognized by lipid-binding domains.

the local concentration of various lipid isoforms in the course of signaling. The activity of these proteins, in turn, is regulated by their abundance, by their post-translational modification, and, in particular, by their subcellular localization. Consequently, these enzymes frequently possess modular lipid- and protein-binding domains, which allow them to respond to changes in the local protein and lipid environment.

One broad class of lipid kinases phosphorylates uncharged lipid signaling mediators, converting them to more highly polar products that also can act as signaling mediators. For example, DAG is converted to PA by DAG kinases. Thus, DAG kinases mediate a switch in local activity from one class of effectors (those activated by DAG binding) to another (those that are regulated by binding to PA). Similarly, sphingosine is converted by sphingosine kinase to S1P, which can leave the membrane and activate GPCRs. The reverse reactions (dephosphorylation of PA and S1P) are mediated by a family of related integral membrane enzymes, the lipid phosphate phosphatases.

The six-carbon inositol sugar head group of PI provides ample scope for the action of lipid kinases and phosphatases (**Figure 7.8**). Hydroxyl groups at positions 2–6 of the ring are available for phosphorylation (the oxygen at position 1 is linked by a phosphodiester bond to the glycerol backbone). As will be described in more detail below, phosphorylation at positions 3–5 is most associated with signaling, creating specific membrane binding sites for modular lipid-binding domains and providing precursors for signaling mediators generated by the action of PLC. In general, different kinases and phosphatases, each of which can have distinct subcellular localization and modes of regulation, are used to modify each position of the inositol ring. This enables very precise spatiotemporal control over the phosphorylation state of the phosphoinositides.

EXAMPLES OF MAJOR LIPID SIGNALING PATHWAYS

We have outlined above how membranes and their lipid constituents can participate in signaling in many ways, both direct and indirect. In this section, we will consider in greater depth several examples where the modification of membrane lipids plays a particularly direct role in the process of signal processing and transmission.

Phosphoinositides can serve as membrane binding sites and as a source of signaling mediators

Phosphoinositides provide a compelling illustration of the two major roles of membrane lipids in signaling: as a source of diffusible signaling mediators, and as specific membrane-localized binding sites that can recruit and activate signaling proteins.

 Akt activation is discussed in Chapter 5.

The lipid $PI(4,5)P_2$ is at the nexus of these two different functions. Cleavage of $PI(4,5)P_2$ by PI-PLC generates DAG and IP_3 (see Figure 6.8), each of which goes on to activate downstream targets (such as PKC and endoplasmic reticulum calcium channels, respectively). However, $PI(4,5)P_2$ itself can directly bind to and recruit to the plasma membrane a variety of signaling proteins, particularly those that regulate localized actin polymerization. $PI(4,5)P_2$ can also be further modified by phosphorylation or dephosphorylation to generate distinct phosphoinositides that can bind and recruit other targets. In particular, $PI(4,5)P_2$ can be phosphorylated by phosphatidylinositol 3-kinase (PI3K) to generate phosphatidylinositol 3,4,5-trisphosphate [$PI(3,4,5)P_3$], a pivotal regulator of cell survival and growth, and many other activities. For example, $PI(3,4,5)P_3$ stimulates downstream signaling by recruiting and activating effectors such as the Akt kinase. $PI(3,4,5)P_3$ has another important property: unlike its precursor $PI(4,5)P_2$, it cannot be cleaved by PI-PLC. Thus, activation of PI3K has the effect of shutting down signaling from $PI(4,5)P_2$, whereas PLC activation, by destroying the substrate for PI3K, has the effect of shutting down PI3K signaling. This mutual inhibition presumably facilitates the precise spatiotemporal regulation of signal output within the cell (**Box 7.1**).

Box 7.1

In signaling, the activity of many lipid-modifying enzymes is regulated, in turn, by other lipids and the enzymes that produce or modify those lipids. These complex regulatory arrangements can have interesting consequences, such as promoting the formation of discrete membrane domains with different lipid composition, or promoting the precise switching from one type of signal to another. Three examples mentioned in the text are illustrated in **Figure 7.9**. PI3K and PI-PLC can both be activated by similar stimuli (for example, receptor tyrosine kinase activation), but the two consume the same substrate, $PI(4,5)P_2$. Thus, in situations where substrate is limiting, whichever enzyme is activated first at a particular site will signal transiently (until it exhausts the substrate) while preventing activation of the latecomer (**Figure 7.9a**). Phosphatidylinositol 5-kinase (PI5K) and PLD co-localize, and each makes a product that activates the other enzyme. Thus the activity of each reinforces the other, leading to high local concentrations of their products. This is an example of *positive feedback* (**Figure 7.9b**). The small G protein Rheb activates both mTOR and PLD. PLD makes a product, PA, which is also necessary for mTOR activity. This is an example of *coherent feedforward* (**Figure 7.9c**). The requirement for PLD could function to limit mTOR activity to specific locations (where PLD is present), or to delay activation of mTOR following Rheb activation (until PLD can generate sufficient PA). Such relationships and their depiction (*network architecture*) are discussed in Chapter 11.

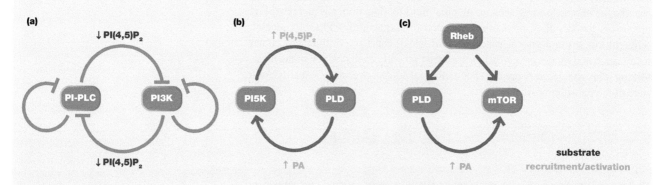

Figure 7.9

Lipid-modifying enzymes involved in (a) mutual inhibition, (b) positive feedback, and (c) coherent feedforward regulation. $PI(4,5)P_2$, phosphatidylinositol 4,5-bisphosphate; PI-PLC, phosphatidylinositol phospholipase C; PI3K, phosphatidylinositol 3-kinase; PI5K, phosphatidylinositol 5-kinase; PLD, phospholipase D; PA, phosphatidic acid.

Two well-understood ways in which PI-PLC can be activated are by GPCRs and by tyrosine kinases. Heterotrimeric G proteins activate the PLC-β family of PLCs. Generally, this is accomplished by binding of the PLC to $G\alpha_q$ subunits, but the βγ subunits may participate in activating some PLC-β isotypes (see Figure 6.9). Examples of G-protein-mediated PLC activation include platelets activated in response to the thrombin receptor, and *Drosophila* retinal cells, where G protein activation is mediated by visual rhodopsin.

By contrast, the PLC-γ family is activated by receptors with intrinsic or associated tyrosine kinase activity, such as mitogenic growth factor receptors or B and T cell receptors. PLC-γ contains SH2 domains, which mediate recruitment and binding to the tyrosine-phosphorylated receptors or associated proteins. This interaction has two consequences: tethering PLC to the membrane where its substrate resides, and promoting phosphorylation of PLC by associated kinases. Tyrosine phosphorylation is important for full catalytic activity of the enzyme.

The primary downstream consequences of PI-PLC activation include release of intracellular Ca^{2+} and activation of PKC, though biological effects have also been attributed to local decreases in $PI(4,5)P_2$ density, or to increases in fatty acids or PA due to further enzymatic modification of the DAG generated by PLC.

Phosphoinositide species provide a set of membrane binding signals

The other major arm of phosphatidylinositol signaling is the selective binding, recruitment, and activation of effectors to various phosphorylated forms of PI. This provides another example of the "writer/eraser/reader" theme introduced in earlier chapters. In this case, the writers are the PI kinases, the erasers are the PIP phosphatases, and the readers are lipid-binding domains on signaling proteins. A number of modular lipid-binding domains (including the PH, PX, and FERM domains) interact with specific phosphoinositides. Other proteins bind to phosphoinositides through less-well-defined basic (positively charged) motifs. The availability of five potential phosphorylation sites on the inositol ring (see Figure 7.8) means a great deal of information can be encoded in one small molecule. In principle, 2^5 or 32 different phosphorylation states are possible, but in practice only the 3, 4, and 5 positions are used in cells, generating eight possible states. Furthermore, the high density of phosphate groups on some phosphoinositides, and their associated negative charge, means that electrostatic attraction can provide considerable binding energy for interacting with positively charged protein surfaces. The major phosphoinositides that participate in signaling, along with examples of their effectors and the kinases and phosphatases that generate them, are shown in **Figure 7.10**.

Different phosphoinositides vary widely in their subcellular distribution, and this plays an important cell-biological role by providing a kind of "identity tag" for different types of membranes. For example, $PI(4,5)P_2$ is almost exclusively found on the plasma membrane, while **phosphatidylinositol 3-phosphate [PI(3)P]** is found on endosomes and PI(4)P on the Golgi (**Figure 7.11**). This is largely due to the subcellular distribution of the enzymes that generate the different phosphoinositides (PI kinases and PIP phosphatases). Key regulatory enzymes include PI(4)P 5-kinases (PIP5Ks), which dominate local generation of $PI(4,5)P_2$, and PI 5-phosphatases, such as synaptojanin. Synaptojanin is associated with the endocytic machinery and degrades $PI(4,5)P_2$ on plasma membrane-derived vesicles after endocytosis, a key step for their proper targeting and recycling. Importantly, phosphoinositide signaling is intimately interconnected with signaling by small G proteins, both in regulating membrane trafficking and in signaling. Not only are many of the guanine nucleotide exchange factors (GEFs) and GTPase-activator proteins (GAPs) for small G proteins regulated by phosphoinositide binding, but the G proteins and phosphoinositides often act cooperatively to bind to and activate downstream factors.

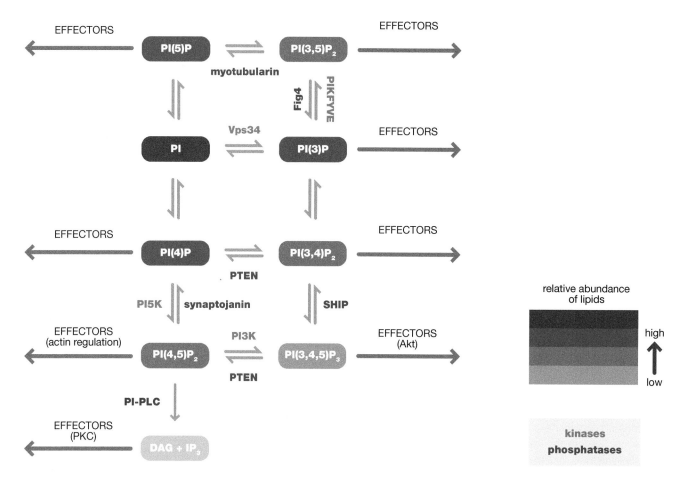

Figure 7.10

Phosphoinositide metabolism in signaling. Major phosphoinositides involved in signaling are shown, and examples of enzymes that metabolize them. Not all possible reactions indicated by arrows have been shown to be biologically significant. PI3K, phosphatidylinositol 3-kinase; PIP5K, phosphatidylinositol(4)P 5-kinase; PI-PLC, phosphatidylinositol-specific phospholipase C; DAG, diacylglycerol; IP$_3$, inositol trisphosphate; PKC, protein kinase C.

There are several different PI3K isoforms activated by distinct upstream stimuli. In terms of signaling, the best understood of these is the heterodimeric class IA subtype, for which the localization and activity of the catalytic subunit are regulated by a regulatory subunit containing SH2 domains. As in the case of PLC-γ described above, the SH2 domains mediate binding to activated tyrosine kinases and their phosphorylated substrates, thereby coupling activation of the PI3K to activation of tyrosine kinases. Class IB subtypes possess a similar heterodimeric organization, but in this case respond to G protein βγ subunits. All class I PI3Ks use PI(4,5)P$_2$ as their preferred substrate, thus their predominant product is PI(3,4,5)P$_3$ (see Figure 7.10). By contrast, the class III PI3K, Vps34, prefers PI as a substrate and thus generates predominantly PI(3)P. Vps34 is found primarily in endosomes and the Golgi, and its major role is to regulate vesicle trafficking.

PI(3,4,5)P$_3$ generated by class I PI3Ks can be dephosphorylated by several distinct phosphatases. PTEN specifically dephosphorylates the 3 position of the inositol

Figure 7.11

Subcellular localization of phosphoinositides. Subcellular localization of (a) PI(4,5)P$_2$, (b) PI(4)P, and (c) PI(3)P, as revealed by the binding of green fluorescent protein (GFP)-tagged lipid-binding domains specific for each lipid. Lipid-binding domains used are indicated beneath each panel. (From G. Di Paolo & P. De Camilli, *Nature* 443:651–657, 2006. With permission from Springer Nature.)

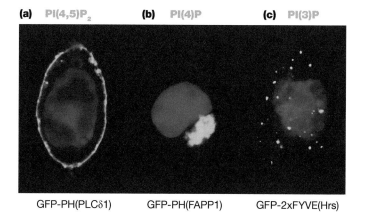

ring, essentially counteracting the effects of the PI3K. Since PI3K and its product generally promote cell growth, proliferation, and survival, it is not surprising that PTEN is a strong *tumor suppressor*, and that loss of PTEN activity is often seen in tumors. Another family of phosphatases that degrades PI(3,4,5)P_3 is the SHIP family, which specifically removes the phosphate from the 5 position to generate PI(3,4)P_2. This lipid has a distinct but overlapping set of effectors compared with PI(3,4,5)P_3. Recruitment of SHIP to activated T cell receptors, for example, is thought to be important for a temporal shift from one class of phosphoinositide-binding effectors to another.

Phospholipase D generates the important signaling mediator, phosphatidic acid (PA)

Phospholipase D cleaves the most abundant membrane phospholipid, PC, into phosphatidic acid (PA) and choline (see Figure 7.7). While free choline has no known signaling functions, PA affects a myriad of cellular events, both through the direct binding and recruitment of effector proteins, as well as through the biophysical effects of PA and its immediate metabolites on the lipid bilayer. There are two highly related PLD isoforms in vertebrates, PLD_1 and PLD_2, both of which require phosphoinositides such as PI(4,5)P_2 for activity. PLD_1 is also regulated by PKC, and by small G proteins including Rheb, RalA, and members of the Rho and Arf families. Less is known about the regulation of PLD_2. Activation of PLD likely involves the combination of membrane recruitment and allosteric activation, also seen for many other lipid-metabolizing enzymes.

One important physiological role for PLD is to cooperate with PI(4,5)P_2 and Rho family G proteins to regulate local organization of the actin cytoskeleton. This is accomplished in part through the physical interaction of PLD with PIP5Ks, which are activated by PA. Since both PLD_1 and PLD_2 are dependent on the product of PIP5K [PI(4,5)P_2], this is an example of a mutually reinforcing positive feedback loop that may help generate discrete, localized membrane domains with distinct signaling properties (see Box 7.1).

Another role for PLD is in regulating the intracellular trafficking of membrane vesicles. PLD is a major effector of the Arf family of small G proteins, which function to regulate the transport and targeting of vesicles. The localized activation of PLD at sites of vesicle fusion and fission (during exocytosis and endocytosis, respectively) is thought to be important for lowering the energy barrier for these events. This has been proposed to be due to direct effects of PA on membrane curvature. The PA generated by PLD tends to contain one unsaturated fatty acid; the kink in the acyl chain generated by the double bond makes the shape of PA rather conical, with a small solvent-accessible head connected to a relatively broad hydrophobic anchor. As a result, high PA density on the inner leaflet of the bilayer promotes negative (concave) curvature in the membrane (**Figure 7.12**). The process of fusion of a vesicle with a membrane (or its reverse, the pinching off of a vesicle from a membrane) involves passing through a transient intermediate stage with strongly negative membrane curvature at the "neck." The localized activation of PLD at such sites can ease the transition through this energetically unfavorable state.

Phospholipase D plays a role in mTOR signaling

PA generated by PLD is also required for activation of the **mechanistic target of rapamycin (mTOR)**, a protein kinase that acts as a master regulator of cell growth, survival, and metabolism (**Figure 7.13**). The activity of mTOR is regulated by environmental inputs such as the levels of mitogens, amino acids, oxygen, and ATP. Activated mTOR, in turn, phosphorylates and regulates the activity of proteins that control translation, such as ribosomal S6 kinase (S6K1) and eukaryotic initiation factor 4E binding protein (4E-BP1). In this way, the rate of protein

Figure 7.12

Effect of PA on membrane curvature.
(a) Space-filling model of PA; note the kink induced by a double bond in one fatty acid chain. (b) The triangular cross-sectional shape of PA induces negative curvature when concentrated on the inner leaflet of a membrane bilayer. (c) Transient areas of extreme negative curvature (*blue arrows*) at sites where a membrane vesicle fuses with or pinches off from a membrane. (a, Adapted from B. Alberts, et al., *Molecular Biology of the Cell*, 5th ed. Garland Science, 2008.)

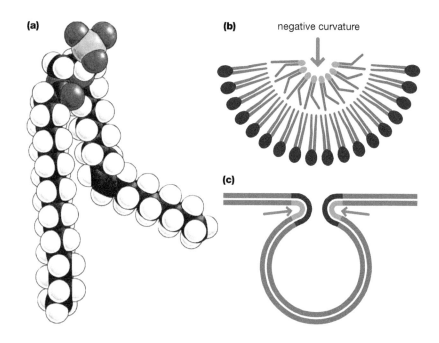

synthesis is coupled to environmental conditions and the metabolic state of the cell. mTOR exists in two multiprotein complexes, termed mTORC1 and mTORC2, which are distinguished by the presence of the regulatory subunits Raptor or Rictor, respectively. The activity of mTORC1 is positively regulated by a small G protein, Rheb, which itself is negatively regulated by the TSC complex, consisting of TSC1 and TSC2. TSC was named for tuberous sclerosis complex, a genetic predisposition to non-malignant tumors, which is due to loss of function mutations in the genes encoding TSC1 or TSC2. It is not surprising that loss of a key negative regulator of the mTOR complex would lead to deregulation of cell growth and survival—a hallmark of tumorigenesis. The TSC complex receives and integrates

Figure 7.13

The mTOR pathway and its regulation by PLD. Activation of the mTOR complex (mTORC1) is mediated by the G protein Rheb, and requires phosphatidic acid (PA). Rheb is inhibited by the TSC1/2 complex. A variety of environmental cues regulate mTOR activity, including the protein kinases AMPK (regulated by cellular ATP levels) and Akt (regulated by signals that promote proliferation, growth, and survival). In addition, phospholipase D (PLD) serves as a central hub for integrating cellular signals and regulating mTOR activity. PI5K, phosphatidylinositol 5-kinase; PIP$_2$, phosphatidylinositol 4,5-bisphosphate.

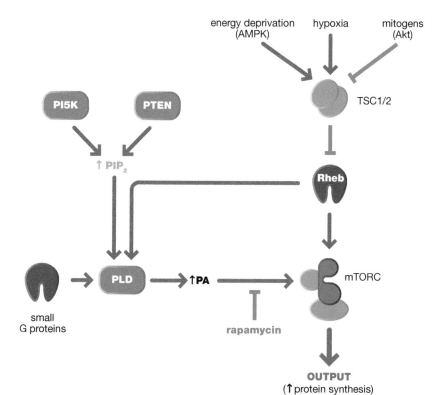

inputs from a number of upstream signals. For example, TSC is phosphorylated and inhibited by the Akt kinase, providing a physiological link between the PI3K–Akt and mTOR pathways in regulating cell growth, proliferation, and survival. Furthermore, mTORC2 (but not mTORC1) phosphorylates Akt at an activating site, providing positive feedback between mTOR and Akt.

PA interacts directly with mTOR and is required for the activation of both mTORC1 and mTORC2. The immunosuppressive drug rapamycin binds to the same site on mTOR as PA, and thus, it is likely that rapamycin inhibits mTOR in mammals by blocking its interaction with PA (mTOR was first isolated as the target of rapamycin). Suppression of PLD activity, like rapamycin treatment, prevents the formation of mTOR complexes, and thus, it is likely that PA facilitates the association of mTOR with the companion proteins Raptor and Rictor. The apparent requirement for PA indicates that PLD plays a critical role in regulating mTOR activity. Consistent with this, PLD activity is frequently elevated in cancer cells and PLD overexpression stimulates mTOR, while experimental inhibition or down-regulation of PLD suppresses mTOR activity and leads to apoptosis in many cancer cell lines.

The details of how PLD activity responds to environmental signals are still under investigation (see Figure 7.13). PLD1 is up-regulated in response to a variety of mitogenic signals, and the mechanism is likely to involve a network of small G proteins that are upstream activators of PLD1, such as RalA and Rheb. PLD activity is elevated in response to amino acids and is dependent on Rheb, and RalA is similarly activated in response to amino acids, directly associates with PLD1, and is required for elevated PLD activity. The dependence of elevated PLD activity on Rheb is an example of *coherent feedforward regulation:* Rheb activates mTOR directly and also activates PLD, which generates a second mTOR activator, PA (see Box 7.1).

A significant conceptual and technical hurdle to clarifying these regulatory relationships is that the PLD–PA–mTOR connection may be specific to mammalian cells, as PA and PLD have not yet been shown to regulate the TOR pathway in more genetically tractable model systems such as yeast and *Drosophila*. Thus, integration of PLD into the regulatory networks controlling mTOR may have been a relatively recent evolutionary innovation.

The metabolism of sphingomyelin generates a host of signaling mediators

The sphingolipid ceramide and its metabolites constitute a large and ever-expanding family of bioactive lipids that regulate a variety of cellular activities. Ceramide can be generated from the abundant membrane lipid sphingomyelin by the action of sphingomyelinase (SMase), or can be generated *de novo* from fatty acids. In turn, ceramide can be broken down by ceramidase (CDase) to sphingosine plus a fatty acid. Both ceramide and sphingosine can directly regulate effector proteins and, furthermore, can be phosphorylated by lipid kinases to generate bioactive derivatives: ceramide 1-phosphate (C1P) and sphingosine 1-phosphate (S1P), respectively (**Figure 7.14**).

This network provides a specific example of the general principle that the levels of individual lipids are intimately interconnected and in dynamic equilibrium. Thus, any disturbance of the equilibrium (for example, activation of a modifying enzyme) will ultimately affect the levels of many different bioactive lipids. In this regard, it is useful to consider the relative abundance of the various players. Sphingomyelin is a rather abundant lipid (comprising up to 30% of plasma membrane phospholipid), and is thus thought to play a largely structural role. Its derivatives are much less abundant, however, so that fairly small changes in the

Figure 7.14

Major ceramide metabolic pathways involved in signaling. Shown are structures of the ceramide lipids, along with the major enzymes that generate them. Reverse reactions also occur in cells but are not shown here. SMase, sphingomyelinase; CDase, ceramidase; SK, sphingosine kinase; CK, ceramide kinase.

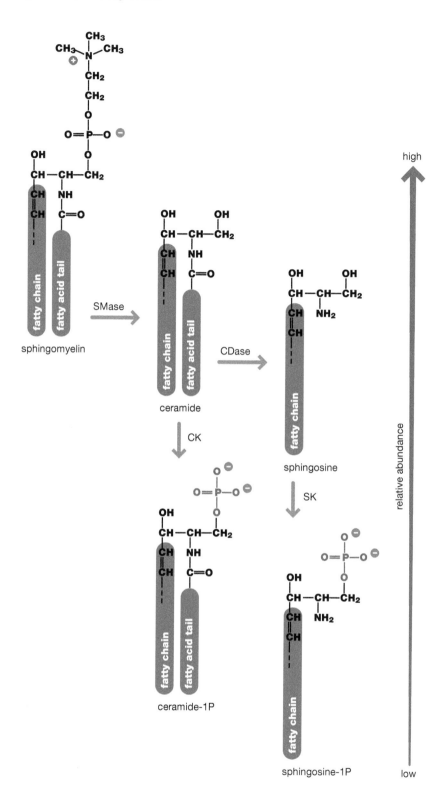

rate of sphingomyelin metabolism can have relatively large effects on the levels of these lipids. The least abundant products, such as S1P, have concomitantly higher potencies; S1P binds to a GPCR with high affinity and thus can activate signaling even at very low (nanomolar) levels.

The signaling activities of each of the products are further constrained by their biophysical properties. Sphingomyelin, with its bulky hydrophobic groups and highly polar head, is essentially confined to just one membrane leaflet. Once the polar head group is removed by SMase, however, ceramide is sufficiently nonpolar that it can readily flip between the inner and outer leaflets. Removal of the fatty acid chain from ceramide by CDase generates compounds with a single aliphatic chain, such as sphingosine and S1P, that are now sufficiently hydrophilic to be able to leave the membrane. Such compounds can exert their effects throughout the cell, or even to other cells throughout the organism.

Sphingolipids play a role in the cellular response to stresses, largely through generation of ceramide and its further metabolites. There are a number of different SMases, which differ in their subcellular localization and specific activation properties. Environmental stresses such as ionizing and ultraviolet radiation, reactive oxygen and nitrogen species, and chemotherapeutic drugs can activate the so-called acid SMases, likely directly through reactive oxygen species and indirectly through phosphorylation by PKCδ. Another class of SMases, the neutral SMases, is activated in response to cytokines such as tumor necrosis factor α (TNFα) and interleukin-1 (IL-1). Downstream targets of the ceramide generated by SMase include the protein phosphatase PP2A and the protease cathepsin D, though many details remain unknown.

S1P, which is generated by the action of sphingosine kinases (SKs) on sphingosine, has emerged as a very potent cell–cell signaling molecule with a variety of developmental and homeostatic roles in metazoans. For example, signaling by a number of cytokines leads to activation of SK and thus to an increase in S1P levels, and this is important for the proinflammatory effects of these cytokines. S1P signaling is also critical for the proper development and function of the endothelial and smooth muscle cells that form blood vessels. Most of the biological effects of S1P in cell–cell signaling are exerted through activation of a class of plasma membrane GPCRs termed the EDG or S1PR family. These receptors are activated by S1P at nanomolar levels, leading to the activation of a variety of heterotrimeric G proteins. S1P may also function intracellularly as a direct signaling mediator, for example in Ca^{2+} homeostasis, but the specific intracellular effectors of S1P are not well characterized.

Phospholipase A$_2$ generates the precursor for a family of potent inflammatory mediators

Inflammation is a physiological response to infection, allergens, or trauma that leads to localized swelling, redness, and pain. These symptoms are caused in large part by the dilation and increased permeability of blood vessels, and by the recruitment of white blood cells (leukocytes) to the site. The eicosanoids are a class of lipid signaling molecules derived from arachidonic acid (AA) that play a critical role in the inflammatory response. They are generated locally by the action of cytosolic PLA$_2$ to generate free AA, which is then further modified by enzymes such as cyclooxygenases and lipoxygenase to generate prostaglandins and leukotrienes, respectively (**Figure 7.15**). These inflammatory mediators act locally (in the cell of origin or nearby cells) through binding to GPCRs, leading to physiological responses. Because of the impact of inflammation on human health, many commonly used drugs target this pathway. For example, aspirin and nonsteroidal anti-inflammatory drugs (NSAIDs) such as ibuprofen and naproxen target the cyclooxygenases (COX1 and COX2) that generate intermediates in prostaglandin synthesis.

Although there is a wide variety of PLA$_2$ isoforms in animal cells, only the cytosolic PLA$_2$α (cPLA$_2$α) form is specific for phospholipid substrates with AA at the *sn*-2 position. The 20-carbon aliphatic chain of AA has four double bonds, and it is the obligate precursor for eicosanoid biosynthesis. Mice lacking cPLA$_2$α are highly deficient in their inflammatory responses, firmly implicating cPLA$_2$α-derived AA in inflammation. In unstimulated cells, cPLA$_2$α is found in the cytosol but rapidly relocalizes to perinuclear membranes upon activation by cytokines, mechanical trauma, or other proinflammatory stimuli. Activation requires binding of Ca^{2+}

Figure 7.15

Generation of inflammatory mediators from arachidonic acid. Regulatory enzymes are shown in *blue*. The action of cytosolic phospholipase A$_2$α (cPLA$_2$α) on phosphatidylcholine generates arachidonic acid and lysophosphatidylcholine (lyso-PC). Arachidonic acid is further processed by the cyclooxygenase (COX) and lipoxygenase (5-LO) pathways to generate potent bioactive lipids (*green*) that signal by binding to G-protein-coupled receptors. Lyso-PC can be further modified to generate platelet-activating factor (PAF), which also signals through a G-protein-coupled receptor. PGH$_2$, prostaglandin H$_2$; LTA$_4$, leukotriene A$_4$.

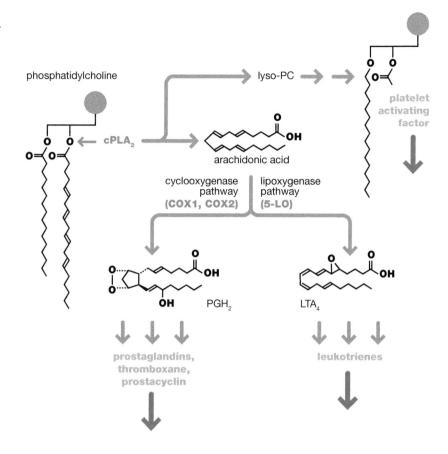

Nuclear receptors are discussed in Chapter 8.

to the C2 domain of cPLA$_2$α, which promotes its association with membranes. cPLA$_2$α is also activated by phosphorylation by MAP kinases such as Erk1/2, and by binding to C1P. Thus, eicosanoid generation is tied into a variety of signal transduction pathways. The other product of PLA$_2$ action on PC is lysophosphatidylcholine, which can then be acetylated to generate another extremely potent inflammatory mediator, platelet-activating factor (PAF), which exerts its biological effects at subnanomolar concentrations by binding to its own GPCR.

In most cell types, cyclooxygenases co-localize with cPLA$_2$α on perinuclear membranes to convert AA into an intermediate, prostaglandin H$_2$ (PGH$_2$), which is then further processed in a cell-type-specific manner to generate a wide variety of specific prostaglandins, prostacyclins, and thromboxanes. In inflammatory cells such as macrophages and mast cells, cPLA$_2$α co-localizes with another enzyme, 5-lipoxygenase (5-LO), to generate the intermediate leukotriene A$_4$ (LTA$_4$), which is further processed to generate a variety of leukotrienes. Specific plasma membrane transporters facilitate efflux of the finished bioactive lipids out of the cell.

The potent biological effects of eicosanoids are due to their binding to a family of more than a dozen related GPCRs, which differ in the spectrum of specific mediators they bind, by their tissue distribution, and by their downstream effectors. Stimulation of these receptors, and the consequent activation of heterotrimeric G proteins, leads to cell-type-specific effects such as contraction of airway smooth muscle cells, vascular leakage, vasodilation, and pain. Interestingly, some eicosanoids have also been shown to bind to nuclear receptors of the PPAR (peroxisome proliferator-activated receptor) class. Although these receptors may be mostly involved in maintaining homeostasis of lipid biosynthetic pathways, it is possible that they may have signaling functions as well.

SUMMARY

The unique chemical properties of lipid molecules cause them to self-organize into lipid bilayers, which form a barrier that is relatively impermeable to most hydrophilic molecules. Lipid species in the membrane can serve as binding sites that are recognized by specific lipid-binding domains. The binding of protein molecules to the lipid bilayer can dramatically enhance their interaction with one another, and many critical signaling interactions take place at membranes. The individual lipid molecules can also serve to encode and transmit signaling information directly. The protein interactions in which lipids participate can be dramatically regulated by modifying enzymes such as lipid kinases and phosphatases, which can covalently modify lipid head groups, thus altering their recognition by distinct proteins. In addition, lipids can be degraded by specific lipases to release products which serve as diffusible signaling mediators that regulate various downstream targets.

QUESTIONS

Answers to these questions can be found online at www.routledge.com/9780367279370

1. What are the possible signaling consequences of a particular membrane lipid being confined to either the inner or outer leaflet of the membrane?

2. Experimental evidence suggests that the plasma membrane is not homogeneous but contains regions with distinct lipid compositions. How might this heterogeneity affect the properties of proteins embedded in or associated with the membrane?

3. What properties make phosphatidylinositol (PI) particularly useful in signaling?

4. Eicosanoids have been shown to bind to G-protein-coupled receptors (GPCRs) and to nuclear receptors (NRs). What are the physical properties of eicosanoids that allow this? How might you experimentally distinguish between GPCR-dependent and NR-dependent effects?

BIBLIOGRAPHY

BIOLOGICAL MEMBRANES AND THEIR PROPERTIES

Cho W (2006) Building signaling complexes at the membrane. *Sci. STKE* 2006(321), pe7.

Groves JT & Kuriyan J (2010) Molecular mechanisms in signal transduction at the membrane. *Nat. Struct. Mol. Biol.* 17, 659–665.

Hancock JF (2006) Lipid rafts: Contentious only from simplistic standpoints. *Nat. Rev. Mol. Cell Biol.* 7, 456–462.

Huang WYC, Boxer SG & Ferrell JE Jr. (2024) Membrane localization accelerates association under conditions relevant to cellular signaling. *Proc. Natl Acad. Sci. UA.* 121(10), e2319491121. doi: 10.1073/pnas.2319491121.

Owen DM, Williamson D, Rentero C & Gaus K (2009) Quantitative microscopy: Protein dynamics and membrane organisation. *Traffic* 10, 962–971.

Sych T, Levental KR & Sezgin E (2022) Lipid-protein interactions in plasma membrane organization and function. *Annu. Rev. Biophys.* 51, 135–156. doi: 10.1146/annurev-biophys-090721-072718.

van Meer G, Voelker DR & Feigenson GW (2008) Membrane lipids: Where they are and how they behave. *Nat. Rev. Mol. Cell Biol.* 9, 112–124.

LIPID-MODIFYING ENZYMES USED IN SIGNALING

Aloulou A, Ali YB & Bezzine S, et al. (2012) Phospholipases: An overview. *Methods Mol. Biol.* 861, 63–85.

Bunney TD & Katan M (2011) PLC regulation: Emerging pictures for molecular mechanisms. *Trends Biochem. Sci.* 36, 88–96.

Burke JE & Dennis EA (2009) Phospholipase A2 structure/function, mechanism, and signaling. *J. Lipid Res.* 50(Suppl), S237–S242.

Hammond GRV & Burke JE. (2020) Novel roles of phosphoinositides in signaling, lipid transport, and disease. *Curr. Opin. Cell Biol.* 63, 57–67. doi: 10.1016/j.ceb.2019.12.007.

Michell RH (2008) Inositol derivatives: Evolution and functions. *Nat. Rev. Mol. Cell Biol.* 9, 151–161.

Suh PG, Park JI, Manzoli L, et al. (2008) Multiple roles of phosphoinositide-specific phospholipase C isozymes. *BMB Rep.* 41, 415–434.

EXAMPLES OF MAJOR LIPID SIGNALING PATHWAYS

Di Paolo G & De Camilli P (2006) Phosphoinositides in cell regulation and membrane dynamics. *Nature* 443, 651–657.

Funk CD (2001) Prostaglandins and leukotrienes: Advances in eicosanoid biology. *Science* 294, 1871–1875.

Hannun YA & Obeid LM (2008) Principles of bioactive lipid signalling: Lessons from sphingolipids. *Nat. Rev. Mol. Cell Biol.* 9, 139–150.

Krauss M & Haucke V (2007) Phosphoinositide-metabolizing enzymes at the interface between membrane traffic and cell signalling. *EMBO Rep.* 8, 241–246.

Ogretmen B (2018) Sphingolipid metabolism in cancer signalling and therapy. *Nat. Rev. Cancer* 18(1), 33–50. doi: 10.1038/nrc.2017.96.

Roth MG (2008) Molecular mechanisms of PLD function in membrane traffic. *Traffic* 9, 1233–1239.

Saxton RA & Sabatini DM (2017) mTOR signaling in growth, metabolism, and disease. *Cell* 168(6), 960–976. doi: 10.1016/j.cell.2017.02.004.

Sun Y & Chen J (2008) mTOR signaling: PLD takes center stage. *Cell Cycle* 7, 3118–3123.

Tei R, Morstein J, Shemet A, et al. (2021) Optical control of phosphatidic acid signaling. *ACS Cent. Sci.* 7(7), 1205–1215. doi: 10.1021/acscentsci.1c00444.

Information Transfer across the Membrane

8

Conveying information from the extracellular environment into the interior of the cell is perhaps the most fundamental hurdle that must be overcome in cell signaling. While the water-impermeable plasma membrane provides an essential function in shielding the cell's contents from the environment, it necessarily imposes a barrier to free communication. The membrane physically isolates the cytosol from many environmental cues, particularly those provided by water-soluble signaling molecules such as peptides and hydrophilic small molecules but also other cues such as those from adjoining cells or the extracellular matrix. However, it is essential for cells to continuously adapt their behavior to their environmental conditions. This is especially true in the case of cells in multicellular organisms, where extensive communication between cells is essential for proper development and function. In this chapter, we will examine the various strategies that have evolved to allow selective signals to be transmitted through the plasma membrane, focusing primarily on the mechanisms used by transmembrane receptors in metazoans.

PRINCIPLES OF TRANSMEMBRANE SIGNALING

Transmembrane signaling depends on the ability of receptors on the cell to receive input from the environment. Receptors are proteins that, when bound to specific signaling molecules (*ligands*), undergo a change in activity that transmits a signal. Because their activity depends on whether or not they are bound to a ligand, they serve to convert information on extracellular ligand concentrations into an intracellular activity. Most receptors are found on the plasma membrane, where they are exposed to the extracellular environment. A few receptors, those that respond to signals that can freely cross the plasma membrane, are located within the cell.

DOI: 10.1201/9780429298844-8

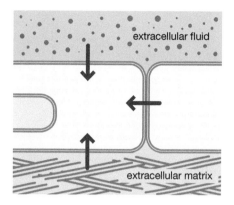

Figure 8.1

Cells receive a variety of environmental cues. Adjacent cells, the extracellular matrix, and soluble factors from the extracellular fluid all provide signals that must be accurately received and interpreted by the cells of multicellular organisms.

The cell must process and respond to a diversity of environmental cues

Before discussing specific signal transduction mechanisms, it will be useful to consider the types of information the cell needs to gather from its environment (**Figure 8.1**). The simplest types of signals, which are important even for free-living unicellular organisms such as yeast, are the levels of nutrients, oxygen, and many other raw materials needed to build and maintain the cell. The levels of such compounds will determine the activity of cellular biosynthetic pathways, and will also affect behavior such as growth and motility, so the organism can adapt to take full advantage of abundant resources while avoiding suboptimal conditions when possible.

The number of critical inputs is substantially larger for multicellular organisms, however. A wide variety of soluble signaling molecules secreted by other cells are present in the extracellular fluid bathing the cell, including factors such as hormones, cytokines, and mitogens. **Hormones** are secreted molecules that regulate cellular activities, often at distant sites in the body (in humans, familiar examples include growth hormone, insulin, and estrogen). In many cases, hormones are secreted by specialized **endocrine** glands and tissues. Hormones can be either polypeptides, which are derived from the processing of larger proteins, or small molecules synthesized from organic building blocks. More specialized types of peptide hormones include **cytokines**, which regulate various aspects of immune cell function, and **mitogens**, which induce the proliferation of cells. Mitogens are often also called growth factors, but this term can lead to confusion between mitogens and true **growth factors**, which specifically promote cell growth (an increase in cell bulk).

In addition to soluble signaling molecules, cells must also respond to more local, fixed cues, such as molecules present on the surface of adjoining cells. During development and in adult tissues, it is essential that cells be able to adapt their behavior in harmony with the properties of surrounding cells. For example, adjoining epithelial cells must make tight contacts with their neighbors to form an epithelial sheet. And during development of the nervous system, the growth cones of elongating axons must interpret attractive or repulsive cues on adjacent cells, in order to make connections to the appropriate target cells. Cells need to be able to detect and react to both the presence of molecules on the surface of adjacent cells, as well as the local density of those molecules.

In addition to cell-surface molecules, cells must also interpret signals from the **extracellular matrix (ECM)** in which they are embedded. The ECM is secreted by cells and consists of fibrillar proteins such as collagen that provide tensile strength, proteoglycans, the polysaccharide hyaluronic acid, and factors that modulate cell adhesion and motility such as fibronectin and its relatives. The ECM plays a structural role in tissues, but also plays a more active role in regulating activities such as cell migration, differentiation, and tissue organization. Some ECM components directly signal to cells through their presence and local density, and the ECM can also serve as a reservoir for binding and presenting soluble hormones and signaling proteins to cells. In addition, the ECM can provide mechanical cues to the cell, through its resistance to being deformed as the cell makes attachments and applies forces to it.

Cells must also respond to a wide variety of other types of physical signals. In multicellular animals, among the most important of these are electrical signals propagated by neurons. These will not be explicitly discussed in this chapter. In addition, specialized cells can respond to sensory stimuli such as light, pressure, odors, and other cues important for the life of multicellular organisms.

Three general strategies are used to transfer information across the membrane

Although there is tremendous diversity in extracellular cues, the cell uses just three general mechanisms to transduce those cues through the plasma membrane to the cytosol (**Figure 8.2**). First, in a relatively few cases, the signal itself can passively cross the plasma membrane. This specialized class of signaling molecules includes the membrane-permeable gas *nitric oxide* (*NO*) and lipid-soluble hormones including steroid hormones and their relatives. For these signals, no specialized mechanism is required to traverse the membrane, and the cytosolic effectors for these signals can bind them directly. The signaling molecules exert their influence by binding to **intracellular receptors**. This is by far the simplest and most direct mechanism of transmembrane signaling.

The second major strategy involves **membrane channels**. These are transmembrane pores, assembled from protein subunits, which selectively pass certain classes of small molecules such as ions. **Gated channels** can be opened and/or closed in response to specific signals, allowing them to function as receptors that transmit information across the membrane. Neurons conduct electrical signals by the rapid opening and closing of channels in response to changes in membrane potential, ion concentrations, and neurotransmitters.

The final broad strategy involves **transmembrane receptors**. These proteins consist of a ligand-binding domain on the extracellular face of the membrane, one or more hydrophobic transmembrane segments, and an intracellular portion that couples to downstream signaling effectors. In many cases, the intracellular part of the receptor itself has intrinsic catalytic activity, for example a protein kinase domain, which is activated by ligand binding. In other cases, the receptor is associated noncovalently with other effector proteins, which relay the information that the ligand-binding domain is occupied. For most cells in multicellular organisms, this is by far the most prevalent and versatile mechanism for transmembrane signaling, coupling a wide variety of extracellular stimuli to changes in a number of intracellular activities.

Many drugs target receptors

The binding sites of receptors that bind to protein ligands (such as peptide hormones) are typical of other protein–protein interaction surfaces (see, for example, the complex between growth hormone and its receptor, Figure 2.4). However, receptors can bind and respond to a wide variety of other ligands, including simple ions and small organic molecules. Regardless of the nature of the ligand, for

(a) membrane permeable signals **(b) gated channels** **(c) transmembrane receptors**

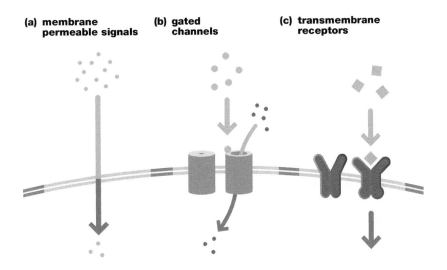

Figure 8.2

Three ways for a signal to cross the plasma membrane. (a) A few signaling molecules can passively diffuse across the membrane into the cytosol. (b) Some signaling molecules bind to and regulate the opening of membrane channels. (c) Many signaling molecules bind to transmembrane receptors, leading directly or indirectly to activation of intracellular enzymes.

binding to convey information, it must lead to physical changes in the receptor itself, either in its conformation or in its association with other molecules.

Many drugs that are used to treat human medical conditions target the ligand-binding sites of cell-surface receptors. This is particularly true for the class of G-protein-coupled receptors (GPCRs), which control a host of medically relevant processes such as heart rate, blood pressure, inflammation, and neurotransmission. Compounds that mimic natural ligands that cause activation of the receptor and transmission of downstream signals are termed **agonists**. Morphine and other opiates are agonists for the opioid receptor. On the other hand, some compounds may bind to a receptor and fail to evoke an activating response—they may, for example, bind to the ligand-binding site without inducing changes needed to transmit a signal. Furthermore, by binding the receptor at a site that is the same as (or overlaps with) the site for natural activating ligands or agonists, they can prevent normal receptor activation. Such compounds are termed **antagonists**. Antagonists are analogous to enzyme inhibitors, in that they inhibit receptor activity in a concentration-dependent fashion. Loratadine (Claritin®), an antihistamine, is a histamine H1 receptor antagonist; so-called beta blockers, used to lower blood pressure and control heart rate, are β-adrenergic receptor antagonists.

TRANSDUCTION STRATEGIES USED BY TRANSMEMBRANE RECEPTORS

For receptors whose ligands cannot pass through the plasma membrane, information transfer depends on somehow conveying to the intracellular part of the receptor whether or not ligand is bound to the extracellular part. This is not a trivial problem, as the two parts of the receptor are separated by the hydrophobic lipid bilayer. The intracellular and extracellular portions of the receptor, which are hydrophilic, cannot move through the membrane to physically interact with each other, and thus, any communication must be through the relatively featureless hydrophobic transmembrane segments. These typically consist of rigid α helices ~20 amino acids in length, with their hydrophobic side chains projecting outward into the plane of the membrane (**Figure 8.3**). With a few exceptions, only two general solutions to this problem are used in biological receptors: concerted conformational changes for those receptors with multiple membrane-spanning segments (multiple-pass receptors), and the dimerization or oligomerization of receptors that span the membrane only once (single-pass receptors).

Receptors with multiple membrane-spanning segments undergo conformational changes upon ligand binding

Many receptors, including the large class of GPCRs and all membrane channels, contain multiple membrane-spanning segments. For GPCRs, binding of ligands to the extracellular portion of the receptor induces concerted conformational changes that involve alterations in the relative packing of the seven transmembrane helices, leading to conformational changes on the intracellular face of the receptor. As discussed in more detail below, these changes increase the affinity of the receptor for G proteins, thus activating downstream signaling. The fact that there are multiple transmembrane helices tightly packed against one another allows a change of conformation (relative orientation of the helices) to be transmitted through the membrane. By contrast, such a change in relative orientation is not possible for a single transmembrane segment, which has no adjacent helices to "push" against.

Gated channels represent the other major class of receptors where ligand binding is converted into conformational change. Membrane channels consist of a number of similar or identical subunits arranged in a ring structure in the plane of the membrane. Each subunit contains multiple membrane-spanning segments, and in the

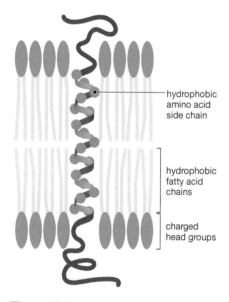

Figure 8.3

hydrophobic amino acid side chain

hydrophobic fatty acid chains

charged head groups

A helical transmembrane segment.
For proteins that traverse the lipid bilayer, the transmembrane portion almost always consists of a rodlike α-helical segment comprised of 20–25 amino acids. The side chains of the amino acids in the transmembrane segments are hydrophobic (*blue circles*), interacting favorably with the hydrophobic fatty acid chains in the interior of the membrane bilayer.

case of ligand-gated channels, a ligand-binding site is present on the side facing the extracellular environment. As we will discuss later in this chapter, ligand binding induces concerted conformational changes that alter the permeability of a pore at the center of the ring. Channel opening can thus lead to very rapid and dramatic changes in the intracellular concentration of the molecules that are allowed to pass through which, in turn, can have widespread effects on intracellular reactions.

Receptors with a single membrane-spanning segment form higher-order assemblies upon ligand binding

For those receptors that span the membrane only once, the extracellular and intracellular domains are connected only by a single, hydrophobic stalk bobbing in a fluid planar membrane. Concerted conformational changes cannot be transmitted across the membrane by single-pass receptors, and so they must adopt a completely different strategy for conveying to the cytosol the binding state of the extracellular domain. For these receptors, signaling depends on changes in the interaction of receptor molecules with each other upon ligand binding (**Figure 8.4**). In many cases, this involves receptor homo- or heterodimerization: unliganded receptors are monomeric or only loosely associated, whereas ligand binding causes two receptor molecules to closely interact with each other. The simplest way that this can be done is if the ligand itself can simultaneously bind to two different receptor molecules. For example, platelet-derived growth factor (PDGF) is a dimer in its native state and so can bind to two receptors. Thus, low concentrations of PDGF will rapidly induce the formation of receptor dimers on the membrane. Receptor dimerization can also be induced by other specific mechanisms; for example, by a single ligand that has two distinct receptor-binding sites (as in the case of growth hormone), by ligands that are physically clustered on the surface of cells or the extracellular matrix, or by ligands that induce changes in the conformation of the receptor that increase the affinity of receptors for each other (as in the case of epidermal growth factor, EGF).

Why does a receptor dimer, in which the intracellular domains of the two receptor molecules are closely apposed to each other, have different activities from the individual receptors in isolation? In general, dimerization or oligomerization facilitates enzymatic reactions that would not be favored in the monomeric state. For example, when the intracellular portion of the receptor has protein kinase activity, dimerization makes it much more likely that the receptors will phosphorylate each other due to their proximity (**Figure 8.5**). An example is the PDGF receptor, which has tyrosine kinase activity. Such **transphosphorylation** can have two important consequences, both of which play a role in downstream signaling. In most cases, transphosphorylation at a specific site in the *activation loop* of the kinase domain is sufficient to stably increase its catalytic activity. Once the receptor is activated, it can then phosphorylate other intracellular targets, including its dimer partner. In addition, transphosphorylation can create binding sites for phosphodependent modular protein binding domains such as SH2 domains, thereby recruiting cytosolic effector proteins containing these domains to the receptor.

Some receptors use a hybrid approach that combines aspects of concerted conformational changes and induced dimerization. For example, the insulin receptor (a receptor tyrosine kinase) consists of two identical sets of subunits covalently linked by disulfide bonds (see Figure 8.8). Thus, insulin binding does not affect the dimerization state *per se*, but instead induces conformational changes in the extracellular domain that are transmitted to the intracellular catalytic domains via changes in the relative orientation of the two transmembrane helices. These changes position the catalytic domains to facilitate transphosphorylation and catalytic activity only when insulin is bound. In the case of the epidermal growth factor receptor (EGFR) family of receptor tyrosine kinases, ligand binding promotes dimerization of the extracellular domains, which in turn causes the

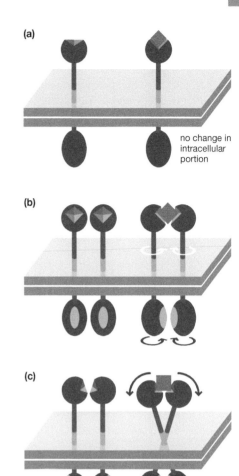

(a)

no change in intracellular portion

(b)

(c)

Figure 8.4

Multiple transmembrane segments allow receptors to transduce information about ligand binding through the membrane. (a) For a receptor that contains a single transmembrane segment, there is no straightforward mechanism by which the conformational changes in the extracellular domain induced by ligand binding can be transmitted to the intracellular portion. (b, c) When a receptor dimerizes or contains multiple transmembrane segments, ligand binding may induce conformational changes in the intracellular portion, for example, by rotation of the transmembrane portions relative to each other.

 Activation of kinases by activation-loop phosphorylation is discussed in Chapter 3.

Figure 8.5

Activation of receptor tyrosine kinases.
Ligand binding induces dimerization of
the receptor. The tyrosine kinase catalytic
domains then rapidly transphosphorylate
each other on the "activation-loop" site on
the kinase domain itself, leading to increased
kinase activity, and on other sites that
serve as docking sites for cytosolic SH2
domain-containing proteins.

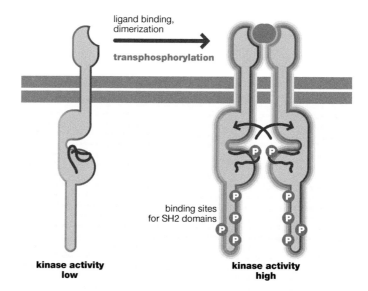

two catalytic domains to form a catalytically active, asymmetric dimer. Another
example is provided by bacterial chemotaxis receptors, which are constitutively
dimerized but transmit information on ligand binding through subtle changes in
the relative orientation of their helical intracellular domains, which couple to intracellular effectors.

Receptor clustering confers advantages for signal propagation

The general strategy of using receptor dimerization or oligomerization to transmit
signals confers a number of interesting properties to the signaling pathways that
they regulate.

The primary effect of dimerization or oligomerization is to greatly increase the
local concentration of the receptor for its partner(s), including any proteins that
may be bound to it. Since the rate of a bimolecular reaction depends on the concentration of the two reactants, increasing the local concentration of receptors can
greatly increase reaction rates when either the receptor itself or receptor-associated
proteins can enzymatically modify each other. Furthermore, this simple property
provides a mechanism to make signal responses more switchlike, responding in an
all-or-none fashion to changes in ligand concentration.

This principle can be illustrated by the case of receptors that are coupled to nonreceptor tyrosine kinases, which include the receptors for a wide variety of cytokines
and hormones, adhesion receptors, and the B cell and T cell receptors on the surface of lymphocytes. As will be discussed later in this chapter, all these different
receptors share the property that the transmembrane receptor itself is noncovalently associated with an intracellular tyrosine kinase (the specific kinase varies
depending on the particular receptor). In the absence of stimulation, these tyrosine
kinases adopt an inactive conformation, but will occasionally "flip" into an active
conformation for a moment before switching back to the more stable inactive conformation. Furthermore, because cytosolic phosphatase activity is relatively high,
any substrate that might be phosphorylated during the brief period when the kinase
is active is likely to be dephosphorylated rapidly. Thus, in the unstimulated state,
both kinase activity and substrate phosphorylation are low.

The situation is quite different when receptors (and associated kinases) dimerize or
associate into higher-order structures upon ligand binding (**Figure 8.6**). Now, whenever a kinase briefly adopts the active conformation, it is situated in very close proximity to another kinase molecule (associated with the dimer partner), as well as the

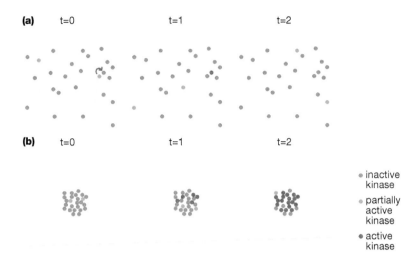

(a) t=0   t=1   t=2

(b) t=0   t=1   t=2

• inactive kinase

• partially active kinase

• active kinase

Figure 8.6

The effect of clustering on kinase activation. Typically, a kinase can exist in three different states: inactive (*pale brown*), partially active (*pale pink*), and phosphorylated and fully active (*pink*). The unphosphorylated kinase rapidly flips between the inactive and partially active states, with the majority of the kinase being in the inactive state at any given time. The partially active kinase can phosphorylate and activate any adjacent kinase molecule during the brief period when it is active (arrow). This phosphorylated kinase remains active until cellular phosphatases remove the phosphate, reverting the kinase back to the inactive state. (a) When the kinase molecules are sparsely and evenly distributed, partially active kinase rarely has the opportunity to activate other kinase molecules, and phosphatases rapidly inactivate the resulting activated kinase before it has a chance to activate other molecules. (b) By contrast, when kinases are clustered together, partially active kinase can phosphorylate and activate multiple adjacent kinases which, in turn, can activate additional adjacent kinases. Soon, most of the kinases in the cluster are phosphorylated and activated. Cellular phosphatases cannot easily inactivate the cluster, because any dephosphorylated kinase is rapidly rephosphorylated by its neighbors. Time steps are indicated by t = 0, t = 1, and t = 2.

second receptor molecule itself. It is therefore very likely that these substrates will be phosphorylated before the catalytic domain reverts to the inactivate conformation. If the kinase manages to phosphorylate the activation-loop site of the second kinase, that second kinase will now be stably activated, and will be highly likely to phosphorylate the first kinase and its associated receptor, thus stably activating the first kinase as well. Furthermore, if either kinase or receptor should be dephosphorylated by phosphatases, it is likely to be rapidly rephosphorylated by its neighbors. Thus, the net effect of dimerization or clustering is high kinase activity and high levels of receptor phosphorylation; both of these states can transmit downstream signals (by phosphorylating other proteins and by binding to downstream effectors, respectively).

Another potential advantage of receptor clustering is that it allows information about the activity of one receptor or receptor dimer to be transmitted to multiple other receptors, allowing a higher order of information processing. In the example above, clustering of kinase-associated receptors ensures that activation of the entire cluster is likely to be relatively stable, even if an individual receptor becomes momentarily inactivated by ligand dissociation or dephosphorylation. More generally, if one receptor is bound to a particular downstream effector, that effector may be able to modify the activity of other receptors within the cluster.

In a broader sense, clustering can promote the formation of phase-separated membrane domains, with high concentrations of receptors, receptor-binding proteins, and their substrates (a type of *biomolecular condensate*). Such domains can have very different properties than the surrounding areas of the membrane. Proximity makes intermolecular interactions and reactions much more efficient than if molecules needed to diffuse throughout the cell to encounter their substrates or binding partners. Some cells contain very specific subcellular compartments where particular signaling components are clustered together even in the absence of stimulation, which minimizes the time needed for components to diffuse and increases the efficiency of interactions. Such specializations are often seen in systems where speed of the response is important, for example, neuromuscular junctions and the photosensitive organelles of photoreceptor cells in the eye.

Clustering also can increase the apparent affinity of receptors for their ligands and for downstream effectors compared to isolated monomers, leading to more efficient binding. When receptors are sparsely distributed, whenever a binding partner dissociates from the receptor, it is likely to diffuse away before it has the opportunity to rebind. However, when there is a very high local concentration of receptors in the vicinity, as when receptors are clustered, the binding partner is much more likely to rebind to another receptor before it has a chance to escape by diffusion (see Figure 2.11). This effect is particularly strong when the overall density of receptors is relatively sparse, and when the fraction of receptors that are bound to ligand is low, both conditions that are common for physiological receptors.

 Biomolecular condensates are discussed in Chapter 5.

G-PROTEIN-COUPLED RECEPTORS

GPCRs are by far the most abundant class of receptors in most eukaryotes, with ~900 distinct GPCRs found in the human genome. All of these receptors share the same overall topology, consisting of an extracellular N-terminus, an intracellular C-terminus, and seven membrane-spanning helices that are connected by hydrophilic loops that project into the cytosol or extracellular environment [thus they are often referred to as seven-transmembrane receptors (7-TMRs) or heptahelical receptors]. GPCRs respond to a wide variety of ligands, including small molecules, polypeptides, and lipids, and they signal by coupling ligand binding to the activation of heterotrimeric G proteins.

G-protein-coupled receptors have intrinsic enzymatic activity

G protein signaling is introduced in Chapter 3.

The broad outlines of GPCR signaling are depicted in **Figure 8.7a.** When an activating ligand binds, concerted conformational changes are transmitted to the intracellular aspect of the receptor. The activated receptor then associates with a GDP-bound

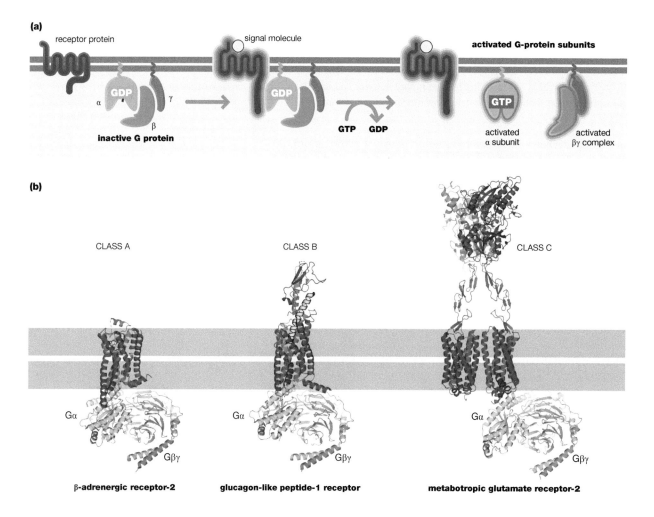

Figure 8.7

Signaling by G-protein-coupled receptors. (a) Binding of ligand to the GPCR leads to nucleotide exchange on the Gα subunit, and its dissociation from the Gβγ subunits. Both the activated Gα and Gβγ subunits can bind to and regulate downstream effector proteins. Gα and Gγ subunits associate with the membrane via covalently linked lipid groups (wavy lines). (b) Examples of high-resolution structures of representative G-protein-coupled receptors (*purple*). Gα is depicted in *green*, Gβ in *blue*, and Gγ in *pink*. Ligand is depicted in *yellow*. Approximate location of the lipid bilayer is indicated by *gray box*. (Structures courtesy of Aashish Manglik; PDB 3SN6, 6X18, 7MTS.)

heterotrimeric G protein and acts as a *guanine nucleotide exchange factor* (*GEF*) to promote release of GDP and binding of GTP to the Gα subunit. Conformational changes in the Gα subunit induced by GTP binding then lead to its dissociation from both the receptor and from the Gβγ subunits. Depending on the specific subunits and effectors, either the Gα or Gβγ subunits then bind to downstream effectors to modulate their activity. Direct effectors include ion channels, phospholipase C (PLC), adenylyl cyclases, phosphodiesterases, and GEFs for Rho family GTPases.

The last decade has seen an explosion of structures of GPCRs, due to remarkable advances both in X-ray crystallography and cryo-electron microscopy (cryo-EM) of membrane proteins. To date, the structures of nearly 100 different GPCRs have been solved, often in complex with various ligands and partner proteins. These structures have given us a far better sense of the wide variety of GPCR structures, especially for the different subfamily classes of GPCRs that were first delineated by sequence similarity. **Figure 8.7b** shows examples of the major classes of GPCRs, including classes A, B, and C.

Class A GPCRs, exemplified by the β-adrenergic receptor-2, form by far the largest family (over 700 members), and the major target for most FDA-approved GPCR-targeted drugs. These receptors are relatively small, lacking significant extracellular regions outside of the membrane. Their natural ligands tend to be hormones, peptides, chemokines, lipids, or other small molecules.

Class B GPCRs come in two major classes, the secretin receptors (class B1) which bind hormones involved in metabolism, and adhesion GPCRs (class B2) which play a role in cell–cell adhesion and cell migration. These receptors have a large extracellular domain. An example of the B1 family, the glucagon-like peptide-1 receptor (GLP-1R), is shown in Figure 8.7b. This is the target of GLP-1 agonist drugs like Ozempic® (semaglutide) that reduce blood glucose levels and obesity.

Class C GPCRs are distinguished by their very large extracellular domain and the requirement for constitutive receptor dimerization. The metabotropic glutamate receptors (neurotransmitter receptors) belong to this family and the mGluR2 family member is shown in Figure 8.7b.

These structures show the remarkable structural variety found in this large family of cellular receptors, explaining in part how they can be used to sense such a wide variety of ligands and physiological inputs. At the same time, these structures also show how well conserved the seven-transmembrane helix core is, as well as the interactions with the intracellular heterotrimeric G protein.

It is important to note that while there are many hundreds of GPCRs, the number of heterotrimeric G proteins to which they couple is much more limited. For example, in humans there are fewer than 20 Gα subunits, divided into four classes based on their effectors. Since there are far more GPCRs than downstream G proteins and effectors, activation of many different GPCRs is likely to lead to the same signal output. Furthermore, many GPCRs can interact with and activate multiple classes of heterotrimeric G proteins. For these reasons, often one cell will contain only a few different GPCRs, and therefore be specialized to respond specifically to a certain class of signaling molecule. This concept is best exemplified in the olfactory system, which mediates our sense of smell. Many hundreds of different olfactory GPCRs are encoded in the vertebrate genome, each of which responds to a different class of odorants. All of these olfactory GPCRs are coupled to Gα$_s$ and the activation of adenylyl cyclase. If a cell were to express many different odorant receptors, it would be unable to discriminate which one was activated because they all couple to the same effectors. For this reason, each sensory cell of the olfactory epithelium expresses just a single odorant receptor, so that it will respond only to a particular class of odorants.

Signaling by GPCRs can be very fast and lead to enormous signal amplification

The speed, magnitude, and duration of downstream signals propagated by GPCRs are dependent on a number of different steps. First, ligand binding must induce conformational changes in the receptor that increase its affinity for heterotrimeric G proteins and render it able to act as a GEF. After G protein binding, nucleotide exchange must occur on the $G\alpha$ subunit, whereupon the G protein subunits can dissociate and bind their effectors. Once the G protein dissociates from the receptor, another G protein can take its place. Repeated cycles of G protein activation allow signal amplification (multiple G proteins activated per activated receptor). Each activated G protein (either the $G\alpha$ or $G\beta\gamma$ subunits) can stably bind to and activate a single effector. But because these effectors are either enzymes or ion channels, a single activated effector can further amplify a signal considerably. After activation, the system then resets back to baseline through several mechanisms. Activated $G\alpha$ subunits eventually hydrolyze bound GTP through their intrinsic GTPase activity, and this can be stimulated considerably by regulator of G protein signaling (RGS) proteins, which act as *GTPase-activator proteins* (*GAPs*) for the $G\alpha$ subunit. The receptor itself can be deactivated either by ligand dissociation or by desensitization pathways involving phosphorylation of the receptor by *G-protein-coupled receptor kinases* (*GRKs*), as discussed in more detail below.

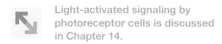 Light-activated signaling by photoreceptor cells is discussed in Chapter 14.

Signaling by GPCRs can be extremely rapid. For example, in the retinal rod and cone cells of the eye, activation of rhodopsin (a specialized GPCR that responds to the light-induced isomerization of bound retinal) can occur within a few milliseconds, and effectors are activated within a few hundred milliseconds. In this case, a single activated receptor can activate up to ~200 G proteins per second. Thus, not only is the signal very fast, but it is also highly amplified, as each activated G protein can go on to activate downstream effectors. This is not surprising, given the enormous evolutionary pressure on organisms to process visual stimuli as rapidly as possible. On the other hand, in more typical GPCR signaling pathways, such as stimulation of the β-adrenergic receptor leading to activation of adenylyl cyclase, activation of $G\alpha_s$ takes ~0.5 seconds, so both the speed and potential signal amplification (the number of molecules of G protein activated per second by activated receptor) are considerably lower.

TRANSMEMBRANE RECEPTORS ASSOCIATED WITH ENZYMATIC ACTIVITY

Except for the ion channels, all transmembrane receptors are directly or indirectly linked to intracellular enzyme activities. For many receptors, the enzymatic activity is encoded in the same polypeptide chain as the ligand-binding activity; such receptors are said to have intrinsic enzymatic activity. These include the GPCRs discussed above, in which ligand binding activates GEF activity, and receptors with protein kinase, protein phosphatase, or guanylyl cyclase activity. Receptors can also be noncovalently associated with enzymes such as protein kinases and proteases.

Receptor tyrosine kinases control important cell fate decisions in multicellular eukaryotes

Receptor tyrosine kinases (RTKs) are transmembrane receptors with intracellular protein tyrosine kinase domains. There are ~50 RTKs in the human genome, which can be divided into families based on their sequence similarity (**Figure 8.8**). These receptors bind polypeptide ligands that regulate cell proliferation, growth, differentiation, or migration; thus, they play a critical role in development and tissue homeostasis in multicellular organisms (metazoans). Most RTKs are activated by dimerization induced by ligand binding, which allows the tyrosine kinase catalytic

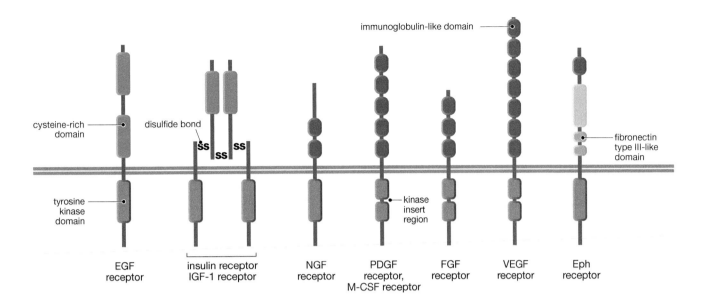

immunoglobulin-like domain

cysteine-rich domain

disulfide bond

fibronectin type III-like domain

tyrosine kinase domain

kinase insert region

EGF receptor

insulin receptor IGF-1 receptor

NGF receptor

PDGF receptor, M-CSF receptor

FGF receptor

VEGF receptor

Eph receptor

Figure 8.8

Receptor tyrosine kinases. The domain structures of the major families of receptor tyrosine kinases. EGF, epidermal growth factor; IGF, insulin-like growth factor; NGF, nerve growth factor; PDGF, platelet-derived growth factor; M-CSF, macrophage-colony-stimulating factor; FGF, fibroblast growth factor; VEGF, vascular endothelial growth factor; Eph, ephrin. (Adapted from R. Weinberg, *The Biology of Cancer.* Garland Science, 2013.)

domains of the dimerized receptors to transphosphorylate each other on the "activation loop" and thus stably activate each other. In some cases, dimerization can also directly induce allosteric changes that increase kinase activity. The specific details of the catalytic activation mechanism vary among the different RTKs, but involve some combination of active-site rearrangement and unblocking of the substrate-binding cleft, leading to a catalytically active kinase domain capable of phosphorylating the dimer partner and other substrates.

In some cases, for example, the EGF and PDGF receptors, autophosphorylation of the receptor on multiple sites is sufficient to recruit effector proteins containing SH2 or PTB domains, and thus transmit downstream signals. These effectors include the regulatory subunit of phosphatidylinositol 3-kinase (PI3K), PLC-γ, and adaptor proteins such as Grb2 and/or Shc that recruit activators of Ras (see **Figure 4.12**). In other cases, such as the insulin receptor and fibroblast growth factor (FGF) receptor, receptor activation leads to the phosphorylation of *scaffold proteins* on multiple sites, and it is these scaffold proteins that then serve as the primary sites for recruitment of downstream effectors. Such scaffolds include IRS1 for the insulin receptor and FRS2 for the FGF receptor. In some cases, the effectors that are recruited are themselves phosphorylated by the receptor, which further increases their activity or recruits additional effectors. But regardless of the specific details, activation of all RTKs leads to increased tyrosine phosphorylation of the receptor and receptor-associated proteins, and to the recruitment of a variety of downstream effector proteins that couple to various signaling pathways including the Ras/Raf/MAPK (mitogen-activated protein kinase), PI3K/Akt, and PLC/Ca^{2+}/protein kinase C pathways.

TGFβ receptors are serine/threonine kinases that activate transcription factors

In metazoans, the transforming growth factor β (TGFβ) receptor family is the only class of receptors with intrinsic serine/threonine kinase activity (other eukaryotes, such as plants, have a wide variety of such receptors however). The TGFβ receptors bind to ligands such as TGFβ, activin, bone morphogenetic proteins (BMPs), and nodal that regulate developmental cell fate and proliferation. In humans, there are ~12 distinct receptors in the TGFβ family, which can be functionally divided into two classes (type I and type II). All have a similar overall structure, with a single membrane-spanning domain and an intracellular serine/threonine kinase domain.

TGFβ receptors couple to transcriptional activation in a rather direct fashion, by phosphorylating and activating the SMAD family of transcription factors (**Figure 8.9**).

Figure 8.9

Signaling by TGFβ receptors. Type I and type II receptors exist as homodimers, which associate into a heterotetrameric complex upon ligand binding. The type II receptor phosphorylates the type I receptor, activating it and increasing its affinity for R-SMADs (*light purple*). Phosphorylation of R-SMAD by the type I receptor leads to its dissociation from the receptor and formation of a heterotrimeric complex with SMAD4 (*light green*). This complex is then free to translocate to the nucleus and regulate transcription of TGFβ-responsive genes.

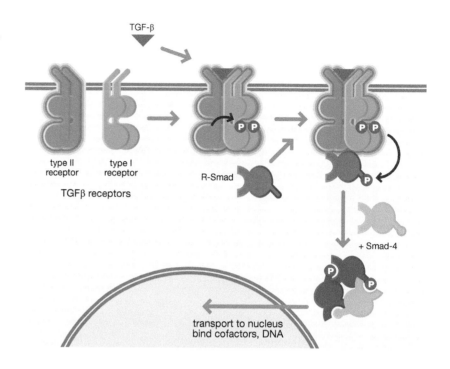

As we have seen for other single-pass receptors, binding of the receptors to ligand induces dimerization, but in this case preexisting dimers of type I and type II receptors associate upon ligand binding, generating a heterotetrameric activated complex. In this complex, the type II receptor, which is constitutively active, can phosphorylate the type I receptor through proximity. This has two important consequences: it greatly increases the affinity of the receptor for a class of SMAD proteins called the R-SMADs, and it increases the kinase activity of the type I receptor itself. The type I receptor then phosphorylates the associated R-SMAD, leading to its dissociation from the receptor. The phosphorylated R-SMAD can then form heterotrimeric complexes with SMAD4, the so-called "co-SMAD," and this complex is then competent to be transported into the nucleus and bind specific chromatin sites where it can activate transcription of TGFβ-responsive genes.

It is interesting to note the very close parallels between the TGFβ receptor signaling mechanism and that used by receptors that couple to tyrosine kinases, despite the dissimilarity of the individual components. In both cases, ligand binding induces receptor dimerization/oligomerization, leading to transphosphorylation. This, in turn, activates the intrinsic kinase activity of the receptor and recruits downstream effectors, which can then be phosphorylated and thus activated by the receptor. In plants, transmembrane receptors with intracellular serine/threonine kinase domains are highly abundant and diverse. Clearly this is a robust and effective signal transduction mechanism that has been used multiple times in the course of evolution.

 · The role of receptor serine/threonine kinases in signaling in plants is discussed in Chapter 13.

Some receptors have intrinsic protein phosphatase or guanylyl cyclase activity

A relatively large number of protein tyrosine phosphatases (PTPs) possess transmembrane and extracellular domains, and thus are presumed to act as receptors, though in most cases their specific ligands are not known. Often the extracellular domains of the receptor tyrosine phosphatases (RPTPs) resemble domains that mediate cell–cell or cell–substrate adhesion, and in at least a few instances, RPTPs have been shown to mediate cell–cell adhesion through homophilic interactions. This raises the possibility that these proteins may play a general role in modulating local protein tyrosine dephosphorylation in response to cell–cell adhesion. This is plausible because many of the proteins that control cell–cell junctions, cell–cell and cell–substrate adhesion, and coupling to the actin cytoskeleton are regulated by tyrosine phosphorylation in metazoans.

The mechanism whereby ligand binding might regulate the phosphatase activity of the RPTPs is not yet fully understood. The cytoplasmic region for most of these receptors consists of two tandem PTP domains, though only the membrane-proximal domain is thought to be capable of catalytic activity (**Figure 8.10**). There is some evidence that dimerization or clustering of the receptor inhibits PTP activity, though the precise mechanism is not known. Thus, the unliganded, monomeric receptor is constitutively active, and ligand binding leads to decreased activity. If this is the case, the net result of ligand binding to RTKs and RPTPs would be the same: a net increase in tyrosine phosphorylation in the vicinity of the receptors.

Another class of receptors with intrinsic enzymatic activity is the membrane guanylyl cyclase receptors (mGCs). There are seven distinct mGCs in mammals, including the receptors for atrial natriuretic peptide (ANP) and its relatives BNP and CNP (B-type and C-type natriuretic peptides), which regulate kidney and smooth muscle function, and for guanylin and uroguanylin, which regulate intestinal water and electrolyte transport. The ligands for other mGCs remain unknown, but the fact that several mGCs are found specifically in the olfactory epithelium or retina suggests roles in sensory transduction.

Regulation of these enzymes by ligand binding is as yet poorly understood. The mGCs all contain a protein kinase-like domain (kinase homology domain, or KHD) located between the transmembrane and guanylyl cyclase (GC) domains (**Figure 8.11**). The KHD does not appear to be capable of catalyzing phosphate transfer itself, but is highly phosphorylated in the basal (unliganded) state. Ligand binding to the mGC receptors apparently shifts the equilibrium toward dimerization, and induces conformational changes in the dimer that promote ATP binding to the KHD and relieve inhibition of the GC domain, which becomes activated. Phosphorylation of the KHD seems to be required for receptor activation, and dephosphorylation by phosphatases leads to its inactivation (desensitization). The kinases and phosphatases that mediate these reactions are not yet known.

Noncovalent coupling of receptors to protein kinases is a common signaling strategy

Many cell-surface receptors do not have intrinsic catalytic domains, but instead their intracellular portions interact noncovalently with proteins with catalytic activity, such as protein kinases or proteases. A wide variety of receptors couple to nonreceptor tyrosine kinases, exemplified by the cytokine receptors, T and B cell receptors, and integrins. Here, we will briefly outline some specific aspects of how these receptors activate their associated kinases, and the downstream effectors of that activation.

The cytokine receptors, which couple to STAT transcriptional activators, have been introduced in Chapter 5. In general, these receptors are associated with nonreceptor tyrosine kinases of the JAK family (so-called "Janus kinases," after the two-faced Roman god, because they contain tandem catalytic domains). Receptor dimerization induced by ligand binding leads to transphosphorylation of the cytoplasmic tails of the receptor; the phosphorylated receptor then serves

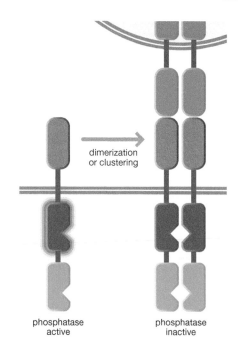

dimerization
or clustering

phosphatase
active

phosphatase
inactive

Figure 8.10

Model for regulation of receptor tyrosine phosphatase activity. In the unliganded state (left), the phosphatase is active. Most RPTPs have two phosphatase homology domains, but only the membrane-proximal domain (*purple*) has enzymatic activity. Upon ligand binding and clustering, for example, by homotypic interactions with RPTPs from an adjacent cell (*orange*), conformational changes lead to inactivation of the phosphatase domain and thus localized increases in tyrosine phosphorylation.

T cell receptor signaling is described in more detail in Chapter 14.

to recruit the SH2 domain-containing *STAT proteins,* which are then phosphorylated by the associated JAK (**Figure 8.12a**). STAT phosphorylation induces conformational changes that promote its dimerization, nuclear transport, and DNA binding, and ultimately leads to changes in the transcription of STAT-responsive genes.

Similar themes are seen in the transduction of signals from immune (B and T cell) receptors, and from adhesion receptors such as integrins. In the case of the T cell receptor—perhaps the best-studied example of a receptor that signals through nonreceptor tyrosine kinases—the CD4 co-receptor of the T cell receptor is associated with the Src family kinase Lck (**Figure 8.12b**). When receptor aggregation is induced by interaction of the T cell receptor with peptide–MHC (major histocompatibility complex) complexes displayed by antigen-presenting cells, Lck becomes activated through transphosphorylation. The proximal effect of Lck activation is the phosphorylation of the receptor ζ chain on pairs of tyrosine residues separated by ~10 amino acid residues, which constitute a recognition site (termed the immunoreceptor tyrosine–based activating motif, or ITAM) for a second nonreceptor tyrosine kinase, ZAP-70. This kinase itself has tandem SH2 domains that engage the doubly phosphorylated ITAM; binding both localizes ZAP-70 to the liganded receptor and activates it so that it can then phosphorylate a number of other receptor-associated proteins. This promotes assembly of a large signaling complex (often termed the "immune synapse"), which leads to activation of a variety of downstream signaling pathways.

Integrins comprise a diverse family of cell-surface adhesion receptors that bind to extracellular matrix- or cell-associated peptides, such as fibronectin, laminin, and fibrinogen. Each integrin is a heterodimer consisting of α and β subunits, each with a large extracellular ligand-binding domain and a small intracellular

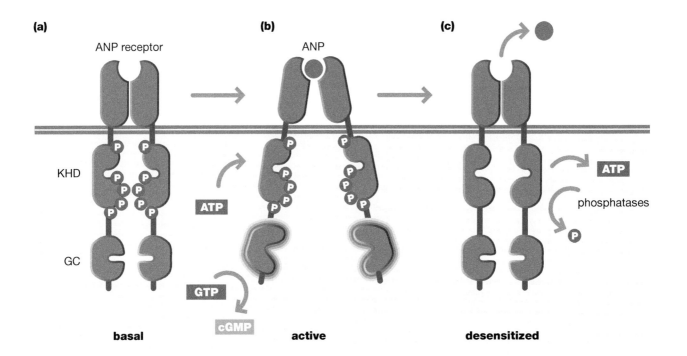

Figure 8.11

Regulation of receptor guanylyl cyclases. (a) In the basal (unactivated) state, receptor guanylyl cyclases such as the atrial natriuretic peptide (ANP) receptor exist as homodimers. The kinase homology domain (KHD) is highly phosphorylated, and the guanylyl cyclase (GC) domain is inactive. (b) Upon ligand binding, conformational changes lead to binding of ATP to the KHD and to activation of the GC domain, leading to increased intracellular cyclic guanosine monophosphate (cGMP) levels. (c) Dephosphorylation of the KHD leads to release of ATP, inactivation of the GC domain, and release of bound ligand. The desensitized receptor cannot be activated until the KHD is phosphorylated again.

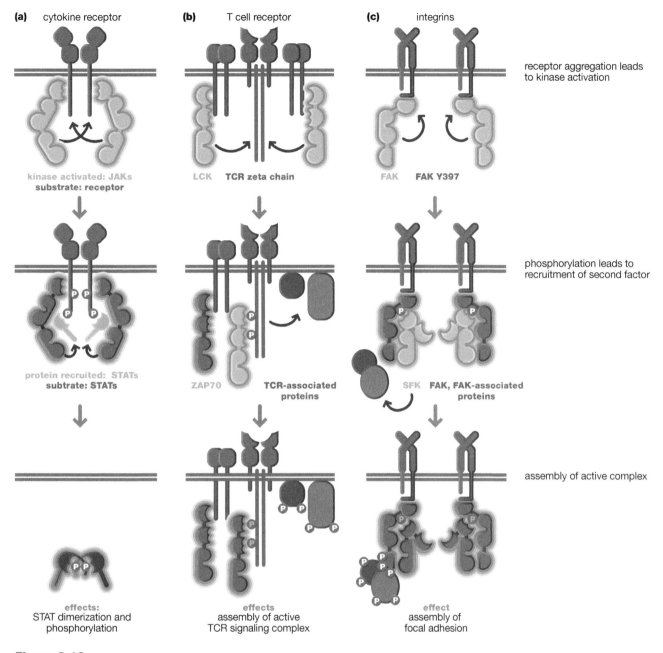

Figure 8.12

Three examples of receptors coupled to activation of nonreceptor tyrosine kinases. Examples depicted are (a) cytokine receptors, (b) T cell receptor (TCR), and (c) integrins. For each receptor, the first step (upper panels) involves increased local concentration and activity of a nonreceptor tyrosine kinase (*green*) associated with the receptor, promoted by proximity in the dimerized/clustered receptor. This leads to increased phosphorylation of a substrate protein (arrows). In the second step (middle panels), this phosphorylated substrate serves to recruit an SH2 domain-containing effector protein (*green*). In the third step (lower panels), phosphorylation of additional substrates leads to downstream biological effects. See the text for details of each pathway. In (b), for clarity, only one set of kinases is shown in the middle and lower panels. FAK, focal adhesion kinase; SFK, Src family kinase.

portion. Integrins play a very important role in coupling cell adhesion to the actin cytoskeleton, so the cell can apply traction force to its surroundings and thereby move. Integrins operate on a number of different levels, functioning as adhesion molecules, as a means to transmit information about the nature of their binding partners, and as a mechanical linkage between the outside environment and the cytoskeleton. Here, we will consider two signaling modes, one of which involves clustering and activation of associated nonreceptor tyrosine kinases and another that involves allosteric changes.

The first level of signaling from integrins is via nonreceptor tyrosine kinases, and is conceptually similar to other examples where receptor dimerization/clustering is coupled to kinase activation. In this case, the integrin-associated kinase is focal adhesion kinase (FAK), a nonreceptor tyrosine kinase with a central catalytic domain flanked by rather large N-terminal and C-terminal regions with a number of protein interaction motifs (**Figure 8.12c**). These interaction domains mediate binding to the integrins and other proteins. Upon integrin clustering, FAK molecules are brought into close proximity, undergo a conformational change and dimerization, and a key tyrosine residue N-terminal to the catalytic domain is phosphorylated. This phosphotyrosine then serves as a docking site for the SH2 domain of Src family kinases (SFKs). Binding of the SFK has the dual consequence of localizing it to the site of integrin engagement, and of activating it by preventing it from adopting the closed, inactive conformation. The activated SFK then phosphorylates a number of proteins, most notably FAK itself and other FAK-associated proteins such as p130Cas, altering their activity and creating binding sites to recruit additional SH2 domain-containing proteins to the nascent focal adhesion.

A second type of signaling by integrins serves to mechanically couple extracellular ligands and the actin cytoskeleton. Integrins undergo conformational changes when their extracellular domains bind to ligand, which alter the binding properties of the intracellular portion by changing the relative orientation of the α and β chains. The effect is to increase the affinity of the intracellular domain for proteins that couple to the actin cytoskeleton, such as talin (**Figure 8.13**). Thus, ligand binding is intimately associated with linking the adhesion receptors to the cytoskeleton. Indeed, the same conformational changes can transmit information in the reverse direction—from inside the cell to the outside (so-called "inside-out signaling"). This is because binding of talin to the intracellular domain of integrin (promoted, for example, by increased local concentrations of talin due to clustering, or by post-translational modifications of talin that increase its affinity for the integrin) converts the integrin to the active conformation, which has higher affinity for extracellular ligands. This interplay between ligands and the actin cytoskeleton will tend to promote the localized clustering of activated integrins into patches. These patches serve as the nucleators for the formation of focal adhesions, highly complex cellular structures that couple sites of cell adhesion to F-actin cables (stress fibers).

Figure 8.13

Bidirectional signaling by integrins.
(a) In the basal state, the unliganded integrin heterodimer has relatively low affinity for extracellular matrix (ECM) components via its extracellular domain, and for talin via its intracellular domain. Conformational changes induced either (b) by binding to high concentrations of ECM components or (c) by binding high concentrations of talin or post-translationally modified talin, lead to (d) a receptor with higher affinity for both ECM and talin. The effect is to generate clusters of activated integrins linking ECM to the actin cytoskeleton. Note, the conformational change in integrin depicted in the figure is exaggerated for clarity.

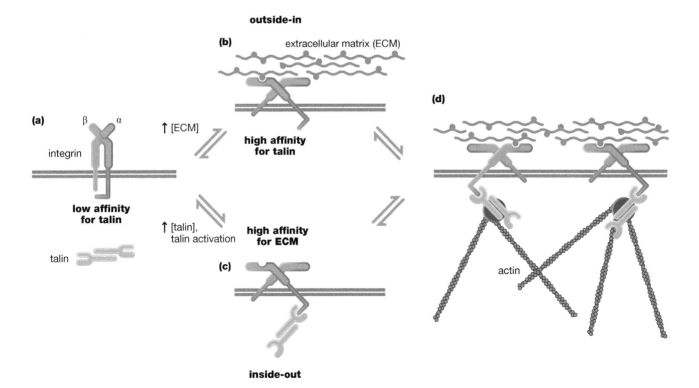

Some receptors use complex activation pathways that involve both kinase activation and proteolytic processing

A number of other receptors couple more indirectly to the activation of protein kinases; the activation mechanisms for such receptors are more complex and involve additional activities, such as regulated proteolysis and the remodeling of elaborate multiprotein complexes. In this section, we will consider the activation of the transcription factor NF-κB by Toll-like receptors; in the following section, we will discuss Wnt and Hedgehog, two other pathways in which transcriptional activators are regulated by a combination of phosphorylation and proteolysis.

The **Toll-like receptors (TLRs)** play an important role in innate immunity by detecting pathogens and stimulating host defenses to fight off the invaders. There are 13 members of the TLR family in mammals, each binding to a distinct pathogen-specific structural motif, such as lipopolysaccharide (LPS), flagellin, or single- or double-stranded RNA. Ligand-bound TLRs indirectly activate the *NF-κB* family of transcription factors. NF-κB is a central mediator of responses to a wide variety of cellular stimuli. It exists in a latent, inactive form in unstimulated cells that can be rapidly mobilized without requiring new transcription or translation. NF-κB is activated by the phosphorylation, ubiquitylation, and proteasome-mediated degradation of inhibitory subunits, termed IκBs, that are bound to the latent NF-κB. Phosphorylation of IκB, which triggers its degradation, is mediated by a heterotrimeric assembly termed the IKK (IκB kinase) complex.

The mechanism whereby TLR ligand binding promotes IκB degradation is quite complex, involving a number of intermediates. The mechanism involves receptor dimerization and conformational changes leading to binding of a class of adaptor proteins containing a TIR domain, exemplified by MyD88 (**Figure 8.14**). These adaptors also contain a so-called *death domain* (*DD*), which then binds to the death domains of the IRAK family of serine/threonine kinases, leading to activation of the IRAKs, presumably through a combination of proximity and conformational changes. Activated IRAKs then promote downstream signaling by dissociating from the receptor and binding to and activating TRAF-6, a ubiquitin E3 ligase. Activated TRAF-6 mediates Lys63-linked polyubiquitylation of itself and other substrates, which in turn leads to the recruitment and activation of yet another kinase, TAK1, in complex with cofactors TAB2 and TAB3. The result of TAK1 activation is, finally, activation of the IKK complex, which then is free to phosphorylate IκB and promote its polyubiquitylation and destruction.

Despite the fearsome complexity of the overall mechanism (which is simplified here for clarity), it is important to note that it involves the same basic mechanisms that have been discussed over and over again in this chapter—receptor aggregation, conformational changes leading to changes in the binding of proteins to the receptor, and activation of associated kinases through proximity.

 NF-κB activation is described in more detail in Chapter 9.

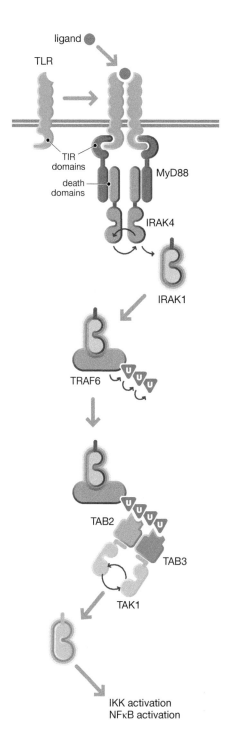

Figure 8.14

Signaling by Toll-like receptors. Dimerization of Toll-like receptors (TLRs) through ligand binding leads to recruitment of the MyD88 adaptor by homotypic interactions of TIR domains. This leads to recruitment and activation of the kinase IRAK4 via homotypic interactions of death domains (*brown oblongs*). IRAK4 activates IRAKI, which binds to TRAF-6, activating its E3 ubiquitin ligase activity which leads to the polyubiquitylation of TRAF-6. The TAB2 and TAB3 adaptors bind to ubiquitylated TRAF-6 via ubiquitin-binding domains, leading to recruitment and transactivation of the kinase TAK1, which activates the IκB kinase complex (IKK), finally leading to activation of NF-κB.

Polyubiquitylation and proteasome-mediated degradation are discussed in more detail in Chapter 9.

Wnt and Hedgehog are two important signaling pathways in development

The **Wnt** signaling pathway also uses a complex mixture of kinase activation and proteolysis (**Figure 8.15**). Members of the Wnt family control many developmental cell fate decisions, and thus play an important role in normal development and in tissue homeostasis. The effector for canonical Wnt signaling (there are other modes of signaling from the Wnt receptors, such as the so-called planar cell polarity pathway, that will not be discussed here) is a latent transcription factor, β-catenin. In unstimulated cells, β-catenin is prevented from entering the nucleus and is rapidly degraded via sequestration in a multiprotein complex (the cytoplasmic destruction complex) that contains β-catenin, two scaffolding proteins (axin and APC), and two serine/threonine kinases (CK1 and glycogen synthase kinase 3, or GSK-3). Within this complex, β-catenin is phosphorylated, leading to recognition by β-TrCP (a specificity factor for a ubiquitin E3 ligase), polyubiquitylation, and ultimately β-catenin is targeted to the proteasome for degradation. Upon Wnt stimulation, however, this complex is disrupted, allowing accumulation of intact, uncomplexed β-catenin, which is then free to translocate to the nucleus and stimulate transcription in association with nuclear coactivators.

Wnt receptors have two components: Frizzled, a member of the GPCR family, and a single-pass receptor called LRP. Wnt apparently binds to both, thereby promoting the close association of the two co-receptors. The receptor heterodimer has two new properties, both of which are important for downstream signaling. First, the tail of LRP becomes accessible for phosphorylation by CK1 and GSK-3, which greatly increases its affinity for axin (note that in this case, CK1 and GSK-3 are not parts of the destruction complex, but comprise a distinct, membrane-associated pool). Second, the liganded Frizzled is able to induce the phosphorylation and recruitment of yet another protein, Disheveled. This may involve the activation of a heterotrimeric G protein by liganded Frizzled. Phosphorylated Disheveled and axin can also associate with each other directly, so assembly of an activated LRP-Frizzled-Disheveled-axin complex on the membrane is likely to be highly cooperative. The assembly of this complex then leads indirectly to the dissociation of the destruction complex and release of free β-catenin. This is likely

Figure 8.15

Wnt signaling. The receptor for Wnt consists of Frizzled, which has the structure of a G-protein-coupled receptor, and LRP. In the absence of Wnt (left), β-catenin is rapidly degraded by a cytosolic destruction complex. The core of this consists of APC and axin, which recruit β-catenin and the kinases GSK-3 and CK1, leading to phosphorylation of β-catenin. Phosphorylated β-catenin is recognized by β-TrCP, which leads to its polyubiquitylation and thus its degradation by the proteasome. In the presence of Wnt, however, the receptor heterodimer assembles (right); Disheveled (DVL) is recruited, and LRP is phosphorylated, leading to axin recruitment. Recruitment of axin to the receptor leads to disruption of the cytosolic destruction complex, through depletion of the pool of cytosolic axin. β-Catenin is no longer phosphorylated, ubiquitylated, and degraded, and is free to translocate to the nucleus and regulate transcription of Wnt-dependent genes.

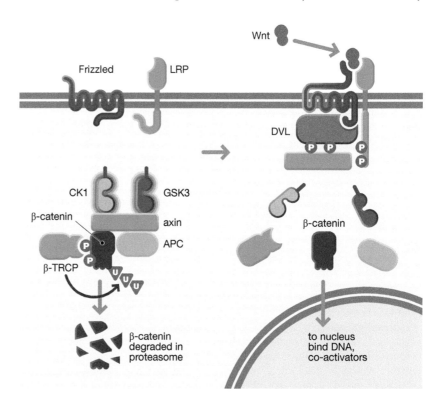

accomplished through simple mass action, by sequestering axin at the membrane and decreasing the pool that is available to form the destruction complex.

The **hedgehog (Hh)** signaling pathway has a number of mechanistic similarities to the Wnt pathway. Hh signaling is important for the normal development and patterning of virtually all tissues in multicellular organisms. Hh was first studied in *Drosophila*, where there is a single Hh ligand; in vertebrates, there are three Hh ligands, the most widely distributed of which is termed Sonic hedgehog (Shh). As in the case of Wnt, the ultimate effect of Hh signaling is transcriptional activation, mediated for vertebrate Hh by members of the Gli family (**Figure 8.16**). In cells not stimulated by Hh ligands, Gli is phosphorylated by protein kinase A, GSK-3, and CK1. This leads to degradation of its C-terminus by the proteasome, generating a truncated form that acts as a transcriptional repressor. Hh ligands cause a change in the processing of Gli, such that a full-length or alternatively processed form that acts as a transcriptional activator predominates. It is thought that the balance in the cell between the repressor and activator forms of Gli determines signal output, allowing graded signals (for example, from spatial gradients of Hh ligands) to be read out by each cell in terms of transcriptional activity.

One of the most interesting aspects of Hh signaling in vertebrate cells is that it is intimately connected to the primary cilium, a specialized filamentous organelle constructed from microtubules. The structure of the primary cilium is very similar to that of the flagella found on single-celled eukaryotes, and in vertebrates, most cells contain a single primary cilium. More and more evidence is emerging that this organelle can serve as a signaling center, by co-localizing receptors, ligands, and cytosolic effectors, and also by actively transporting various components within the cilium via microtubule-based motor proteins. It is interesting to note that while the primary cilium is absolutely essential for Hh signaling in vertebrates, it appears to be unnecessary in *Drosophila*, where most cells are not ciliated. It is not known whether the dependency of the pathway on the primary cilium was lost in *Drosophila*, or whether it evolved *de novo* in the vertebrate lineage.

How is Gli processing affected by Hh ligands? The receptor for Hh ligands is a protein with 12 membrane-spanning segments termed Patched (Ptc). Ptc normally functions to negatively regulate the activity of a seven-transmembrane protein of the GPCR family termed Smoothened (Smo). The precise nature of this negative regulation is still under investigation, but it appears not to be due to simple protein–protein interaction. One possibility is that Ptc (which is related to a family of bacterial membrane transporters) regulates the influx or efflux of a small-molecule ligand for Smo. Whether Smo actually functions as a GPCR is also still under investigation, but what is clear is that its conformation dramatically changes upon binding of Hh ligands to Ptc. This causes Smo to become heavily phosphorylated on its C-terminus and to associate with other proteins such as arrestin (which normally functions to down-regulate and direct some aspects of downstream signaling from GPCRs, as discussed later in this chapter). In addition, in vertebrates, binding of Hh to Ptc leads to its relocalization from the primary cilium, where it is replaced by Smo. These changes lead to a rearrangement of the complexes that process Gli, such that Gli escapes phosphorylation and C-terminal processing and can thus enter the nucleus to activate transcription of its target genes.

A variety of receptors couple to proteolytic activities

After protein kinases, proteases are the enzymes most often used by receptors to convey information to the cell interior. Above, we discussed several examples where proteolysis plays a supporting role in receptor signaling. We will now consider two examples where proteolysis is the predominant mechanism. Proteolysis is unlike many other signaling transactions, in that it is essentially irreversible; resetting the system requires synthesis of new proteins and/or further degradation

 The roles of proteases in signaling are discussed in more detail in Chapter 9.

Figure 8.16

Hedgehog signaling. Signaling in vertebrates by Sonic hedgehog (Shh) occurs at the primary cilium. (a) When Shh is absent, the Shh receptor Patched (Ptc) is localized on the primary cilium and inhibits the activity of Smoothened (Smo). Gli transcription factors undergo phosphorylation by protein kinase A (PKA), glycogen synthase kinase 3 (GSK-3), and CK1, leading to recognition by β-TrCP, polyubiquitylation, and partial proteolysis by the proteasome. The remaining fragment of Gli (GliR) has transcriptional repressor activity. (b) In the presence of Shh, Ptc inhibition of Smo is relieved, and Ptc relocalizes from the primary cilium. Smo becomes activated and undergoes conformational changes that allow it to be phosphorylated by a G-protein-coupled receptor kinase (GRK) and other kinases. Phosphorylated Smo binds to arrestin and is transported up the primary cilium with the assistance of motor proteins such as Kif7. Gli is also transported up the cilium and avoids phosphorylation. Full-length, unphosphorylated Gli translocates to the nucleus where it can activate gene transcription.

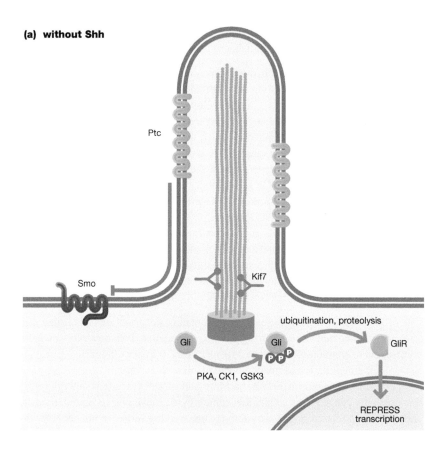

(a) without Shh

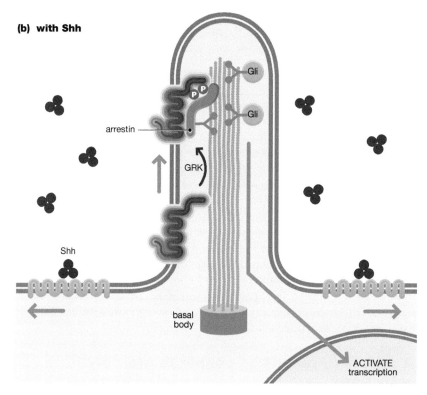

(b) with Shh

of signaling mediators generated by proteolysis. Thus, this mechanism is most likely to be used for systems that must respond strongly and decisively to a signal, but for which temporal control on short time scales (that is, the ability to turn off or otherwise regulate the signal once initiated) is less important.

The *Notch* signaling pathway was introduced in Chapter 5. Notch receptors are proteolytically processed upon binding to their ligands (such as Delta), which are presented on the surface of a neighboring cell (the "sender" cell). In the "receiver" cell,
ligand-induced cleavage liberates the Notch intracellular domain (NICD), which
then translocates to the nucleus and promotes transcription of Notch-responsive
genes. Ligand engagement first stimulates cleavage by an extracellular metalloprotease of the ADAM/TACE family at the "S2 site" of Notch, releasing the Notch
extracellular domain (NECD) (**Figure 8.17**). The remaining receptor fragment is
then cleaved within the hydrophobic transmembrane segment (at the "S3 site") by
the γ-secretase complex to release the active NICD into the cytosol. Endocytosis of
the Notch ligand by the sender cell is required for optimal signaling in the receiver
cell, and it is likely that pulling forces generated by endocytosis of the ligand
in the sender cell expose the S2 cleavage site on the receptor. This represents a
rare example of signal transmission by a single-pass transmembrane receptor that
does not require dimerization or clustering. A number of transmembrane proteins,
including the amyloid precursor protein (APP), whose cleavage products accumulate in neurons in Alzheimer's disease, are cleaved in the membrane in a two-step
process similar to Notch. Thus, the Notch pathway may be only one specific example of a more widespread but poorly understood signaling mechanism.

Another example of receptors linked directly to proteolytic activities is provided by the **death receptors**, which induce *apoptosis*, or programmed cell death,
when activated by their ligands. These receptors include the tumor necrosis factor receptor (TNFR), TRAIL receptor, and Fas/CD95. Apoptosis is caused by the
activation of a class of proteases termed *caspases*; the immediate consequence of
death-receptor engagement is caspase activation. The ligands for death receptors,
such as tumor necrosis factor (TNF) and TRAIL, are all homotrimeric in structure, and their transmembrane receptors are also constitutively trimeric, even in the
absence of ligand. When ligand binds to the extracellular domain of the receptors,
however, conformational changes are induced which expose protein interaction
motifs on the intracellular portion of the receptor. All death receptors contain a
modular domain termed the **death domain (DD)**, which can dimerize with DDs in
other proteins, as mentioned above for TLR signaling.

In the case of Fas, an archetype for death receptors, the DDs that are exposed upon
receptor activation bind to the DDs of an adaptor protein, FADD (**Figure 8.18**). In
addition to its DD, FADD contains another modular dimerization motif, the **death**

Apoptosis and caspases are
discussed in more detail in
Chapter 9.

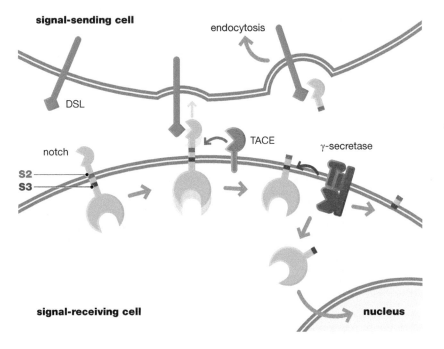

Figure 8.17

Notch signaling. DSL (delta-serrate-Lag2)
ligand on the sender cell engages Notch
receptors on the receiver cell. Pulling
forces generated by endocytosis of DSL
by the sender cell expose the S2 site
of Notch for proteolytic cleavage by the
metalloprotease TACE. Cleavage at the
S2 site allows access to the S3 site by the
γ-secretase complex, which then cleaves
Notch at the intramembrane S3 site. The
Notch intracellular domain (NICD) is then
free to relocalize to the nucleus to regulate
transcription of Notch-responsive genes in
the signal-receiving cell.

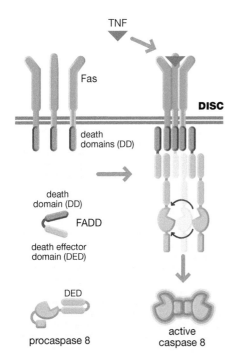

Figure 8.18

Signaling by death receptors. Death receptors such as Fas exist as constitutive trimers. Binding of trimeric ligands such as Fas ligand (FasL) induces conformational changes in the receptor that promote homotypic interactions of the death domains (*light brown oblongs*) of the receptor with those of adaptors such as FADD. This exposes the death effector domains (DEDs, *green oblongs*) of FADD, which interact with the DEDs of procaspase-8. The resulting large assembly, termed the death-inducing signaling complex (DISC), promotes autocatalytic processing of procaspase-8, likely by a combination of proximity and conformational change. Active caspase-8 is then free to diffuse throughout the cell to cleave substrates and effect the apoptotic cell death program.

effector domain (**DED**). Once the DD of FADD is engaged by the receptor, the DED is free to bind the DED of one of the so-called *initiator caspases*, caspase-8 or caspase-10. This, in turn, serves to activate the caspase, inducing its proteolytic processing from an inactive proenzyme form to an active holoenzyme. The supramolecular complex containing the receptor, adaptor proteins such as FADD, and initiating caspases such as caspase-8 is termed the death-inducing signaling complex (DISC). Recent evidence shows that engagement of FADD by the activated receptor can nucleate the formation of long filaments of caspase-8, mediated by DED oligomerization. This presumably serves to amplify the number of caspases that can be activated by a single activated receptor.

Caspase activation by the receptor can be explained at least in part by proximity—bringing multiple procaspases together in the DISC increases the likelihood that one will cleave the other into the activated form, initiating a cascade in which all bound procaspases will rapidly be activated. This is directly analogous to the mechanism of activation of tyrosine kinases by aggregation, discussed above. Structural studies have also shown that the association of procaspase molecules into large supramolecular filaments by receptor-mediated aggregation induces conformational changes that are likely to facilitate activation and proteolytic processing.

GATED CHANNELS

A further major class of signaling receptors is the gated ion channels, which respond to ligands or other environmental cues by altering the permeability of the membrane to ions or other small molecules. Such gated channels play a fundamental role in neurotransmission, opening in response to neurotransmitters, membrane depolarization, or other stimuli. These channels allow specific ions to flood rapidly across the membrane, causing massive and very rapid changes in the electrical properties of the membrane, the basis for electrical signals. Because of their importance for neurotransmission, such channels are the targets of a number of familiar drugs. For example, local anesthetics such as procaine and lidocaine block voltage-gated sodium channels, while calcium channel blockers (CCBs) are used to treat high blood pressure. Gated channels also regulate intracellular signaling, as in the case of calcium channels on the endoplasmic reticulum that open when bound to inositol trisphosphate, leading to the release of intracellular calcium stores. In this section, we will not deal with the electrophysiology of neuronal signaling, but will consider a few examples of gated channels that illustrate the mechanisms of their selectivity and their regulation by stimuli.

Gated channels share a similar overall structure

All gated channels, irrespective of the stimuli that regulate them or the solutes that they conduct, share similarities in overall topology. All are composed of multiple similar or identical subunits arranged in a ring structure in the plane of the membrane (**Figure 8.19a**). Depending on the specific channel, this ring may be composed of two to five subunits (with four or five subunits being the most common arrangement), though in some cases all subunits are combined in a single polypeptide chain. Each subunit can consist of anywhere from two to six hydrophobic α helices that span the membrane, connected by loops of varying lengths. In the center of the ring is a narrow aqueous pore, which allows the selective passage of certain molecules. This pore is lined with helices, arrayed perpendicular to the membrane like barrel staves. The specific properties of the pore determine which molecules can pass: the size of the opening, and in particular its electrostatic properties, determine which anions or cations will be attracted into the pore, and those that will be repelled or sterically excluded (**Figure 8.19b**).

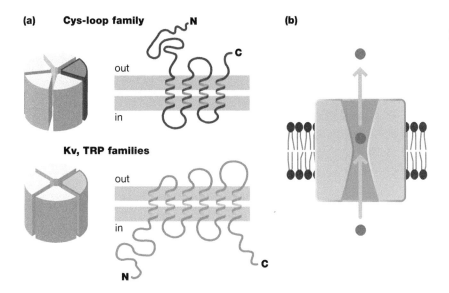

(a) **Cys-loop family**

out

in

Kv, TRP families

out

in

(b)

Figure 8.19

The structure of gated ion channels.
(a) Diagrammatic depiction of the organization and subunit structure of gated ion channels. Pentameric subunit composition, in which each subunit has four transmembrane helices (top), is typical of a variety of ligand-gated ion channels of the Cys-loop superfamily involved in neurotransmission. A tetrameric subunit composition, where each subunit has six transmembrane helices, is found in voltage-gated potassium (Kv) channels and TRP (transient receptor potential) channels. (b) Diagrammatic cross section of a channel embedded in the lipid bilayer. Ions with particular properties (size and charge) can selectively pass through the pore of the open channel.

In addition to the physical structure of the pore itself, the gated channels must also be able to control the opening and closing of the pore in response to ligands or other signals (such as changes in voltage across the membrane, changes in temperature, or changes in extracellular pH). These chemical or physical changes must be converted into mechanical energy, to drive changes in the relative orientation of the helices that line the pore. In this way, a channel that in its resting state is sterically "closed" is converted to an open form that allows passage of small molecules across the membrane. We will briefly consider two families of gated channels for which structural data are available.

The voltage-gated potassium channel provides clues to mechanisms of gating and ion specificity

Voltage-gated potassium (Kv) channels play an important role in propagating action potentials in neurons. They respond to membrane depolarization by opening and allowing the passage of K^+ out of the cell down its concentration gradient, thereby repolarizing the membrane to its resting state. Kv channels are one example of a large family of channels consisting of four identical subunits, each of which has six transmembrane helices. Pioneering work by the group of Rod MacKinnon has provided high-resolution X-ray crystal structures of a number of Kv channels in different states, providing perhaps the best view yet of how ion selectivity and gating can be achieved.

The pore of the Kv channel contains a narrow constriction roughly in the middle that serves as a selectivity filter. This filter is lined by 20 oxygen atoms, arranged in four potential K^+-binding sites in which their partial negative charges electrostatically shield the positively charged potassium ions (**Figure 8.20**). The way in which the oxygen atoms coordinate to the K^+ is very similar to how the "hydration shell" of water molecules shields the ion when it is dissolved in water. Thus, there is little if any energetic cost to stripping the hydration shell from the ion as it passes through the filter. The actual size of the pore is a good fit for K^+ (1.33 Å), while it would be too loose for Na^+ (0.95 Å), the other abundant singly charged cation in the cell. There is evidence that the binding of multiple K^+ ions to the selectivity filter induces a slight conformational change in the helices lining the pore, thus lowering the affinity of the ion for the filter (because some of the binding energy is used for the rearrangement). This, combined with electrostatic repulsion between K^+ ions in the filter, ensures that the affinity is not so high that ions would have difficulty escaping from the filter, leading to low conductance (the number of ions that can pass per unit time). Similar principles are used in other channels to generate pores specific for various ions depending on their size and charge.

Figure 8.20

The selectivity filter of the K⁺ channel.
(a) Open-channel model of the bacterial KcsA K⁺ channel, showing the large aqueous pore and the selectivity filter near the extracellular (upper) face of the membrane. Potassium ions are shown as *green* spheres. (b) X-ray crystal structure of the KcsA selectivity filter. Potassium ions are shown as *green* spheres, and water molecules as *red* spheres. Note that K⁺ ions on either side of the selectivity filter are hydrated, but the ions passing through the filter have been dehydrated. Carbonyl groups from the amino acids lining the filter coordinate to the K⁺, replacing water molecules. The size of the pore is a perfect fit for K⁺. (a, Adapted from E. Gouaux & R. MacKinnon, *Science* 310:1461–1465, 2005. With permission from AAAS; b, adapted from Y. Zhou, et al., *Nature* 414:43–48, 2001. With permission from Springer Nature.)

The modular design of signaling proteins is discussed in Chapter 10.

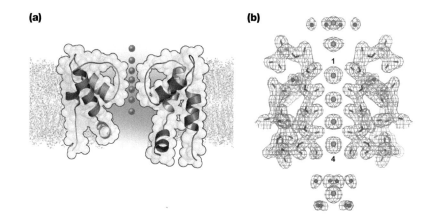

(a) **(b)**

Much has also been inferred about the gating mechanism of Kv channels from crystal structures. While many of the details still are under investigation, it is clear that a number of positively charged residues in a so-called "voltage sensor" (consisting of helices 3 and 4 of the channel) are physically translocated within the membrane in response to depolarization (**Figure 8.21a**). In the resting state, the external side of the membrane carries a slight positive charge relative to the inside, whereas when the membrane depolarizes, the inside becomes positively charged relative to outside due to the rapid influx of sodium ions. Because of simple electrostatic forces, the positively charged residues in the voltage sensor will always be attracted to the more negatively charged side of the membrane and repelled by the more positively charged side. In the crystal structures, which correspond to the open conformation, the positively charged residues of the voltage sensor are accessible to solvent on the extracellular side of the membrane. Presumably in the closed conformation, at normal membrane resting potential in which the inside is negatively charged relative to outside, those positively charged residues would be oriented differently so that they are accessible to the cytosol.

Clusters of negatively charged residues are present on both sides of the channel to help stabilize the positively charged residues in either of the two orientations, and the space in between these two negatively charged clusters is entirely hydrophobic. Thus, only two conformations are likely to be stable, and the voltage sensor is unlikely to get "stuck" halfway open. This switchlike property, in which only the fully closed and fully open states are stable, is very important for channel function. Remarkably, it has been shown that the voltage sensor can be transferred to other channels to confer voltage gating, once again demonstrating the modularity of signaling proteins.

Structural and experimental studies have led to a model for how the translocation of the voltage sensor would affect the packing of the helices lining the pore. In this model, the pore is constricted when the sensor is oriented toward the cytosol, and opens when the sensor is oriented toward the extracellular environment, as expected from the known electrophysiology of the channel (**Figure 8.21b**).

A number of other channels share the same overall topology of the Kv channel, but are gated by other stimuli. An important group of channels for signaling is the TRP (transient receptor potential) family. These channels are permeable to Ca²⁺ (in addition to other cations, with varying degrees of specificity), and are often activated by multiple signaling inputs, including small molecules, lipids and lipid metabolites, heat or cold, and voltage. Such channels can act in signaling as input integrators or coincidence detectors. Several interesting examples include those that respond to heat and cold (these channels are how sensory neurons in the body detect temperatures that are uncomfortably cold or hot). TRPV1, the so-called vanilloid receptor, opens in response to heat, and also to the binding of the compound capsaicin, which is what gives chili peppers their fiery taste. By contrast, the TRPM8 channel opens in response to cool temperatures, and also to the binding of compounds such as menthol that provide a sensation of cooling.

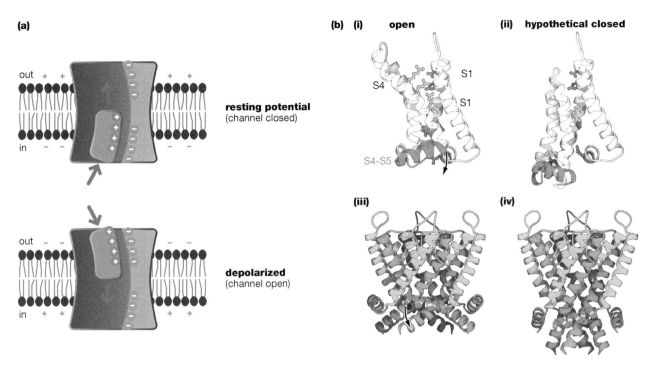

Figure 8.21

Mechanism of gating of the voltage-gated K⁺ channel. (a) Schematic representation of the "voltage sensor" of the voltage-gated K⁺ (Kv) channel, based on crystal structures of bacterial and bacterial–vertebrate hybrid Kv channels. The voltage sensor is depicted in *light pink*; for clarity only one sensor is shown, while the tetrameric channel contains a total of four sensors arrayed around the perimeter of the channel. The voltage sensor has four positive gating charges (*blue* + signs), which are attracted to the negatively charged side of the membrane, thus driving movement of the sensor depending on the membrane potential. The gating charges move relative to two clusters of negative charges on the main body of the channel (*pink* – signs). The area between the two clusters of negative charges is hydrophobic, thus only two positions ("open" and "closed") for the voltage sensor are stable. (b) Top left (i) is a representation of the voltage sensor and S4–S5 linker helix in the open conformation from the crystal structure. Helices are drawn as ribbons. The view is from the pore looking out, with the extracellular solution "above" and the intracellular solution "below." The positive gating charges are shown as *blue* sticks. Negatively charged residues in the external and internal clusters are *pink*; the hydrophobic phenylalanine in the middle is *green*. Top right (ii) is a depiction of a hypothetical closed conformation of the voltage sensor. The positive gating charges now reach toward the intracellular solution, and are stabilized through interactions with the internal negative cluster. The inward displacement of the S4 helix pushes down on the N-terminal end of the S4–S5 linker helix (*orange*), causing it to tilt toward the intracellular side and to close the pore. Bottom left (iii) is a depiction of the open conformation of the S4-S5 linker helices and pore from the crystal structure. Note that the voltage sensor (above) is rotated 180° relative to the pore as depicted here. Bottom right (iv) is a hypothetical model of the S4–S5 linker helices and pore in a closed conformation based on the crystal structure of a closed K⁺ channel pore (KcsA). (b, Adapted from S.B. Long, et al., *Nature* 450:376–382, 2007. With permission from Springer Nature.)

Ligand-gated ion channels play a central role in neurotransmission

Signal transmission at synapses involves the release of neurotransmitters such as acetylcholine, glutamate, and γ-aminobutyrate (GABA) from the signaling (presynaptic) cell, which bind to *ligand-gated ion channels* on the receiving (postsynaptic) cell. The opening of these channels allows ions to flow across the membrane, either initiating an action potential in the case of cation channels, or inhibiting it in the case of anion channels. Because of their central role in all aspects of the nervous system, the ligand-gated ion channels are the targets of a number of drugs. The Cys-loop superfamily of ligand-gated ion channels has been studied intensively, and includes the nicotinic acetylcholine receptor (nAChR) and the receptors for GABA, glycine, and serotonin. These channels share a common overall topology, consisting of five nonidentical subunits, each with four transmembrane helices (see Figure 8.19a). Ligand-binding sites have been mapped to the large N-terminal extracellular domain, in the cleft between adjacent subunits.

Figure 8.22

Structure of a pentameric ligand-gated ion channel. (a) Left, ribbon representation of GLIC (a bacterial cation channel with the same topology as mammalian ligand-gated ion channels) viewed from within the membrane with the extracellular solution above. This channel is in the open conformation. Right, structure of the pentameric channel viewed from the extracellular side. Each channel subunit is depicted in a different color. (b) Left, view of the α2 helices of GLIC defining the pore region. The front subunit is removed for clarity. The molecular surface is shown as *gray* mesh. Upper right, intracellular part of the pore region. Shown are the inferred positions of Cs+ (*gray*), Rb+ *(blue)*, and Zn2+ *(pink)* from crystal structures. Bottom right, schematic representation of the pore-opening mechanism. The α2/α3 helices of two subunits in the closed (left) and open (right) conformation are shown. The ion-coordinating glutamate residues are shown in *pink*, the permeating ions in *blue*. (Adapted from R.J.C. Hilf & R. Dutzler, *Nature* 457:115–118, 2009. With permission from Springer Nature.)

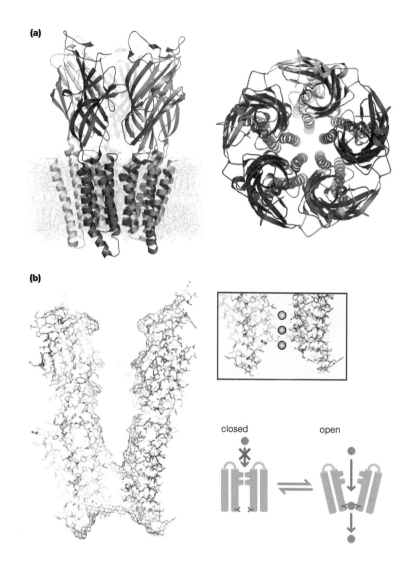

High-resolution structures of vertebrate ligand-gated ion channels are now available, but selectivity and gating mechanisms of such channels were first elucidated in the structures of prokaryotic channels with sequence and structural homology to vertebrate channels (**Figure 8.22a**). The structures of two closely related cation channels show the receptor in the closed and open conformation. In the open conformation, a fairly wide funnel-shaped chamber lined with hydrophilic residues leads to a relatively narrow transmembrane pore (**Figure 8.22b**). At its narrowest point, at the intracellular membrane border, a ring of negatively charged glutamate residues presumably acts as a specificity filter, helping to attract and shield the cation after its hydration shell has been stripped away, and ensuring that only positively charged solutes of a certain size can pass. This ring is highly conserved in other ion channels that conduct cations, such as the nAChR.

In the closed conformation, the top part of the pore, near the extracellular membrane border, is sterically blocked by a plug consisting of bulky hydrophobic side chains. The difference between open and closed conformations appears to be due to a rotation of the pore-forming helices perpendicular to the plane of the membrane, which opens the pore on the extracellular side and narrows it on the intracellular side (Figure 8.22b). This rotation must be caused by ligand binding, and the structures reveal close physical linkages between the ligand-binding sites in the extracellular domain and the transmembrane helices that could translate conformational changes between the two.

MEMBRANE-PERMEABLE SIGNALING

MEMBRANE-PERMEABLE SIGNALING

The simplest and most straightforward way for information to cross the plasma membrane is for signaling molecules to diffuse passively through the membrane itself (see Figure 8.2a). Two types of molecules function this way: gases (nitric oxide, oxygen, and, to a lesser extent, carbon monoxide) and hydrophobic small molecules such as steroid hormones. It is the small size of these signaling molecules and their solubility in both aqueous and hydrophobic environments that gives them the relatively unusual ability to traverse freely and rapidly between the extracellular fluid and the cytosol. The receptors for these two classes of molecules are found in the cytosol and thus are fundamentally different from all other receptors, which must span the plasma membrane to function.

Nitric oxide mediates short-range signaling in the vascular system

As introduced in Chapter 6, **nitric oxide** (**NO**) is an important signaling molecule that regulates physiological processes such as blood flow through the vasculature. NO is a simple diatomic gas that can pass freely through cell membranes. The chemical reactivity of NO with heme or oxygen is very high—its half-life in tissues is only several seconds—and so its range of action is relatively short. This is an example of a **paracrine** signaling molecule, meaning that it primarily functions to signal between adjacent or nearby cells. In the best-understood physiological context, NO is synthesized by NO synthase in vascular endothelial cells; the newly generated NO then diffuses to adjacent vascular smooth muscle cells, where it causes relaxation and thus dilation of the blood vessels.

The receptor for NO in the target cell is **soluble guanylyl cyclase** (**sGC**), an enzyme that converts GTP into cyclic GMP (cGMP) and pyrophosphate. sGC is cytosolic and is distinct from the transmembrane receptor guanylyl cyclases (mGCs) discussed above. NO binding can activate sGC by several hundredfold, thus coupling increases in cytosolic NO levels to increases in cytosolic cGMP. cGMP then acts as a small, soluble signaling mediator to regulate the activity of a variety of downstream targets, notably cGMP-dependent protein kinase isoforms (**Figure 8.23a**). It is these downstream effectors of cGMP that ultimately mediate the cellular responses to NO, such as smooth muscle relaxation.

Soluble guanylyl cyclase is a heterodimer that contains a regulatory domain with a heme functional group (a ring structure with a single central iron atom). In the absence of NO, this regulatory domain represses the activity of the catalytic cyclase domain of the enzyme. Binding of NO to the ferrous iron of the heme group displaces a histidine side chain from the heme, inducing conformational changes that relieve repression of

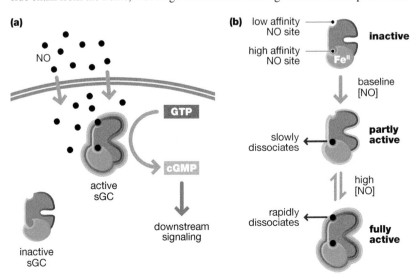

(a)

NO

GTP

cGMP

active sGC

downstream signaling

inactive sGC

(b)

low affinity NO site

high affinity NO site

inactive

Fe^II

baseline [NO]

slowly dissociates

partly active

high [NO]

rapidly dissociates

fully active

Figure 8.23

Signaling by nitric oxide. (a) Nitric oxide (NO) generated by one cell diffuses into a neighboring cell where it binds to and activates soluble guanylyl cyclase (sGC). Activation of sGC leads to increased intracellular levels of cGMP which, in turn, regulates other signaling proteins. (b) sGC has two NO-binding sites. The high-affinity site contains a heme group (Fe^II), and is bound even at relatively low NO concentrations. Binding to this site partially activates sGC. At higher concentrations, NO binds to the low-affinity site, leading to full activation of sGC. When NO levels return to resting levels, NO rapidly dissociates from the low-affinity site and sGC activity decreases.

the catalytic domain. Once bound to heme, dissociation of NO is relatively slow, with a half-life of approximately 2 or 3 minutes. For maximal activity, however, a second molecule of NO must also bind to sGC; this can rapidly dissociate if NO concentrations go down, leaving a partially active "sensitized" form. It is thought that basal levels of NO in the vasculature are sufficient to maintain this sensitized form, setting the tonic level of vascular constriction (the K_d for binding of NO to the heme of sGC is in the picomolar range, while basal NO levels are in the nanomolar range). Acute signaling (the release of a bolus of NO from adjoining endothelial cells) leads to binding of NO to the second, lower-affinity site, leading to maximal sGC activity which returns to the basal level rapidly as tissue NO levels diminish (**Figure 8.23b**). Thus, this receptor system is set up to maintain relatively stable basal levels of activity, and to rapidly respond to local changes in ligand concentration.

Carbon monoxide (CO), which has rather similar chemical properties to NO, can also bind to and activate sGC, though the affinity is much lower and the maximal activation of sGC is much less (two- to fourfold, as opposed to many hundredfold for NO). The physiological significance of CO signaling is not yet fully established, but CO levels can rise in tissues in response to stress. This has the potential to modulate sGC activity, either by moderately activating it or by acting as a partial antagonist of NO signaling.

O_2 binding regulates the response to hypoxia

Molecular oxygen (O_2) can act as a signaling molecule in addition to its fundamentally important role in cellular respiration. Decreases in tissue O_2 levels (hypoxia) induce short-term physiological responses, such as the shutdown of inessential activities that consume ATP, and an increased rate of anaerobic glycolysis. In the longer term, hypoxia induces transcriptional changes leading to tissue-level responses such as increased **angiogenesis** (the generation of new blood vessels). Transcriptional responses to hypoxia are mediated by a transcription factor, HIF-1α (hypoxia inducible factor), which under normal O_2 levels is rapidly degraded. Binding of O_2 activates heme-containing proline hydroxylase domain (PHD) proteins, which hydroxylate two prolines in HIF-1α. This modification allows binding of VHL, a ubiquitin E3 ligase, leading to HIF-1α ubiquitylation and targeting to the proteasome. By contrast, under hypoxic conditions, the PHD proteins are inactivated, and HIF-1α begins to accumulate in the nucleus to promote transcription of hypoxia-dependent genes (**Figure 8.24**).

Figure 8.24

Signaling by oxygen. (a) Under normal tissue oxygen levels, molecular oxygen (O_2) binds to the heme group (Fe^{II}) of proline hydroxylase (PHD) and activates it. PHD hydroxylates two prolines in the transcription factor HIF-1α (hypoxia inducible factor), leading to its recognition by the VHL ubiquitin ligase. VHL adds long chains of ubiquitin (U) to HIF-1α, targeting it for degradation by the proteasome. (b) In hypoxic conditions, tissue oxygen concentrations are too low to activate PHD. HIF-1α is not recognized by VHL, and is free to bind DNA and regulate transcription of target genes.

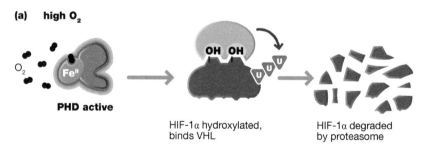

(a) high O_2

O_2

PHD active

OH OH

U U U

HIF-1α hydroxylated, binds VHL

HIF-1α degraded by proteasome

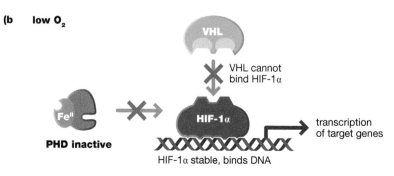

(b) low O_2

VHL

VHL cannot bind HIF-1α

Fe^{II}

PHD inactive

HIF-1α

transcription of target genes

HIF-1α stable, binds DNA

VHL takes its name from a genetic cancer predisposition, Von Hippel-Lindau syndrome. Afflicted individuals carry only one functional copy of VHL; the second has an inactivating mutation. Whenever a cell loses the functional allele through somatic mutation, HIF-1α is stable and constitutively active in that cell and its descendants. This leads to increased angiogenesis and altered metabolic activity, both of which promote tumor cell growth.

The receptors for steroid hormones are transcription factors

The other class of signaling molecules that can easily pass through the plasma membrane is comprised of hydrophobic hormones and related compounds, which include the steroid hormones (such as estrogen, progesterone, testosterone, and hydrocortisone), vitamins A and D, thyroid hormone, and retinoic acid. The receptors for these compounds are transcription factors of the **nuclear receptor** (**NR**) superfamily. In humans, there are ~48 members of the NR family. The endocrine NRs bind to their specific ligands with high affinity, generally in the nanomolar range, consistent with the low circulating levels of their hormone and vitamin ligands. A number of other NRs bind to specific classes of lipids and lipid metabolites with relatively lower affinity, and are thought to function primarily in cellular lipid homeostasis. Finally, a number of NRs are considered **orphan receptors**, as their physiological ligands have not yet been identified.

The NRs all bind to specific DNA sites as homo- or heterodimers via an N-terminal DNA-binding domain (**Figure 8.25a**). The ligand-binding domain is situated in the C-terminus of the molecule. The glucocorticoid receptor (GR) serves as a paradigm for NR action. When not bound by its ligand, GR is sequestered in the cytosol in an inactive complex with chaperone proteins such as hsp90, which presumably function to prevent the aggregation and/or degradation of the relatively unstable (and thus partially unfolded) unliganded receptor. Upon ligand binding, the receptor undergoes a conformational change, chaperone proteins and other co-repressors are released, and the receptor is now free to dimerize, relocate to the nucleus, and bind palindromic NR-binding sites on chromatin (**Figure 8.25b**). The DNA-bound form can activate transcription by binding to a host of coactivator proteins, such as chromatin remodeling complexes including histone acetyl transferase (HAT). The net effect is to increase transcription from a battery of hormone-responsive genes. Thus, these receptors represent the most simple and straightforward mechanism of transcriptional regulation by an outside signal: direct activation of a transcription factor by its ligand.

 The role of protein acetylation in transcriptional regulation is described in Chapter 4.

The simple model described above does not capture all of the variety and subtlety of regulation by different members of the NR family. For example, some of these receptors are constitutively bound to specific sites on the chromatin even in the absence of ligand. In these cases, the unliganded receptor acts as a transcriptional repressor by interacting with corepressor proteins such as histone deacetylase complex (HDAC). When ligand binds, the conformational changes in the ligand-binding domain lead to release of the co-repressors and binding of a new set of coactivator proteins, such as HAT, which remodel the chromatin and convert the erstwhile repressor into a transcriptional activator (**Figure 8.25c**). Note that while the specific effects of ligand binding are quite different from those described above for GR, the immediate consequence of ligand binding is the same; that is, to induce conformational changes in the ligand-binding domain that alter the binding partners and thus the biological activity of the NR.

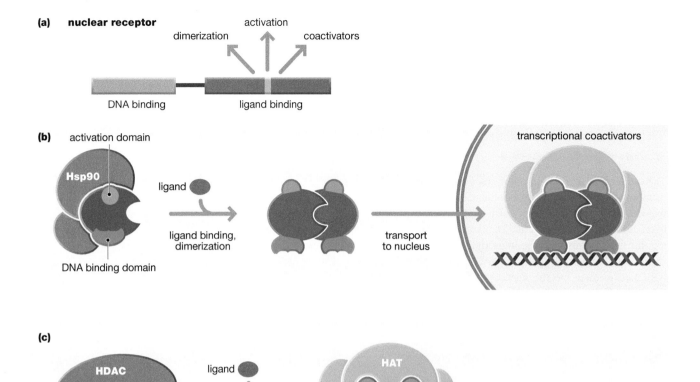

Figure 8.25

Signaling by nuclear receptors. (a) All nuclear receptor (NR) family members consist of an N-terminal DNA-binding region (*orange*), and a C-terminal ligand-binding region (*purple*). (b) When their ligand is not present, NRs such as glucocorticoid hormone receptors reside in the cytosol in complex with hsp90 and other proteins. Ligand binding leads to dissociation of hsp90, conformational changes, and dimerization. The receptor dimer is imported into the nucleus, binds DNA, and associates with transcriptional coactivators. (c) Some NRs are constitutively bound to DNA. In the absence of ligand, they are associated with transcriptional co-repressors such as histone deacetylase (HDAC). Ligand binding induces conformational changes that lead to dissociation of co-repressors and binding of transcriptional coactivators such as histone acetyl transferase (HAT).

DOWN-REGULATION OF RECEPTOR SIGNALING

While receptor activation is the basis for signal transmission through the membrane, the ability to turn receptor signaling off, or to adjust its sensitivity, is also critical for cells to respond to their environment. As we discussed in Chapter 2, very high-affinity complexes necessarily have long half-lives (on the order of several minutes or longer), and transmembrane receptors tend to have very high affinity for their ligands, which are present at relatively low levels in the environment. Thus, the inactivation of a ligand-bound receptor by dissociation of the ligand is a relatively slow process. To circumvent this constraint, cells have adopted a number of other mechanisms to terminate the signal, or to allow the signal output to adapt to different levels of ligand. Here, we provide a few examples of specific mechanisms used to down-regulate receptor signaling. It should be emphasized that this is not meant to be a comprehensive list; indeed, it is likely that there are nearly as many ways to down-regulate receptor-mediated signals as there are ways to initiate the signal itself.

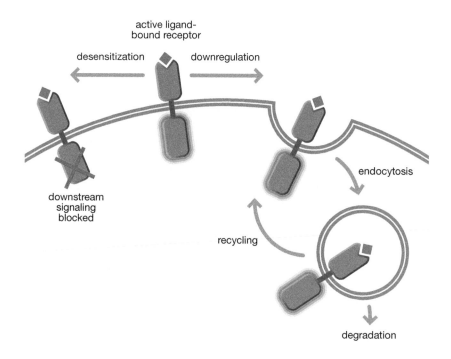

active ligand-
bound receptor

desensitization   downregulation

endocytosis

downstream
signaling
blocked

recycling

degradation

Figure 8.26

The fate of activated receptors. Active, liganded receptor (center) is subject to down-regulation and desensitization. Down-regulation involves internalization of the receptor–ligand complex, which can then be recycled to the cell surface or degraded in the lysosome. Desensitization involves blocking downstream signaling from the liganded receptor.

In general, there are two ways in which the signals emanating from receptors can be modulated (**Figure 8.26**). The first and most straightforward is removing the receptor itself from the cell surface, often termed receptor down-regulation. Not only does this reduce the number of receptors that can respond to the signal over time, but activated receptors already bound to ligand can also be targeted for degradation once they are internalized. Receptor internalization is a relatively slow process, operating on time scales of a few minutes. The second general mechanism is **desensitization**, or attenuation, in which the ligand-bound receptor becomes less active in transmitting downstream signals. Desensitization can be quite rapid, on the order of milliseconds. This is the case for ligand-gated ion channels, where relatively brief stimulation can cause the channel to reversibly adopt a closed (desensitized) conformation refractory to further stimulation. Many other receptors undergo some form of desensitization, such as GPCRs as discussed below. Desensitization can allow cells to adjust the baseline level of signal output to the current level of input signal, so they can respond to a much wider range of signal strength than would be possible otherwise.

Ubiquitylation regulates the endocytosis, recycling, and degradation of cell-surface receptors

In the absence of ligand, most cell-surface receptors undergo relatively slow cycles of endocytosis, sorting, and recycling back to the plasma membrane. Once receptors are activated by ligand binding, however, the dynamics of this process usually change dramatically: the rate of endocytosis increases, and many of the internalized receptor–ligand complexes are targeted for degradation in the lysosome. The net effect is to rapidly (within a few minutes) clear the activated receptor from the cell surface and extinguish its ability to transmit downstream signals. While this process is common to all types of cell-surface receptors, we will consider the specific example of the epidermal growth factor receptor (EGFR), an RTK that has been particularly well studied.

Internalization of the activated EGFR can occur by a variety of specific mechanisms (see Figure 5.15). Of these, clathrin-mediated endocytosis is the most rapid, and is likely to be dominant under physiological conditions of relatively low receptor number and ligand concentration. A number of studies have shown

Figure 8.27

Recruitment of Cbl to the activated EGF receptor. Upon ligand binding, the epidermal growth factor (EGF) receptor is activated and autophosphorylates on a number of sites. Some sites bind the SH2 domain of Cbl directly; others bind the SH2 domain of the Grb2 adaptor which, in turn, binds to the proline-rich tail of Cbl (*blue*) via its SH3 domains. Cbl has ubiquitin E3 ligase activity and, when recruited to the receptor, ubiquitylates the receptor and receptor-associated proteins, leading to engagement of the endocytic machinery.

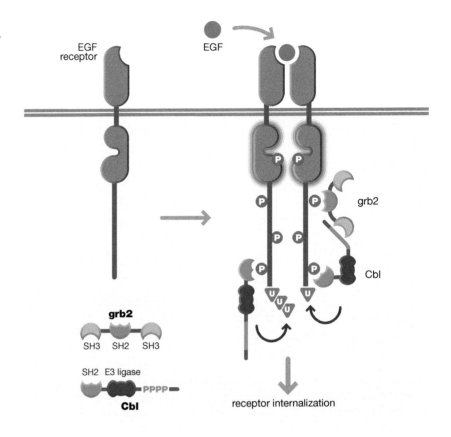

 Ubiquitin and the enzymes that add and remove it from proteins are discussed in detail in Chapter 4.

that ligand-stimulated endocytosis through the clathrin-mediated pathway requires RTK activity and, in particular, receptor autophosphorylation. For down-regulation, the key consequence of receptor autophosphorylation is the recruitment of Cbl, a ubiquitin E3 ligase. Cbl contains an SH2 domain that can bind directly to phosphorylated sites on the receptor. In addition, Cbl also binds to the adaptor Grb2, which itself has an SH2 domain that binds to different phosphorylated sites on the receptor (**Figure 8.27**). In this way, Cbl recruitment is tightly coupled to receptor activation.

The association of Cbl with the activated receptor leads to receptor ubiquitylation (either monoubiquitylation or Lys63-linked polyubiquitylation) and to rapid internalization via clathrin-coated vesicles. In other cases, ubiquitin is attached to receptor-associated adaptor proteins rather than the receptor itself, but in all cases the ubiquitylated receptor complex is rapidly recognized by the clathrin-dependent endocytic machinery, incorporated into clathrin-coated vesicles, and internalized. Once removed from the cell surface, the receptor–ligand complex (which still retains signaling activity) is transported to an organelle called the **multivesicular body**, where the decision is made whether to send it to the lysosome for degradation or recycle it back to the surface (**Figure 8.28**). This sorting process is accomplished by the so-called **ESCRT complexes**, which consist of a number of proteins containing ubiquitin-recognition motifs as well as lipid- and membrane-binding domains. Those proteins targeted for degradation (for example, those that are Lys63-polyubiquitylated) then pinch off from the external limiting membrane of the multivesicular body to form internalized vesicles. Because these vesicles are now topologically isolated from the cytosol, any receptor–ligand complexes found there can no longer signal. Ultimately, the multivesicular body fuses with a lysosome and the internalized vesicles, lipid and protein alike, are degraded.

While the specific details might vary for different receptors, this overall scheme of down-regulation is likely to be common for the vast majority of ligand-activated cell-surface receptors. Novel properties induced by activation, such as conformational changes or phosphorylation, lead to the recruitment of ubiquitin ligases,

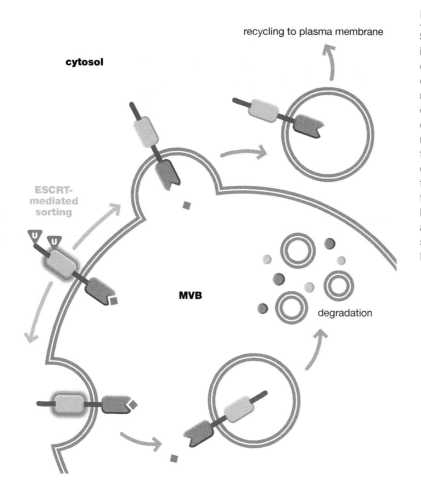

cytosol

recycling to plasma membrane

ESCRT-mediated sorting

MVB

degradation

Figure 8.28

Sorting of receptor–ligand complexes in the multivesicular body. Upon endocytosis, activated receptor–ligand complexes are transported to the multivesicular body (MVB). On the outer membrane of the MVB, ESCRT complexes, which contain ubiquitin- and membrane-binding domains, sort receptors for degradation or recycling. Those to be degraded are transported to vesicles inside the MVB, which are physically isolated from the cytosol so receptors can no longer signal. Receptors to be recycled are transported by vesicles back to the cell surface. Ultimately, the MVB fuses with a lysosome and its contents are degraded.

which mark the receptor for internalization and target it for recycling or degradation. The wide variety of ubiquitin ligases and adaptor proteins that can be involved, as well as variations in the number of ubiquitin units conjugated and the linkage of polyubiquitin chains, provide ample scope for regulation of this process to adapt to the specific requirements of the situation. The key role of this pathway in modulating signal output is highlighted by the fact that a mutant version of Cbl was originally isolated as a viral oncogene. The oncogenic form of Cbl lacks ubiquitin ligase activity but still binds to activated receptors, thereby preventing the binding of normal endogenous Cbl. Thus, activated receptors such as the EGFR cannot be efficiently down-regulated, leading to unrestrained signaling and inappropriate cell proliferation.

G-protein-coupled receptors are desensitized by phosphorylation and adaptor binding

GPCRs can be activated very rapidly and their activation can result in tremendous signal amplification, so for this class of receptors it is particularly important to be able to control signal output by desensitizing activated receptors. Desensitization of GPCRs is accomplished by two families of proteins that specifically associate with the activated form of the receptor: the GPCR kinases (GRKs) and the arrestins. GRKs associate with activated GPCRs and phosphorylate them, generally on the C-terminus, to generate high-affinity binding sites for arrestins. Typically, once arrestin is bound to the phosphorylated receptor the binding of G proteins to the receptor is blocked, preventing activation of additional G proteins. Furthermore, the arrestin can mediate association with the endocytic machinery and ubiquitin-conjugating enzymes, thus promoting endocytosis and, in some cases, degradation of the liganded receptor (**Figure 8.29**).

Figure 8.29

GPCR desensitization by GRK and arrestin. Conformational changes induced by ligand binding to GPCRs (*purple*) recruit and activate GPCR kinases (GRKs), which phosphorylate the C-terminal tail of the receptor. The phosphorylated receptor then recruits and binds to arrestin, which typically prevents the receptor from activating G proteins, targets it for endocytosis, and also serves as a scaffold for assembly of G-protein-independent signaling complexes.

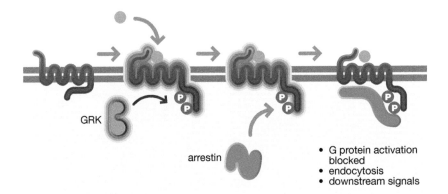

- G protein activation blocked
- endocytosis
- downstream signals

There are seven GRKs in humans, differing in their tissue distribution and specific domain structures. All have an N-terminal RGS homology (RH) domain that can interact with specific G protein α subunits, and a central serine/threonine kinase catalytic domain. The C-terminal region mediates association with membranes, either via a lipid-binding PH domain or via sites for covalent lipid attachment, and, in some cases, can also interact specifically with G protein βγ subunits. Recent structures have shown that in its basal state, the GRK catalytic domain is held in an inactive conformation through interactions with the N-terminal RH domain; upon engagement of the RH domain with active GPCRs, the catalytic domain undergoes conformational changes that activate it and position it to phosphorylate the GPCR C-terminus.

For receptor down-regulation, the key property of the arrestins is their ability to bind specifically to the phosphorylated form of activated GPCRs. Binding to the receptor induces major concerted conformational changes in arrestin which, in turn, unmask binding sites for a variety of other proteins that promote endocytosis and ubiquitylation of the arrestin–receptor complex, as well as promoting downstream signaling (**Figure 8.30a** and **b**). In most cases, the immediate consequence of arrestin binding is to block the activated receptor from interacting with and thus activating G proteins. Receptor-bound arrestin also associates with clathrin and the AP-2 clathrin adaptor, and thus very efficiently promotes internalization of the complex via clathrin-coated vesicles.

The fate of the internalized receptor then depends on specific details of the receptor and arrestin involved, and their mode of interaction. In some situations, arrestin rapidly dissociates from the receptor after internalization, allowing receptor dephosphorylation and recycling back to the cell surface (resensitization). In other situations, arrestin remains stably associated with the receptor, the arrestin and receptor are ubiquitylated by associated ubiquitin ligases, and the complex is targeted for degradation in the lysosome. Finally, it has recently been shown that in some cases, arrestin does not bind to the central, G protein binding cleft of the activated receptor, but instead binds exclusively to the phosphorylated C-terminal tail of the GPCR. In this case, although arrestin remains flexibly tethered to the receptor, G proteins can still bind to and become activated by the receptor, even after the receptor has been internalized via endocytosis. Structures of this so-called "megaplex" (containing the receptor, G protein, and arrestin) demonstrate that arrestin binding need not result in desensitization of the bound receptor (**Figure 8.30c**). Overall, multiple structures of arrestin bound to various GPCRs have revealed that the specific orientation, or "pose," of the arrestin molecule relative to the GPCR exhibits considerable variation. It is possible that the specific GPCR sites that are phosphorylated by GRKs affect this orientation, and thus influence the fate of the GPCR–arrestin complex.

What might the role of the megaplex be in GPCR signaling? One interesting property of arrestins is their ability to initiate signaling on their own once they have bound to activated GPCRs. Activated arrestins can associate with and activate nonreceptor tyrosine kinases of the Src family, and can serve as scaffolds for activation of mitogen-activated protein (MAP) kinase cascades and other signaling

(a)   (b)

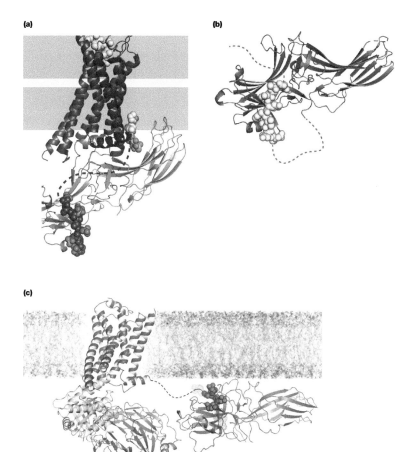

(c)

Figure 8.30

Binding of arrestin to GPCRs. (a) Cryo-EM structure of the neurotensin receptor (*purple*) complexed with Arrestin 2 (*orange*). The activating ligand neurotensin is shown in *yellow*. The phosphorylated C-terminus of the receptor that engages arrestin is shown in space-filling mode. Note how arrestin occupies the central cytosolic cavity of the GPCR normally bound by G protein, thus preventing G protein activation. Phosphatidylinositol (4,5)-bisphosphate (PIP_2) that helps stabilize arrestin binding is shown in space-filling mode. In this structure, the flexible C-terminus of arrestin was deleted. Approximate position of residues from the C-terminus of the receptor not resolved in the structure are indicated by *dotted purple line*. Position of membrane is indicated by *gray rectangle*. (b) X-ray crystal structure of unactivated Arrestin 2. C-terminal arrestin segment that occupies the site bound by phosphorylated GPCR C-terminus in activated arrestin is shown in *yellow* in space-filling mode (compare panels a and b). Approximate position of residues not resolved in the structure is indicated by dotted lines. (c) Cryo-EM structure of the "megacomplex" containing chimeric GPCR (β_2-adrenergic receptor with the C-terminal tail of the arginine vasopressin type 2 receptor [β_2V_2R], *purple*) bound to Arrestin 2 (*orange*) and G_s protein. For G_s, the α subunit is *green*, the β subunit is *blue*, and the γ subunit is *pink*. Note how in this structure, arrestin is bound only by the receptor C-terminal tail, allowing G protein binding and activation (compare panel a with panel c). The phosphorylated C-terminus of the GPCR is shown in space-filling mode; residues of the receptor not resolved in the structure are indicated by a dotted line. (a, Courtesy of John Janetzko and Brian Kobilka; PDB 6UP7. b, Courtesy of John Janetzko and Brian Kobilka; 1G4M. c, Courtesy of Anthony Nguyen and Robert Lefkowitz; PDB 6NI2 and 6NI3.)

mediators. For example, binding of arrestin to the Smo GPCR family member is thought to play an important role in Hh signaling, as described earlier in this chapter. These arrestin-dependent signals can, in some cases, cooperate with signals that are directly dependent on the G protein activated by the receptor, and can oppose the G-protein-mediated signaling in other cases. Thus, the GRK–arrestin system is not only a very efficient means to regulate the activity and dynamics of GPCR signaling, but also plays a larger role in sculpting the precise downstream signaling output from the activated receptor.

SUMMARY

Transmitting extracellular signals through the plasma membrane to the cell interior is one of the most fundamental challenges of cell signaling. Cells use a very limited number of mechanisms to accomplish this task. By far, the most common mechanism involves transmembrane receptors, where ligand binding to the extracellular portion leads to altered enzymatic activity of the receptor or its associated proteins in the cytosol. The other mechanisms involve ligand-gated ion channels, which open or close in response to stimuli, and in a few cases the use of signaling molecules that can passively cross the membrane and exert their effects directly in the cytosol. For transmembrane receptors, information about ligand binding is conveyed to the cytosolic portion of the receptor either by concerted conformational changes in the case of receptors with multiple membrane-spanning segments, or dimerization or oligomerization in the case of receptors with a single membrane-spanning segment. Protein phosphorylation and proteolysis are the most common activities that are coupled to receptor binding. Cells have also evolved a number of mechanisms to down-regulate the activity of receptors, providing an additional level of control over the extent and dynamics of signaling output.

QUESTIONS

1. What general classes of extracellular signals do cells need to detect?

2. What are the three main ways for signals to cross the cell membrane?

3. You are studying a transmembrane receptor with an intracellular kinase domain, and observe by microscopy that it clusters into membrane microdomains upon stimulation by its ligand. Discuss the potential mechanisms by which clustering of such a receptor might significantly enhance phosphorylation and activation of its kinase domain.

4. The human genome encodes hundreds of G-protein-coupled receptors (GPCRs), but an individual cell usually expresses only a handful of GPCRs. What aspects of GPCR signaling may limit the number of GPCRs that can be simultaneously used in a cell?

5. There are dozens of receptor tyrosine kinase (RTK) receptors and ligands in humans. By contrast, there is only a handful of ligands and receptors for Wnt and Hedgehog signaling. Why might this be the case?

6. Many drugs act as antagonists of transmembrane receptors. If a small-molecule G-protein-coupled receptor (GPCR) antagonist inhibits its target by 50% at a concentration of 2×10^{-8} M, what concentration would be needed in the bloodstream to inhibit receptor activity by 99%? If the compound has a molecular mass of 500 Da, how much of the drug would be needed to achieve this concentration in the bloodstream (assume that all of the drug enters the bloodstream, which has a volume of 5 L)?

7. Receptors that have intrinsic or associated kinase activity depend on dimerization and/or clustering in order to transmit downstream signals. A key step in activating the associated kinase activity is usually input-dependent phosphorylation on the kinase activation loop. However, there are a few receptor-associated kinases that do not require activation-loop phosphorylation for signaling. How might it be possible to transmit signals without activation-loop phosphorylation? Similarly, consider the role of cellular phosphatase activity. In what ways would a much higher or lower phosphatase activity affect the signaling properties of the receptor?

8. In most cells, the amounts of Gα and Gβγ subunits are closely matched. What effect might you expect if you were able to highly overexpress G protein β and γ subunits in the cell? Propose a mechanism whereby the cell normally keeps the amounts of different subunits in balance.

9. What properties of nuclear receptors make them particularly well suited for regulating cellular lipid metabolism? In what way is this regulatory role similar to and different from their role as receptors for extracellular signals?

10. Signaling by G-protein-coupled receptors (GPCRs) is generally very rapid (half-time for changes in downstream signaling mediators can be milliseconds), while signaling by tyrosine kinases is generally quite a bit slower (half-time for recruitment of SH2 effector proteins of several minutes). What molecular mechanisms might explain this difference in response dynamics?

11. The erythropoietin (Epo) receptor of a famous long-distance runner is sequenced, and it is found to have a mutation at a Ser residue known to be phosphorylated upon stimulation with Epo. The Ser residue is converted to Ala in the mutant allele. How might this mutation be linked to this person's athletic endurance?

BIBLIOGRAPHY

TRANSDUCTION STRATEGIES USED BY TRANSMEMBRANE RECEPTORS

Cooper JA & Qian H (2008) A mechanism for SRC kinase-dependent signaling by noncatalytic receptors. *Biochemistry* 47, 5681–5688.

Lemmon MA, Schlessinger J & Ferguson KM (2014) The EGFR family: Not so prototypical receptor tyrosine kinases. *Cold Spring Harb. Perspect. Biol.* 6(4), a020768.

Oh D, Ogiue-Ikeda M, Jadwin JA, et al. (2012) Fast rebinding increases dwell time of Src homology 2 (SH2)-containing proteins near the plasma membrane. *Proc. Natl Acad. Sci. USA* 109, 14024–14029.

G-PROTEIN-COUPLED RECEPTORS

Lohse MJ, Hein P, Hoffmann C, et al. (2008) Kinetics of G-protein-coupled receptor signals in intact cells. *Br. J. Pharmacol.* 153(1), S125–S132.

Oldham WM & Hamm HE (2008) Heterotrimeric G protein activation by G-protein-coupled receptors. *Nat. Rev. Mol. Cell Biol.* 9, 60–71.

Rasmussen SG, DeVree BT, Zou Y, et al. (2011) Crystal structure of the β2 adrenergic receptor-Gs protein complex. *Nature* 477, 549–555.

Seven AB, Barros-Álvarez X, de Lapeyrière M, et al. (2021) G-protein activation by a metabotropic glutamate receptor. *Nature* 595(7867), 450–454. doi: 10.1038/s41586-021-03680-3.

Weis WI & Kobilka BK (2018) The molecular basis of G protein-coupled receptor activation. *Annu. Rev. Biochem.* 87, 897–919. doi: 10.1146/annurev-biochem-060614-033910.

Wettschureck N & Offermanns S (2005) Mammalian G proteins and their cell type specific functions. *Physiol. Rev.* 85, 1159–1204.

Zhang X, Belousoff MJ, Zhao P, et al. (2020) Differential GLP-1R binding and activation by peptide and non-peptide agonists. *Mol. Cell* 80(3), 485–500.e7. doi: 10.1016/j.molcel.2020.09.020.

TRANSMEMBRANE RECEPTORS ASSOCIATED WITH ENZYMATIC ACTIVITY

Bachmann M, Kukkurainen S, Hytönen VP & Wehrle-Haller B (2019) Cell adhesion by integrins. *Physiol. Rev.* 99(4), 1655–1699. doi: 10.1152/physrev.00036.2018.

Bray SJ (2016) Notch signalling in context. *Nat. Rev. Mol. Cell Biol.* 17(11), 722–735. doi: 10.1038/nrm.2016.94.

Fitzgerald KA & Kagan JC (2020) Toll-like receptors and the control of immunity. *Cell* 180(6), 1044–1066. doi: 10.1016/j.cell.2020.02.041.

Hata A & Chen YG (2016) TGF-β signaling from receptors to Smads. *Cold Spring Harb. Perspect. Biol.* 8(9), a022061. doi: 10.1101/cshperspect.a022061.

Lemmon MA & Schlessinger J (2010) Cell signaling by receptor tyrosine kinases. *Cell* 141(7), 1117–1134. doi: 10.1016/j.cell.2010.06.011.

Mohebiany AN, Nikolaienko RM, Bouyain S & Harroch S (2013) Receptor-type tyrosine phosphatase ligands: Looking for the needle in the haystack. *FEBS J.* 280(2), 388–400. doi: 10.1111/j.1742-4658.2012.08653.x.

Rim EY, Clevers H & Nusse R (2022) The Wnt pathway: From signaling mechanisms to synthetic modulators. *Annu. Rev. Biochem.* 91, 571–598. doi: 10.1146/annurev-biochem-040320-103615.

Schmierer B & Hill CS (2007) TGFβ-SMAD signal transduction: Molecular specificity and functional flexibility. *Nat. Rev. Mol. Cell Biol.* 8, 970–982.

Shi M, Zhang P, Vora SM & Wu H. (2020) Higher-order assemblies in innate immune and inflammatory signaling: A general principle in cell biology. *Curr. Opin. Cell Biol.* 63, 194–203.

Zhang Y & Beachy PA (2023) Cellular and molecular mechanisms of Hedgehog signalling. *Nat. Rev. Mol. Cell Biol.* 24(9), 668–687. doi: 10.1038/s41580-023-00591-1.

GATED CHANNELS

Gouaux E & MacKinnon R (2005) Principles of selective ion transport in channels and pumps. *Science* 310, 1461–1465.

Hilf RJC & Dutzler R (2009) Structure of a potentially open state of a proton-activated pentameric ligand-gated ion channel. *Nature* 457, 115–118.

Kim DM & Nimigean CM (2016) Voltage-gated potassium channels: A structural examination of selectivity and gating. *Cold Spring Harb. Perspect. Biol.* 8(5), 029231. doi: 10.1101/cshperspect.a029231.

Long SB, Tao X, Campbell EB & MacKinnon R (2007) Atomic structure of a voltage-dependent K+ channel in a lipid membrane-like environment. *Nature* 450, 376–382.

MEMBRANE-PERMEABLE SIGNALING

Billas I & Moras D (2013) Allosteric controls of nuclear receptor function in the regulation of transcription. *J. Mol. Biol.* 425(13), 2317–2329. doi: 10.1016/j.jmb.2013.03.017.

Liu L & Simon MC (2004) Regulation of transcription and translation by hypoxia. *Cancer Biol. Ther.* 3, 492–497.

Lundberg JO & Weitzberg E (2022) Nitric oxide signaling in health and disease. *Cell* 185(16), 2853–2878. doi: 10.1016/j.cell.2022.06.010.

Sonoda J, Pei L & Evans RM (2008) Nuclear receptors: Decoding metabolic disease. *FEBS Lett.* 582, 2–9.

DOWN-REGULATION OF RECEPTOR SIGNALING

Goh LK & Sorkin A (2013) Endocytosis of receptor tyrosine kinases. *Cold Spring Harb. Perspect. Biol.* 5(5), a017459. doi: 10.1101/cshperspect.a017459.

Huang W, Masureel M, Qu Q, et al. (2020) Structure of the neurotensin receptor 1 in complex with β-arrestin 1. *Nature* 579(7798), 303–308.

Nguyen AH, Thomsen ARB, Cahill 3rd TJ, et al. (2019) Structure of an endosomal signaling GPCR-G protein-β-arrestin megacomplex. *Nat. Struct. Mol. Biol.* 26(12), 1123–1131.

Premont RT & Gainetdinov RR (2007) Physiological roles of G protein-coupled receptor kinases and arrestins. *Annu. Rev. Physiol.* 69, 511–534.

Seyedabadi M, Gharghabi M, Gurevich EV & Gurevich VV (2021) Receptor-Arrestin Interactions: The GPCR Perspective. *Biomolecules* 11(2), 218. doi: 10.3390/biom11020218.

Shenoy SK & Lefkowitz RJ (2011) β-Arrestin-mediated receptor trafficking and signal transduction. *Trends Pharmacol. Sci.* 32, 521–533.

von Zastrow M & Sorkin A (2021) Mechanisms for regulating and organizing receptor signaling by endocytosis. *Annu. Rev. Biochem.* 90, 709–737.

Regulated Protein Degradation

As we have seen in Chapter 4, the post-translational modification of proteins is one of the most common ways of transmitting signals in the cell. The most drastic form of post-translational modification is **proteolysis**, the breaking of the peptide bonds that form the backbone of proteins. Proteolysis can be used to break down or degrade proteins for the purpose of eliminating their activity, or it can serve as an essential step in generating active enzymes or signaling proteins from longer, inactive precursors. Regulated proteolysis is central to two of the most fundamental activities of the cell: control of the cell cycle and the decision to undergo programmed cell death (apoptosis). This chapter will explore some of the special properties of signaling pathways that involve proteolysis, and discuss in greater detail the ubiquitin–proteasome pathway of protein degradation, and the role of proteases in apoptosis and other regulated forms of cell death.

GENERAL PROPERTIES AND EXAMPLES OF SIGNAL-REGULATED PROTEOLYSIS

The most fundamental way in which proteolysis differs from other post-translational modifications is that it is essentially irreversible—the only way to regenerate the intact, uncleaved substrate is through the relatively slow and energetically costly process of translating a new polypeptide chain. This contrasts with other post-translational modifications, for example, phosphorylation or histone acetylation, where the action of the writers (kinases and histone acetyl transferases) is opposed by erasers (phosphatases and histone deacetylases). Most other types of signaling inputs, such as the opening of membrane channels to allow ions to flood in or out, are also rapidly reversible (in the case of channels, by channel closing and the action of ion pumps). The unique irreversibility of proteolysis is useful for signaling situations where it is not desirable for the system to revert easily to the starting state (**Figure 9.1**).

DOI: 10.1201/9780429298844-9

Figure 9.1

Signaling by proteases is irreversible.
(a) The consequences of the activity of proteases (either the activation of enzymes from inactive zymogens, or the proteolytic destruction of proteins) are irreversible, and the system can be reset only by synthesizing new proteins. (b) Proteases can act like a ratchet-and-pawl mechanism, preventing a signaling pathway from easily reverting to its previous state.

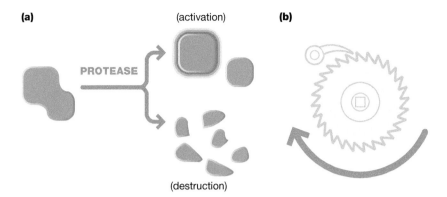

(a) (activation) **(b)**

PROTEASE

(destruction)

Two important examples of processes characterized by an irreversible commitment step are the cell cycle and apoptosis, both of which are discussed in more detail later in this chapter. In the case of the cell cycle, it is essential that the process proceed in an orderly fashion from one step to another—DNA synthesis must occur once before mitosis, for example, and mitosis must occur once before a new round of DNA synthesis. If the cycle were to run in reverse (for example, if mitosis occurred before the genomic DNA could be replicated), the consequences for the cell and organism could be disastrous. Similarly, once the process of programmed cell death begins, it must go to completion to avoid damage to surrounding cells and tissues.

Proteases are a diverse group of enzymes

Proteases, the enzymes that hydrolyze the peptide bonds of proteins, are an evolutionarily ancient and highly diverse group. The peptide bond is inherently quite stable, so a variety of proteases have evolved to accomplish housekeeping functions such as breaking down the proteins in food (to generate energy and provide raw materials for synthesis of new biomolecules) or eliminating cellular proteins that are damaged or no longer needed. Proteases can be classified on the basis of their reaction mechanism into the metalloproteases, serine proteases, cysteine proteases, and aspartyl proteases.

Most proteases exhibit pronounced preferences for the amino acid residues that flank the bond to be cleaved. Substrate specificity can be determined not only by the catalytic site itself but also, in many cases, by additional interactions with substrates outside of the catalytic site. This specificity is particularly important for those proteases that have been harnessed for signal transduction.

Most cellular proteases are synthesized in an inactive precursor form, termed a **zymogen**. This is quite sensible, as it would be potentially dangerous to have fully active proteases rampaging about the cell as soon as they emerged from translation. Proteolytic cleavage of the zymogen at specific sites is needed to release the catalytically active enzyme. This prevents damage to the synthesizing cell, and permits precise activation at the time and place needed. Such an arrangement also allows the system to greatly amplify an initial signal, particularly in cases where a single activated protease can go on to cleave and thereby activate many more zymogen molecules of the same type. This sort of positive feedback loop is common where a rapid and explosive burst of activity is needed. Thus proteolysis confers distinct dynamic properties to signaling—a proteolytically regulated process can switch from one state to another very rapidly, but it can only be reversed slowly through new protein synthesis.

Blood coagulation is regulated by a cascade of proteases

The principles of zymogen activation and signal amplification are illustrated in the blood coagulation or clotting cascade of vertebrates. Clotting is, of course, important

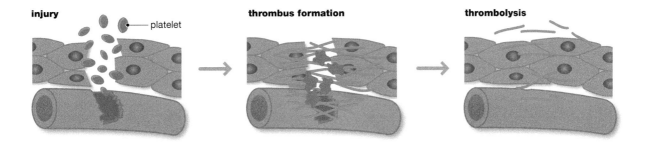

to prevent catastrophic loss of blood from damaged blood vessels after injury. The amount of clotting must be strictly controlled, however; too much clotting can cause blockage of blood vessels leading to tissue death, as in stroke, and is potentially life-threatening. This is a tricky problem for the organism, and so the clotting cascade has evolved under intense pressure to plug leaks rapidly and efficiently, to self-limit to only the areas where it is needed, and to remodel and ultimately break down over time to allow the repair and replacement of damaged tissue.

The molecular machinery that drives this elaborate process consists of a cascade of serine proteases that are specifically activated by damaged tissue and by blood platelets that adhere to the site of damage. The ultimate effector of the clotting pathway is fibrin, a protein that self-assembles into a tangled fibrillar network (a thrombus) that physically fills the breach in the damaged tissue (**Figure 9.2**). Fibrin is generated from a soluble precursor, fibrinogen, which is a major component of serum (present at a concentration of ~3 mg/mL). It is converted by the proteolytic action of thrombin from its native soluble conformation to the conformation that self-assembles into fibrils. Thrombin also cleaves and activates a transglutaminase, Factor XIII, that cross-links the fibrin network in the clot. Thrombin is present at relatively high concentrations in the blood in a zymogen form, termed prothrombin, and must itself be proteolytically processed to become active.

Thrombin activation must be very carefully controlled, and thus is dependent on a number of factors (**Figure 9.3**). One of these, an accessory factor (Factor V),

Figure 9.2

The physiological role of the clotting cascade. Tissue injury causing damage to blood vessels leads to contact between the serum and damaged cells, and to the binding and aggregation of platelets to damaged tissue. This activates the clotting cascade, leading to the deposition of a tangled network of fibrin (a thrombus, or clot; *blue*) that prevents the escape of blood cells from the damaged vessel. Once the injury has been repaired, the clot is disassembled (thrombolysis).

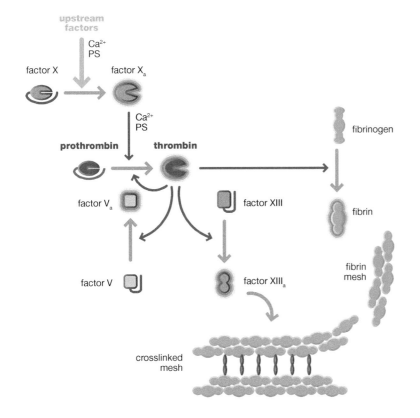

Figure 9.3

The clotting cascade. The pivotal event of clotting is the activation of thrombin by cleavage of its zymogen, prothrombin. Activation of thrombin can be accomplished either by another molecule of activated thrombin or by a different protease, Factor X_a. Factor X is similarly activated by other upstream proteases. Once activated, thrombin cleaves and activates fibrinogen, leading to formation of fibrin filaments. Thrombin generates more thrombin directly by cleaving prothrombin, and indirectly by cleaving and thus activating the cofactor Factor V. Thrombin also activates Factor XIII, which cross-links the fibrin network to generate a more stable, hard clot. Many steps require Ca^{2+} and negatively charged lipids such as phosphatidylserine (PS).

is itself cleaved by thrombin to generate the active factor (Factor V_a), providing one positive feedback loop. The fact that thrombin itself can cleave and activate more molecules of prothrombin provides a second positive feedback loop. Other essential factors for prothrombin cleavage are Ca^{2+} ions and negatively charged membrane lipids. As we saw in Chapter 7, negatively charged lipids such as phosphatidylserine are almost always present only on the inward (cytosolic) face of cell membranes, so their exposure to the blood signals that ruptured cells and tissue injury are present.

The very first molecules of thrombin that are activated to initiate clot formation are generated by another protease, Factor X, which must, in turn, be activated by zymogen cleavage, a process again facilitated by Ca^{2+} ions, negatively charged phospholipids, and additional accessory factors. The entire multistep cascade, initiated either by negatively charged surfaces (the so-called "intrinsic pathway") or by exposed tissue proteins (the "extrinsic pathway"), will not be further discussed, but is described in great detail in standard biochemistry texts. Once the process is initiated, however, a number of factors contribute to limiting the extent of clot formation to the immediate site of injury. One is blood flow, which constantly dilutes the activated proteases and accessory factors at the site of injury; another is the presence of circulating protease inhibitors such as antithrombin, which dampen protease activity away from the immediate site of injury.

Finally, once tissues have had a chance to repair themselves, the clot must be removed, a process termed thrombolysis. Perhaps not surprisingly, this involves yet another serine protease, termed plasmin, that specifically cleaves fibrin and thereby dissolves the clot. The zymogen form of plasmin is plasminogen, which, in turn, is activated by other proteases such as tissue plasminogen activator (t-PA) and is inhibited by additional factors. Because blood clots are a significant health risk (causing strokes and heart attacks, two of the main causes of death for humans), drugs that inhibit coagulation such as coumarin, which prevents the interaction of prothrombin with membranes, or those that enhance clot lysis (such as t-PA) have saved many lives.

Regulated proteolysis by metalloproteases can generate signaling molecules and alter the extracellular environment

Proteases play a number of roles on the extracellular surface of the cell, which include degrading extracellular matrix proteins to regulate cell motility, generating bioactive signaling molecules from larger precursors, and cleaving receptors as an essential step in their activation or down-regulation. Many of these activities are performed by metalloproteases such as the matrix metalloproteases (MMPs) and disintegrin metalloproteases (ADAMs, A Disintegrin And Metalloprotease). These proteases all require a metal ion, generally zinc, for catalytic activity. They are synthesized as inactive precursors and transported through the secretory pathway to the cell surface, where most remain associated via a C-terminal transmembrane segment or phospholipid anchor; some lack a membrane anchor and are released from the cell. The domain structures of the major families of metalloproteases are depicted in **Figure 9.4**.

ADAMs regulate signaling pathways by cleaving membrane-associated proteins

ADAMs are also known as "sheddases" for their ability to cleave membrane-associated proteins from the cell surface, a process termed ectodomain shedding. Such an activity can either promote signaling (for example, where a bioactive ligand is released) or inhibit it (such as when the extracellular ligand-binding region of a cell-surface receptor is cleaved off). Two particularly well-understood cases

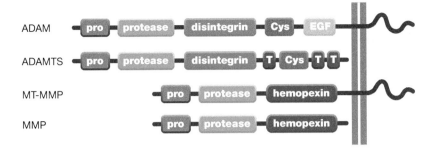

Figure 9.4

Structure of extracellular metalloproteases. A typical ADAM protease consists of a pro-domain (pro), a metalloprotease catalytic domain (protease), a disintegrin domain that mediates binding to cell-surface integrins, a cysteine-rich element (Cys), and an epidermal growth factor-like motif (EGF), followed by a transmembrane segment and cytoplasmic tail. ADAMTS proteins are distinguished by multiple thrombospondin repeats (T) and lack the EGF motif and transmembrane and intracellular segments. A typical matrix metalloprotease (MMP) contains a pro-domain and catalytic domain followed by a hemopexin-like domain. While most MMPs are secreted, the membrane-type MMPs (MT-MMPs) are anchored to the membrane either by a transmembrane segment (depicted here) or by a glycosylphosphatidylinositol lipid group.

involve the processing by ADAM-17 of tumor necrosis factor α (TNFα) and the epidermal growth factor (EGF) family of soluble mitogens. In fact, ADAM-17 was first characterized as TNFα converting enzyme (TACE) for its role in processing the membrane-bound precursor form of TNFα into its soluble, active form, though later studies demonstrated that ADAM-17 is a sheddase with a particularly broad range of substrates. Gene knockout studies have shown that the biological phenotype of mice lacking ADAM-17 closely parallels that of mice with mutations in the EGF receptor (EGFR) family or their ligands. Since ADAM-17 cleaves a variety of EGFR ligands from the membrane, this provides convincing evidence that receptor signaling activity can be regulated at the level of ligand processing. Such a mechanism is probably useful in generating localized signals, where a ligand exerts its influence in the immediate vicinity of its shedding (this is termed **juxtacrine signaling**).

Eph–ephrin signaling illustrates another way in which proteolysis can help sculpt signaling responses at the cell surface. Eph receptors are receptor tyrosine kinases; binding of the membrane-anchored ephrin to its corresponding Eph receptor on another cell generates bidirectional signals impacting both cells. In the developing nervous system, Eph–ephrin signaling generally leads to repulsion of the two cells, an important cue in guiding axons to their proper targets. ADAM-10 specifically recognizes the Eph–ephrin complex, and binding induces a conformational change in the protease that increases its activity toward the bound ephrin, cleaving it from the cell surface (**Figure 9.5**). This allows the two interacting cell membranes to separate, an obvious requirement if two cells are to be repelled from each other. ADAM-10 also plays a major role in signaling by Notch receptors (discussed below).

The potent biological activity of ADAMs raises the question of how they might be regulated by other signals. Control can be exerted by the cell at a variety of levels, including transcription, proteolytic activation, trafficking to the cell surface, and internalization. Most intriguing, however, are cases where activation of other signaling pathways (by protein kinase C agonists or Ca²⁺ ionophores, for example) modulates ADAM activity. In many cases, regulation is likely to involve the intracellular domains of ADAMs, which typically contain potential phosphorylation

Figure 9.5

Role of ADAM proteases in Eph/ephrin-mediated repulsive interactions. The EphA3 receptor (*blue*) interacts with ADAM-10 (*brown*) on the surface of one cell. Upon binding of EphA3 to ephrin A5 (*orange*) on the surface of a second cell, ADAM-10 is activated and cleaves the associated ephrin. This allows the surfaces of the two cells to separate from each other. Binding of ephrin to Eph also activates the intracellular tyrosine kinase domain of the Eph receptor.

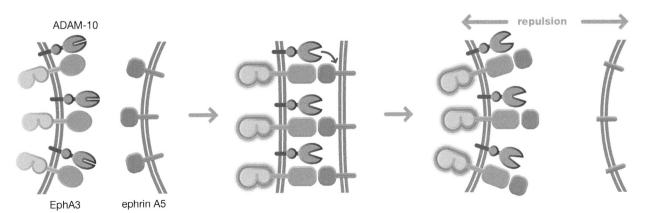

sites and proline-rich regions that can bind proteins containing SH3 domains. In one example, G-protein-coupled receptors (GPCRs) have been shown to stimulate EGFR signaling in an ADAM-dependent fashion. This cross-talk involves activation of ADAM proteases by the GPCR, most likely via a pathway involving phosphorylation of their cytoplasmic domains. ADAM-mediated cleavage of EGFR ligands then leads to activation of their receptors on the same or nearby cells.

MMPs participate in remodeling the extracellular environment

Another important class of metalloproteases is the matrix metalloproteases (MMPs). As their name suggests, MMPs target extracellular matrix proteins such as collagen and fibronectin, in addition to a number of other substrates. They participate in a variety of normal events that involve remodeling of the extracellular environment, such as wound healing, ovulation, and angiogenesis. For example, the first MMP to be isolated degrades connective tissue in the tail of the tadpole during its metamorphosis into a frog. MMPs also play a role in a number of diseases such as cancer, where they are important for the ability of tumors to invade adjoining tissues and to spread to distant sites (metastasis), and also in arthritis, where they contribute to the degeneration of connective tissue in the joints. MMPs are also a key component of **podosomes**, actin-rich cell-surface protrusions that mediate adhesion and invasion. These structures are found both in normal cells that invade tissues, such as macrophages and osteoclasts, and in cancer cells, where they are termed invadopodia (**Figure 9.6**). MMPs secreted by podosomes and invadopodia are thought to allow cells to eat their way through dense protein networks that would otherwise be impassible.

Regulation of MMPs can occur at a number of levels, though in most cells their production is controlled transcriptionally and occurs only under specific conditions.

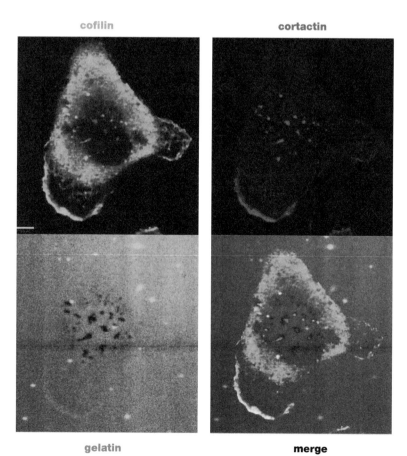

Figure 9.6

Invadopodia are sites of matrix degradation. MDA-MB-231 breast cancer cells expressing fluorescently tagged wild-type cortactin (*red*) were plated on a matrix of gelatin labeled with Alexa-405 (*cyan*) for 4 hours and fixed. The cells were immunostained for cofilin (*green*). Cortactin and cofilin are two proteins that regulate the organization of the actin cytoskeleton. Mature invadopodia are characterized by cortactin puncta that co-localize with cofilin and degrade the underlying gelatin matrix (shown by a lack of gelatin fluorescence). Invadopodia seen on tumor cells are structurally and functionally similar to podosomes seen on normal invasive cells. (Images courtesy of Marco Magalhaes and John Condeelis.)

Like other metalloproteases, they are produced as inactive zymogens and must be proteolytically processed, either by active MMPs or by other proteases. Again, as in the thrombin cascade, complex regulatory interrelationships allow both fine control of local activation and the potential for amplification of the signal. MMPs (and other metalloproteases) are also regulated by specific factors, including the tissue inhibitors of metalloproteases (TIMPs). These proteins act as potent inhibitors by forming very high-affinity 1:1 complexes with their targets. Each TIMP differs in the spectrum of MMPs, ADAMs, and other metalloproteases that it can inhibit, contributing additional specificity that can be used to control the localized activation of metalloproteases.

Proteolysis activates the thrombin receptor

Extracellular proteases, in some cases, directly participate in receptor activation. The GPCRs activated by thrombin and other proteases provide one example. These receptors, termed protease-activated receptors (PARs), play an important role in the cellular responses to vascular injury. They provide a direct mechanism for the proteases activated in the clotting cascade, such as thrombin, to signal directly to surrounding cells and to platelets. During normal clotting, for example, platelets respond to thrombin by becoming activated, a process that involves shape change, increased adhesiveness, and release of a variety of inflammatory mediators, mitogens, and procoagulant factors.

The PARs have the same seven-transmembrane topology typical of all GPCRs, but differ in that their N-termini contain a peptide sequence that can bind to and potently activate the same receptor in *cis*. In the unstimulated state, however, this autoactivating ligand is held in a conformation that prevents interaction with its binding site (**Figure 9.7**). Protease cleavage releases the conformational constraints on the autoactivating ligand, allowing it to bind to the ligand-binding site and activate the receptor. As in all GPCRs, this then leads to activation of heterotrimeric G proteins and downstream signaling. Because the PARs are irreversibly activated by proteolysis (unlike typical GPCRs), their down-regulation and recycling take on particular importance. Before activation, PARs undergo continuous cycles of internalization and re-insertion into the plasma membrane, but once activated, they are rapidly targeted for lysosomal degradation upon internalization. This mechanism is apparently distinct from the typical GPCR down-regulation mechanisms that involve arrestin binding and monoubiquitylation.

Regulated intramembrane proteolysis (RIP) is an essential step in signaling by some receptors

Another way in which proteases directly participate in receptor activation is illustrated by the Notch signaling pathway. As outlined in **Figure 8.17**, activation of Notch by its ligands involves multiple proteolytic steps: an initial "S2" cleavage liberates the extracellular domain of the receptor, allowing a second protease, the presenilin/γ-secretase complex, to cleave the receptor in the plane of the membrane, liberating the Notch intracellular domain (NICD) which thereupon translocates to the nucleus to regulate transcription. The S2 cleavage is performed by ADAM proteases, often ADAM-10. Indeed, in mice, ADAM-10 knockouts phenocopy Notch pathway mutations, strongly suggesting that ADAM-10 is primarily responsible for Notch activation *in vivo*.

The process of ADAM-mediated ectodomain cleavage followed by further processing within the membrane by the γ-secretase complex is also used for a number of other proteins, and has been termed **regulated intramembrane proteolysis (RIP)** (**Figure 9.8**). Other proteins subject to RIP include Notch ligands such as Delta, the amyloid precursor protein (APP), and the EGFR family member ErbB4. Cleavage of APP can generate amyloid β, a hydrophobic extracellular peptide that aggregates into amyloid fibrils, which are the major component of the senile

GPCR signaling is discussed in Chapter 8.

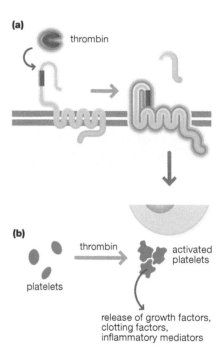

Figure 9.7

Activation of protease-activated receptor (PAR) by thrombin. (a) PAR has the seven-transmembrane topology of a typical G-protein-coupled receptor (GPCR). The activating ligand peptide (*purple*) cannot interact with its binding site until the receptor is cleaved by thrombin. After cleavage, ligand peptide binds the receptor intramolecularly and induces the active conformation. (b) Signaling by the thrombin-activated PAR on platelet plasma membranes leads to platelet activation. Activated platelets adhere to cells and to each other, and secrete a wide variety of factors that regulate clotting, inflammation, and tissue repair.

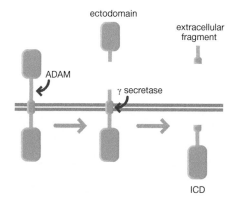

ectodomain

extracellular fragment

ADAM

γ secretase

ICD

Figure 9.8

Regulated intracellular proteolysis (RIP). In the first step of RIP, activated ADAM protease cleaves the extracellular portion of a generic transmembrane protein, shedding the ectodomain. The membrane-associated product is then a substrate for the γ-secretase complex, which cleaves within the transmembrane segment. This results in a soluble intracellular domain (ICD), which can signal in the cytosol and nucleus (as in the case of Notch and ErbB4), and a small extracellular fragment. In the case of amyloid precursor protein, this extracellular fragment, amyloid β, can aggregate into fibrils and is associated with Alzheimer's disease.

plaques found in the brains of Alzheimer's disease patients. Normally, however, APP is likely to serve as an adhesion receptor and signal to the cytosol, both in its intact form and after RIP-mediated cleavage.

Another interesting example of RIP-mediated signaling is ErbB4/HER4, one of the four members of the EGFR family of receptor tyrosine kinases. Binding of ErbB4 to its ligand, neuregulin-1, induces RIP via ADAM-17 and the γ-secretase complex. The products include the shed ectodomain, which may act as a "decoy" to bind ligand and sequester it from activating additional ErbB4 molecules, and the intracellular domain (ICD), which retains tyrosine kinase activity and has been reported to phosphorylate substrates in the nucleus such as Mdm2, a regulator of the tumor suppressor p53. The ErbB4 ICD has been shown to be essential for some biological activities, such as the formation of astrocytes in the developing mouse brain. It apparently performs this function by binding to other proteins, such as transcriptional coactivators or co-repressors, and helping escort them to the nucleus where they exert their effects. Neuregulin-1, the ligand that activates ErbB4, is also processed by the RIP pathway and generates an ICD that may have its own signaling outputs.

UBIQUITIN AND THE PROTEASOME DEGRADATION PATHWAY

The cell must be able to degrade cytosolic proteins rapidly and efficiently in some situations, for example, when a protein is unable to fold correctly or has become damaged. The accumulation and aggregation of such damaged proteins can be highly toxic to the cell. To avoid this problem, eukaryotes have evolved an elaborate system that specifically tags damaged proteins for destruction and delivers them to the proteasome, a molecular machine that digests them into short peptides. This system for targeting specific proteins for removal from the cytosol has also been exploited extensively for signal transduction, providing a rapid and irreversible way to change the protein constituents of the cell in response to various regulatory inputs. In this section, we will consider the machinery involved in this process, and discuss how it is used to control the cell cycle and other signaling events.

The proteasome is a specialized molecular machine that degrades intracellular proteins

While the cell has a need to hydrolyze damaged or misfolded proteins, it must simultaneously avoid the potential dangers of unrestrained proteolytic activity. For this reason, sites of proteolysis must be sequestered from other components of the cytosol. *Lysosomes*, which mostly mediate the degradation of proteins derived by endocytosis from the cell surface or extracellular environment, are shielded from the cytosol by a lipid membrane. The proteolytic destruction of most cytosolic proteins, however, is mediated by a very large multisubunit protein structure termed the **proteasome**. This consists of a hollow cylinder lined on the inside with proteases, and capped at each end by a "lid" structure that controls access to the proteolytic machinery of the inner chamber (**Figure 9.9**). Proteins targeted for destruction are specifically bound by the lid of the proteasome, unfolded, and threaded inside, where they are reduced to small peptides that are released into the cytosol. In this way, the cytosol is shielded from the potent proteolytic activities of the proteasome, and only specifically targeted proteins are ushered into the chamber for destruction.

In eukaryotes, the intact proteasome (called the 26S proteasome, for its sedimentation behavior upon centrifugation) consists of a cylindrical core (the 20S core particle) and two 19S regulatory particles capping the ends of the cylinder. The core

(a)

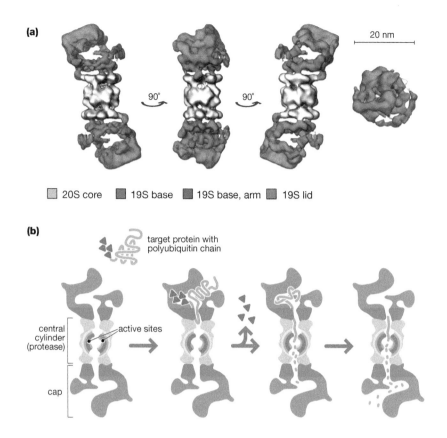

20 nm

☐ 20S core   ☐ 19S base   ☐ 19S base, arm   ☐ 19S lid

(b)

target protein with polyubiquitin chain

central cylinder (protease)

active sites

cap

Figure 9.9

The proteasome. (a) Three-dimensional reconstruction of the 26S proteasome from electron microscope images. Side views and the view from the capped end of the proteasome complex (right) are shown, with the approximate location of proteasome components indicated by colors. (b) Processive protein digestion by the proteasome. The proteasome cap recognizes a substrate protein marked by a polyubiquitin chain, and subsequently translocates it into the proteasome core, where it is digested. (a, Adapted from P.C. da Fonseca & E.P. Morris, *J. Biol. Chem.* 283:23305–23314, 2008. With permission from The American Society for Biochemistry and Molecular Biology, published under CC BY 4.0; b, adapted from B. Alberts, et al., *Molecular Biology of the Cell*, 5th ed. Garland Science, 2008.)

consists of 28 subunits stacked into four rings, and contains three distinct protease activities, each specific for different peptide substrates. Thus, any polypeptide that enters the core is likely to be efficiently cleaved into small oligopeptides by the combined action of multiple proteases. For the isolated core particle, entry to the inner chamber of the proteasome is sterically blocked. Entry is mediated by the 19S regulatory particle that, upon interaction with the core particle, forms a narrow channel just large enough to accommodate a polypeptide chain. Energy from ATP hydrolysis is required to unfold targeted proteins and to thread the unfolded chains into the inner chamber.

The regulatory particle also performs the important task of selecting which proteins will be passed through the proteasome and thereby destroyed. In eukaryotes, most proteasome targets bear a distinctive post-translational mark, the covalent addition of long chains of Lys48-linked polyubiquitin. These chains are added by cellular ubiquitin ligases in a process described in more detail in Chapter 4. The regulatory particle contains several proteins with ubiquitin-binding domains; some of these are integral components of the regulatory particle, while some are accessory adaptor proteins that help recruit specific targets to the proteasome. The preference for substrates with long polyubiquitin chains is probably due to two factors. First, the presence of multiple ubiquitin-binding domains on the proteasome means that polyubiquitylated targets will bind much more tightly than singly or lightly ubiquitylated targets due to avidity. Second, there are deubiquitinase (DUB) activities associated with the regulatory particle that can remove ubiquitins from the end of the polyubiquitin chain; thus, lightly ubiquitylated substrates (with relatively short chains) may be deubiquitylated and released from the proteasome before the protein can be unfolded and stuffed into the inner chamber. In any case, once a protein is committed to destruction, it begins to unfold and other DUB activities remove the polyubiquitin chain at its base, allowing the ubiquitin to be recycled.

The cell cycle and its regulation are described in Chapter 14.

The cell cycle is controlled by two large ubiquitin-conjugating complexes

The cell-cycle machinery ensures that each step in the cell cycle—mitosis, G₁, S phase (where the genome is replicated), G₂, mitosis again—occurs only at the appropriate time and after the previous steps have been completed. Two of the most fateful decisions that a cell must make are whether to initiate DNA replication (and thus commit the cell to divide), and when to segregate the chromosomes and other cell contents into two daughter cells through the process of mitosis followed by cytokinesis. For multicellular organisms in particular, mistakes here can be disastrous, leading to abnormal development or to diseases such as cancer. Not surprisingly, an elaborate mechanism has evolved to regulate these processes tightly. This involves the sequential activation of a series of *cyclin-dependent kinases* (*CDKs*), serine/threonine kinases consisting of a CDK catalytic subunit and a regulatory *cyclin* subunit that is needed for activity. Two large, multisubunit ubiquitin ligase complexes work along with the CDKs to control progression through the cell cycle. Through their action, components necessary for the previous stage of the cycle are polyubiquitylated and targeted for destruction by the proteasome, ensuring that the cycle progresses to the next step and cannot run backward.

The two ubiquitylating complexes are the **anaphase-promoting complex** (**APC**), which regulates the timing of mitosis, and the **SCF complex**, which acts at a number of steps in the cell cycle. Each of these complexes has a core architecture consisting of a RING-type E3 ubiquitin ligase subunit, a cullin scaffold subunit, and one or more adaptor subunits that, in turn, interact with specificity-determining subunits that bind and recruit substrates (**Figure 9.10**). Thus, for these complexes, at least three interacting proteins together perform the tasks normally performed by a single E3 ligase—recruiting the E2–ubiquitin and the substrate, and facilitating the transfer of ubiquitin to the substrate. For both the APC and the SCF complexes, multiple specificity-determining subunits can interact with the same core machinery, allowing each complex to be targeted to specific classes of protein substrates depending on the subunit bound.

Figure 9.10

Structure of the SCF complex and the APC. (a) The *Saccharomyces cerevisiae* SCF ubiquitin ligase complex: cullin scaffold subunit (Cul1, *green*), RING-type E3 ubiquitin ligase (Rbx1, *pink*), adaptor subunit (Skp1, *blue*), specificity component (Skp2, an F-box protein, *purple*), and an E2 ubiquitin ligase (*orange*). The active-site cysteine (Cys) of the E2, where ubiquitin would be covalently attached, is shown in space-filling representation (*blue*). (b) Structure of the human APC derived from cryoelectron microscopy, in the absence (left) or presence (right) of the Cdh1 specificity subunit (*purple*). The cullin subunit Apc2 is colored *green*, and the Doc1 subunit, which is involved in substrate recognition, is colored *yellow*. Note that parts (a) and (b) are rendered at different scales. (a, Adapted from N. Zheng, et al., *Nature* 416:703–709, 2002. With permission from Macmillan Publishers Ltd.; b, adapted from F. Herzog, et al., *Science* 323:1477–1481, 2009. Courtesy of Holger Stark and Jan-Michael Peters.)

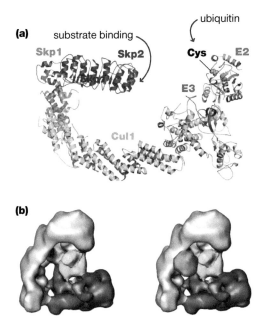

SCF recognizes specific phosphorylated proteins, targeting them for destruction

The core components of the SCF are the E3 (Rbx1), a cullin (Cul1), and the Skp1 adaptor. The specificity components of the SCF complex are called F-box proteins, of which at least 50 are known in the human genome. Many F-box proteins have a generic WD40 repeat domain structure, but a key and defining characteristic of most is that they specifically bind to phosphorylated target proteins, many of which are substrates of specific cyclin-CDKs or other kinases regulated in the cell cycle. Recognition of a phosphorylated site by the SCF complex becomes the kiss of death—the protein is rapidly polyubiquitylated and consigned to destruction by the proteasome. In this way, a normally transient and reversible post-translational modification—phosphorylation—results in the permanent and irreversible elimination of the modified protein (**Figure 9.11**). This is but one of the many ways that CDKs and ubiquitin ligases work together in regulating cell-cycle progression.

The large number of F-box proteins that can bind to the SCF complex, and the variety of substrates recognized by each F-box protein, make it difficult to generalize about the role of the SCF complexes in the cell cycle—virtually every step in the process is regulated in some way by SCF-mediated proteolysis. For example, the F-box protein β-TrCP is important in targeting both inhibitors of the cell cycle (for example, Wee1, a kinase that phosphorylates and thereby inhibits CDKs) and stimulators (for example, Cdc25, the phosphatase that removes the Wee1-mediated inhibitory phosphorylation). Which substrates are targeted by β-TrCP in a given situation depends on their phosphorylation state, which is mediated by yet other kinases and by the CDKs themselves. Another F-box protein, Skp2, specifically targets CDK inhibitors, which include p21 and p27. The proteolytic destruction of these inhibitors is a key step in progressing from the G_1 to S phase of the cell cycle.

Two APC species act at distinct points in the cell cycle

The APC is an even more elaborate molecular assembly than the SCF complex: in addition to RING and cullin subunits, it contains roughly ten more proteins. In contrast to the SCF, however, it associates mainly with just two specificity subunits, Cdc20 and Cdhl. Thus, there are only two major species of APC (APC^{Cdc20} and APC^{Cdh1}), each of which has a distinct role in the cell cycle. APC^{Cdc20} is critical in driving the events of mitosis, and in particular in triggering the metaphase-to-anaphase transition (when sister chromatids physically separate so they can migrate to opposite poles of the mitotic spindle, ultimately to be incorporated into the two daughter cells). Once anaphase is triggered, APC^{Cdh1} takes over, and its

Figure 9.11

SCF couples CDK activity to protein degradation. (a) Substrate is phosphorylated by an activated cyclin–cyclin-dependent kinase (CDK) complex. Upon phosphorylation, it is targeted by the F-box protein (FB) of the SCF complex and polyubiquitylated. The ubiquitylated substrate is then destroyed by the proteasome. (b) In such a system, the concentration of substrate is inversely related to the level of CDK activity.

(a)

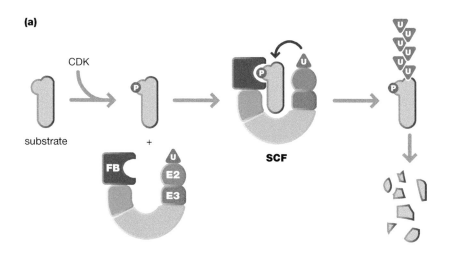

(b)

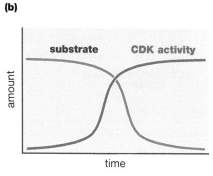

activity is critical to maintaining a stable G_1 phase and repressing the initiation of a new round of DNA synthesis until appropriate signals are received (**Figure 9.12**).

In late G_2 phase of the cell cycle, mitotic cyclin–CDK complexes accumulate and become activated, leading to phosphorylation of APC core subunits. This phosphorylation greatly increases the affinity of Cdc20 for the APC complex, leading to formation of active APC^{Cdc20}. Among the most important substrates for the active complex are the mitotic cyclins and a protein called securin. Polyubiquitylation and degradation of the mitotic cyclins allows progression from mitosis, but also serves to inactivate the kinase that activates APC^{Cdc20}—a negative feedback that leads ultimately to loss of APC^{Cdc20}. Degradation of securin leads to cleavage of the bonds that hold sister chromatids together, allowing them to move to opposite poles of the mitotic spindle. This is the most critical step in mitosis, because premature separation (before all chromosomes have had a chance to align on the mitotic spindle, with sister chromatids attached to opposite spindle poles) can lead to chromosome

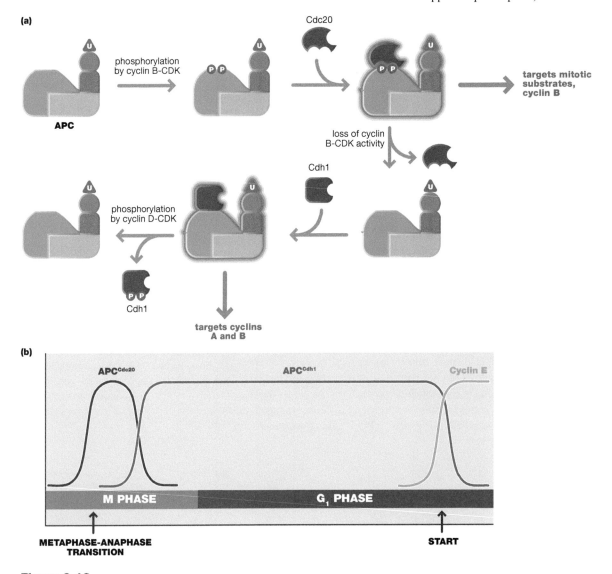

Figure 9.12

Regulation of APC activity by phosphorylation. (a) During M phase, the cyclin B-cyclin-dependent kinase (CDK) complex phosphorylates the anaphase-promoting complex (APC), increasing its affinity for the specificity subunit, C^{dc20}. The now-active APC^{Cdc20} targets substrates such as securin and cyclin B. Decreasing cyclin B-CDK activity leads to dephosphorylation of APC and dissociation of Cdc20. Cdh1 now binds and targets S-phase and mitotic cyclins (cyclins A and B, respectively) to maintain a stable G_1 phase. If G_1/S cyclin (cyclin E) begins to accumulate, Cdh1 is phosphorylated and no longer binds APC, inactivating it until the next M phase. (b) APC^{Cdc20} activity (*purple*), APC^{Cdh1} activity (*pink*), and cyclin E accumulation (*green*) are plotted during M and G_1 phases. Metaphase–anaphase transition (where chromosomes separate in mitosis) and *start* (the point at which a cell is committed to undergoing DNA replication and cell division; also called the *restriction point*) are indicated.

loss or breakage. An elaborate control system, termed the spindle assembly checkpoint, ensures that APC^{Cdc20} is not activated and securin is not polyubiquitylated and degraded until all chromosomes have assumed their proper positions.

Once APC^{Cdc20} has been activated and has caused mitotic cyclins to be degraded, its activity wanes due to loss of APC phosphorylation. At this point, Cdc20 can no longer bind and the second specificity subunit, Cdhl, can associate with the APC. APC^{Cdh1} targets not only the cyclins necessary for M phase (cyclin B), but also those needed to trigger S phase (cyclin A), thus preventing premature entry into a new cell cycle. It does not, however, ubiquitylate the cyclins needed to initiate exit from G_1. When transcription of these cyclins (such as cyclin E) is stimulated by mitogens and other proliferative signals, the activity of the G_1/S cyclin–CDK complex increases, leading to the phosphorylation of Cdhl. Once phosphorylated, it can no longer bind to APC, allowing S-phase cyclins to begin to accumulate and initiate a new round of DNA synthesis. This event marks the commitment of a cell to replicating its DNA and ultimately dividing into two cells, termed the *restriction point* in mammalian cells, and *start* in budding yeast (where much of the cell-cycle machinery was first elucidated).

NF-κB is controlled by regulated degradation of its inhibitor

Members of the nuclear factor κB (**NF-κB**) family of transcription factors are activated in response to a wide variety of signals, particularly in the innate and adaptive immune responses. NF-κB is present in a latent, inactive form in the cytosol of unstimulated cells, from where it can be mobilized rapidly and in the absence of new protein synthesis. The activation of NF-κB depends on the regulated degradation of inhibitory subunits or domains which, in turn, is mediated by polyubiquitylation and proteasome-mediated proteolysis.

The NF-κB family consists of five related proteins, which have in common a Rel homology domain that mediates dimerization and DNA binding (**Figure 9.13**). Three of these proteins [p65 (RelA), c-Rel, and RelB] have transcriptional transactivation domains, whereas the other two (p105/p50 and p100/p52) lack the transactivation domain but instead contain a C-terminal inhibitory domain consisting of a number of *ankyrin repeats*. For these subunits, proteolytic removal of the inhibitory domain is required for activity. A variety of heterodimers are possible, with distinct but overlapping transcriptional activation profiles. The "canonical" NF-κB species consists of a p65/p50 heterodimer.

The activity of NF-κB is held in check in unstimulated cells by association with inhibitory subunits, termed IκBs. There are several distinct species of IκB, all of which consist of multiple ankyrin repeat domains; for the canonical NF-κB pathway, the predominant inhibitory subunit is IκBα. These inhibitors function by retaining the ternary NF-κB/IκB complex in the cytosol, through the combined action of a nuclear export signal on IκB and steric blocking of nuclear localization signals present on the other subunits. The key to unleashing the transcriptional activity of NF-κB, therefore, is the physical removal of the inhibitory subunit.

 Activation of IKK by receptors is described in Chapter 8.

This is accomplished through signal-induced phosphorylation of the IκB, followed by recognition of the phosphorylated site by the SCF family E3 ligase, SCF$^{β\text{-}TrCP}$ (already introduced above). Of course, this raises the question of how

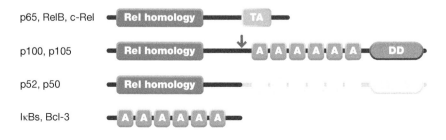

Figure 9.13

The NF-κB family. The Rel homology domain, transcriptional transactivation domain (TA), ankyrin repeats (A), and death domain (DD) are depicted. p52 and p50 are derived by proteolysis of p100 and p105, respectively.

phosphorylation of IκB is regulated. This is the culmination of a rather complicated series of signaling events ultimately leading to activation of the IKK (IκB kinase) complex, which consists of two catalytic subunits, IKKα and IKKβ, and a regulatory subunit, NEMO (**Figure 9.14**). It is interesting to note that the activation of IKK involves a number of proteins of the TRAF family that are themselves RING-family ubiquitin E3 ligases. However, the TRAFs mediate Lys63-linked polyubiquitylation, and thus do not target proteins for proteasomal degradation. Instead, the Lys63-linked modification of substrates such as TRAF itself seems to function primarily in helping to recruit and assemble signaling complexes that ultimately lead to the activation of IKK. Another interesting aspect of NF-κB regulation is that NF-κB activity strongly promotes the transcription of IκBα, which serves as a negative feedback loop to shut down NF-κB signaling after an initial burst of activity.

In addition to the canonical NF-κB pathway, a noncanonical pathway involves activation of heterodimers containing p100 and RelB. Both p100 and p105 contain C-terminal domains that function as IκBs, so proteolytic processing of the precursor form is required for activation of complexes containing these subunits. In the case of p105, processing can be either constitutive or inducible, and the role of polyubiquitylation in processing is still not settled. In the case of p100, however, it is clear that p100–RelB heterodimers can be activated by a subset of NF-κB activators, in a process involving phosphorylation of p100 by IKKα (in the absence of IKKβ and NEMO). Once phosphorylated, p100 is recognized by SCF$^{β-TrCP}$, polyubiquitylated, and targeted to the proteasome. In this case, however, the entire protein is not degraded, just the C-terminal IκB-like region. Such partial degradation by the proteasome is quite unusual, and for p100 and p105 (and a few other known examples, such as the Gli transcription factors that are effectors of Hedgehog signaling) specialized sequence elements are responsible. These consist of a relatively unstructured glycine-rich region, which may form a hairpin loop and insert into the proteasome, juxtaposed to a very stably folded region, which presumably thwarts the proteasome's normal unfolding machinery and prevents further degradation. In the case of the noncanonical NF-κB pathway, partial degradation leads to the release of a p52-RelB heterodimer, which can then translocate to the nucleus and promote transcription of a unique set of NF-κB targets.

Figure 9.14

Canonical and noncanonical pathways of NF-κB activation. In the canonical pathway, a latent heterotrimeric complex consisting of two Rel subunits and an IκB subunit (here, p65, p50, and IκBα) is phosphorylated by the heterotrimeric IκB kinase (IKK) complex. Phosphorylated IκB is targeted by SCF$^{β-TrCP}$ and polyubiquitylated, leading to proteasomal degradation. The released dimeric NF-κB translocates to the nucleus and induces transcription through its transactivation domain (TA). In the noncanonical pathway, a latent p100–RelB heterodimer is phosphorylated by IKKα. The phosphorylated tail of p100 is polyubiquitylated and subjected to partial proteolysis in the proteasome, generating an active p52–RelB heterodimer. Colors correspond to those in Figure 9.13.

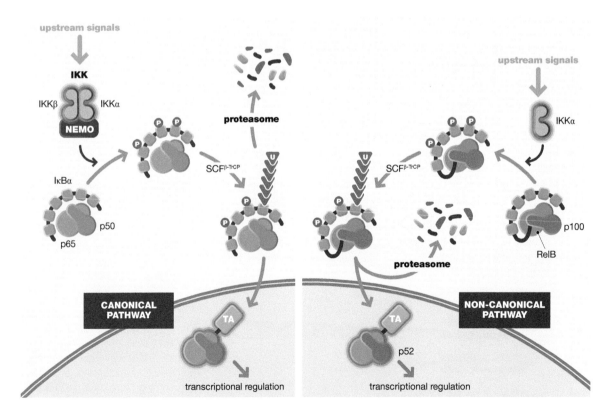

PROTEASE-MEDIATED CELL DEATH PATHWAYS

In the development and day-to-day life of metazoans (multicellular organisms), occasions arise that necessitate the death of individual cells for the benefit of the whole organism. The contents of lysed cells can be quite toxic to surrounding tissues, and so a specialized physiological process termed **apoptosis** has evolved to ease cells through a relatively neat and tidy process of programmed cell death, minimizing the effects on neighboring cells and thus the organism as a whole. The decision to initiate this self-destruct program is obviously important both for the individual cell and for the organism—literally a matter of life and death. At the heart of the apoptotic machinery are the *caspases*, a specialized class of proteases, which execute the cell death program by cleaving key cell proteins. More recently, a broader role for caspases in other physiological responses, such as immunity, has emerged. In this section, we will discuss the regulation and action of caspases, focusing primarily on their role in apoptosis. We will also briefly describe the process of *autophagy*, in which cells subjected to nutrient deprivation and other stresses target their intracellular contents for digestion via the lysosomal pathway.

Apoptosis is an orderly and highly regulated form of cell death

For free-living unicellular organisms, cell death may be a tragedy for the individual cell but has little impact on the larger population. In metazoans, however, the fate of each cell is tied to the fate of the organism as a whole. Normal development requires that some cells die. For example, in *Caenorhabditis elegans*, a simple roundworm consisting of roughly 1000 cells, more than 100 cells undergo programmed cell death in the process of generating the finished worm. Furthermore, cells that have been damaged, particularly those that have sustained extensive or irreparable damage to the genomic DNA, pose a considerable danger to multicellular organisms—cells that are allowed to pass on a damaged genome to their progeny may develop into cancers that can threaten the survival of the whole organism. Apoptosis provides an efficient way to remove such cells without generating cell debris that could initiate inflammatory or immune responses which could pose additional dangers to the surrounding cells.

Cell death in eukaryotes can take a variety of forms, depending on the cause and the cellular context. In almost all cases (with the exception of acute trauma), the process is actively regulated by signaling pathways that recognize specific triggers, and then execute an orderly program of self-destruction (**Table 9.1**). This process is termed **regulated cell death**. At one end of the spectrum is **apoptosis**, a highly orchestrated form of cell death initiated by specific signals and leading to characteristic biochemical and morphological changes (**Figure 9.15**). The chromatin of a cell undergoing apoptosis condenses, and nucleases are activated that degrade the

Table 9.1

Mechanisms of Regulated Cell Death. Examples of different mechanisms of cell death are compared. "Morphology" describes the physical manifestations of cell death, whether apoptotic, necrotic, or having properties of both (see Fig. 9.15). Note that there is considerable overlap between different regulated cell death pathways, and triggering one mechanism can lead to activation of secondary mechanisms. MOMP, mitochondrial outer membrane permeabilization; MLKL, mixed-lineage kinase-like pseudokinase; RIPK3, receptor-interacting serine/threonine protein kinase 3; ROS, reactive oxygen species; GPX4, glutathione peroxidase 4; PARP1, poly-ADP ribose polymerase 1

	Trigger	Key effectors	morphology
Extrinsic apoptosis	Death receptor activation	Caspases	apoptotic
Intrinsic apoptosis	Stress/homeostasis failure, leading to MOMP	Caspases	apoptotic
Necroptosis	Death receptors, pathogen recognition receptors	MLKL/RIPK3	necrotic
Ferroptosis	Excess ROS, iron	GPX4 inactivation, lipid oxidation	necrotic
Pyroptosis	Innate immunity	Caspases, Gasdermin-mediated membrane pores	necrotic/mixed
Parthanatos	DNA damage	PARP1 hyperactivity	apoptotic/mixed
Autophagy-dependent cell death	Developmental cues, starvation	Autophagy machinery	autophagic

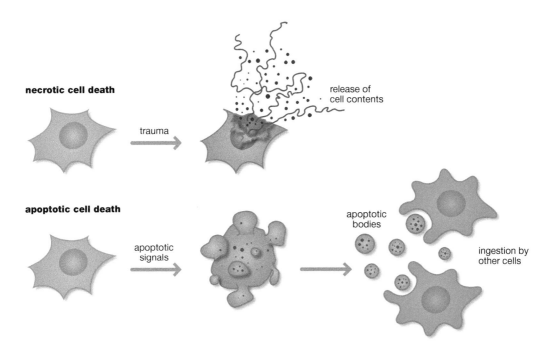

Figure 9.15

Necrotic and apoptotic cell death. In necrotic cell death, the cell ruptures and its contents are released, leading to localized tissue damage and inflammation. By contrast, a cell undergoing apoptosis dies by a programmed sequence of events, leading to its fragmentation into membrane-enclosed apoptotic bodies that are engulfed and digested by surrounding cells.

DNA into oligonucleosome-sized fragments (this "laddering" of degraded DNA, visible upon electrophoresis, is a hallmark of apoptosis). The nuclear membrane breaks down, as do other membrane-bound organelles, and cytoskeletal changes lead to cell rounding and membrane out-pocketing or "blebbing." The resulting blebs ultimately pinch off, creating membrane-enclosed packets of cell contents, termed apoptotic bodies, which can be ingested by surrounding cells or by professional phagocytes such as macrophages. These cell fragments are then completely digested by lysosomes within the engulfing cell. In this way, the contents of the apoptotic cell, including its nucleic acids, can be recycled without ever being exposed to the extracellular environment. Apoptosis contrasts with cell death induced by physical trauma or acute stress (a process called **necrosis**), in which cells spill their contents into their surrounding tissues. This tends to evoke strong inflammatory responses, which may be harmful to surrounding cells. However, such responses can also be important in mounting an effective defense against pathogenic invaders, or in the remodeling and repair of damaged tissues.

Apoptosis is initiated either by signals from the extracellular environment, acting through plasma membrane receptors, or by signals from within the cell itself, such as physiological stresses or DNA damage. As discussed in more detail below, these *extrinsic* and *intrinsic* pathways differ in important respects, but both lead to the activation of a specialized group of cysteine proteases, termed the **caspases**, that specifically cleave peptide bonds C-terminal to aspartic acid in their protein targets. Most caspase-mediated signaling pathways involve a hierarchy of two classes of caspases, initiator caspases and effector or executioner caspases. The **initiator caspases** are directly activated by apoptotic signals, as discussed below. Once activated, they amplify the resulting signal by proteolytically activating the **effector caspases**, which go on to cleave cell proteins that mediate the physiological manifestations of the cell death program. As we have seen many times already, cascades of sequentially activated proteases are commonly used to generate widespread, irreversible physiological changes.

The activity of caspases is tightly regulated

There are 13 distinct caspase genes in humans, which can be roughly divided into two classes (initiator and effector) based on their function, activation mechanism, and structural features (**Figure 9.16**). All caspases have a similar overall structure, consisting of an N-terminal pro-domain followed by large and small catalytic

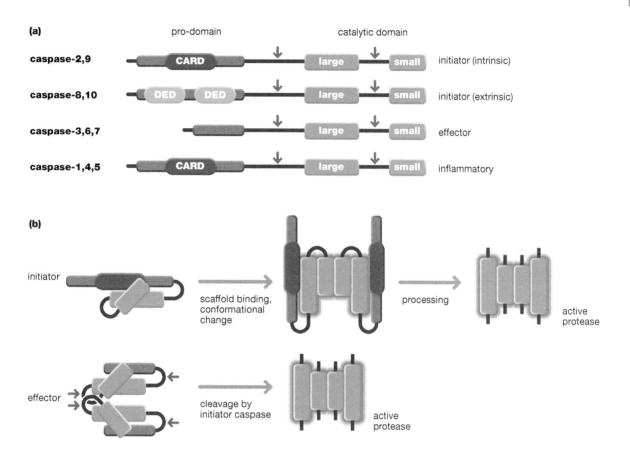

Figure 9.16

Caspase domain structure and activation mechanism. (a) All caspases have an N-terminal pro-domain followed by large and small protease catalytic domain subunits (*blue*). Sites of cleavage by caspases are indicated by *pink arrows*. The caspase recruitment domain (CARD) and death effector domain (DED) mediate homotypic protein interaction with scaffold and adaptor proteins during activation. (b) Activation of initiator and effector caspases. Interaction with scaffold complexes induces conformational changes that activate initiator caspases, while effector caspases must be activated by cleavage by upstream initiator caspases (*pink arrows*).

subunits. These three domains are connected by linker sequences that contain sites for caspase cleavage. All caspases are initially expressed as inactive zymogens, though the mechanism of their activation is different for the two classes. Initiator caspases (caspases 2, 8, 9, and 10) have an extended N-terminal pro-domain and are activated predominantly by dimerization or assembly into higher order structures, whereas effector caspases (caspases 3, 6, and 7) have a shorter pro-domain and are activated by cleavage by upstream initiator caspases. Inflammatory caspases (caspases 1, 4, and 5) are similar in structure and activation mechanism to initiator caspases. In all cases, the mature, active caspase is a tetrameric complex consisting of two large and two small catalytic subunits.

Initiator caspases must respond to and become activated by diverse upstream signaling inputs. This is accomplished by stimulus-induced aggregation, the same mechanism underlying many receptor-mediated signaling strategies. Recruitment of inactive, monomeric initiator caspases into a multimeric structure or scaffold leads to dimerization of the precursors, along with conformational changes that rearrange the catalytic domain and allow cleavage of the dimer partner into its mature form. As discussed below, the multimeric complex that mediates the activation of the extrinsic pathway is the *death-inducing signaling complex (DISC)*, whereas the intrinsic pathway is triggered by assembly of the *apoptosome*. More specialized signal-induced scaffolds also exist. For example, caspase-2 is activated in the nucleus in response to genotoxic stimuli by a structure termed the PIDD osome (PIDD is a major transcriptional target of p53, which is activated by DNA damage, cell-cycle checkpoints, and other stress responses). Similarly, pathogens activate caspases 1, 4, and 5 through the assembly of a scaffolding complex termed the inflammasome, which plays an important role in the innate inflammatory response (see below).

By contrast, the effector caspases exist normally as inactive dimers. Their activation requires cleavage of the linker between the two catalytic segments, allowing rearrangement of the catalytic site into its fully active conformation. This strategy makes the activation of effector caspases absolutely dependent on the levels of activity of the upstream initiator caspases. In lymphocytes, another unrelated protease, the serine protease granzyme B, can also cleave and activate effector caspases, but this seems to be an exception.

In vitro studies showed that caspases are highly specific for peptide substrates with an aspartic acid (Asp) residue preceding the bond to be cleaved, followed by an amino acid with a small, uncharged side chain. Proteomic studies have shown that hundreds of different proteins are cleaved in cells where effector caspases are activated. Some of these are clearly important for the physical execution of apoptosis. For example, the caspase-activated DNase (CAD) and the Rho-associated kinase (ROCK) are activated by caspase-mediated cleavage, causing DNA degradation and reorganization of the actin cytoskeleton, respectively, while the lamins that provide structural integrity to the nuclear envelope are inactivated by caspase-mediated cleavage. However, many of the substrates might be "innocent bystanders" that play no role in the physical events of apoptosis. It is interesting to note that, in many cases, cleavage by caspases modifies or stimulates the activity of a substrate, rather than simply inactivating it by degradation. In this way, proteolysis can promote the irreversible activation of pathways that lead to the orderly death of the cell.

Since inadvertent activation of caspases would have dire consequences for the cell, mechanisms are needed to keep basal activity levels low in the absence of strong apoptotic stimuli. A family of caspase inhibitors termed IAPs (inhibitor of apoptosis proteins) serves to repress activated caspases under most conditions. These proteins use two mechanisms to inhibit their targets. First, they can bind directly to and inhibit activated caspases though protein interaction modules termed BIR domains. Second, most IAPs also have a RING-type ubiquitin E3 ligase activity, which can mono-and polyubiquitylate the caspase and either inactivate it or target it for proteasome-mediated degradation. IAPs can themselves be down-regulated in order to sensitize cells to apoptotic stimuli. One way this is accomplished is by the translocation or increased expression of IAP antagonists such as Smac/DIABLO, which bind to and sequester IAPs and thereby prevent their binding to caspases.

The extrinsic pathway links cell death receptors to caspase activation

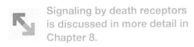

Signaling by death receptors is discussed in more detail in Chapter 8.

The **extrinsic apoptotic pathway** is induced by a number of ligands that interact with specific cell-surface receptors. These include Fas ligand (FasL) and its receptor Fas, tumor necrosis factor (TNF) and its receptor TNFR, and TRAIL and its receptors DR4 and DR5. These receptors are collectively referred to as *death receptors*. Binding of the corresponding trimeric ligand to the death receptor induces conformational changes that expose *death domains (DDs)* on the intracellular portion of the receptor. Using the prototypic death receptor Fas as an example, the activated receptor then recruits an adaptor, FADD, through homotypic DD-mediated interactions. Receptor binding exposes the *death effector domain (DED)* of FADD, which then recruits initiator caspases (caspase-8 and to a lesser extent caspase-10) through homotypic interaction with the DED domain located in the caspase pro-domain (see Figure 8.18). Collectively, these interactions generate the active **death-inducing signaling complex (DISC)**, which serves as a platform for the dimerization and allosteric activation of the bound caspase. Upon activation, bound caspases cleave each other to generate soluble, heterotetrameric caspase that is free to activate caspase-3, the downstream effector caspase. In some but not all cell types, simultaneous activation of the intrinsic pathway (discussed further below) is also necessary for efficient induction of apoptosis. This is accomplished by caspase-8-mediated cleavage and activation of BID, a key promoter of the intrinsic pathway. This also illustrates a more general point, that activation of one regulated cell death program often engages some elements of other cell death programs, especially in the later stages.

Other death receptors use variations on this overall strategy for apoptotic activation (**Figure 9.17**). There are five receptors for the pro-apoptotic ligand TRAIL, but only two of these (DR4 and DR5) can activate caspases. The others can bind ligand, but because they lack intracellular DDs, they are incapable of downstream signaling. Therefore, they act as "decoys" to soak up and sequester TRAIL, thus blocking apoptosis in cells where they are expressed at high levels. In the case of the TNFR, the major signaling output under most conditions is activation of NF-κB, not apoptosis. NF-κB is activated by the recruitment of a different DD-containing adaptor, TRADD, to the activated receptor. TRADD, in turn, recruits TRAF2 and the IKK complex, ultimately leading to the phosphorylation, ubiquitylation, and destruction of IκB. The active NF-κB that is released strongly suppresses apoptosis by inducing transcription of various apoptotic inhibitors, such as IAPs and a DED-containing decoy protein FLIP, which competes with initiator caspases to prevent their recruitment to the DISC. Only when NF-κB signaling is for some reason blocked is a second, pro-apoptotic complex formed. This complex contains TRADD, FADD, and caspase-8, and functions analogously to the DISC formed in response to TRAIL or FasL. In this case, apoptosis is a "fail-safe" mechanism that kicks in under circumstances where the primary response, NF-κB induction, is ineffective. TNFR signaling is just one example of many instances where the pathways leading to activation of NF-κB and to apoptosis intersect.

Infection of cells with bacterial or viral pathogens can activate an alternative DISC-like complex termed the inflammasome (**Figure 9.18**). The result of activation of this pathway is typically not apoptosis, but instead the processing and secretion of a host of proinflammatory cytokines. The caspases that are activated include caspase-1 and caspase-5, which contain in their pro-domains another protein interaction module, termed the caspase recruitment domain or CARD domain, in place of the DED. Activation is achieved by assembly of a multiprotein complex consisting of a sensor protein (such as NALP1, NALP3, or IPAF) that is allosterically activated by pathogen components. Activation induces oligomerization

Figure 9.17

Signaling by various death receptors.
Fas ligand (FasL) binding to Fas leads to aggregation of the adaptor FADD, and recruitment/activation of initiator caspase. Some TRAIL receptors (DR4, DR5) signal through FADD in the same fashion, whereas other "decoy" receptors for TRAIL do not contain intracellular death domain (DD) motifs and thus cannot activate caspases. The primary result of binding of tumor necrosis factor α (TNFα) to its receptor (TNFR) is the aggregation of the TRADD adaptor, leading to recruitment of TRAF2, IκB kinase (IKK) activation, and activation of NF-κB. A secondary pathway leads to formation of a cytosolic complex containing TRADD and FADD, which can recruit and activate initiator caspases. This pathway is normally strongly inhibited by NF-κB; thus, it is only activated under conditions where NF-κB activity is suppressed. The domain structures of FADD and TRADD are indicated in the inset.

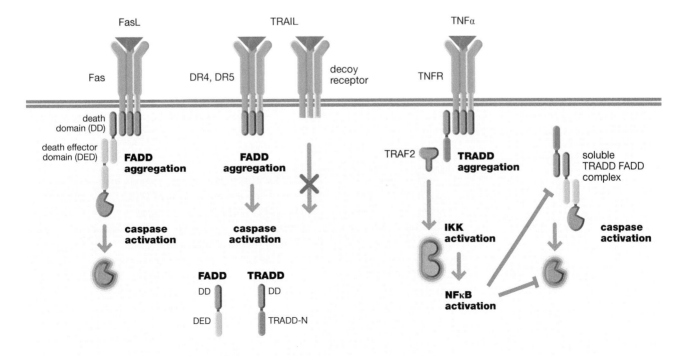

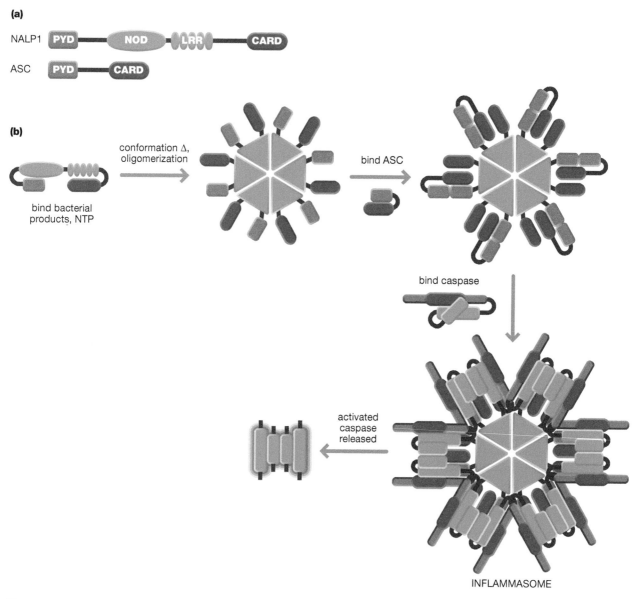

Figure 9.18

Assembly of the NALP1 inflammasome. (a) Domain structure of NALP1 and ASC. NALP1 contains an N-terminal Pyrin domain (PYD), a domain that mediates nucleotide binding and oligomerization (NOD), leucine-rich repeats (LRR) that bind microbial products such as cell-wall proteoglycans, and a CARD domain (caspase recruitment domain). The ASC adaptor contains Pyrin and CARD domains. (b) Binding of bacterial products and nucleotide (NTP) to NALP1 induces conformational changes leading to oligomerization. NALP1 now binds ASC and recruits inflammatory caspases such as caspase-1, which are activated by conformational change and dimerization. Processed caspase is then released from the complex. The organization of subunits shown is diagrammatic and not based on physical structural data.

of the sensor, and exposes other protein interaction domains that can directly or indirectly (through adaptor proteins) interact with the CARD domain in the caspase pro-domain. In the case of the NALP1 inflammasome, activation exposes its PYD domains, which recruit and bind to PYD domains of an adaptor such as ASC. This exposes the CARD domain of the adaptor, thus recruiting caspases through CARD–CARD interactions. This scheme is conceptually very similar to death receptor DISC formation, with the PYD substituting for the DD and the CARD for the DED. Activated caspase-1 can directly cleave and activate proinflammatory cytokines such as interleukin-1β, and indirectly induce the processing and secretion of others. It is interesting to note that plants use a very similar mechanism to respond to pathogen-derived foreign substances, where the sensor proteins are *nucleotide-binding and leucine-rich repeat receptors (NLRs)* (see Figure 13.13).

Under some conditions, particularly in macrophages and related cells, activation of inflammatory caspases can induce a distinct regulated cell death program termed *pyroptosis*. In this case, activated caspases cleave the substrate gasdermin D (GSDMD), which releases an N-terminal domain (GSDMD-N) that translocates to the plasma membrane, oligomerizes, and forms a pore. The resulting membrane permeabilization leads to the release of cell contents and necrotic cell death. It is thought the dead cell corpses that result from pyroptosis may serve as "pore-induced intracellular traps" (PITs) that sequester bacteria until the dead cells can be ingested by phagocytes.

Mitochondria orchestrate the intrinsic cell death pathway

In vertebrates, the **intrinsic apoptotic pathway** responds to signals generated from within the cell, most of which fall into the general category of "stress responses." For these pathways, the mitochondrion plays a surprising role in integrating the welter of potentially conflicting information on the status of the cell and, where necessary, executing the early phases of the apoptotic program. Central to this formidable task is a group of proteins termed the **Bcl2 family**, of which at least 18 are found in humans. The balance between pro-apoptotic and anti-apoptotic Bcl2 family members determines the fate of the cell. When pro-apoptotic Bcl2 family members predominate, pores form in the mitochondrial outer membrane that increase its permeability (termed mitochondrial outer membrane permeabilization, or MOMP), leading to the release of cytochrome *c* and other components from the mitochondria. Ultimately, it is these released mitochondrial components that nucleate the assembly in the cytosol of the *apoptosome*, yet another large cytosolic complex similar to the DISC, that serves as a scaffold for the recruitment and activation of caspases.

There are three major classes of Bcl2 proteins, defined by their biological activities and their structural elements. All contain at least one of four conserved sequence motifs, termed Bcl homology (BH) domains, and many also have a C-terminal transmembrane helix that can anchor them to cellular membranes, particularly the mitochondrial outer membrane (**Figure 9.19**). These three categories are the anti-apoptotic Bcl2 proteins, which include Bcl2 and Bcl-X$_L$; the pro-apoptotic or effector Bcl2 proteins, BAX and BAK; and the pro-apoptotic BH3-only proteins, which include BAD, BID, BIM, NOXA, and PUMA. Many of these Bcl2 family members can interact with each other, usually via insertion of the helical BH3 segment of one into a hydrophobic pocket on the surface of its partner (described below). It is the stoichiometry of the interactions among these classes of proteins that determines whether apoptosis will be activated. Under normal conditions, there is an excess of anti-apoptotic proteins, and the activity of the pro-apoptotic proteins BAX and BAK is held in check. Cellular stress signals, however, induce either increased expression or post-translational modification of the BH3-only class of proteins, and this shifts the equilibrium such that the pro-apoptotic Bcl2 proteins can aggregate on the outer membrane of the mitochondrion and induce MOMP (**Figure 9.20**).

Figure 9.19

Bcl2 family of apoptotic regulators. Representative structures are shown for the three major groups. Positions of Bcl2 homology domains (BH1, BH2, BH3, and BH4) and transmembrane helix (TH) are indicated. Names of family members in each group are provided at left. Note that other, less-well-characterized family members exist.

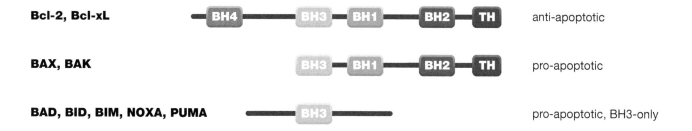

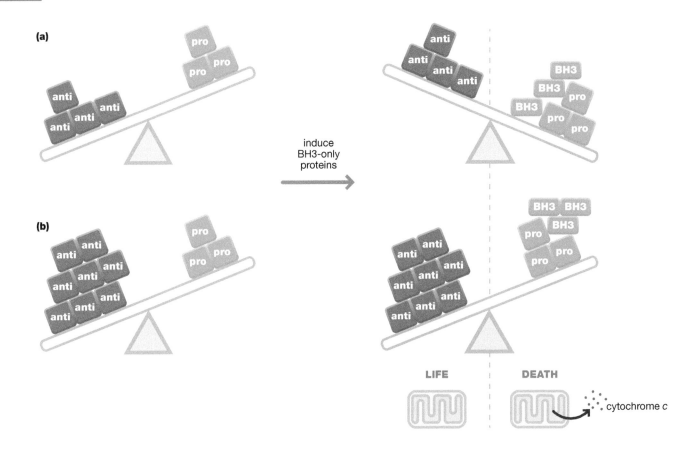

Figure 9.20

Induction of apoptosis depends on the balance of Bcl2 proteins. (a) Normally, anti-apoptotic Bcl2 proteins are in excess over pro-apoptotic Bcl2 proteins. When BH3-only protein activity or abundance increases, the balance shifts and pro-apoptotic proteins are in excess; MOMP is induced and cytochrome *c* and other pro-apoptotic factors are released from the mitochondrial intermembrane space. When levels of anti-apoptotic proteins greatly exceed those of pro-apoptotic proteins, as when Bcl2 is overexpressed, then levels of BH3-only proteins that would normally induce apoptosis have no effect.

BAX and BAK can exist either in soluble or membrane-associated forms. In the soluble form, the C-terminal transmembrane helix (TH) occupies a hydrophobic groove on the surface. Once inserted into the mitochondrial outer membrane via its C-terminal TH segment, however, this hydrophobic groove in BAX or BAK is available to interact with BH3 segments from other Bcl2 family members (**Figure 9.21**). Transient interaction with BH3-only proteins induces profound conformational changes in BAX or BAK that lead first to homodimerization, then to oligomerization and pore formation. It is thought that anti-apoptotic Bcl2 proteins such as Bcl2 normally bind to BAX and BAK on the mitochondrial outer membrane and prevent their dimerization; excess BH3-only proteins displace the anti-apoptotic Bcl2 proteins either directly or indirectly, leading to pore formation, release of mitochondrial contents, and ultimately to mitochondrial fragmentation and loss of mitochondrial function.

The primary level at which the intrinsic pathway is induced is via increased activity or abundance of the BH3-only proteins. Different BH3-only proteins act as sensors for a wide variety of cellular stresses. For example, the transcription of PUMA and NOXA is induced by p53, and BIM transcription is stimulated by mitogen and growth factor deprivation and endoplasmic reticulum (ER) stress (excessive levels of unfolded proteins in the ER). Changes in post-translational modification can also lead to activation of BH3-only proteins, as in the case of loss of phosphorylation of BAD by Akt in the absence of survival signals. As already mentioned, cleavage of BID into its active form by caspase-8 allows extrinsic pathway activity to be amplified by engaging the intrinsic pathway. Expression levels of the anti-apoptotic Bcl2 proteins can also be regulated by survival signals and stress, thereby resetting the threshold for induction of apoptosis (see **Figure 9.20**). Indeed, Bcl2 itself was first discovered as an oncogene whose expression was elevated in B cell lymphomas; it promotes the uncontrolled overgrowth of B cells by inhibiting their apoptosis.

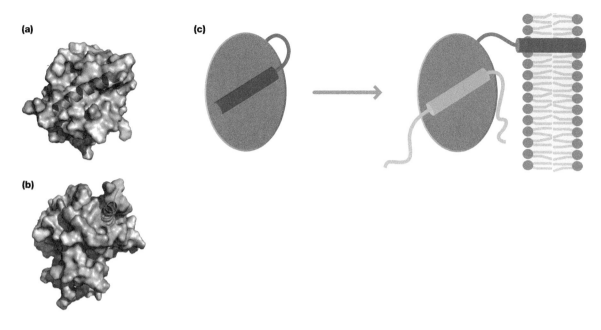

Figure 9.21

The transmembrane helix (TH) segment of pro-apoptotic Bcl2 proteins binds to the same site as BH3-only proteins. (a and b) X-ray crystal structure of full-length BAX, a pro-apoptotic Bcl2 protein. In its soluble state, the C-terminal TH segment of BAX (*brown*) is sequestered in a hydrophobic pocket. The structure in (b) is rotated 90° from that in (a). (c) Diagrammatic representation of binding of BH3-only protein (*green*) to BAX inserted in the mitochondrial outer membrane. The BH3 helix binds to the same hydrophobic pocket previously occupied by the TH segment (*brown*). Transient BH3 binding in this context leads to conformational changes in BAX, and subsequent dimerization, oligomerization, and pore formation. (a and b, Adapted from R.J. Youle and A. Strasser, *Nat. Rev. Mol. Cell Biol.* 9:47–59, 2008. With permission from Springer Nature.)

Once MOMP occurs, cell death is induced through the assembly of the **apoptosome**, which serves as a platform for activation of the initiator caspase-9. Apoptosome assembly depends on cytochrome c that is released from the intermembrane space of the mitochondria. In the cytosol, cytochrome c binds to the adaptor protein Apaf1 and, in concert with nucleotide binding, induces conformational changes that lead to the oligomerization of Apaf1 into a structure reminiscent of a seven-spoked wheel (the "wheel of death") (**Figure 9.22**). Each spoke of the wheel consists of an Apaf1 molecule, oriented with the bound cytochrome c on the outside and with its CARD domain located near the hub. As in the case of

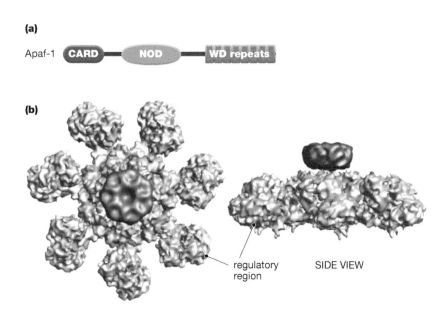

Figure 9.22

The apoptosome. (a) Domain structure of Apaf1. Cytochrome c released from mitochondria binds to the WD repeats, inducing conformational changes, nucleotide binding, and oligomerization of the NOD (nucleotide-binding and oligomerization domain). (b) Three-dimensional reconstruction of the human apoptosome [Apaf1 in complex with cytochrome c and the CARD domain from procaspase-9] from cryoelectron microscopy images. Seven Apaf1 subunits assemble into a wheel-like structure. The CARD domains of Apaf1 and caspase-9 form a disc (*purple*) without visible connections with the rest of the structure. The regulatory regions, consisting of the Apaf1 WD repeats bound to cytochrome are located at the end of each spoke. (b, Adapted from S. Yuan, et al., *Structure* 18:571–583, 2010. With permission from Elsevier.)

the inflammasome, the exposed CARD domains serve to recruit initiator caspase (caspase-9 in this case) through homotypic CARD–CARD interactions. Bound caspase then is activated through conformational changes, undergoes autocleavage, and then is free to activate downstream caspases (caspase-3 and caspase-7).

Cytochrome *c* is not the only pro-apoptotic component released from mitochondria during MOMP. Smac/DIABLO and related IAP antagonists promote apoptosis upon their release by sequestering and inhibiting the IAP proteins, which normally function to inhibit caspases. Some of the released mitochondrial proteins can also function to promote cell death in a caspase-independent fashion. Apoptosis inducing factor (AIF), a flavoprotein, translocates to the nucleus and induces chromatin condensation and DNA fragmentation. EndoG is a nuclease that cleaves chromatin between nucleosomes upon its release from mitochondria. Finally, loss of the electrogenic potential of the mitochondria upon MOMP contributes to cell death by depriving the cell of the means to generate energy.

Autophagy is a mechanism used by cells to digest themselves, which can lead to cell death

Autophagy is a highly regulated process whereby cells engulf their own cytosol and organelles and target them for degradation in the lysosome, allowing the raw materials to be recycled. Autophagy is an element of normal cell homeostasis, providing a means of removing old and damaged organelles and insoluble aggregates of misfolded proteins; it can also be an acute response to nutrient starvation or energy depletion, allowing the cell to survive until conditions are more favorable. For these reasons, autophagy usually enhances cell survival. There are some conditions, however, in which autophagy continues unabated until the cell digests so much of itself that it can no longer survive. This process, which can be part of normal developmental programs or the result of external stress, is termed *autophagy-dependent cell death (ADCD)*.

As might be expected, autophagy is controlled by an elaborate set of signaling mechanisms (**Figure 9.23**). Much of the early work to define the mechanism was done in yeast, where a number of key genes (termed Atg genes) were identified through genetic screens. The process is normally initiated by activation of a multi-component serine/threonine kinase termed the ULK1 complex. Various stresses couple to ULK1 activity; for example, mTORC1 (which is active when conditions are appropriate for cell growth) binds to and represses ULK1, while AMPK (which is activated when ATP levels are low) activates it. Activation of the ULK1 complex triggers the formation of a specialized region of the endoplasmic reticulum termed the Phagophore Assembly Site (PAS). ULK1-mediated phosphorylation leads to the recruitment and activation of another protein complex containing the Class III PI 3-kinase VPS34 and its regulatory subunit Beclin-1, generating areas of high PI(3)P density on the ER membrane.

The PI(3)P-enriched membranes next recruit factors such as WIPI, which in turn recruit a number of ATG gene products that mediate the covalent coupling of several proteins, including ATG8 homolog LC3, to the lipid phosphatidylethanolamine (PE) on the phagophore membrane. This conjugation process is mechanistically very similar to the coupling of ubiquitin and its relatives to proteins, and the ATG proteins include examples with E1, E2, and E3 activities, along with ubiquitin-like proteins (ATG8 homologs and ATG12).

The developing phagophore also recruits endosomes and other membranes from the cell to assemble a crescent-shaped membrane structure that sequesters cell components for digestion. These contents may be general (that is, consisting of all cytosolic contents) or in some cases specific cytosolic structures may be targeted. In the case of damaged mitochondria, targeting involves the polyubiquitylation of mitochondrial outer membrane proteins and the phosphorylation of these

The ubiquitin conjugation machinery is discussed in Chapter 4.

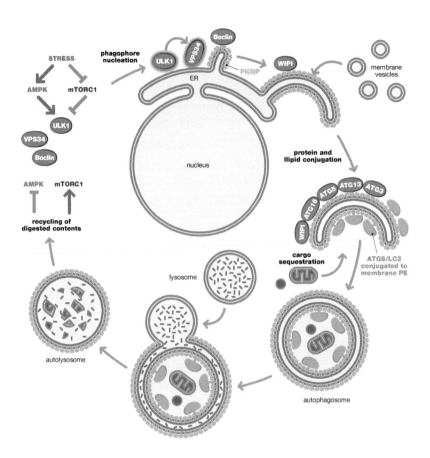

Figure 9.23

Autophagy. The major steps in autophagy are depicted. Autophagy is initiated by activation of the ULK1 kinase complex, which localizes to the Endoplasmic Reticulum (ER) and phosphorylates and activates the VPS34 Class III PI 3-kinase complex. Localized high concentrations of PI(3)P recruit WIPI and other proteins, which in turn recruit a number of ATG proteins along with membrane vesicles from the cytosol to enlarge the growing phagophore. The ATG proteins mediate the covalent conjugation of ATG8 family member LC3 to the membrane lipid PE. Cytosolic cargo is sequestered either passively, or in case of targeted degradation via direct or indirect interactions with LC3. The membranes eventually seal to form an autophagosome surrounded by a double lipid membrane. This fuses with lysosomes, leading to digestion of the contents and their release into the cytosol. Under normal conditions, the process is terminated by relief of nutritional stress and the inactivation/degradation of ULK1 and VSP34. (Adapted from Dikic I & Elazar Z. *Nat Rev Mol Cell Biol*. 19(6):349–364, 2018. With permission from Springer Nature.)

ubiquitin chains; these modified proteins are directly or indirectly recognized by proteins such as LC3 on the developing autophagosome membrane. Ultimately, the autophagosome seals into a double membrane-bounded vesicle containing the cytosolic components slated for destruction, which is accomplished by fusion of the autophagosome with lysosomes. Proteases and other digestive enzymes derived from the lysosome degrade the contents, which are then released for recycling.

Specific mechanisms that lead to unrestrained autophagy and thus to ADCD are not yet completely understood. There is evidence that ADCD can result from unrestrained initiation of autophagy (through sustained hyperactivation of AMPK, for example), or through failure of the feedback mechanisms that normally restore autophagy to resting levels, such as the polyubiquitinylation and proteosome-mediated degradation of key components such as ULK1, VPS34, and Beclin-1.

SUMMARY

Proteolysis is unlike other post-translational modifications in that its effects cannot be reversed, other than by the relatively slow process of new protein synthesis. Most proteases are initially made as catalytically inactive precursors (zymogens), which are activated at the appropriate time and place by proteolytic processing. The regulated activation of proteases plays a key role in a number of signaling pathways, including extracellular processes such as blood clotting and in transmembrane signaling. The targeted degradation of cytosolic proteins is achieved by the proteasome, a molecular machine that recognizes and destroys proteins that have been tagged by polyubiquitin chains. The coupling of protein ubiquitylation to proteasome-mediated destruction is central to many signaling mechanisms, including those that regulate the orderly progression through the cell cycle. Proteolysis also plays a key role in various types of regulated cell death. Most notably, the activation of a specialized class of proteases, the caspases, serves both to initiate and execute the events of programmed cell death (apoptosis).

Answers to these questions can be found online at www. routledge.com/9780367279370

QUESTIONS

1. Discuss the differences between phosphorylation and degradation as modes of regulation. What types of processes might be better regulated by one or the other? How does the functional coupling of phosphorylation and ubiquitylation (by ubiquitin ligases that recognize phosphorylated substrates, such as the SCF complex) affect signaling output from the phosphorylated substrates?

2. Almost all proteases are activated by cleavage of an inactive precursor. In some cases, one activated protease molecule can cleave and activate other molecules of the same type, while in other cases, activation must be performed by a different (upstream) protease. Compare the effects on activation dynamics and localization in these two different cases.

3. What mechanisms prevent the entire blood supply from clotting after a minor injury?

4. Many viruses alter regulation of the infected host cell. For example, several viruses induce the degradation of antiviral factors. Propose strategies that could be used by the virus to accomplish this. In addition, many viruses have evolved mechanisms to inhibit cellular apoptotic pathways. Why might this be the case? Propose several distinct strategies that could be used by viruses to accomplish this.

5. Proteasome inhibitors are increasingly used for cancer therapy. Why might such compounds be effective in preferentially targeting tumor cells?

BIBLIOGRAPHY

GENERAL PROPERTIES AND EXAMPLES OF SIGNAL-REGULATED PROTEOLYSIS

Drag M & Salvesen GS (2010) Emerging principles in protease-based drug discovery. *Nat. Rev. Drug Discov.* 9, 690–701.

Edwards DR, Handsley MM & Pennington CJ (2008) The ADAM metalloproteinases. *Mol. Aspects Med.* 29, 258–289.

UBIQUITIN AND THE PROTEASOME DEGRADATION PATHWAY

Finley D (2009) Recognition and processing of ubiquitinprotein conjugates by the proteasome. *Annu. Rev. Biochem.* 78, 477–513.

Gao M & Karin M (2005) Regulating the regulators: Control of protein ubiquitination and ubiquitin-like modifications by extracellular stimuli. *Mol. Cell* 19, 581–593.

Hayden MS & Ghosh S (2008) Shared principles in NF-κB signaling. *Cell* 132, 344–362.

Komander D (2009) The emerging complexity of protein ubiquitination. *Biochem. Soc. Trans.* 37, 937–953.

Liu T, Zhang L, Joo D & Sun SC (2017) NF-κB signaling in inflammation. *Signal Transduct Target Ther.* 2, 17023. doi: 10.1038/sigtrans.2017.23.

Yamano H (2019) APC/C: Current understanding and future perspectives. *F1000Res* 8, F1000 Faculty Rev-725. doi: 10.12688/f1000research.18582.1.

PROTEASE-MEDIATED CELL DEATH PATHWAYS

Bialik S, Dasari SK & Kimchi A (2018) Autophagy-dependent cell death - Where, how and why a cell eats itself to death. *J. Cell Sci.* 131, jcs215152. doi: 10.1242/jcs.215152.

Cosentino K & García-Sáez AJ (2017) Bax and bak pores: Are we closing the circle? *Trends Cell Biol.* 27, 266–275. doi: 10.1016/j.tcb.2016.11.004.

Dikic I & Elazar Z (2018) Mechanism and medical implications of mammalian autophagy. *Nat. Rev. Mol. Cell Biol.* 19, 349–364. doi: 10.1038/s41580-018-0003-4.

Galluzzi L, Vitale I, Aaronson SA, et al. (2018) Molecular mechanisms of cell death: Recommendations of the Nomenclature Committee on Cell Death 2018. *Cell Death Differ.* 25, 486–541. doi: 10.1038/s41418-017-0012-4.

Jin Z & El-Deiry WS (2005) Overview of cell death signaling pathways. *Cancer Biol. Ther.* 4, 139–163.

Pop C & Salvesen GS (2009) Human caspases: Activation, specificity, and regulation. *J. Biol. Chem.* 284, 21777–21781.

Youle RJ & Strasser A (2008) The BCL-2 protein family: Opposing activities that mediate cell death. *Nat. Rev. Mol. Cell Biol.* 9, 47–59.

Yuan S & Akey CW (2013) Apoptosome structure, assembly, and procaspase activation. *Structure* 21, 501–515.

The Modular Architecture and Evolution of Signaling Proteins

10

Proteins involved in signaling pathways are frequently composed of multiple domains, each of which has a distinct biochemical activity. A **domain** represents a polypeptide sequence that has the ability to fold independently into a functional unit, and is typically between 35 and 250 amino acids in length. Thirty-five residues represents a minimal size required to specify a stable three-dimensional fold, whereas 250 amino acids likely approaches an upper limit for what can be folded without error. Thus, one reason for the multidomain structure of eukaryotic proteins may be the inability of cells to support the folding of domains beyond a relatively modest size. Typically, signaling proteins consist of several different domains, joined by less ordered linker sequences. Because such domains can be assembled together into larger proteins while retaining their distinct structure and activity, they are sometimes referred to as modules or **modular domains**.

Protein domains generally have one of two primary functions: to mediate interactions with other molecules within the cell, or to catalyze enzymatic reactions. We will refer to these as **interaction domains** and **catalytic domains**, respectively. In Chapter 3, we considered catalytic domains such as those in kinases, phosphatases, GEFs, and GAPs. In this chapter, we will focus primarily on the many classes of interaction domains and their role in assembling signaling pathways. We will also examine the more complex signaling properties that can emerge from the linking of multiple domains, both within a single polypeptide chain and in multiprotein complexes. Finally, we will explore how modules can be recombined in evolution to create new functions, or in other cases lead to disease.

DOI: 10.1201/9780429298844-10

(a)

proline-rich peptide ligand

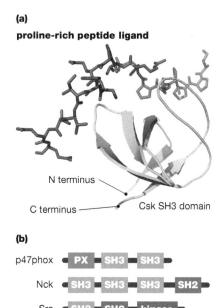

N terminus

C terminus — Csk SH3 domain

(b)

p47phox — PX – SH3 – SH3

Nck — SH3 – SH3 – SH3 – SH2

Src — SH3 – SH2 – kinase

Figure 10.1

Interaction domains form compact, globular modules. (a) The Csk SH3 domain (*green*) bound to a proline-rich peptide (*pink*), illustrating how the N- and C-termini of the domains are located close together in space and on the face opposite the ligand-binding site. (b) Examples of different SH3 domain-containing proteins. SH3 domains can occur in varying numbers and combinations with other domains. (a, Adapted from T. Pawson & P. Nash, *Science* 300:445–452, 2003. With permission from AAAS.)

MODULAR PROTEIN DOMAINS

Protein domains, and the shorter peptide sequences that bind to them, are key building blocks of signaling proteins. By understanding the individual properties of these modules, we can often infer a great deal about a protein containing them: its enzymatic activities, its localization, its binding partners, and its mode of regulation. Here, we look more closely at the general properties of these building blocks and how they are identified.

Protein domains usually have a globular structure

Interaction and catalytic domains typically fold into a globular structure, which is stabilized by a tightly packed core composed largely of hydrophobic amino acids. Residues involved in binding or catalysis are exposed on the surface of the domain, or positioned in surface-accessible pockets. As we shall see, signaling proteins are constructed from a relatively limited number of domain types or families, each of which can be found in many different proteins and in various combinations with other domains. The amino acids that form the hydrophobic core of a domain are usually highly conserved between related members of a domain family, as are surface residues that play an essential function in binding or catalytic activity. More variable residues, involved in the detailed specificity of binding or substrate recognition, are often found in loops on the domain's surface.

In most interaction domains, the N- and C-termini of the domain are close together in space and are located on the face of the domain opposite from the ligand-binding site (**Figure 10.1**). In principle, this arrangement allows an interaction domain to be easily inserted into an existing protein, while leaving its ligand-binding surface exposed to the solvent. This likely facilitated the evolution of multidomain proteins, as we shall discuss further in this chapter.

Bioinformatic approaches can identify protein domains

The genome typically encodes many related copies of a protein domain, which likely arose from a single ancestral gene that was duplicated and modified in the course of evolution. As we have discussed, the members of a domain family possess a number of residues that are particularly important because they are required for the domain to fold correctly, or are essential for a critical function such as ligand recognition or catalysis. These residues are therefore conserved as a domain family gains new members during evolution. Practically, this means that protein domains within the same family will usually show at least 15% identity in their primary amino acid sequences. This feature can therefore be used to discover new domains by computationally aligning the sequences of multiple proteins and looking for regions of 35–250 amino acids that show significant similarities. Indeed, many domain families were originally discovered through this bioinformatic approach, and are frequently termed "homology domains" to denote their homologous sequences. Pleckstrin homology (PH) domains, for example, were originally discovered because there are two PH domain sequences in the protein

Figure 10.2

Identification of domains by database analysis. The sequencing of the human genome has revealed around 20,000 protein-coding genes. Databases of the amino acid sequences encoded by all these genes can be searched to identify groups of conserved residues that constitute a domain. In this example, pleckstrin contains two copies of the pleckstrin homology (PH) domain, which has also been identified in many other proteins, three of which are illustrated here. *Light brown boxes* denote other conserved domains.

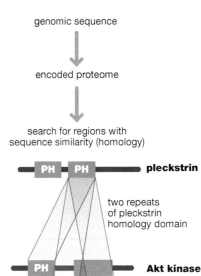

identification of other proteins with pleckstrin homology domains

pleckstrin, and related sequences were subsequently discovered in many other proteins. (**Figure 10.2**). If the presence of a domain is suggested by a sequence, the predicted domain can be expressed using recombinant DNA techniques. The putative domain can then be analyzed to see if it has a folded structure, as well as a biochemical function such as catalytic or binding activity.

Once the characteristic conserved "consensus sequence" signature is defined for a domain family, this allows the identification of new family members. The amino acid sequence of a protein of interest can thus be compared with the conserved sequences of the library of previously established domains to identify its domain composition and organization. Since different members of a domain family usually have similar functions, this approach often suggests the possible biochemical properties of an individual protein. Global analysis of particular domains in an organism's genome also allows a preliminary estimate of the number of gene products that an organism devotes to a particular signaling activity. For example, sequence analysis indicates that the human genome encodes ≈518 protein kinase domains and ≈120 SH2 domains. Approximately 70% of human proteins have one or more recognizable domains, and this percentage is likely to increase as more proteins are subjected to intensive investigation.

Domains can be composed of several smaller repeats

While most signaling domains fold as a single unit, there are a few classes of domains that are composed of several repeats of a smaller conserved unit, which by itself forms an element of secondary structure. This is a versatile arrangement, as the properties of such a domain can potentially be altered by varying both the amino acid sequences of the nonconserved loops and the number of the repeats. For example, WD40 repeats form β strands, which then are organized into a circular structure resembling a propeller with each repeat forming one blade (**Figure 10.3a**). The various specificity subunits that recruit substrate proteins to the SCF ubiquitin ligase complex are examples of WD40 repeat proteins.

In several other cases, such as armadillo repeats, each repeated unit has an α-helical structure. Multiple linked repeats form a twisted, superhelical structure with an extended binding surface, which potentially can bind several different proteins. This device is used by the β-catenin protein (**Figure 10.3b**). β-Catenin has a structural role at cell–cell junctions and is also a central component of the Wnt signaling pathway, in which it binds distinct protein ligands in the cytoplasm and nucleus. Thus, β-catenin must bind a variety of different binding partners, depending on its localization within the cell; this may be facilitated by the versatile properties of its armadillo repeat domain.

 The role of SCF complexes in regulating the cell cycle is discussed in Chapters 9 and 14.

(a)

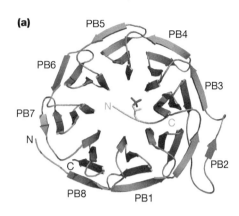

(b)

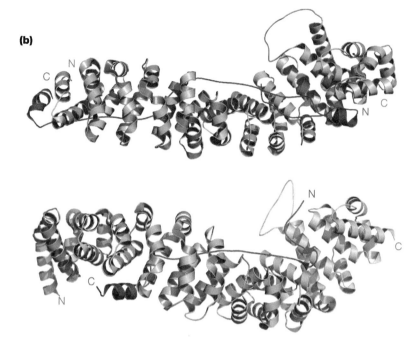

Figure 10.3

Domains can be formed from smaller repeats. (a) The β-propeller from Cdc4, the substrate-binding component for a form of the SCF ubiquitin ligase complex in yeast. The propeller is composed of eight WD40 repeats (labeled PB1–8), which are each composed of four antiparallel β strands. This protein recognizes a specific phosphorylated peptide motif (*green*, with phosphate group highlighted in *pink*).
(b) β-Catenin consists of 12 α-helical armadillo (Arm) repeats, which together form an extended superhelix. A positively charged groove that spans the entire superhelical repeat region forms the binding surface for the majority of β-catenin's interaction partners. The structure of the armadillo repeat domain of β-catenin is illustrated complexed to the cytoplasmic domain of the cell adhesion protein E-cadherin (top) and to the catenin-binding domain (CBD) of the transcription factor Tcf3 (bottom). The β-catenin domain is shown in *blue* and the bound peptides in *pink*. (a, Adapted from S. Orlicky, et al., *Cell* 112:243–256, 2003. With permission from Elsevier; b, adapted from H.J. Dyson & P.E. Wright, *Curr. Opin. Struct. Biol.* 12, 54–60, 2002. With permission from Elsevier.)

Protein domains often act as recognition modules

Many protein domains act as interaction modules that recognize and bind to short peptide sequences, sometimes in a fashion that requires a post-translational modification such as phosphorylation. These peptide sequences or **motifs** are typically located in unstructured regions of their host proteins (*intrinsically disordered regions*, or *IDRs*). When removed from their normal location, the isolated peptides usually retain their ability to act as ligands for interaction domains, as enzyme substrates, or both; for this reason, they can be viewed as modular units of protein function. Signaling proteins often contain a combination of folded domains with enzymatic or binding properties as well as unstructured regions containing peptide motifs. **Table 10.1** summarizes a number of interaction domains involved in signal transduction.

Domains that are related by sequence similarity often have a similar recognition function. For example, SH2 domains primarily bind phosphotyrosine-containing sites, and so the identification of an SH2 domain sequence is strongly suggestive of its primary function. Despite this, however, there are instances where identifying a domain by computational or structural methods may not necessarily reveal its actual function. WD40 repeat domains, for example, have a wide range of binding properties, including binding to phosphorylated or methylated peptide motifs, and are therefore not easy to categorize.

There are also examples of domains with very different sequences and biochemical activities that fold into similar three-dimensional structures. As we will discuss in more detail below, the fold (the overall arrangement of secondary structural elements and how they are connected) originally identified in PH domains, which often bind phospholipids, is also used by other domains, including phosphotyrosine-binding (PTB) domains and EVH1 domains that bind proline-rich sequences (**Figure 10.4**). The PH domain fold therefore represents a common structural framework with versatile binding properties. Conversely, a variety of domains with entirely distinct sequences and structures can converge on very similar biochemical activities. For example, a number of different domain types bind phospholipids such as phosphoinositides, and several distinct classes of domains, using different catalytic mechanisms, have protein phosphatase activity.

Table 10.1

Interaction domains involved in signal transduction

Domain	Binding target	Cellular processes	Example protein
14-3-3	Phosphoserine/phosphothreonine	Signal transduction, subcellular localization	14-3-3
ANK	Repeat domain, diverse binding partners	Diverse	53BP2
ANTH/CALM	Phospholipids	Clathrin-coated pit formation	AP180
ARM	Repeat domain, diverse binding partners	Diverse	β-Catenin
BAR, I-BAR	Dimerization, lipids, and curved surfaces	Endocytosis and cytoskeletal regulation	Amphiphysin
BEACH	Phospholipids	Vesicle trafficking, membrane dynamics, and receptor signaling	Neurobeachin
BH1–BH4	Dimerization	Apoptosis	Bcl2
BIR	Repeat domain, caspases	Apoptosis	XIAP
BRCT	Phosphoserine/phosphothreonine	DNA damage response and cell-cycle regulation	BRCA1
Bromo	Acetyl-lysine	Chromatin regulation	Gcn5p
BTB/POZ	Homodimerization and heterodimerization	Chromatin regulation and protein degradation	Mel26
C1	Diacylglycerol or phorbol ester	Plasma membrane recruitment	c-Raf
C2	Phospholipids (calcium dependent)	Membrane targeting, signal transduction, and vesicle trafficking	PKC
CARD	Homotypic interactions	Apoptosis	RAIDD
Coiled coil (CC)	Oligomerization	Diverse	BCR
CH	Actin	Cytoskeletal regulation	β-Spectrin
Chromo	Methyl-lysine	Chromatin regulation and gene expression	HP1
Chromo-shadow	Hydrophobic pentapeptide motif (binds as a dimer)	Chromatin regulation and gene expression	HP1
cue	Ubiquitin	Protein degradation and sorting	Cue1
Death (DD)	Homotypic interactions	Apoptosis	Fas
DED	Homotypic interactions	Apoptosis	Procaspase-8
DEP	Membranes, G-protein-coupled receptors	Signal transduction, protein targeting, and stabilization	Dsh
EH	Peptides with core NPF motifs	Endocytosis and vesicle trafficking	Eps15
EF-hand	Calcium	Calcium signaling	Calmodulin
ENTH	Phospholipids	Clathrin-dependent endocytosis and cytoskeletal regulation	Epsin
EVH1	Proline-rich sequences	Cytoskeletal regulation, postsynaptic signal transduction	Mena
F-box	Ubiquitin ligase substrates (phosphoserine/phosphothreonine)	Ubiquitylation	Cdc4
FCH	Actin, microtubules	Cytoskeletal regulation	Fes
FERM	Phospholipids	Cytoskeletal regulation and membrane dynamics	PTLP1
FF	Phosphoserine/phosphothreonine	Transcription, splicing	CA150
FH2	Actin, homotypic interactions	Cytoskeletal regulation	mDia
FHA	Phosphoserine/phosphothreonine	DNA repair, signal transduction, vesicular trafficking, protein degradation	MDC1
FYVE	Phospholipids	Signal transduction, vesicular trafficking	Hrs
gat	Ubiquitin (and other binding partners)	Vesicle trafficking and protein sorting	GGA1
gel	Actin	Cytoskeletal regulation	Gelsolin
GK	Phosphoserine/phosphothreonine	Scaffolding	PSD-95

(Continued)

Table 10.1 (*Continued*)

Interaction domains involved in signal transduction

Domain	Binding target	Cellular processes	Example protein
glue	Phospholipids	Vesicular trafficking	Vps36
GRAM	Phospholipids	Vesicular trafficking	MTM1
grip	Arf/Arl family of small GTPases	Golgi targeting	Golgin-97
GYF	Proline-rich sequences	Signal transduction, splicing	CDBP2
HEAT	Repeat domain, diverse binding partners	Vesicular trafficking, protein translation	Importin β1
HECT	Ubiquitin, E2 ubiquitin-conjugating enzymes	Ubiquitylation	E6AP
IQ	Calmodulin	Calcium signaling	Ras-GRF
LIM	Diverse binding partners, other LIM domains	Diverse including gene expression and cytoskeleton organization	hCRP
LRR	Repeat domain, diverse binding partners	Diverse	Rna1p
MBT	Repeat domain, methyl-lysine	Chromatin regulation	CGI-72
MH1	DNA and transcription factors	Transcription	SMAD2
MH2	Phosphoserine, homo-oligomerization	Signal transduction	SMAD2
MIU	Ubiquitin	Vesicular trafficking	RNF168
NZF	Ubiquitin	Ubiquitin-dependent processes	RanBP2
PAS	Diverse binding partners	Signal sensor domain, detecting oxygen tension, redox potential, or light intensity	PASK
PB1	Heterodimerization	Signal transduction	p67phox
PDZ	C-terminal peptide motifs	Diverse, scaffolding	PSD-95
PH	Phospholipids	Membrane recruitment, vesicular trafficking, signal transduction, cytoskeletal regulation	Akt
Polobox	Phosphoserine/phosphothreonine	Cell cycle	Plk1
PTB	Phosphotyrosine	Tyrosine kinase signaling	Shc
Pumilio	Repeat domain, RNA	Gene expression	Pumilio
PWWP	Methyl-lysine, DNA	DNA methylation, DNA repair, transcription	WHSC1
PX	Phospholipids	Protein sorting, vesicular trafficking, hospholipid metabolism	p40phox
RGS	GTP-binding pocket of Ga proteins	Signal transduction	RGS-4
RING	Ubiquitin, E2 ubiquitin-conjugating enzymes, transcription factors	Diverse, ubiquitylation, transcription	Cbl
SAM	Homotypic and heterotypic oligomerization, RNA	Diverse	Ste11
SH2	Phosphotyrosine	Tyrosine kinase signaling	Src
SH3	Proline-rich sequences	Diverse, cytoskeletal regulation	Src
SNARE	Components of SNARE complex	Vesicle-membrane fusion	Syntaxin
SOCS box	Ubiquitin ligase substrates	Ubiquitylation	Socs-1
SPRY	Diverse binding partners	Diverse including cytokine signaling and retroviral defense	RanBPM
START	Lipids	Lipid transport, transcription	StAR
SWIRM	Acetyl-lysine	Chromatin regulation and gene expression	SMARC2
TIR	Homotypic and heterotypic interactions	Cytokine and immune signaling	TLR4
TPR	Repeat domain, diverse binding partners	Scaffolding function in diverse processes	p67phox
TRAF	Components of TNF signaling pathways	Cell survival, protein processing, and ubiquitylation	TRAF-1
TUB	DNA and phospholipids	Metabolism, transcription	Tulp-1

(Continued)

Table 10.1 (*Continued*)

Interaction domains involved in signal transduction

Domain	Binding target	Cellular processes	Example protein
Tudor	Methyl-lysine and methyl-arginine	Chromatin regulation and gene expression	SMN
UBA	Ubiquitin	Ubiquitylation	HHR23A
UEV	Ubiquitin and Pro-Thr/Ser-Ala-Pro peptide	Protein sorting	TSG101
UIM	Ubiquitin	Ubiquitylation	Hrs
VHL	Hydroxy-proline	Ubiquitylation	VHL
VHS	Ubiquitin	Endocytosis and protein sorting	GGA
WD40	Repeat domain, phosphoserine/phosphothreonine, dimethyl-lysine, others	Diverse including cell cycle and ubiquitylation	βTRCP
WW	Proline-rich sequences, phosphoserine/phosphothreonine	Diverse signaling processes	YAP

Figure 10.4

Domains with the same structure can recognize different ligands.
EVH1, PH, and PTB domains (*green*) have the same overall fold (as indicated by the very similar arrangement of α helices and β sheets in the three domains), but have entirely different binding properties and use different surfaces to engage their ligands (*pink*). (a) The EVH1 domain from the actin regulatory protein Mena is shown bound to a proline-rich peptide, (b) the PH domain from the ArfGEF GRP1 interacts with a phosphoinositol lipid head group, and (c) the PTB domain from the scaffold protein IRS1 binds to a tyrosine-phosphorylated peptide.

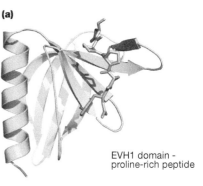

(a) EVH1 domain - proline-rich peptide

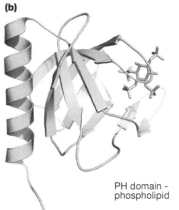

(b) PH domain - phospholipid

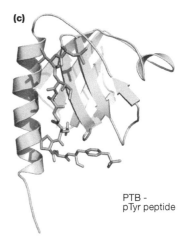

(c) PTB - pTyr peptide

Below, we will discuss three major classes of modular interaction domains: those that recognize peptides or proteins with post-translational modifications, those that recognize specific unmodified peptide or protein motifs, and those that recognize specific phospholipid species. Signaling proteins also employ modular domains that recognize small-molecule signaling mediators, such as calcium and cAMP, but these will not be discussed further here.

INTERACTION DOMAINS THAT RECOGNIZE POST-TRANSLATIONAL MODIFICATIONS

Post-translational modifications, such as phosphorylation of a serine, threonine, or tyrosine residue, can produce binding sites for interaction domains. As first discussed in Chapter 4, the recognition of modified residues provides a relatively simple molecular device through which the cell can respond to the activities of signaling enzymes such as protein kinases. In the case of protein phosphorylation, the post-translational modification is "written" by a kinase and "erased" by a protein phosphatase. A modular interaction domain that recognizes the modified site essentially functions as a "reader" module that interprets the modification and causes downstream changes in function. In addition to phosphorylation, modular interaction domains recognize other types of post-translational modifications, including ADP-ribosylation, acetyl- or methyl-lysine, methyl-arginine, hydroxy-proline, and ubiquitylated or sumoylated lysine (**Figure 10.5**).

If the interaction domain and the modified peptide site are on two separate polypeptide chains, then the binding event will induce formation of a complex between the two proteins. Alternatively, the modified site and interaction domain may be in the same protein, in which case the modification will result in an intramolecular interaction. Such an intramolecular association will suppress the ability of the

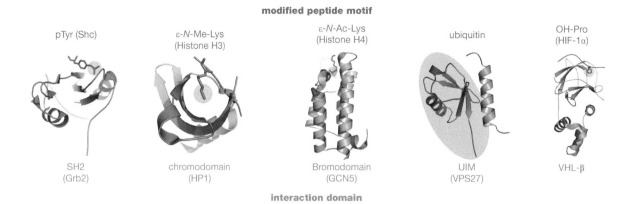

Figure 10.5

Examples of domains that recognize modified peptide motifs. The SH2 domain of Grb2 recognizes a phosphorylated tyrosine residue on Shc. The chromodomain of HP1 recognizes a methylated lysine residue on histone H3, whereas the bromodomain of Gcn5 recognizes an acetylated lysine residue on histone H4. Ubiquitin is bound by the ubiquitin interacting motif (UIM) of Vps27. The Von Hippel-Lindau β protein (VHL-β) recognizes a hydroxylated proline residue on HIF-1α. In each case, the post-translational modification is highlighted in *pink*. Note that different examples are not shown at the same scale. (Adapted from B.T. Seet, et al., *Nat. Rev. Mol. Cell Biol.* 7:473–483, 2006. With permission from Springer Nature.)

 The regulation of Src family kinases by intramolecular interactions is discussed in Chapters 1 and 3.

interaction domain to bind sites on other proteins. It may also induce a conformational change that alters the activity of other domains within the same protein, as we have seen for regulation of the Src tyrosine kinase, in which the interaction of the SH2 domain with a phosphotyrosine site in the C-terminal tail leads to inhibition of the intervening kinase domain.

SH2 domains bind phosphotyrosine-containing sites

The **Src Homology 2 (SH2) domain** was the first modular interaction domain whose binding was shown to be dependent on post-translational modification. This domain was first recognized in protein tyrosine kinases including Src, the product of a viral oncogene that causes sarcomas in chickens. SH2 domains are found in over a hundred different human proteins, and in almost all cases are thought to bind specifically to tyrosine-phosphorylated peptides.

SH2 domains are approximately 100 amino acids in length and fold into a compact structure with a central β sheet that separates the domain into two binding pockets (**Figure 10.6**). One of these is highly conserved among all SH2 domains and serves primarily to bind the phosphotyrosine, while the other—the "specificity pocket"—is more variable and binds the side chains of the adjacent amino acids. Thus, the affinity with which a particular peptide binds a given SH2 domain depends, in part, on its phosphorylation (about half of the binding energy comes from recognition of the phosphotyrosine) and also on the fit of other peptide residues with the specificity pocket of the SH2 domain. Individual SH2 domains typically bind their preferred phosphorylated peptide motifs with a K_d of ~1 μM, although in some cases the interactions can be tighter (~100 nM). However, they also show relatively rapid on- and off-rates, indicating that SH2 domain–phosphopeptide interactions are highly dynamic, allowing signaling to be rapidly initiated and terminated.

Binding to the phosphotyrosine is through a bidentate ionic interaction (such an interaction is illustrated in **Figure 3.11b**) with an essential arginine residue (the fifth residue in β strand B, and hence termed ArgβB5). ArgβB5 sits at the base of the relatively deep pocket, and is precisely positioned to interact with the phosphate group of a phosphotyrosine. The shorter side chains of phosphoserine or phosphothreonine

Figure 10.6

Structure of the SH2 domain of Src. (a) The central β sheet divides the domain into two binding pockets: a highly conserved pocket that is responsible for binding to the phosphotyrosine, and a more variable "specificity" pocket that interacts with the side chains of adjacent amino acids. Tyrosine-phosphorylated peptide ligand is depicted in *yellow*; phosphate group in *pink*. (b) Schematic of an SH2 domain; the presence of distinct phosphotyrosine and specificity pockets is reminiscent of a two-holed socket.

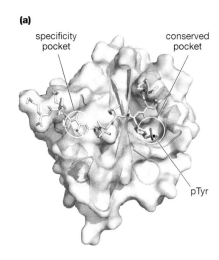

(a) specificity pocket / conserved pocket / pTyr

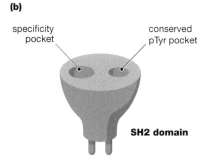

(b) specificity pocket / conserved pTyr pocket / SH2 domain

would be unable to project far enough into the pocket to encounter this arginine, explaining the selectivity of SH2 domains for phosphotyrosine. In the absence of tyrosine phosphorylation, the interaction between a peptide and an SH2 domain is typically very weak (K_d in the mM range), and so is not relevant in cells. However, phosphorylation of the peptide leads to a 1000-fold increase in its affinity for the interaction domain. Phosphorylation therefore acts as a switch to elicit complex formation.

The sequence of amino acids flanking the phosphotyrosine sites strongly influences which SH2-containing proteins are recruited. Phosphopeptides typically bind an SH2 domain in an extended conformation, so that they run across the central β sheet (see **Figure 10.6**), thereby positioning the C-terminal amino acids to interact with the specificity pocket. Many SH2 domains engage only three C-terminal residues (in the +1 to +3 positions relative to the phosphotyrosine), while others bind more extended phosphorylated motifs up to eight amino acids in length, and can engage residues both N- and C-terminal to the phosphotyrosine. Different SH2 domains prefer distinct amino acids in the +1 to +3 positions of the phosphopeptide. For example, the Src SH2 domain can accommodate an isoleucine at position +3 into a hydrophobic specificity pocket. The Grb2 SH2 domain specifically binds phosphopeptides with asparagine in position +2, due to favorable hydrogen-bonding and the β-turn conformation favored by asparagine. The SH2 domains of the p85 adaptor subunit of phosphatidylinositol 3-kinase (PI3K) prefer a methionine at the +3 position, and those of phospholipase C-γ (PLC-γ) prefer a run of hydrophobic amino acids following the phosphotyrosine (**Figure 10.7**). The preference of a given domain for particular flanking residues has been investigated by probing degenerate peptide libraries in which phosphotyrosine is flanked by random combinations of amino acids. The consensus binding motif suggested by such an analysis can then be compared with experimentally determined phosphorylation sites and used to predict potential binding sites.

Some SH2 domains are elements of larger binding structures

There are some cases in which SH2 domains work with other adjacent domains to display somewhat atypical recognition properties. For example, in the ubiquitin E3 ligase Cbl, the SH2 domain is followed by a four-helix bundle and an EF-hand domain. These three domains actually fold together into an integrated structural unit, sometimes called the TKB domain (**Figure 10.8**). This larger domain binds phosphotyrosine through its SH2 subunit, but can also engage residues unusually distant from the phosphorylated sites. For example, peptide residues in the –5 and –6 positions, relative to the phosphotyrosine, are recognized by the four-helix bundle subunit. Thus, the phosphopeptide-binding properties of the Cbl SH2 domain are altered and rendered more complex.

Another example of an interaction domain with novel recognition properties is the tandem SH2 module from the ZAP-70 tyrosine kinase, which functions in immune cell activation. ZAP-70 has two N-terminal SH2 domains that function cooperatively to engage an unusual doubly tyrosine-phosphorylated sequence (the

 T cell signaling is discussed in more detail in Chapter 14.

Figure 10.7

SH2 domain selectivity. (a) The SH2 domain of the Src tyrosine kinase binds to a phosphopeptide with the sequence pYEEI, which is accommodated because the isoleucine in the +3 position fits into a hydrophobic pocket in the variable surface of the SH2 domain. (b) The SH2 domain of the adaptor protein Grb2 selects phosphotyrosine sites with an asparagine at the +2 position; the bulky tryptophan (*pink*) in the SH2 domains forces the binding peptide to form a β-turn structure, rather than adopting an extended conformation. (c) The phospholipase C-γ (PLC-γ) SH2 domain prefers peptides with hydrophobic residues following the phosphotyrosine, particularly an isoleucine at +1. The positively charged phosphotyrosine pocket is indicated in *blue*; peptide ligands are in *yellow*.

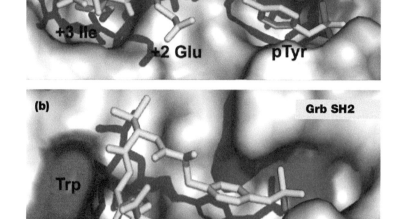

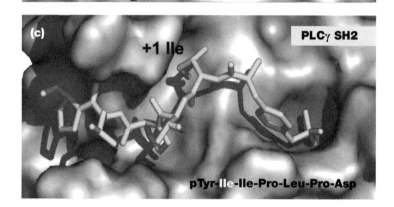

Figure 10.8

Interaction domains can be joined to form new structural folds with novel properties. In Cbl, the SH2 domain (*blue*), four-helix bundle (*green*), and EF-hand (*yellow*) domains form a combined structural unit termed the TKB domain, which interacts with phosphorylated tyrosine residues on receptors. A tyrosine-phosphorylated peptide from the endocytic adaptor APS is depicted in *pink*. Shown are (a) a backbone structure and (b) a surface representation of the TKB domain. (Adapted from J. Hu & S.R. Hubbard, *J. Biol. Chem.* 280:18943–18949, 2005. Published under CC BY 4.0.)

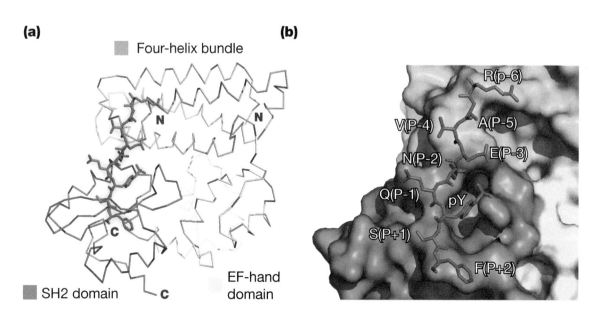

immunoreceptor tyrosine-based activation motif, or *ITAM*) found in the signaling subunits of the T cell antigen receptor and other immune receptors. This interaction requires that the two SH2 domains be tightly coupled, because residues from the C-terminal SH2 domain contribute to the phosphotyrosine-binding pocket of the N-terminal SH2 domain, which is therefore nonfunctional unless the two SH2 domains are correctly oriented in space (**Figure 10.9**). Similar types of modified recognition properties have been observed in tandem domains that belong to other families besides SH2 domains.

Several different types of interaction domains recognize phosphotyrosine

Other types of domains with completely different structural folds can also bind tyrosine-phosphorylated peptide motifs. The interaction of PTB domains with phosphotyrosine-containing sites differs in several features from the interaction involving SH2 domains. First, PTB domains have a completely different fold from SH2 domains, with two orthogonal β sheets forming a β sandwich that is capped by a C-terminal α helix (see **Figure 10.4c**). As a consequence, the mode of peptide recognition is different: the peptide ligand contacts the β-5 strand and C-terminal α helix of the PTB domain, and, in effect, adds an antiparallel strand to one of the β sheets. N-terminal to phosphotyrosine, the peptide forms a type I β turn, which is favored by the peptide motif NPxY (where "x" can be any amino acid); this element anchors the peptide and is the hallmark of most ligands for PTB domains. Stable peptide binding to the PTB domain of the scaffolding protein Shc depends on phosphorylation of the tyrosine in the NPxY motif. The bound phosphotyrosine is coordinated by three basic residues (two arginines and a lysine) of the PTB domain, which form a network of hydrogen bonds with the phosphate group. Unlike SH2 domains, many PTB domains bind with high affinity to unphosphorylated peptides (often with the NPxY consensus). Therefore it is likely that phosphotyrosine recognition evolved relatively late from an existing peptide-binding module.

The PTB domains of Shc and other scaffolding proteins such as IRS1, Dok1, and FRS2 recognize phosphorylated NPxY motifs, and are followed by an unstructured region with several sites for tyrosine phosphorylation. Autophosphorylated receptor tyrosine kinases recruit these proteins via their PTB domains, and then phosphorylate them on multiple tyrosine motifs, which consequently bind the

Figure 10.9

Domains can act cooperatively to regulate signaling. (a) Binding of MHC-bound peptide antigen (*orange dot*) to the T cell receptor (TCR, *light brown*) results in the phosphorylation of ITAMs (immunoreceptor tyrosine-based activation motifs) on the TCR. The tandem SH2 domains of ZAP-70 can then interact with a doubly phosphorylated ITAM on the TCR. This activates the kinase domain of ZAP-70, allowing it to phosphorylate substrates, such as the adaptor protein LAT (not shown). (b) Further details on ZAP-70 activation. In the inactive state (left), interactions between the inter-SH2 region, linker sequence, and kinase domain serve to inhibit kinase activity and also keep the SH2 domains separated, preventing them from engaging phosphorylated tyrosines. Upon ITAM phosphorylation, the SH2 domains orient themselves to dock with the phosphorylated motif. This releases the intramolecular interactions, allowing the kinase to be activated, and also positions the kinase in the correct orientation for phosphorylation of substrates. This extended conformation of ZAP-70 is further stabilized by phosphorylation of tyrosine residues (Tyr315 and Tyr319) within the linker region.

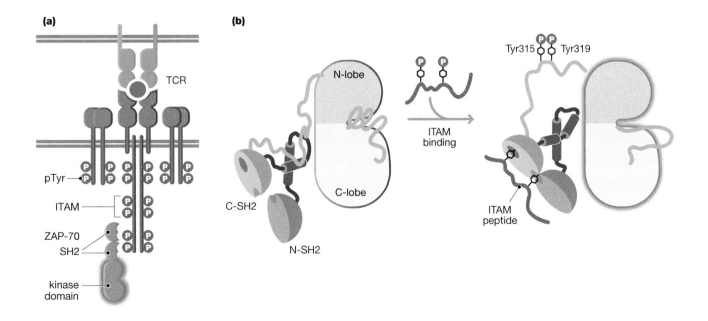

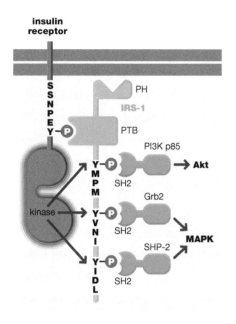

Figure 10.10

Binding of one interaction domain can provide further targets for a catalytic domain. The PTB domain of IRS1 docks to a phosphorylated tyrosine on the insulin receptor. The insulin receptor kinase then phosphorylates multiple tyrosine residues on IRS1, which in turn form docking sites for specific SH2 domain-containing proteins, such as phosphatidylinositol 3-kinase (PI3K), the adaptor Grb2, and the tyrosine phosphatase Shp2. The binding of these SH2 domain-containing proteins leads to the activation of corresponding downstream signaling pathways.

Figure 10.11

Structure of 14-3-3 proteins. The structure of the 14-3-3ζ dimer bound to the phosphopeptide RSHpSYPA. The individual 14-3-3 monomers are shown in *green* and *blue* and the phosphopeptide in *yellow*, with phosphates highlighted in *pink*.

SH2 domains of other regulatory signaling proteins (**Figure 10.10**). PTB domain-containing proteins therefore extend and amplify a receptor's ability to recruit cytoplasmic targets.

The C2 domain is another interaction domain that can, in at least one case, recognize phosphotyrosine sites. Most C2 domains bind to phospholipids. However, the C2 domain of PKCδ, a protein serine/threonine kinase, binds a specific phosphotyrosine-containing motif in the cytoplasmic tail of a transmembrane protein (CDCP1), again through a distinct structural mechanism. This directly links a phosphotyrosine signal, generated by Src-mediated phosphorylation of CDCP1, to serine/threonine phosphorylation. Thus, at least three separate families of interaction domains—SH2, PTB, and C2 domains—have independently converged on the selective recognition of phosphotyrosine sites.

Multiple domains recognize motifs phosphorylated on serine/threonine

The recognition of phosphorylated sites by specific interaction domains is a common feature of signaling pathways, and phosphoamino acids other than phosphotyrosine are also recognized. Like tyrosine, serine and threonine are hydroxyl amino acids, but they lack tyrosine's bulky phenolic ring. Consequently, domains that recognize phosphotyrosine generally do not have the correct shape to accommodate the much smaller phosphoserine or phosphothreonine side chains. Serine and threonine are chemically and structurally very similar, however, so there are many domains that specifically bind to both phosphoserine and phosphothreonine motifs.

At least ten different domain types, including FHA, WW, BRCT, Polobox, MH2, and WD40 domains, can bind phosphothreonine- and/or phosphoserine-containing motifs, and are found in the context of multidomain proteins. There are many more types of phosphoserine/phosphothreonine-binding domains than phosphotyrosine-binding modules, which may reflect the fact that phosphorylated serine/threonine residues are much more abundant in mammalian cells than phosphotyrosine (the ratio is around 100:1). These modules bind selectively to peptide motifs in which a threonine or serine is phosphorylated (often more strongly to phosphothreonine), and in which the phosphorylated residue is flanked by a particular set of amino acids that are preferentially accommodated by the domain's ligand-binding site.

14-3-3 proteins recognize specific phosphoserine/phosphothreonine motifs

The binding of the 14-3-3 family of proteins to phosphorylated sites represents a major mechanism through which cells respond to the effects of basophilic kinases (kinases that phosphorylate serine/threonine sites with nearby basic residues). The 14-3-3 proteins bind serine/threonine phosphorylated motifs with the general consensus RSxpSxP, RxY/FxpSxP, or SWpTx (the latter being a C-terminal sequence). Mammalian 14-3-3 proteins are very abundant and can bind at least 200 phosphorylated proteins involved in many different aspects of cellular regulation, including signaling pathways activated by cell-surface receptors, cell-cycle progression, apoptosis, transcriptional control, cytoskeletal organization, metabolism, and protein trafficking.

Each ~30 kD 14-3-3 polypeptide has nine α helices that form an amphipathic peptide-binding channel (**Figure 10.11**) that has a conserved, positively charged basic pocket composed of a lysine and two arginines, which directly bind the phosphate group. Unlike other phosphopeptide-binding modules, 14-3-3 proteins are never found as components of larger proteins with additional domains, but they do interact noncovalently with one another to form homodimers or heterodimers. Each 14-3-3 dimer therefore has two phosphoserine/phosphothreonine-binding pockets and can interact simultaneously with two different phosphorylated sites; typically, these are located on the same polypeptide chain, although a 14-3-3 dimer might potentially link two distinct phosphorylated proteins.

Binding to a 14-3-3 dimer can modify the function of a phosphorylated protein in one of several ways. One consequence of 14-3-3 binding can be to interfere with the ability of a phosphorylated protein to bind another protein. For example, in the absence of survival signals, the pro-apoptotic protein BAD associates with the pro-survival protein Bcl-X$_L$, inhibiting the survival effects of Bcl-X$_L$ and thus inducing cell death. Extracellular stimuli that promote cell survival induce the phosphorylation of BAD on serine residues that consequently bind 14-3-3 dimers, dislodging BAD from Bcl-X$_L$. Thus, when Bcl-X$_L$ is in a complex with BAD, its anti-apoptotic activity is suppressed; 14-3-3 binding to BAD relieves this inhibition and promotes cell survival.

Association with 14-3-3 proteins can also alter the subcellular location of a phosphorylated protein, typically by anchoring it in the cytoplasm by blocking its nuclear localization signal. For example, association with 14-3-3 proteins can restrict the FOXO transcription factor to the cytoplasm. FOXO normally induces the expression of genes that antagonize cell-cycle progression and induce apoptosis. Thus, preventing its entry into the nucleus leads to increased cell proliferation and survival (**Figure 10.12**).

Interaction domains recognize acetylated and methylated sites

Phosphorylation is but one of several post-translational modifications that are selectively recognized by specific interaction domains. Lysine residues can be methylated or acetylated on the flexible N- or C-terminal tails of histones, for example, leading to changes in chromatin organization and the epigenetic control of gene expression. An ever-increasing number of non-chromatin proteins have also been shown to be subject to lysine methylation and/or acetylation. As with phosphopeptide recognition, peptide motifs with an acetylated lysine flanked by a defined peptide sequence can be selectively recognized by particular domains, of which **bromodomains** are the cardinal example (see Figure 10.5). These are frequently components of proteins involved in chromatin remodeling, such as the histone acetyl transferases themselves, and are therefore intimately involved in controlling gene expression. **Chromodomains** recognize specific peptide motifs in which the lysine is methylated, rather than acetylated. Chromodomains occur in proteins such as the heterochromatin protein 1 (HP1), which binds to histone H3 methylated at Lys9 and modifies chromatin structure to repress gene expression (**Figure 10.13**).

Just as different types of interaction domains recognize phosphorylated sites, there are several different types of interaction domains that bind methylated lysines, including subsets of WD40, tudor, malignant brain tumor (MBT), and PHD finger domains. These different domains can use distinct structural mechanisms for binding to methylated lysines. Lysine methylation is more complex than protein phosphorylation, as a lysine residue can be mono-, di- or trimethylated. Interaction domains can bind preferentially to lysine residues carrying different degrees of methylation, increasing the potential sophistication of this type of binding interaction. For example, chromodomains and tudor domains employ conserved aromatic residues to surround the ε-methyl groups of a methylated lysine residue with a hydrophobic cage (**Figure 10.13b**). By contrast, the WD40 repeat domain of WDR5, a subunit of histone H3 Lys4 methyltransferase, preferentially forms hydrogen bonds to dimethylated Lys4 in histone H3; this interaction also depends

The role of Bcl2 family proteins in triggering programmed cell death is discussed in Chapter 9.

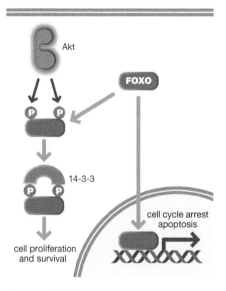

Figure 10.12

14-3-3 proteins can regulate subcellular localization. The FOXO transcription factor regulates the expression of genes leading to cell-cycle arrest and apoptosis. The serine/threonine kinase Akt phosphorylates multiple sites on FOXO, which form docking sites for 14-3-3 proteins. Once bound to 14-3-3, FOXO is sequestered in the cytoplasm. This promotes cell survival and proliferation.

The roles of methylation and acetylation in regulating chromatin structure are discussed in Chapter 4.

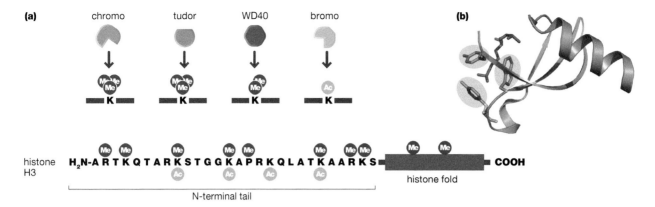

Figure 10.13

Domains that bind modified histone residues. (a) A number of different interaction domains, including chromo, tudor, WD40, and bromo domains, bind to histone lysine or arginine residues that have been methylated or acetylated. The binding of these domains regulates chromatin structure and gene expression. (b) Structure of the chromo domain from *Drosophila* HP1, showing the aromatic residues (*green*) that form a "cage" around the dimethylated lysine of the histone H3 tail peptide (*pink*). (a, Adapted from B.T. Seet, et al., *Nat. Rev. Mol. Cell Biol.* 7:473–483, 2006. With permission from Macmillan Publishers Ltd; b, adapted from A. Brehm, et al., *Bioessays* 26:133–140, 2004. With permission from John Wiley & Sons, Inc.)

Figure 10.14

Ubiquitin-binding domains. (a) Most ubiquitin-binding domains (*blue*), including the UIM, MIU, and UBA domains shown here, recognize the hydrophobic patch centered on Ile44 of ubiquitin (colored *green* on the crystal structures). (b) The specificity of ubiquitin-binding domains depends on the ubiquitin chain length and linkage. For example, the K63-linked di-ubiquitin (top) has a more extended structure than the K48-linked di-ubiquitin (below). Lysines are depicted in space-filled mode. (a, Adapted from J.H. Hurley, et al., *Biochem. J.* 399:361–372, 2006. With permission of Portland Press; b, adapted from K. Newton, et al., *Cell* 134:668–678, 2008. With permission from Elsevier.)

strongly on recognition of an arginine residue at the −2 position relative to the modified lysine.

Ubiquitylation regulates protein-protein interactions

Lysine residues can also be modified by addition of the 76-amino-acid protein ubiquitin, via an isopeptide linkage between the ε-amino group of lysine and the C-terminus of ubiquitin. Furthermore, ubiquitin can itself be ubiquitylated on one of its own lysines (for example, Lys48 or Lys63) or its N-terminus to form a poly-ubiquitin chain. Ubiquitin is in essence a transferable interaction domain that, once linked to its target protein, is recognized by binding modules (collectively termed "ubiquitin-binding domains" or UBDs). There are at least 11 structurally distinct families of UBDs, although these mostly bind the same hydrophobic patch on ubiquitin, centered on Ile44 (**Figure 10.14**). Although UBDs are referred to as "domains," many of these are more akin to linear peptide motifs that engage the ligand-binding surface of ubiquitin.

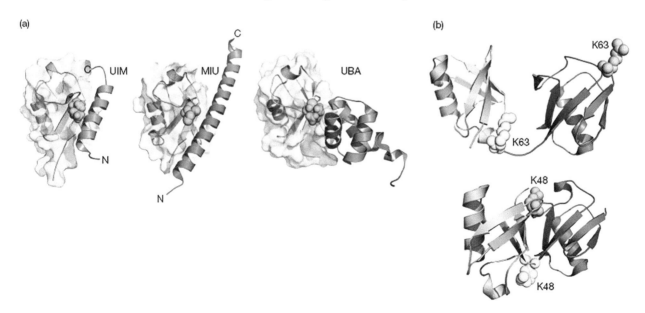

Ubiquitylation regulates a range of cellular processes, including proteasome-mediated proteolysis, protein trafficking to endosomes, post-replicative DNA repair, and signaling downstream of receptors leading to the activation of kinases that control the NF-κB transcription factor. In each of these pathways, the receptors for ubiquitylated proteins have one or more UBD. Since the affinity of ubiquitin for a UBD is typically rather weak, it is likely that multivalent interactions are important for the association of ubiquitylated proteins and their binding partners. Furthermore, there is often a close interplay between protein-protein interactions mediated by phosphorylation and ubiquitylation, in the sense that the substrate-binding domains of E3 protein-ubiquitin ligases often bind their target proteins in a phosphorylation-dependent manner. As a result, the target is only ubiquitylated after it has been phosphorylated (see Figure 4.9).

Ubiquitylation and its consequences are discussed in more detail in Chapters 4 and 9.

INTERACTION DOMAINS THAT RECOGNIZE UNMODIFIED PEPTIDE MOTIFS OR PROTEINS

Thus far, we have discussed interaction domains that bind short peptide sequences that have been post-translationally modified. There are also a number of domain families that recognize unmodified peptide ligands. Here we focus on two domains of this sort—SH3 domains and PDZ domains—and their respective peptide ligands. Both of these domains are found in ~300 copies in the human proteome, making them among the most commonly used modules in signaling proteins. We will also consider domains that interact with each other to form homo- or heterodimers or larger oligomeric structures.

Proline-rich sequences are favorable recognition motifs

A number of interaction domains, including SH3, WW, EVH1, and GYF, bind short, proline-rich peptide motifs. Proline is unique among naturally occurring amino acids in that its side chain is fused to the nitrogen of the peptide backbone, forming a five-member ring. Due to the resulting conformational constraints, proline-rich sequences tend to form a left-handed helix with three residues per turn (a **polyproline type II or PPII helix**). In a PPII helix, both the side chains and the backbone carbonyls project outward from the axis of the helix, and so are available to contact an interaction domain (**Figure 10.15a**).

A number of properties of the PPII helix make it ideal for mediating protein interactions. As opposed to more flexible peptides that are only locked into a single conformation upon binding, the PPII helix forms spontaneously; this reduces the entropic penalty involved in binding. Other non-proline residues can be incorporated without disrupting the helix, and these can contribute to selective binding. In addition, a PPII helix has twofold rotational pseudosymmetry, meaning it has the potential to bind in either orientation (N- to C-terminal or C- to N-terminal) to a domain such as an SH3 domain. Proline-rich sequences were also likely selected as

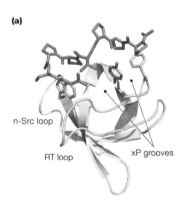

(a)

n-Src loop

RT loop

xP grooves

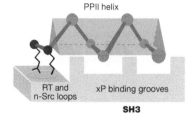

PPII helix

RT and n-Src loops

xP binding grooves

SH3

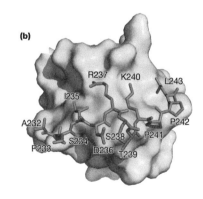

(b)

R237   K240   L243
I235
A232   P242
P233   S234   S238   P241
D236   T239

Figure 10.15

SH3 domains bind polyproline helices. (a) A Sem-5 SH3 domain (*green*) is shown bound to a proline-rich peptide (*orange*). The cartoon below shows the mechanism of polyproline recognition. The core recognition surface of the SH3 domain has two grooves formed by aromatic amino acids (shown in *blue*) that each accommodate an xP peptide motif. Adjacent to this are two variable loops (RT and n-Src) that make contacts with residues flanking the PxxP motif, forming a "specificity" pocket (*green*). (b) The GADS SH3 domain (*green*) bound to a peptide motif from SLP-76 (*orange*) that is centered on RxxK, rather than PxxP. (a, Adapted from A. Zarrinpar, et al., *Science STKE* 179:re8, 2003. With permission of AAAS; b, adapted from B.T. Seet, et al., *EMBO J.* 26:678–689, 2007. With permission from Springer Nature.)

ligands for interaction domains because their structure is incompatible with secondary structure elements (α helices and β sheets) that comprise the core of folded domains. Thus, they are usually exposed on the surface of a protein, or located in intrinsically disordered regions, and so are readily accessible to a binding partner. Furthermore, the proline ring is relatively hydrophobic, which is unusual for side chains exposed on the surface of proteins, so shielding it from the solvent by binding to another protein is energetically favored.

SH3 domains bind proline-rich motifs

PPII helices that contain the consensus motif PxxP are bound by **SH3 domains**, which contain two antiparallel β sheets positioned at right angles to one another. The peptide-binding surface of the SH3 domain has two grooves, each of which accommodates one proline residue and an adjacent amino acid that is usually hydrophobic. Two variable SH3 domain loops (called "RT" and "n-Src" for historic reasons) make numerous contacts with residues that flank the PxxP core motif and can be viewed as forming a specificity pocket. Many SH3 domains bind ligands containing the sequences R/KxxPxxP or PxxPxR/K (where R/K denotes either arginine or lysine), which bind in opposing orientations and therefore make similar contacts with SH3 domains (**Figure 10.15**). In addition, these peptides frequently contain more variable sequences that extend beyond the basic residue and contribute to the specificity of SH3 domain interactions. As we have seen with other interaction modules, SH3 domains are relatively versatile in their binding properties, and in some cases bind peptides that lack a PxxP motif; one example is the C-terminal SH3 domain of the Grb2-like adaptor protein GADS, which serves a key role in signaling downstream of the T cell receptor (TCR) by linking two docking proteins, LAT and SLP-76. The GADS SH3 domain binds with high affinity and specificity to a peptide motif on SLP-76 that is centered on an RxxK motif (**Figure 10.15b**).

PDZ domains recognize C-terminal peptide motifs

As with proline-rich sequences, the C-terminus of proteins is usually exposed and contains unique chemical features. These properties have made it a preferred site for recognition by a specific class of interaction modules termed PDZ domains. These domains have a β-sheet structure with a carboxylate-binding loop, which forms a pocket for the side chains of hydrophobic C-terminal residues such as valine (**Figure 10.16**). Indeed, the great majority of interactions mediated by PDZ domains involves the recognition of such C-terminal sequences, and only a few PDZ domains recognize motifs located within a protein. The key distinguishing feature of a PDZ domain-binding site is therefore the C-terminal hydrophobic residue, which conceptually has the same defining role as phosphotyrosine in binding to an SH2 domain, or a PxxP motif in recognition by an SH3 domain. As with these other domain-peptide interactions, the residues adjacent to the core feature, in this case residues located N-terminal to the C-terminal amino acid, provide a degree of specificity in determining which PDZ domains are recruited to a particular motif. The amino acid at the –2 position (that is, two amino acids from the C-terminus, which is denoted as the "0" position) is especially important, but significant contributions can be made by residues more distant from the C-terminus.

Many transmembrane receptors, such as receptor tyrosine kinases, G-protein-coupled receptors, ion channels, and adhesion proteins, have C-terminal PDZ-binding motifs, as do many cytoplasmic proteins involved in functions such as cell polarity and the regulation of Rho family GTPases. Furthermore, PDZ proteins frequently have multiple tandem PDZ domains (up to 13 in the case of the MUPP1 protein), and can therefore simultaneously bind several different proteins with appropriate C-terminal motifs. PDZ domain proteins therefore commonly serve as scaffolds

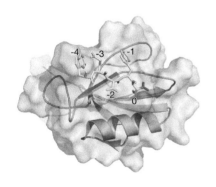

Figure 10.16

PDZ domains bind C-terminal motifs. The PDZ domain (*green*) of Erbin bound to its optimal ligand (*yellow*), WETWV-COOH. The C-terminal hydrophobic residue (valine in this case, with the terminal carboxylate group shown in *pink*, numbered 0) is the key distinguishing feature of PDZ ligands. Adjacent residues (tryptophan and threonine in this example) provide a degree of specificity. The amino acid at the –2 position (threonine) is of particular importance. (Courtesy of Megan McLaughlin and Sachdev Sidhu.)

to co-localize transmembrane receptors and cytoplasmic signaling proteins at a specific site within the cell.

Protein interaction domains can form dimers or oligomers

Most protein interactions discussed above involve a folded protein domain interacting with a peptide ligand. By contrast, some protein domains mediate dimerization or oligomerization as part of their role in the assembly of functional signaling complexes. Dimerization can occur between two identical proteins (homodimerization) or two distinct domains from the same family (heterodimerization), which can bring two different proteins into a common complex to regulate a signaling pathway. Higher-order oligomeric complexes are also possible. We have already introduced how dimerization of 14-3-3 proteins plays an important role in their ability to bind to multiply phosphorylated partners (see Figure 10.11).

The SAM (sterile alpha motif) domain is an example of a module that can form a head-to-tail dimer as well as self-associate into an extended oligomer. SAM domains are found in a variety of proteins ranging from receptor tyrosine kinases and cytoplasmic signaling proteins to transcription factors and polypeptides that regulate chromatin. SAM domains usually interact through two distinct surfaces, termed EH (for end-helix) and ML (for mid-loop); the EH surface of one domain interacts with the ML surface of a second domain (**Figure 10.17**). In addition to forming dimers, a SAM domain with mutually compatible EH and ML sites can assemble into long polymers, as has been observed for the human transcription factor TEL, a member of the Ets family that mediates transcriptional repression. A similar open-ended polymer is formed by the SAM domain of the *Drosophila* protein polyhomeotic, a member of the polycomb group of proteins that maintain chromatin in a transcriptionally repressed state. In both cases, the polymers formed by SAM domains may transmit a signal for transcriptional repression over an extended region of chromatin.

This type of domain-mediated oligomerization can also be dynamically regulated, as in the case of the *Drosophila* SAM domain-containing transcriptional repressor Yan. Receptor tyrosine kinase activation promotes the heterodimerization of Yan with a second SAM domain protein, Mae. Binding to Mae leads to the depolymerization of Yan associated with chromatin, its phosphorylation, and its export from the nucleus. In this way, relief of transcriptional repression can be coupled to extracellular signals during development (**Figure 10.18**).

INTERACTION DOMAINS THAT RECOGNIZE PHOSPHOLIPIDS

We have discussed how extracellular signals can induce the dynamic post-translational modification of intracellular signaling proteins, and that these modifications frequently exert their effects by creating binding sites for the interaction domains of target proteins. However, signaling information can also be carried by non-protein molecules, such as lipids or small molecules, which are modified in response to a stimulus. For example, external signals can induce the phosphorylation of the exposed head groups of phospholipids embedded in membranes through their fatty acid side chains. In the same way that interaction domains can recognize phosphorylated sites on proteins, there are a number of interaction domains that selectively bind the head groups of phospholipids, particularly members of the phosphoinositide family. In this section, we will discuss several examples of domains that recognize specific membrane phospholipids, thereby playing a critical role in regulating the interaction of signaling proteins with membranes.

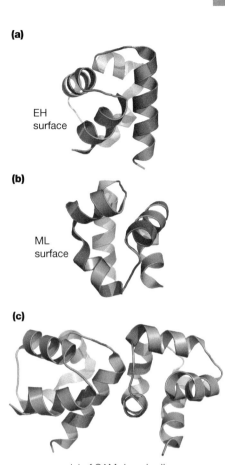

(a) EH surface

(b) ML surface

(c)

model of SAM domain dimer

Figure 10.17

SAM domains can form polymers.
SAM domains have distinct (a) EH and (b) ML surfaces, which can (c) dimerize in a "head-to-tail" fashion and also form longer oligomers. Residues implicated in dimer–dimer interactions are colored *pink*. (Adapted from J.J. Kwan et al., *J. Mol. Biol.* 342:681–693, 2004. With permission from Elsevier.)

 Membrane lipids and their role in signaling are discussed in detail in Chapter 7.

Figure 10.18

Multimerization of SAM domains regulates the transcription factor Yan.
In the absence of receptor tyrosine kinase activation, the majority of Yan molecules (*green*) form a SAM domain-mediated homopolymer, which has a strong affinity for DNA. Binding of Yan to DNA prevents the binding of transcriptional activators, resulting in the repression of target genes. A small amount of Yan forms a SAM–SAM heterodimer with the regulatory protein Mae (*orange*). Receptor tyrosine kinase stimulation leads to the activation of MAP kinase (MAPK), which can dock with Mae and phosphorylate Yan. Phosphorylated Yan is exported to the cytoplasm. To maintain the equilibrium, Yan polymers dissociate from the DNA, allowing transcription to occur. Since *Mae* expression is regulated by Yan, a positive feedback cycle is initiated, ensuring full de-repression of target genes upon MAPK activation. (Adapted from F. Qiao, et al., *Cell* 118:163173, 2004. With permission from Elsevier.)

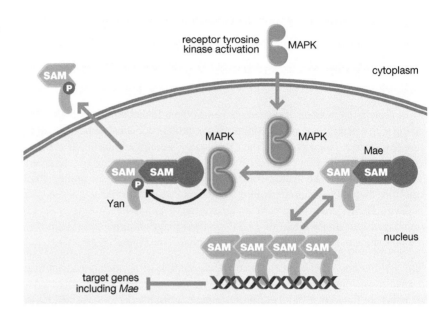

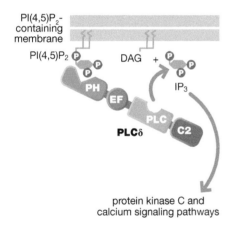

Figure 10.19

The PLC-δ PH domain is recruited to regions of the membrane rich in PI(4,5)P₂. After recruitment to the membrane, the enzymatic activity of PLC-δ then converts PI(4,5)P₂ to inositol 1,4,5-trisphosphate (IP₃) and diacylglycerol (DAG), which activate protein kinase C (PKC) and calcium signaling pathways.

Akt activation is described in Chapter 5.

PH domains form a major class of phosphoinositide-binding domains

Phosphoinositide-binding domains, such as **PH domains**, can selectively recognize one or more forms of the phosphorylated inositol head group of phosphoinositides. A protein can therefore be localized to a region of the membrane enriched in a phospholipid preferentially recognized by its PH domains. These domains can target a protein to a particular subcellular organelle or, by recognizing one of the signaling phosphoinositides, can induce translocation of a protein to a membrane (usually the plasma membrane) following activation of a cell-surface receptor, such as a receptor tyrosine kinase or G-protein-coupled receptor. Such phosphoinositide-dependent localization therefore positions enzymes or adaptors with appropriate phospholipid-binding domains in the proximity of their targets. The enzyme phospholipase C-δ (PLC-δ), for example, has a PH domain that selectively binds phosphatidylinositol 4,5-bisphosphate [PI(4,5)P₂], which is also the substrate for its catalytic domain. This interaction therefore concentrates PLC-δ at regions of the membrane that are enriched for its substrate (**Figure 10.19**). In contrast, the PH domains of the serine/threonine kinase Akt and its activator PDK1 bind selectively to PI(3,4,5)P₃ and PI(3,4)P₂. PDK1 and Akt are cytosolic until PI(3,4,5)P₃ levels in the plasma membrane rise—in response to an external signal that activates PI 3-kinase—providing docking sites for their PH domains. Once at the membrane, PDK1 phosphorylates and activates Akt, which has numerous substrates involved in the control of cell growth and survival.

As mentioned earlier in this chapter, PH domains have the same overall fold as PTB domains, consisting of two orthogonal β sheets capped by a C-terminal a helix (see Figure 10.4). The phosphoinositide-binding pocket of PH domains is on the opposite surface from the α helix, and is typically formed by basic residues located on the loop between the first two β strands, which can make up to 19 hydrogen bonds with the phosphates of the inositol ring. As a consequence, the affinity with which phosphoinositides bind PH domains can be relatively high (K_d as low as 30 nM). Subtle differences in the binding pocket can alter the specificity of phosphoinositide recognition, and some PH domains use a second, distinct binding pocket adjacent to the canonical site to engage phosphoinositides or phosphorylated sphingolipids. It should be noted that the majority of PH domains identified on the basis of sequence similarity cannot be shown to bind with high affinity to phosphoinositides, leaving open the possibility that they bind to other, currently unknown targets.

FYVE domains are phospholipid-binding domains found in endocytic proteins

Just as multiple different classes of domains bind phosphopeptide sites in proteins, there are several distinct types of interaction domains that bind phosphoinositides in membranes. One of these is the FYVE domain, which is stabilized by two zinc-binding clusters and has a shallow, positively charged pocket that interacts with phosphatidylinositol 3-phosphate [PI(3)P] on endosomes (**Figure 10.20a**). Because they make fewer hydrogen bonds with bound phosphoinositides than do PH domains, the affinity of FYVE domains for soluble forms of PI(3)P is rather weak. However, additional features of FYVE domains greatly strengthen their interactions with membranes. First, they insert nonpolar residues into the interior of the membrane, adding binding energy to that provided by selective phosphoinositide recognition. Second, FYVE domain proteins can form dimers and oligomers, increasing their avidity for PI(3)P-rich membranes.

Examples of FYVE domain proteins include Hrs, which is localized to endosomes through binding of its FYVE domain to PI(3)P. Hrs also has a ubiquitin interaction motif and thus it acts as a receptor on endosomes for endocytosed, ubiquitylated receptors. SARA, in contrast, is a scaffold that associates with the TGFβ receptor serine/threonine kinase and R-SMAD, and thereby localizes the activated receptor complex to endosomes through its FYVE domain. This prevents degradation of internalized TGFβ receptors and stimulates signaling through the SMAD2/4 pathway. SARA binding also distinguishes signaling-competent TGFβ receptors internalized through clathrin-coated pits from those that have been internalized through a caveolin-mediated pathway, which are targeted to lysosomes for degradation (see also Figure 5.16).

BAR domains bind and stabilize curved membranes

In addition to PH and FYVE domains, many other protein domains bind negatively charged lipids, and these are particularly prevalent in proteins involved in endocytosis and in protein/vesicle trafficking. These domains can have more complex properties. BAR domains, for example, form dimers of coiled-coil sequences that adopt a banana-shaped structure (**Figure 10.20b**). The concave surface of this dimer is typically positively charged, and thus has some affinity for biological membranes (which tend to have a net negative surface charge due to the phosphate groups of phospholipids). Their curved shape means that they bind with the

Figure 10.20

Interaction domains that bind phospholipids. (a) A model of the interaction of a FYVE domain homodimer with the membrane. FYVE domains typically bind endosomal membranes enriched in PI(3)P; the domain is stabilized by the binding of zinc (*orange spheres*). Basic residues that interact with the phosphorylated lipid head group are shown in *blue*. Note the potential insertion of hydrophobic side chains from the FYVE domain (*yellow*) into the membrane (approximate position indicated by horizontal line). For crystallization, the water-soluble inositol 1,3-bisphosphate [Ins(1,3)P_2] head group was used as an analog for the insoluble membrane lipid PI(3)P. (b) BAR domains interact strongly with curved membranes due to their shape and positively charged binding surface. Binding to a curved membrane increases electrostatic interactions between the domain and the negatively charged membrane, and is energetically favored. BAR domains can thus stabilize and/or promote membrane curvature, for example in the formation of endocytic vesicles. (a, Adapted from M. Lemmon, *Nat. Rev. Mol. Cell Biol.* 9:99–111, 2008. With permission from Macmillan Publishers Ltd; b, adapted from H.T. McMahon & J.L. Gallop, *Nature* 438:590–596, 2005. With permission from Springer Nature.)

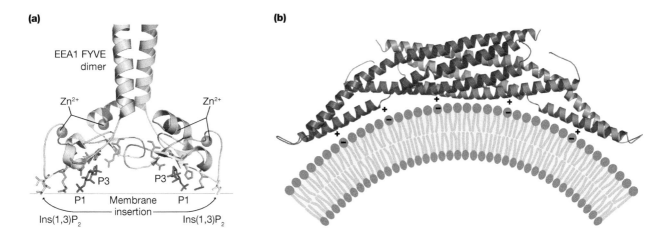

(a) EEA1 FYVE dimer

Zn²⁺   Zn²⁺

P3   P3

P1   Membrane   P1
— insertion —

Ins(1,3)P_2 Ins(1,3)P_2

(b)

highest affinity (lowest energy) to membranes with a specific degree of curvature, matching the curvature of the domain. BAR domains thus can stabilize or even induce membrane curvature, and thereby promote the formation of vesicular or tubular structures of the sort required for endocytosis. The related I-BAR domain has a convex instead of concave surface, and thus can stabilize or induce membrane protrusion.

CREATING COMPLEX FUNCTIONS BY COMBINING INTERACTION DOMAINS

Thus far, we have primarily discussed individual binding domains—the minimal sequence necessary to fold independently and recruit the appropriate ligand. However, it is quite common, through the process of evolution, for distinct domains to recombine and become linked with other domains. In this section, we discuss how such domain combinations can lead to proteins and complexes with more sophisticated signaling behaviors.

Recombination of domains occurs through evolution

Although members of a family of related domains generally share similar biochemical activities, they can be found in proteins containing a wide variety of other kinds of functional domains with either binding or catalytic properties. The range of combinations in which a single class of domains is found suggests that evolutionary diversification has occurred by shuffling domains in different ways. Clearly this provides a ready mechanism for generating new signaling behaviors through the repeated re-use and recombination of a limited set of modular functional units of protein structure.

We can see this combinatorial diversification by examining the modular domain architectures of SH2 domain-containing proteins. **Figure 10.21** illustrates a selection of proteins that contain SH2 domains, and their varied functions. These include regulating the GTP-binding state of small G proteins, phospholipid metabolism, the dynamic reorganization of the actin cytoskeleton, ubiquitylation, protein tyrosine phosphorylation, and transcription. While this figure shows only a small sample of the complement of SH2-containing proteins found in humans, the

Figure 10.21

SH2 domains are found in numerous proteins with varied roles in signal transduction. The domain arrangements of a subset of the ~110 human SH2 domain-containing proteins are depicted. The primary signaling activity of each protein is indicated on the right. Note that in PLC-γ, both the phospholipase C catalytic domain and a PH domain are split into two halves, indicated by n and c.

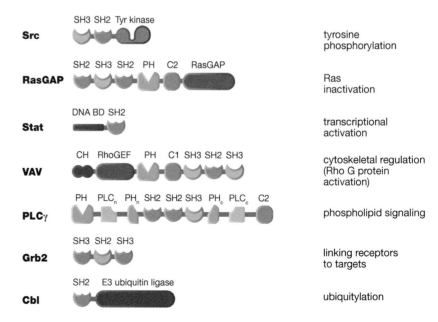

combinatorial diversity of the family is apparent. Similar combinatorial diversity is observed for many other modular domain families.

In some proteins, SH2 domains are linked exclusively to other interaction domains, such as SH3 domains that bind proline-rich sequences. These types of proteins act as **adaptors** that link phosphotyrosine-containing receptors to downstream targets that bind the adaptor's SH3 domains (see below). SH2 domains can also be directly linked to enzymatic domains. Proteins such as phospholipase C-γ have both catalytic domains and several different interaction domains that confer binding to both peptide and phospholipid ligands; the activities of such proteins are therefore potentially regulated by multiple inputs. The cellular response to changes in tyrosine phosphorylation is therefore determined largely by the ability of the resulting phosphotyrosine motifs to associate with specific SH2 domain proteins, and also the downstream pathways that these proteins activate or repress.

Combinations of interaction domains or motifs can be used as a scaffold for the assembly of signaling complexes

One of the major classes of multidomain signaling proteins is the scaffold proteins, which contain multiple interaction domains or peptide motifs within the same polypeptide. In general, such proteins can be used to coordinate the assembly of multiprotein complexes, either in a static way, or in a way that is dynamically modulated by signaling.

One of the simplest examples of creating new functions is illustrated by proteins such as Grb2, which are constructed from two different recognition domain types, linked together to form an adaptor protein. Grb2 contains a single SH2 domain flanked by two SH3 domains. The Grb2 SH2 domain specifically recognizes phosphorylated motifs, such as the pYSNA peptide motif that is generated in the activated PDGF receptor, while the SH3 domains from Grb2 recognize specific proline-rich peptide motifs, such as those in Sos, a GEF for the small G protein Ras. Thus, Grb2 functions as an adaptor because it takes a phosphotyrosine-based input and converts it into a specific output: activation of Ras on the plasma membrane (**Figure 10.22a**). Additional outputs are also possible, as the Grb2 SH3 domains can bind to effectors other than Sos—for example, Cbl (a ubiquitin E3 ligase) or dynamin (a protein involved in endocytosis). Further flexibility is provided because, in principle, different adaptor proteins can be used to convert the same input signal into different potential outputs, depending on the domain combinations found in the adaptor protein (**Figure 10.22b** and **10.22c**). Below, we discuss how other types of domain combinations can lead to even more complex signaling behavior.

Figure 10.22

Flexibility of adaptor protein function. (a) The Grb2 adaptor couples activated receptor tyrosine kinases such as the PDGF receptor (PDGFR) to downstream effectors such as the Ras activator Sos. (b) Grb2 can bind multiple effector proteins with distinct activities through its SH3 domains. Recruitment of Sos promotes Ras signaling, while recruitment of the ubiquitin ligase Cbl promotes receptor ubiquitylation and down-regulation. (c) Members of the Dok family of adaptors contain a PH domain that binds to phosphoinositides, a PTB domain that binds to tyrosine-phosphorylated peptides, and potential tyrosine phosphorylation sites (Y). Thus Dok family adaptors can integrate signals from lipid kinases that generate phosphoinositides such as PIP, and tyrosine kinases that generate phosphotyrosine (pY).

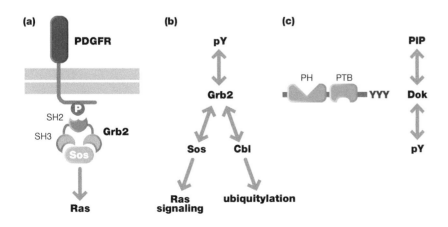

Scaffold proteins containing PDZ domains organize cell–cell signaling complexes such as the postsynaptic density

Proteins containing multiple interaction domains can function as scaffolds to organize complex signaling pathways. Examples include the PDZ domain-containing scaffolds that organize cell–cell signaling junctions. In multicellular organisms, specific signaling must often occur between adjacent cells, and highly specialized complexes may be assembled at the junctions between cells. For example, neurons connect with each other at junctions called synapses. The postsynaptic side contains a specialized cellular substructure known as the **postsynaptic density** (**PSD**), named because it is so densely packed with proteins that it can be easily imaged by electron microscopy. The PSD is typically found on protruding actin-rich structures known as dendritic spines, and it is where neurotransmitter receptors and other signaling proteins are concentrated in the cell.

Synaptic scaffolding proteins are critical for organizing the PSD. Some of these proteins, exemplified by the synaptic scaffold protein PSD95, contain multiple PDZ domains, as well as SH3 and phosphoserine/phosphothreonine-binding guanylyl kinase-like (GK) domains, which all form different interactions (**Figure 10.23**). Most of the neurotransmitter receptors in the PSD have a C-terminus that is recognized by these PDZ domains. In addition, the other domains form interactions that anchor the complex to the cytoskeletal structure of the dendritic spines. Thus these scaffolds, in effect, organize a large supramolecular complex in which the receptors and their downstream signaling effectors form a preassembled structure

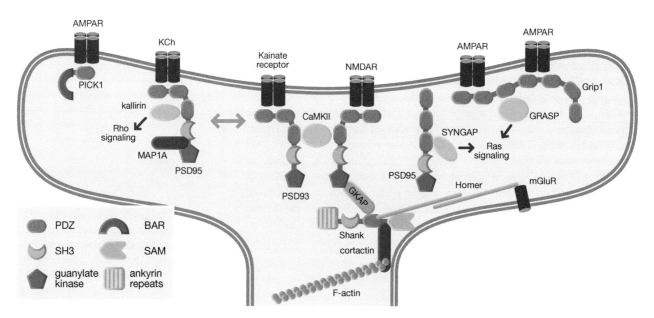

Figure 10.23

PDZ-containing proteins at the postsynaptic density. Many of the proteins located at the postsynaptic density (PSD) contain PDZ domains. These form supramolecular complexes required for proper synapse structure and function. Some of these interactions are illustrated in this figure. The major PDZ-containing protein, PSD95 (also known as Dlg4), can link to many receptors and ion channels such as the NMDA receptor (NMDAR), AMPA receptor (AMPAR), and potassium channel (KCh). PSD93 (also known as Dlg2) also interacts with receptors, such as the kainate receptor, and can form heteromultimers with PSD95. PSD95 links to the PDZ domain-containing Shank protein via the adaptor GKAP. Shank also interacts with the metabotropic glutamate receptor (mGluR) via the protein Homer and connects with the actin cytoskeleton via cortactin. PSD95 connects with microtubules via binding to MAP1A. PICK1, a PDZ and BAR domain-containing protein, can also connect with receptors and ion channels. Grip1, with seven PDZ domains, can bind to AMPARs and regulates the localization of these and other receptors to PSD membranes. Grip1 can bind GRASP to regulate Ras signaling, which can also be affected by SYNGAP binding to PSD95. Kalirin also binds PSD95 and stimulates Rho signaling. CaMKII phosphorylates a number of PSD proteins including PSD95. AMPAR, AMPA (α-amino-3-hydroxyl-5-methyl-4-isoxazole propionic acid) receptor; CaMKII, calcium/calmodulin-dependent kinase II; NMDAR, NMDA (N-methyl-D-aspartate) receptor. (Adapted from W. Feng and M. Zhang, *Nat. Rev. Neurosci.* 10:87–99, 2009. With permission from Springer Nature.)

properly localized in the cell to allow for an efficient and rapid response to neuro-transmitters released from the presynaptic cells. Notably, other cell–cell junctions, such as the tight junctions between adjacent epithelial cells, also use related PDZ scaffolds for organization. As we saw in Chapter 5, multivalent interactions such as those mediated by PDZ scaffolds often mediate formation of phase-separated *molecular condensates* that can make reactions more efficient and specific.

Proteins with multiple phosphotyrosine motifs function as dynamically regulated scaffolds

As has been discussed above, there are many examples of proteins that contain multiple phosphotyrosine motifs (see Figure 10.10). Such proteins can act as scaffolds that transiently organize larger signaling complexes in response to phosphorylation. The scaffolding activity of such proteins is dynamically regulated by the activities of tyrosine kinases and phosphatases, in contrast to the constitutive activity of scaffolds such as the PDZ-containing proteins discussed above (**Figure 10.24**). Examples include the receptor tyrosine kinases—such as the PDGF and EGF receptors, which contain multiple tyrosine motifs that are phosphorylated upon receptor activation—and receptor-associated scaffold proteins such as IRS1 (see Figure 4.12). Further examples are provided by immune signaling molecules like LAT and SLP-76, which contain multiple tyrosine motifs that are phosphorylated by the tyrosine kinase ZAP-70 upon its recruitment to the activated T cell receptor (TCR). In this case, phosphorylation of LAT and SLP-76 transiently creates SH2 binding sites, nucleating the formation of a large scaffold complex involving LAT, SLP-76, the SH2/SH3 adaptor GADS, the SH2-containing kinase ITK, and the SH2-containing enzyme PLC-γ. Formation of this complex is required for proper activation of PLC-γ, which in turn is critical for inducing the overall TCR response.

RECOMBINING INTERACTION AND CATALYTIC DOMAINS TO BUILD COMPLEX ALLOSTERIC SWITCH PROTEINS

The catalytic domains involved in signaling must often be regulated by specific upstream signals. The modular architecture of signaling proteins provides a flexible solution to this problem: coupling catalytic domains with specific interaction domains can generate allosteric molecules in which upstream regulatory signals, localization, and catalytic activity are coordinately regulated (**Figure 10.25**). Such molecules can be termed **allosteric switch proteins**. Below, we describe several examples of complex multidomain switches to illustrate the diversity of mechanisms and regulatory relationships that can emerge from using these components in new combinations.

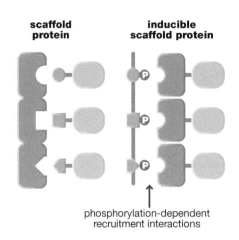

scaffold protein   inducible scaffold protein

phosphorylation-dependent recruitment interactions

Figure 10.24

Properties of scaffold proteins. Scaffolds have binding sites for multiple partners and function to nucleate assembly of large multiprotein complexes. An inducible scaffold requires signaling input to promote binding activity. For example, binding may be dependent on phosphorylation by an upstream kinase.

The role of scaffold proteins in TCR activation is illustrated in Chapter 14.

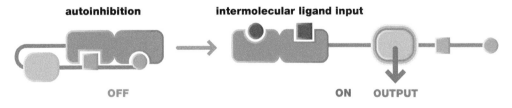

autoinhibition   intermolecular ligand input

OFF   ON   OUTPUT

Figure 10.25

Allosteric switch proteins. Many signaling enzymes are regulated by modular interaction domains (*orange*). In the example illustrated here, activity of the catalytic domain (*green*) is repressed by intramolecular interactions in the basal (inactive) state. Disruption of these intramolecular interactions, for example by ligands for the interaction domains (*purple*), leads to activation of the catalytic domain.

 Regulation of PKA is discussed in Chapter 6.

Many signaling enzymes are allosteric switches

We have already seen several examples of this type of regulation for protein kinases. For example, we saw in Chapters 1 and 3 that intramolecular interactions mediated by SH2 and SH3 domains lock the catalytic domain of Src family kinases in an inactive conformation. Src is therefore an allosteric switch protein, in which kinase activity is specifically induced by disruption of the autoinhibitory interactions, either through dephosphorylation of the SH2-binding motif, or by binding to competing SH2 and SH3 ligands (see Figure 1.9). We have also seen how in PKA the kinase subunit is noncovalently associated with a regulatory (R) subunit made up of two cAMP-binding domains and a pseudosubstrate peptide inhibitor. This multidomain complex results in an inhibited, inactive kinase that can be activated by binding of cAMP to the R subunits. Thus we can see how this kind of modular allosteric regulation can be achieved, not only through intramolecular autoinhibitory interactions, but also through noncovalent interactions with other molecules. These types of modular, autoinhibitory schemes are observed for many classes of catalytic functions, including kinases, phosphatases, GEFs, GAPs, and many others. There is even growing evidence that multidomain scaffold proteins can exist in inactive states involving intramolecular interactions.

14-3-3 Protein regulates the Raf kinase by coordinately binding two phosphorylation sites

Above, we described how association with 14-3-3 proteins can alter a phosphorylated target protein's localization or can occlude its interactions with other partners. However, 14-3-3 domains can also be used to regulate a protein's catalytic activity in a phosphorylation-dependent manner. In the Erk-MAP kinase pathway, the c-Raf serine/threonine protein kinase (a MAPKKK) is activated by the small G protein Ras and, in turn, stimulates signaling through the MAP kinase pathway. In the basal state, 14-3-3 proteins bind to phosphorylated c-Raf, inhibiting its activity. In this complex, each protomer of a 14-3-3 dimer binds a distinct phosphorylation site in an inactive c-Raf molecule: one N-terminal to the kinase domain (Ser259 in human c-Raf) and a second at the C-terminus (Ser621) (**Figure 10.26**). This two-point interaction acts as a clamp to stabilize an autoinhibited conformation in which kinase activity is suppressed and the Ras-binding and membrane-binding domains are blocked. Dephosphorylation of the N-terminal site has two important

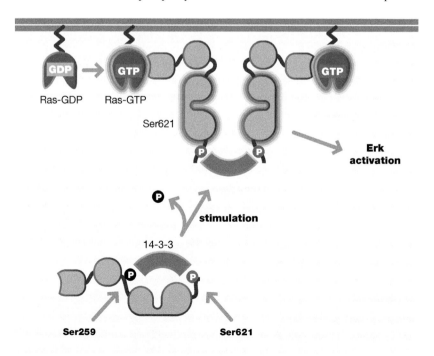

Figure 10.26

Regulation of Raf kinase by 14-3-3 proteins. 14-3-3 proteins hold Raf in an inactive state by binding simultaneously to phosphorylated Ser259 and Ser621. Dephosphorylation of Ser259 leads to 14-3-3-mediated Raf dimerization, association of Raf with activated Ras and the plasma membrane, and activation of the MAPK signaling pathway. Note that Raf typically is pre-associated with its substrate MEK in cells, not shown here for clarity.

consequences: it relieves steric inhibition of the catalytic and regulatory domains, and it promotes c-Raf dimerization by binding of a 14-3-3 dimer to the C-terminal phosphosites of two c-Raf molecules. These changes lead to stable activation of c-Raf on the membrane.

Mutation of Pro261 of c-Raf to Ser has been observed in Noonan syndrome, a human genetic disorder associated with heart defects, facial abnormalities, short stature, and other features. This mutation prevents Ser259 phosphorylation and thus 14-3-3-mediated inhibition of c-Raf activity. The aberrantly active c-Raf then signals through the Erk-MAP kinase pathway, ultimately leading to disease. This example underscores the modular nature of signaling proteins, the regulation of a key signaling kinase by its association with a 14-3-3 dimer, and the severe phenotypic consequences when this inhibitory interaction is lost.

 MAP kinase cascades are described in Chapter 3.

Certain plant protein kinases are regulated by modular light-gated domains

An example of the range and general utility of these modular systems is provided by plants, where some serine/threonine protein kinases can be activated by light. Such a mechanism is particularly useful in plants, which derive their energy from light and must therefore adjust their physiology and orientation depending on its availability. In one well-studied case, the ability to respond to blue light is provided by the LOV domain, a light-gated interaction module. LOV domains tightly bind flavin mononucleotide, and are found in a diverse set of light-regulated signaling proteins in plants. In the case of kinases, the LOV domain inhibits the kinase catalytic domain through an intramolecular interaction when in the resting (dark) state. Blue light creates a covalent adduct between the flavin and a conserved cysteine in the LOV domain, leading to a conformational change that releases the LOV domain from the catalytic domain, and thereby stimulates kinase activity (**Figure 10.27**). The modularity of this regulatory mechanism has been demonstrated in the laboratory, as investigators have been able to engineer recombinant proteins containing LOV domains that can be regulated in cells by light (further discussed below and in Chapter 15).

Regulation of the neutrophil NADPH oxidase by modular interactions

One of the most fascinating molecular machines that is allosterically regulated by modular interaction domains is the neutrophil NADPH oxidase. This enzyme is activated during the phagocytosis of bacteria to produce the toxic superoxide anion (O_2^-) within specialized internal vesicles termed phagosomes, which are used to kill the bacteria. Given the toxicity of superoxide, it is particularly critical that the oxidase catalytic activity only be induced at the right place and time. Fittingly, a complex assembly reaction is required to form the active enzyme. The NADPH oxidase has multiple components, including two transmembrane proteins—p91phox and p22phox—that form a heterodimeric flavocytochrome that mediates electron transfer to molecular oxygen, and three cytosolic regulatory proteins—p47phox, p67phox, and p40phox. Upon stimulation by phagocyte receptors, these three cytosolic proteins translocate to the membrane and associate with the integral membrane proteins to yield a functional oxidase complex (**Figure 10.28**).

Each of the cytosolic proteins has multiple interaction domains, primarily SH3 domains, through which they interact with one another, and with the integral membrane phox proteins, following a phagocytic signal. However, in unstimulated cells, these domains are sequestered through intramolecular interactions that prevent adventitious oxidase activation. For example, two adjacent SH3 domains of p47phox bind to the tail of p47phox, which blocks the ability of these SH3 domains to associate in *trans* with another protein. Activating signals, which

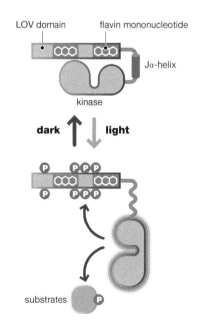

Figure 10.27

Activation of kinases with LOV domains by light. Light interacts with flavin mononucleotides of the LOV domain. This triggers a conformational change, including disruption of a critical α helix (Jα), that leads to dissociation and activation of the kinase domain and phosphorylation of substrates, including the LOV domain-containing regulatory region. (Adapted from J.M. Christie, *Annu. Rev. Plant Biol.* 58:21–45, 2007. With permission from *Annual Reviews*.)

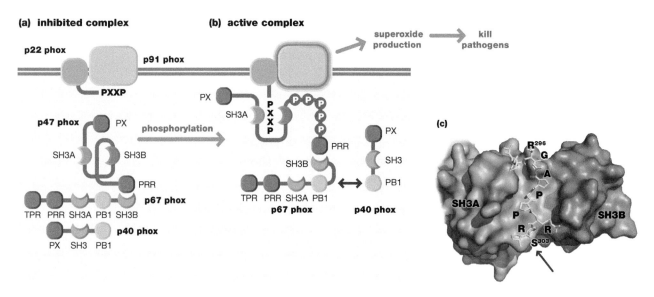

Figure 10.28

Multiple interaction domains regulate NADPH oxidase activation. (a) The NADPH oxidase complex consists of multiple subunits: p22phox and p91phox are transmembrane proteins, whereas p67phox, p47phox, and p40phox are cytosolic regulatory components. The p47phox subunit is held in an inactive state by its two SH3 domains (*blue* and *orange*) being locked together via interaction with a C-terminal polybasic motif. Other domains indicated are the TPR repeat domain, a proline-rich region (PRR), PX domain, and PB1 domain. (b) Phosphorylation of the polybasic motif of p47phox releases the SH3 domains, allowing them to interact with the polyproline motif (PxxP) of p22phox. This allows interaction of additional domains, thus activating the complex. (c) Crystal structure of the SH3A and SH3B domains of p47phox, showing interaction with the polybasic motif (depicted in *yellow*). Phosphorylation of residues in this region, including Ser303 (arrow), disrupts this intramolecular interaction. (a and b, Adapted from S.S.-C. Li, *Biochem. J.* 390:641–653, 2005. With permission of Portland Press; c, from Y. Groemping, et al., *Cell* 113:343355, 2003. With permission from Elsevier.)

include binding of the small G protein Rac and phosphorylation of multiple serine residues in p47phox, result in disruption of the intramolecular SH3-binding interaction. The released p47phox SH3 domains are now free to bind to an alternative proline-rich sequence in the membrane-associated subunit p22phox, yielding a functional oxidase that is essential for limiting microbial infections. In this system, the two tandem SH3 domains from p47phox pack together to form a single binding surface that recognizes a binding motif that is longer and more specific than typically seen for a single SH3 domain. This likely confers particularly tight control over the enzyme activity (**Figure 10.28c**).

CREATING NEW FUNCTIONS THROUGH DOMAIN RECOMBINATION

We have seen how evolution has harnessed the amazing flexibility of modular catalytic and interaction domains to build complex signaling devices and to create new behaviors and phenotypes. It is easy to envision how the molecular processes that drive evolution, such as gene duplication, exon shuffling, and point mutation, could lead to the generation of new signaling functions

Of course, in most cases, we can only infer such events indirectly. There are, however, instances in which the generation of new functions through domain rearrangements can be directly demonstrated. We discuss below several examples from human disease, and from directed efforts to engineer new functions in the laboratory.

The evolution of cell signaling machinery is discussed in more detail in Chapter 13.

Some modular domain rearrangements can lead to cancer

Although domain rearrangement provides a powerful mechanism for generating new functions, there is a potential trade-off in this functional flexibility, in that genetic rearrangements that lead to novel domain combinations can also lead to misfunction and disease in individuals. A random domain rearrangement may in some cases offer a fitness advantage, while in most other instances it is likely to confer a fitness disadvantage, either at the level of the individual mutated cell, or at the level of the organism. Here, we discuss two examples of how *oncogenes*—genes that can cause cancer when mutated or overexpressed—can be generated by chromosome translocations. Oncogenes generally lead to the uncontrolled proliferation or increased survival of cells in which they are expressed, so these mutations provide a fitness advantage for the cell, but clearly are a disadvantage for the organism, as the resulting tumor might ultimately lead to its death.

Above, we introduced the propensity of SAM domains to oligomerize into extended complexes. This behavior can also induce aberrant signaling. In some human leukemias, for example, a chromosomal translocation leads to the fusion of the self-oligomerizing SAM domain from the gene *TEL* to the catalytic domain of the nonreceptor tyrosine kinase Abl (**Figure 10.29a** and **b**). The resulting chimeric protein is constitutively oligomerized, resulting in persistent activation of the tyrosine kinase domain and oncogenic transformation (remember that dimerization or clustering is the most common mechanism for activating tyrosine kinases, as it promotes transphosphorylation and stabilization of the active conformation of the catalytic domain). Several other oncogenic fusion proteins (involving tyrosine kinases other than Abl) are similarly activated by the oligomerizing function provided by the SAM domain of TEL.

Activation of the Abl tyrosine kinase by chromosome translocation is also seen in patients with chronic myelogenous leukemia (CML). In this case, a coiled-coil oligomerization domain derived from the *BCR* gene on chromosome 22 is fused to the tyrosine kinase domain and C-terminus from the *Abl* gene on chromosome 9

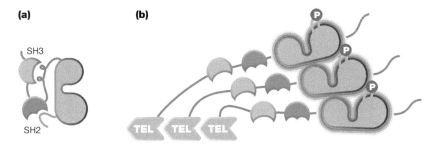

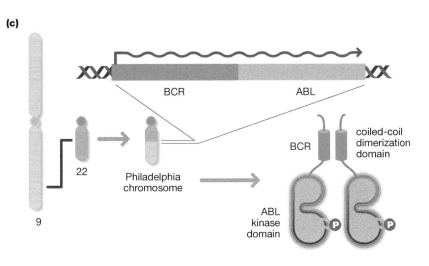

Figure 10.29

Activation of Abl by fusion with oligomerization domains. (a) Abl is normally held in an inactive state by intramolecular interactions, mediated primarily by its SH2 and SH3 domains. (b) When fused to TEL, SAM domains of the fusion protein oligomerize, leading to autophosphorylation and activation of the Abl catalytic domain. (c) The Philadelphia chromosome is generated by translocation of the tip of chromosome 9 (encoding Abl) to chromosome 22 (encoding BCR). The translocation generates a hybrid protein containing a portion of BCR fused to the N-terminus of Abl. Dimerization or oligomerization of the coiled-coil domains from BCR leads to constitutive activation of the Abl kinase domains.

(**Figure 10.29c**). The characteristic 9;22 chromosomal translocation that generates this fusion protein is termed the "Philadelphia chromosome," and was one of the first specific genetic abnormalities to be directly correlated with human cancer. A functional coiled-coil domain in the BCR portion has been shown to be required for oncogenic transformation by the BCR–Abl fusion. Thus, as in the case of the TEL-Abl oncogene, constitutive dimerization or clustering of BCR–Abl is thought to lead to persistent kinase activity and oncogenic transformation. BCR–Abl is well known as the target for the drug imatinib (Gleevec®), which specifically inhibits the Abl tyrosine kinase domain. The effectiveness of imatinib in treating CML was the first dramatic success in the effort to develop rational cancer therapies that target specific signaling proteins.

Modules can be recombined experimentally to engineer new signaling behaviors

Work in the emerging field of **synthetic biology**, which is aimed at using natural biological components to build novel functional systems, has demonstrated that signaling modules can be used to generate non-natural, custom-designed signaling proteins, pathways, and cellular behaviors.

 The role of scaffold proteins in controlling MAP kinase output is discussed in Chapter 3.

Many eukaryotic cells use scaffold proteins to properly wire their MAP kinase pathways. Most cells contain several of these closely related pathways that mediate distinct responses; scaffolds are thought to assemble the correct partners into a single complex, thus promoting their efficient interaction, and preventing their interaction with incorrect but related partners elsewhere in the cell. Recombination of individual domains from MAP kinase scaffold proteins has been used to construct chimeric scaffold proteins that assemble novel combinations of MAPK components (**Figure 10.30**). These chimeric scaffolds can reroute signaling in a living cell so that a specific input now produces a normally unrelated response. For

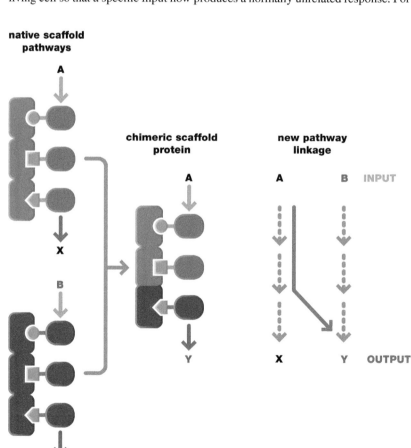

Figure 10.30

Chimeric scaffolds can rewire signaling pathways. In this example (modeled on MAP kinase cascades), two scaffolds assemble distinct protein complexes, such that stimulus A leads to output X and stimulus B leads to output Y. A chimeric, engineered scaffold protein can redirect signal output such that stimulus A now leads to output Y.

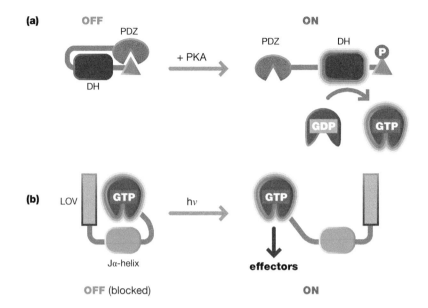

(a)

OFF

PDZ

DH

+ PKA

ON

PDZ   DH

P

GDP   GTP

(b)

LOV

GTP

Jα-helix

hν

GTP

effectors

OFF (blocked)   ON

Figure 10.31

Engineering allosteric switch proteins that respond to novel signaling inputs.
(a) Coupling Rac activation to protein kinase A (PKA) activity. DH domains act as GEFs for Rho family G proteins. In this example, activity of the DH domain is repressed by an engineered intramolecular interaction between a PDZ domain and the C-terminus. Phosphorylation of the C-terminus by PKA disrupts this intramolecular interaction, leading to GEF activity and Rac activation. (b) Activation of Rac by light. In this example, a constitutively GTP-bound form of Rac is fused to a LOV domain, thus blocking its interaction with downstream effectors. Light induces conformational changes that release the LOV domain, leaving Rac-GTP free to bind effectors.

example, a yeast cell can be reprogrammed such that stimulation with a mating pheromone induces the response normally associated with high osmotic stress.

Engineered domain recombination involving catalytic and interaction domains can also be used to build new allosteric signaling proteins and new cellular behaviors. For example, the small G protein Rac is a master regulator of actin polymerization, which drives membrane protrusion and cell movement. Recombination of a Rac-specific GEF domain (a Dbl homology or DH catalytic domain) with different interaction domains can result in novel allosteric switches in which Rac activation and actin polymerization are controlled by novel, non-natural inputs. For example, recombination with an autoinhibitory intramolecular PDZ domain interaction that can be relieved by PKA phosphorylation leads to a PKA-induced GEF (**Figure 10.31a**). Alternatively, recombination of Rac itself with a light-gated LOV domain module from plants generates a novel light-induced Rac (**Figure 10.31b**).

These kinds of engineered signaling proteins clearly illustrate the functional flexibility of modular domains. Such engineered proteins also have several potential uses. First, examples like light-controlled signaling proteins can be used as powerful research tools to activate pathways in a spatially and temporally controlled manner using specific patterns of activating light. Second, these engineered signaling proteins might, in the longer term, allow reprogramming of cells so that they carry out custom-designed therapeutic sensing/response behaviors, such as those involved in detecting and eliminating diseased cells or pathogens.

SUMMARY

The modular architecture of signaling proteins, coupled with the binding properties of interaction domains, provides a flexible solution to the problem of how to control signaling protein activities, and to coordinately couple them to upstream regulatory signals and downstream targets. As we have seen, this is achieved not only by the association of proteins with one another, but also through interaction domains that recognize a range of biomolecules such as phospholipids.

Modular interaction domains can either recognize unmodified peptide motifs or specifically bind only to peptide motifs after they have been post-translationally modified. In the latter case, such domains serve to couple post-translational modification to the localization or activity of downstream effector proteins. Proteins with multiple modular interaction domains or binding motifs can serve as scaffolds to assemble signaling proteins into larger molecular complexes. When combined with catalytic domains, modular interaction domains can be used to generate allosteric switches that can be regulated by a variety of inputs. The modular architecture of signaling proteins also provides a ready mechanism to explain the evolution of more complicated signaling behaviors from a relatively limited toolkit of modular functional units.

QUESTIONS

Answers to these questions can be found online at www.routledge. com/9780367279370

1. Most eukaryotic signaling proteins have modular structures composed of several domains that each carry out a distinct subfunction. What evolutionary constraints or advantages may have led to this type of organization?

2. A particular modular protein interaction domain is often found in many different signaling proteins in a given organism. Hypothesize how a modular protein interaction domain family may have expanded over the course of evolution. What issues may arise as such a domain family expands in size?

3. You are analyzing an SH2-phosphotyrosine peptide interaction and find that the K_d for the unphosphorylated peptide is 1 mM, while the K_d for the phosphorylated form is 100 nM. This peptide is found on the C-terminal tail of a receptor tyrosine kinase that is expressed at low levels. What is the approximate concentration range for the SH2-containing protein in the cell?

4. Protein interaction modules, such as SH2 domains or PDZ domains, often recognize a key physical feature in their cognate peptides, such as a phosphotyrosine side chain (SH2) or a free C-terminus (PDZ). How do individual members of the domain family establish distinct specificities?

5. Describe the general ways in which protein interaction domains can be combined with catalytic domains to regulate their activity and function.

6. How can fusions between interaction and catalytic domains lead to disease? How can they be harnessed to engineer new cellular behavior? If modular recombination of signaling domains can lead to diseases such as cancer, why would these features not have been selected against by natural selection?

7. The human genome encodes over 100 different SH2 domains, most of which bind specifically to phosphotyrosine peptides. By contrast, the budding yeast has a single SH2 domain, which binds to RNA polymerase II in a serine/threonine phosphorylation-dependent fashion. Yeast has no dedicated tyrosine-specific kinases and few tyrosine-specific phosphatases. Based on this information, provide a model for the evolution of the SH2 domain family and tyrosine kinase signaling.

8. Most signaling proteins contain multiple functional domains (either binding or catalytic domains). In many cases, expression in the cell of a mutant form of the protein, in which one of these domains has been inactivated, can act as a dominant negative mutation; that is, the mutant protein can inhibit the activity of the normal, endogenous protein. Explain how this can occur. In other cases, mutation of one domain can sometimes lead to the opposite effect—constitutive activity. When might this be the case?

9. In the laboratory, how might you create a modular domain that has a novel binding activity?

BIBLIOGRAPHY

MODULAR PROTEIN DOMAINS

Dyson HJ & Wright PE (2002) Coupling of folding and binding for unstructured proteins. *Curr. Opin. Struct. Biol.* 12, 54–60.

Mayer BJ (2015) The discovery of modular binding domains: Building blocks of cell signalling. *Nat. Rev. Mol. Cell Biol.* 16(11), 691–698. doi: 10.1038/nrm4068.

Orlicky S, Tang X, Willems A, et al. (2003) Structural basis for phosphodependent substrate selection and orientation by the SCFCdc4 ubiquitin ligase. *Cell* 112, 243–256.

Pawson T & Nash P (2003) Assembly of cell regulatory systems through protein interaction domains. *Science* 300, 445–452.

INTERACTION DOMAINS THAT RECOGNIZE POST-TRANSLATIONAL MODIFICATIONS

Au-Yeung BB, Deindl S, Hsu LY, et al. (2009) The structure, regulation, and function of ZAP-70. *Immunol. Rev.* 228, 41–57.

Brehm A, Tufteland KR, Aasland R & Becker PB (2004) The many colours of chromodomains. *Bioessays* 26, 133–140.

Hu J & Hubbard SR (2005) Structural characterization of a novel Cbl phosphotyrosine recognition motif in the APS family of adapter proteins. *J. Biol. Chem.* 280, 18943–18949.

Hurley JH, Lee S & Prag G (2006) Ubiquitin-binding domains. *Biochem. J.* 399, 361–372.

Newton K, Matsumoto ML, Wertz IE, et al. (2008) Ubiquitin chain editing revealed by polyubiquitin linkage-specific antibodies. *Cell* 134, 668–678.

Pascal SM, Singer AU, Gish G, et al (1994) Nuclear magnetic resonance structure of an SH2 domain of phospholipase C-gamma 1 complexed with a high affinity binding peptide. *Cell* 77, 461–472.

Seet BT, Dikic I, Zhou MM & Pawson T (2006) Reading protein modifications with interaction domains. *Nat. Rev. Mol. Cell Biol.* 7, 473–483.

INTERACTION DOMAINS THAT RECOGNIZE UNMODIFIED PEPTIDE MOTIFS OR PROTEINS

Kwan JJ, Warner N, Pawson T & Donaldson LW (2004) The solution structure of the *S. cerevisiae* Ste11 MAPKKK SAM domain and its partnership with Ste50. *J. Mol. Biol.* 342, 681–693.

Qiao F, Song H, Kim CA, et al. (2004) Derepression by depolymerization; structural insights into the regulation of Yan by Mae. *Cell* 118, 163–173.

Seet BT, Berry DM, Maltzman JS, et al. (2007) Efficient T-cell receptor signaling requires a high-affinity interaction between the Gads C-SH3 domain and the SLP-76 RxxK motif. *EMBO J.* 26, 678–689.

Zarrinpar A, Bhattacharyya RP & Lim WA (2003) The structure and function of proline recognition domains. *Sci. STKE* 2003(179):re8.

INTERACTION DOMAINS THAT RECOGNIZE PHOSPHOLIPIDS

Kutateladze TG (2006) Phosphatidylinositol 3-phosphate recognition and membrane docking by the FYVE domain. *Biochim. Biophys. Acta* 1761, 868–877.

Lemmon MA (2008) Membrane recognition by phospholipid-binding domains. *Nat. Rev. Mol. Cell Biol.* 9, 99–111.

McMahon HT & Gallop JL (2005) Membrane curvature and mechanisms of dynamic cell membrane remodelling. *Nature* 438, 590–596.

CREATING COMPLEX FUNCTIONS BY COMBINING INTERACTION DOMAINS

Chothia C, Gough J, Vogel C & Teichmann SA (2003) Evolution of the protein repertoire. *Science* 300, 1701–1703.

Christie JM (2007) Phototropin blue-light receptors. *Annu. Rev. Plant Biol.* 58, 21–45.

Feng W & Zhang M (2009) Organization and dynamics of PDZ-domain-related supramodules in the postsynaptic density. *Nat. Rev. Neurosci.* 10, 87–99.

Kim E & Sheng M (2004) PDZ domain proteins of synapses. *Nat. Rev. Neurosci.* 5, 771–781.

Li SS (2005) Specificity and versatility of SH3 and other proline-recognition domains: Structural basis and implications for cellular signal transduction. *Biochem. J.* 390, 641–653.

Peisajovich SG, Garbarino JE, Wei P & Lim WA (2010) Rapid diversification of cell signaling phenotypes by modular domain recombination. *Science* 328, 368–372.

Vogel C, Bashton M, Kerrison ND, et al. (2004) Structure, function and evolution of multidomain proteins. *Curr. Opin. Struct. Biol.* 14, 208–216.

RECOMBINING INTERACTION AND CATALYTIC DOMAINS TO BUILD COMPLEX ALLOSTERIC SWITCH PROTEINS

Fu H, Subramanian RR & Masters SC (2000) 14-3-3 proteins: Structure, function, and regulation. *Annu. Rev. Pharmacol. Toxicol.* 40, 617–647.

Groemping Y, Lapouge K, Smerdon SJ & Rittinger K (2003) Molecular basis of phosphorylation-induced activation of the NADPH oxidase. *Cell* 113, 343–355.

Light Y, Paterson H & Marais R (2002) 14-3-3 antagonizes Ras-mediated Raf-1 recruitment to the plasma membrane to maintain signaling fidelity. *Mol. Cell. Biol.* 22, 4984–4996.

Lim WA (2002) The modular logic of signaling proteins: Building allosteric switches from simple binding domains. *Curr. Opin. Struct. Biol.* 12, 61–68.

Park E, Rawson S, Li K, et al. (2019) Architecture of autoinhibited and active BRAF-MEK1-14-3-3 complexes. *Nature* 575(7783), 545–550. doi: 10.1038/s41586-019-1660-y.

CREATING NEW FUNCTIONS THROUGH DOMAIN RECOMBINATION

Bashor CJ, Helman NC, Yan S & Lim WA (2008) Using engineered scaffold interactions to reshape MAP kinase pathway signaling dynamics. *Science* 319, 1539–1543.

Bashor CJ, Horwitz AA, Peisajovich SG & Lim WA (2010) Rewiring cells: Synthetic biology as a tool to interrogate the organizational principles of living systems. *Annu. Rev. Biophys.* 39, 515–537.

Chau AH, Walter JM, Gerardin J, et al. (2012) Designing synthetic regulatory networks capable of self-organizing cell polarization. *Cell* 151, 320–332.

Dueber JE, Yeh BJ, Chak K & Lim WA (2003) Reprogramming control of an allosteric signaling switch through modular recombination. *Science* 301, 1904–1908.

Golub TR, Goga A, Barker GF, et al. (1996) Oligomerization of the ABL tyrosine kinase by the Ets protein TEL in human leukemia. *Mol. Cell. Biol.* 16, 4107–4116.

Karginov AV, Ding F, Kota P, et al. (2010) Engineered allosteric activation of kinases in living cells. *Nat. Biotechnol.* 28, 743–747.

Kim CA, Phillips ML, Kim W, et al. (2001) Polymerization of the SAM domain of TEL in leukemogenesis and transcriptional repression. *EMBO J.* 20, 4173–4182.

Lim WA (2010) Designing customized cell signalling circuits. *Nat. Rev. Mol. Cell Biol.* 11, 393–403.

McWhirter JR, Galasso DL & Wang JY (1993) A coiled-coil oligomerization domain of Bcr is essential for the transforming function of Bcr-Abl oncoproteins, *Mol. Cell. Biol.* 13, 7587–7595.

Park SH, Zarrinpar A & Lim WA (2003) Rewiring MAP kinase pathways using alternative scaffold assembly mechanisms. *Science* 299, 1061–1064.

Wu Y, Frey D, Lungu OI, et al. (2009) A genetically encoded photoactivatable Rac controls the motility of living cells. *Nature* 461, 104–108.

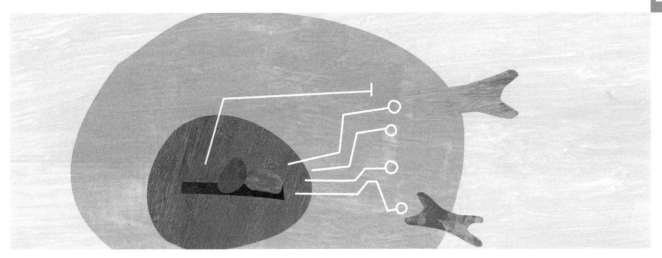

Information Processing by Signaling Devices and Networks

11

A living cell must take in a vast amount of information about its environment and internal state, and use this information to respond in various ways. What genes should be turned on or off? Should it grow, divide, differentiate, or die? Should it move or change its shape? Thus, from a signaling perspective, cells are complex information-processing systems, akin to computers and other human-made devices. How then do cellular components process information and respond appropriately?

The previous chapters have introduced the fundamental molecular parts and mechanisms that make up cell signaling systems. In this chapter, we try to understand how these molecular parts can be used to build devices that are capable of executing particular information-processing tasks. Our understanding of signaling design principles is still relatively primitive—we are just starting to reach beyond a descriptive understanding of signaling systems to grasp the fundamental principles of how molecular systems execute complex information-processing tasks. In part, our knowledge is limited because relatively few signaling systems have been quantitatively analyzed at high resolution. However, in recent years, computational systems biology and synthetic biology—approaches that combine analysis, simulation, and construction of molecular systems—have become powerful tools in understanding how signaling systems achieve complex behaviors.

In this chapter, we will examine a series of specific information-processing tasks that are common to many cellular processes, and provide examples of how molecular systems are able to accomplish these tasks. This will not be an encyclopedic examination of all possible cellular behaviors, but instead we will focus on a few commonly used design principles. We consider how cell signaling devices process various **input** stimuli, including input amplitude, duration, and combinations (multi-input control), and how they control **output** responses, including output amplitude and duration. We will examine how this diverse array of processing behaviors can be generated using signaling molecules as the basic components.

DOI: 10.1201/9780429298844-11

SIGNALING SYSTEMS AS INFORMATION-PROCESSING DEVICES

In order to discuss the general properties of information processing in cells, it is useful to step back from the seemingly overwhelming complexity of the biological system to a more abstract perspective. The fields of computer science and communications, with their focus on the design of highly complex information-processing devices and systems, provide a helpful theoretical framework for such a discussion.

Signaling devices can be considered as state machines

All information-processing systems are built from **state machines**—devices that can exist in multiple discrete states, and which can change their state in response to specific instructional inputs (**Figure 11.1a**). If these states have different functional properties (what we will refer to as outputs), then the state machine will serve as an input/output device. It is easy to understand how human-made machines serve as input/output devices (**Figure 11.1b**). Take, for example, an automatic door, which can exist in the open or closed state; the open state has a distinct functional output—it is the only state in which a person can pass through the door. This automatic door is an input/output device because a particular input—in this case, motion detected by the door's sensor—serves as an instruction to convert from the closed state to the open state. Thus, the input of motion results in the functional output of allowing a person to walk through the door.

Another example of a human-made state machine is the transistor—the component that enables most of our digital electronic systems. A transistor can switch between a high-resistance state and a low-resistance state. The application of a high voltage to the transistor (the input) switches it to the low-resistance state, which in turn allows current to flow through the transistor (the output). When linked in complex ways, these simple input/output devices can yield immensely complex decision-making machines, such as computers.

Living cells can also be considered as state machines (**Figure 11.1c**). For example, a cell can exist in distinct states, such as a quiescent (survival) state, a growing/dividing (proliferation) state, and a dying (apoptotic) state. These different fates are controlled by environmental inputs (such as growth factors, metabolic state, and mechanical inputs) that serve as the instructions that induce changes in state. Often, it is a complex combination of signals that serves as the instructions to change state in cellular systems.

At a more microscopic scale, the individual proteins and pathways within a cell can also act as state machines. For example, a simple allosteric kinase can exist in an active or inactive conformation, and this conformation might be controlled by the input of phosphorylation/dephosphorylation. In another example, a transmembrane receptor involved in detecting a mitogen is also a state machine—its input is the concentration of mitogen, while its output is the activated state of the receptor.

Signaling devices are organized in a hierarchical fashion

Information-processing systems, both biological and human-made, are often organized in a hierarchical fashion. State machines can exist at a variety of scales, and often a larger-scale state machine is constructed from many smaller-scale state machines. Take the example of an automatic door—while this system can be viewed as a single state machine, if one looks deeper into the inner workings of the door's sensor and processor, one will find many smaller state machines (including transistors) that are linked together to yield the whole system.

(a) generic state machine

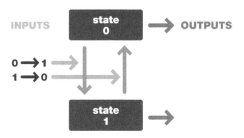

(b) man-made devices

(c) biological devices

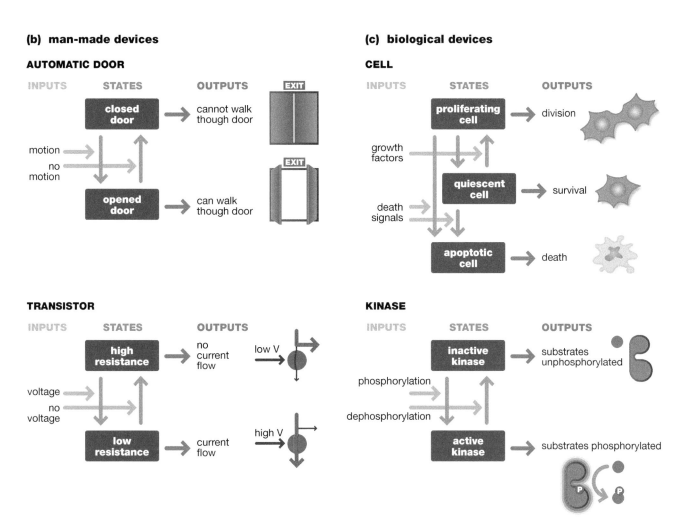

Figure 11.1

Signaling devices as state machines. (a) A generic state machine switches between two different states (0 and 1) when provided with specific instructions (inputs). Output depends on the state. (b) Examples of human-made devices that function as state machines. An automatic door will open only when it detects motion. A transistor will allow current to flow only when an external voltage is applied. (c) Examples of biological state machines involved in cell signaling. A cell can be viewed as a device that can exist in many distinct states, such as dividing (proliferating), quiescent, and apoptotic, with specific signals providing instructions to switch from one state to another. A signaling molecule, such as a kinase, can be viewed as a state machine, with specific inputs (for example, phosphorylation) serving as instructions to switch from one state to another.

Cellular information-processing machines are also built in a hierarchical manner. For example, although the whole cell can be considered as one state machine, it is clear that the individual proteins (such as receptors and kinases) that make up a growth-control pathway are themselves smaller-scale state machines (**Figure 11.2**). Similarly, multiple proteins and pathways interact with each other to yield a larger network capable of more complex signal processing.

Figure 11.2

Signaling input/output devices are organized in hierarchical fashion. A single signaling molecule is an input/output device when its activity is controlled by specific inputs. This molecule may be part of a larger multiprotein pathway, which itself functions as a modular input/output device. Finally, this pathway device may be a component of a larger, more complex signaling network at the cellular level. Important signaling behaviors can be mediated by events that span this whole range of scales. In this example, input from binding of ligands to cell-surface receptors leads to the output of transcriptional regulation of specific genes.

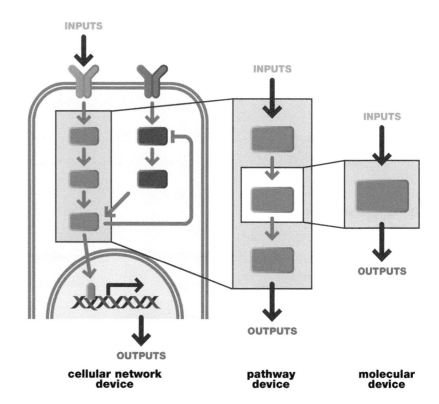

cellular network device · **pathway device** · **molecular device**

In summary, cells use a range of scales and different types of solutions to solve information-processing tasks. In some cases, a signal-processing device can comprise just a single signaling protein, whereas in other cases, a larger network of interacting molecules is involved. We will explore how particular tasks can be accomplished by devices that operate both at the molecular scale and at the network scale. In the case of networks, many of the systems we will examine also incorporate transcriptional elements, which lead to changes in the abundance of different proteins. This is because at a systems biology level, it is often this whole ensemble of integrated molecular elements that yields a particular functional cellular behavior.

Signaling devices face a variety of challenges in input detection

If one considers the multitude of different stimuli that bombard a cell, it is natural to wonder how the cell is able to use this information to respond appropriately. To accomplish this, cellular devices, no matter what specific physiological role they play, must perform a range of common tasks. At the most basic level, input stimuli must be detected and transformed into a specific response output. More specifically, often the amplitude (strength) of the input must be accurately measured, as well as how the input changes over time (**Figure 11.3**). In many cases, a cellular

Figure 11.3

Signaling systems respond to different types of input changes. Signaling systems respond to the changes in the amplitude of inputs, which may be sustained or transient. Most signaling systems must monitor and respond to changes in the amplitude of many different inputs simultaneously.

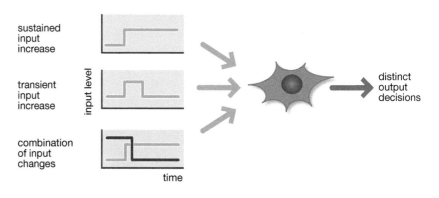

sustained input increase

transient input increase

combination of input changes

input level

time

distinct output decisions

device must measure multiple inputs and make a coordinated response that integrates these multiple signals. While the exact molecules that receive and transmit information will be different for a particular response in a particular cell, one often uncovers similar solutions to accomplish these common tasks.

Despite these commonalities, all signaling systems must be highly adapted to their specific roles in the cell. Depending on their function, for example, many signaling systems have evolved to respond only to a specific type of input stimulation, effectively ignoring the many other possible inputs vying for attention. Some signaling systems show responses that are proportional to the input level, whereas others only become activated above a given threshold. Some signaling systems respond immediately upon stimulation, whereas others require sustained input before switching on—if, for example, the response is very costly and the cell needs to ensure that it does not trigger the output in response to random fluctuations in stimuli. Throughout this chapter, we consider some of the ways that signaling systems have evolved to address each of these challenges.

Proteins can function as simple signaling devices

How are signaling molecules built to function as input/output devices? Throughout the previous chapters in this book, we have reviewed various mechanisms by which a molecule can change its functional output based on input stimuli (**Figure 11.4**). Many signaling proteins act as a simple *allosteric switch*, whereby the active conformation of the protein is stabilized in the presence of a stimulatory covalent modification (for example, phosphorylation) or ligand binding (**Figure 11.4a**). For example, the kinase domain of the insulin receptor is activated by phosphorylation of its activation loop, which in turn leads to repositioning of the activation loop to allow binding of peptide substrate. Allosteric control mechanisms like this are common to many protein kinases.

Regulation of protein kinases by activation-loop phosphorylation is discussed in Chapter 3.

Some proteins achieve input/output control by acting as a modular allosteric switch. In these types of proteins, the catalytic domain of the protein itself may not function as an allosteric switch. However, other regulatory domains (usually protein–protein interaction domains) can provide switching function by participating in autoinhibitory interactions with the catalytic domain (**Figure 11.4b**). For example, in the tyrosine phosphatase Shp2, the protein tyrosine phosphatase (PTP) catalytic domain is by itself constitutively active. However, the intact protein contains two SH2 domains. In the folded tertiary structure, the N-terminal SH2 domain binds to the PTP domain, blocking its active site—the SH2 domain autoinhibits the phosphatase activity. This multidomain protein then acts as an input/output device because binding of tyrosine-phosphorylated ligands to the Shp2 SH2 domains leads to the release of the SH2–PTP domain autoinhibitory interaction, resulting in activation of the phosphatase activity.

Multiple polypeptides can also interact to form a protein-complex allosteric switch. For example, protein kinase A (PKA) has a catalytic kinase domain and a regulatory domain (**Figure 11.4c**; see also Figure 6.7). The regulatory domain binds to the catalytic domain, blocking its substrate-binding site with a pseudo-substrate sequence, thus inhibiting its activity. This complex now acts as a switch that can be activated by the input of the small signaling mediator, cAMP. cAMP specifically binds to the regulatory subunit, which causes the subunit to release the catalytic domain, which is now active. This mechanism is not that different from the autoinhibitory mechanism described above, except that regulation is achieved by domains on a separate polypeptide chain.

In addition to changes in conformation or interactions, another simple and common way to change protein activity in response to inputs is to change its subcellular localization (**Figure 11.4d**). For example, the transcription factor Pho4 is normally localized in the cytoplasm, where it is inactive because it cannot interact

Regulated changes in subcellular localization are discussed in Chapter 5.

with the promoters that it can regulate. However, signal input (dephosphorylation of Pho4) leads to its import into the nucleus, where it can now exert its output activity on transcription. Similarly, phosphorylation of FOXO transcription factors by kinases such as Akt leads to their export from the nucleus and sequestration in the cytosol, leading to transcriptional changes.

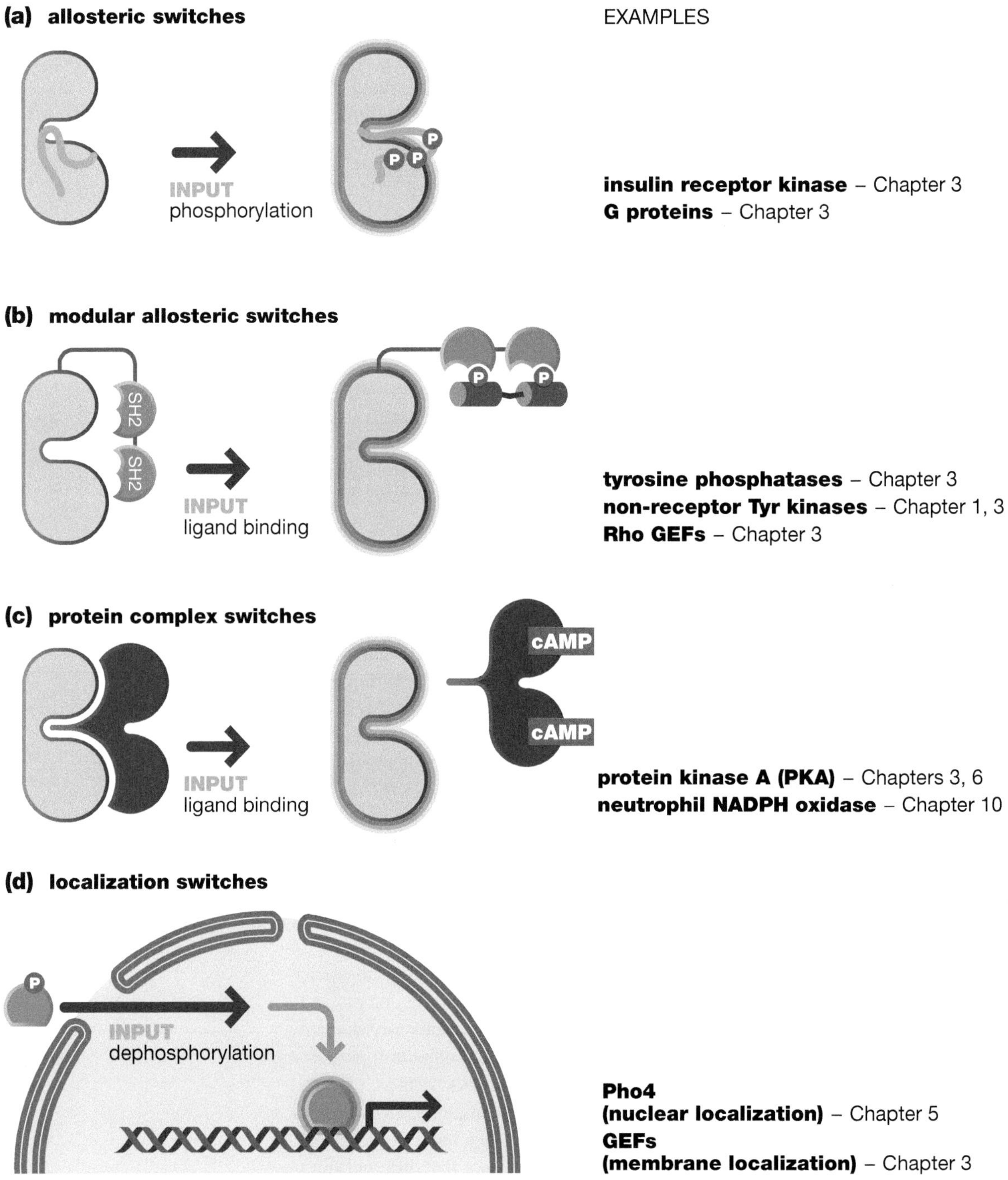

(a) allosteric switches

EXAMPLES

INPUT
phosphorylation

insulin receptor kinase – Chapter 3
G proteins – Chapter 3

(b) modular allosteric switches

SH2
SH2

INPUT
ligand binding

tyrosine phosphatases – Chapter 3
non-receptor Tyr kinases – Chapter 1, 3
Rho GEFs – Chapter 3

(c) protein complex switches

INPUT
ligand binding

cAMP

cAMP

protein kinase A (PKA) – Chapters 3, 6
neutrophil NADPH oxidase – Chapter 10

(d) localization switches

P

INPUT
dephosphorylation

XXXXXXXXXXXXXX

**Pho4
(nuclear localization)** – Chapter 5
**GEFs
(membrane localization)** – Chapter 3

Figure 11.4

Examples of molecular input/output devices. One example is illustrated for each type; other examples, and chapter numbers where examples are discussed in more detail, are listed on the right. (a) Allosteric switch proteins are activated by conformational changes in response to inputs such as post-translational modification or ligand binding. (b) Modular allosteric switches are regulated by intramolecular interactions involving modular binding domains. (c) Protein-complex switches are regulated by interactions between different protein subunits. (d) Localization switches are driven by changes in subcellular localization.

INTEGRATING MULTIPLE SIGNALING INPUTS

In both cellular and electronic signaling, many devices are more complex than those described above. For example, cellular responses are highly dependent on the integration of multiple inputs. In responding to inputs, the cell's ultimate output must take account of a wide range of conditions—not just a single signal, but often multiple external signals or stresses, as well as internal states (such as the energy status of the cell or what stage of the cell cycle it is in). It would be hard for a cell to function if it did not have complex signaling systems capable of monitoring and integrating multiple signals. Thus, it is not surprising that the majority of signaling proteins and networks function as multi-input state machines. They are capable of responding to many different inputs, some of which act in a positive manner (activate output), and some of which act in a negative manner (repress output) (**Figure 11.5**).

Cells use many different strategies to build signal-integrating devices. In this section, we explore several examples of signaling devices that can integrate multiple signals.

Logic gates process information from multiple inputs

To understand how information from multiple inputs can be processed, it is helpful to draw parallels between cellular signaling systems and human-made digital control systems, which depend on **logic gates**. Logic gates specify outputs depending on the combination of two inputs; they come in a variety of types, such as AND, OR, NOR, or XOR gates, depending on their input/output behavior (**Figure 11.6**). For example, in a simple two-input AND gate, the output is switched on only when both inputs 1 and 2 are present, and not in the presence of either input 1 or input 2 alone. In a NOR gate, output is switched on only in the absence of both inputs 1 and 2. Either input alone (or together) is sufficient to switch output off. When these types of digital control gates are linked together, the resulting circuits can be capable of highly complex responses—the microprocessor chips in computers and other electronic devices, for example, can contain millions of logic gates.

Many signaling proteins or networks show behavior analogous to logic gates. For example, some proteins will be strongly activated only in the presence of two different inputs and weakly activated in the presence of either individual input—such proteins are analogous to an AND gate. Of course, most signaling proteins do

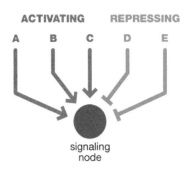

Figure 11.5

Most signaling proteins integrate many different inputs. These include combinations of activating and repressing signals.

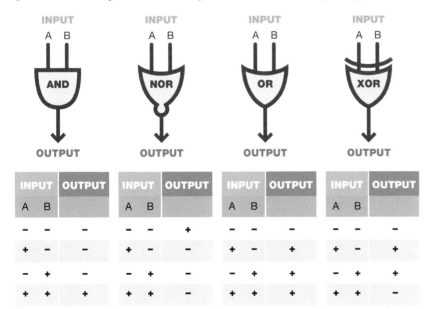

INPUT		OUTPUT
A	B	
–	–	–
+	–	–
–	+	–
+	+	+

INPUT		OUTPUT
A	B	
–	–	+
+	–	–
–	+	–
+	+	–

INPUT		OUTPUT
A	B	
–	–	–
+	–	+
–	+	+
+	+	+

INPUT		OUTPUT
A	B	
–	–	–
+	–	+
–	+	+
+	+	–

Figure 11.6

Signal integration by multi-input logic gates. Examples are shown of various types of digital two-input gates. For each type of gate, the standard symbol used to represent the gate in schematic drawings is indicated above, and how output depends on each input (A and B) is shown below.

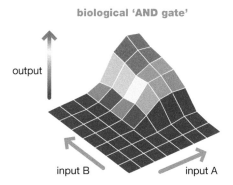

biological 'AND gate'

output

input B   input A

Figure 11.7

A biological AND gate. A biological signaling device, although not truly digital in response, approximates an AND gate if high output activity is only obtained when two inputs (A and B) are both present at high levels. Note that there is no output activity when either of the two inputs is absent.

 The consequences of multiple protein modifications are discussed in more detail in Chapter 4.

not show the absolute on–off properties of digital systems (**Figure 11.7**), because their physiological inputs (the concentrations of proteins, lipids, and other biomolecules, and the activity of enzymes) usually vary continuously over a range of values, as opposed to being either present or absent. Despite this, logic gates provide a useful analogy for information processing by signaling proteins and networks, which is critical for precise cellular decision-making.

Simple peptide motifs can integrate multiple post-translational modification inputs

One simple but effective way for signaling proteins to integrate multiple inputs is to have multiple post-translational modifications within the same peptide. The functional output of having a single modification versus two modifications could be very different, especially if this peptide serves as a modification-dependent binding site for downstream effectors. The combination of multiple possible modifications greatly expands the possible states of a protein, providing the means for more sophisticated regulation.

An example of antagonistic modifications occurs at the N-terminal tail of histone H3 (**Figure 11.8**). When Lys9 is methylated by histone methyl transferases, the modified site acts as a signal to recruit the chromodomain of the protein HP1, which acts to silence nearby transcription by altering chromatin structure. However, during mitosis, the aurora-B kinase can phosphorylate the nearby Ser10 residue of histone H3. This modification blocks HP1 binding, even if Lys9 is methylated. Thus, this simple histone-tail peptide can act as a sophisticated signal-integration point—its recruitment of the regulatory protein HP1 is activated by methylation but then can be overridden by phosphorylation.

This is only one example of how the interplay between post-translational modifications can be used to integrate signals. Such relationships can be antagonistic, as described in the example above, or cooperative or sequential (that is, modification A is a priming modification which is required for modification B to occur). For example, the protein kinase GSK-3 can only efficiently phosphorylate substrates that have been previously phosphorylated by a second kinase, such as CK1 or CK2.

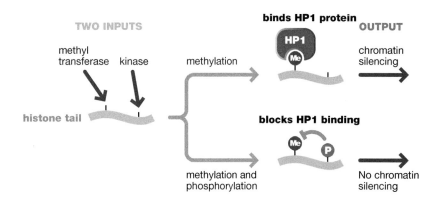

TWO INPUTS

methyl transferase   kinase   methylation

binds HP1 protein   **OUTPUT**

HP1 / Me

chromatin silencing

histone tail

blocks HP1 binding

methylation and phosphorylation

No chromatin silencing

INPUT		OUTPUT
methylation	phosphorylation	silencing
−	−	−
+	−	+
−	+	−
+	+	−

Figure 11.8

Peptide motifs can integrate multiple post-translational modifications. The tail of histone H3 can be trimethylated on Lys9 by a histone methyl transferase, which leads to recognition by the HP1 protein. In turn, HP1 binding leads to chromatin remodeling and silencing. However, phosphorylation of Ser10 in the histone H3 peptide by the aurora-B kinase blocks HP1 recognition. This behavior is analogous to another type of two-input logic gate, an AND NOT gate.

Cyclin-dependent kinase is an allosteric signal-integrating device

An allosteric kinase such as a cyclin-dependent kinase (CDK) responds to multiple inputs, including post-translational modifications and ligand binding (**Figure 11.9**). These input-integration properties are critical for the central role of CDKs in controlling progression in the cell cycle—many distinct inputs must be weighed before CDK activation can proceed. More specifically, to be activated, Cdk2 must both be phosphorylated on the activation loop (Thr160 in human Cdk2) by the CAK complex, and also bind to a cyclin partner. These two positive inputs are both required for full CDK activity, and thus mimic AND-gate control. At a mechanistic level, these two inputs work together to properly position the C-helix and activation loop of CDK, to allow for catalytic activity (see Figure 3.14).

The activity of CDK can be overridden by several negative inputs, however. These inputs act like safety mechanisms to prevent activation before the appropriate time in the cell cycle, or to halt cell-cycle transitions when checkpoints have indicated that something is amiss and must be corrected. Phosphorylation at two other sites within the CDK (residues Thr14 and Tyr15 in Cdk2) by the kinase Wee1 and its relative Myt1 shuts off the kinase activity. In addition, binding of CDK inhibitor proteins, such as p27, to the CDK–cyclin complex can also shut off kinase activity by causing large conformational changes in the kinase domain, thus blocking ATP binding, and also by occupying the substrate-docking-site groove on the cyclin subunit. Thus, the negative phosphorylation input and negative regulator input have a NOR-gate relationship—either is sufficient to prevent activation of the CDK–cyclin complex.

Modular signaling proteins can integrate multiple inputs

Modular proteins can also serve as signal integrators. An example is the signaling protein WASP and its homologs such as N-WASP, which are key regulators of the actin cytoskeleton (**Figure 11.10**). N-WASP has a catalytic domain that can interact with and activate the actin-related protein 2/3 (Arp2/3) complex—a seven-protein machine that nucleates new actin filaments that grow as branches from existing filaments. Activation of N-WASP (and subsequently Arp2/3) must occur with high spatial and temporal precision in order to yield specific actin-driven morphological changes, such as cell movement, endocytosis, and phagocytosis.

 Cell-cycle control is explored in Chapter 14.

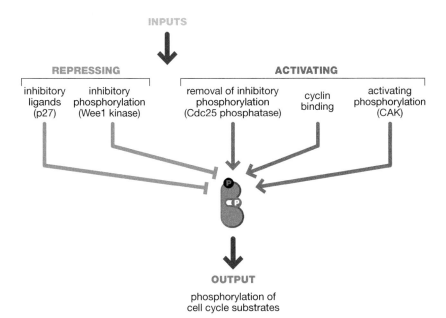

Figure 11.9

Cyclin-dependent kinase (CDK) acts to integrate a wide variety of both activating and repressing inputs.
Activating inputs include binding of the cyclin subunit, phosphorylation by CAK, and dephosphorylation by Cdc25. Repressing inputs include phosphorylation by Wee1 and binding of inhibitory ligands such as p27.

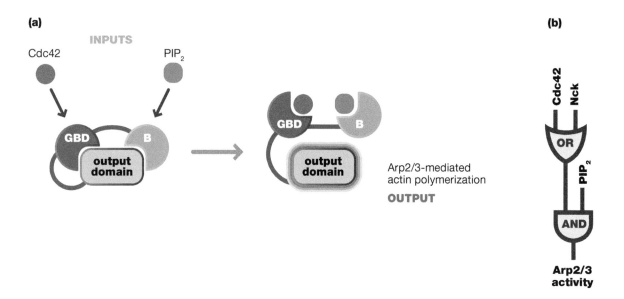

(a)

INPUTS

Cdc42   PIP$_2$

GBD   B

output domain

GBD   B

output domain

Arp2/3-mediated actin polymerization

OUTPUT

(b)

Cdc42   Nck

OR

PIP$_2$

AND

Arp2/3 activity

Figure 11.10

Modular signaling proteins act as switches that integrate multiple inputs.
(a) Two domains of the actin regulator N-WASP are involved in autoinhibition of the output domain. Relief of autoinhibition requires the cooperative binding of two inputs: GTP-bound Cdc42 to the G protein binding domain (GBD), and the phospholipid phosphatidylinositol 4,5-bisphosphate (PIP$_2$) to the basic domain (B). Each input alone is a poor activator. (b) The adaptor Nck can substitute for Cdc42. N-WASP thus behaves like a more complex input/output device containing both AND and OR gates.

Not surprisingly, N-WASP is regulated by multiple inputs. These inputs include proteins, such as the active (GTP-bound) form of the small G protein Cdc42, and lipids, such as phosphatidylinositol 4,5-bisphosphate (PIP$_2$). N-WASP is normally only strongly activated when both Cdc42 and PIP$_2$ are present: thus, the two inputs have an AND relationship. This dual control is thought to target actin polymerization precisely to sites on the membrane that contain both Cdc42 and PIP$_2$.

This AND-gate control is only observed in the intact N-WASP protein. As with many other examples, N-WASP uses an autoinhibitory mechanism to achieve this complex input control. N-WASP has multiple modular domains, including a small G protein binding domain that can bind to activated Cdc42, and a basic domain that can bind to PIP$_2$. In the intact protein in its basal state, these two domains participate in autoinhibitory interactions that repress the ability of the N-WASP catalytic domain to stimulate actin nucleation. These autoinhibitory interactions are cooperative; the complex is tightly held in the OFF state, and it is very difficult for either single input to release the autoinhibited state. However, if both Cdc42 and PIP$_2$ are present, they can cooperate to release autoinhibition, resulting in much higher activity when both ligands are present.

Examined more closely, regulation of the activity of WASP family proteins is even more complex. For example, SH3 domains from proteins such as the Nck adaptor can bind to sites within a proline-rich region of N-WASP and stimulate its catalytic activity. Nck seems to be able to substitute for Cdc42; either of these two inputs, together with PIP$_2$, promotes activation. This three-input system contains both OR and AND gates: the output is dependent on (Cdc42 OR Nck) AND PIP$_2$ (**Figure 11.10b**). Furthermore, N-WASP can also be regulated by tyrosine phosphorylation and other inputs. We can see that, much as in the case of complex electronic circuits, increasingly complex biological behaviors can be derived by linking together simple input/output relationships.

This type of general structure—a switch composed of a modular output catalytic domain combined with autoinhibitory input domains—is highly flexible and is observed in many different signaling proteins. Such a modular allosteric framework can, in principle, lead to integration of many different inputs and many different types of integrating relationships. The evolutionary flexibility of this system has been demonstrated in the laboratory by swapping input domains in N-WASP, resulting in novel synthetic proteins that respond to different input stimuli specific for the new autoinhibitory domains.

Transcriptional promoters can integrate input from multiple signaling pathways

T cell activation is described in more detail in Chapter 14.

Processing of multiple inputs can also be achieved at a network level by transcriptional promoters that integrate the activities of multiple transcription factors. For example, when T cells are activated by antigen-presenting cells, one of the important outputs is activation of the interleukin-2 (IL-2) promoter and the subsequent production and secretion of the cytokine IL-2. IL-2 promotes survival and proliferation of the activated T cell. The IL-2 promoter serves as a critical integration point, acting as an AND gate that detects two distinct branches of the T-cell-activation pathway (**Figure 11.11**).

One of the pathways activated upon T-cell-receptor stimulation is the Erk MAP kinase (MAPK) pathway. Activated Erk phosphorylates and activates the transcription factor AP-1. However, this factor alone is not sufficient to activate transcription from the IL-2 promoter. T-cell-receptor activation also activates a second pathway branch involving Ca^{2+} signaling. Release of Ca^{2+} from intracellular stores activates the phosphatase calcineurin, which dephosphorylates the transcription factor NFAT, allowing it to be transported into the nucleus. The IL-2 promoter has binding sites for both AP-1 and NFAT, and only when these factors bind simultaneously is transcription activated. The binding of the two factors is cooperative: the complex binds DNA much more stably than either AP-1 or NFAT alone. Thus, the IL-2 promoter functions as an AND gate that confirms that both the MAPK and Ca^{2+} branches of the T-cell-signaling network have been activated, something that will only occur with robust and sustained T cell activation.

RESPONDING TO THE STRENGTH OR DURATION OF AN INPUT

In addition to simply detecting the presence of an input, it can be very important for a cell to measure its amplitude (strength) or duration. In this section, we will look at different ways that signal output can be related to the strength or timing of input signals. We will also introduce ways of describing how different components of a signaling system (usually different signaling proteins) interact with each other and modify each other's behaviors to process information. The specific nature and wiring of these interrelationships—the **network architecture**—can provide the basis for surprisingly sophisticated and complex signaling devices (see **Box 11.1** for a primer on network architecture).

Throughout this chapter, we describe examples of network architectures that are capable of performing specific information-processing tasks. However, it is important to note that there is not a simple one-to-one relationship between network architecture and network function— a given architecture might actually perform a number of different processing functions, depending largely on the exact

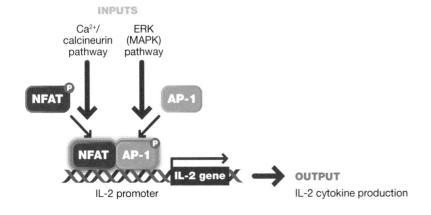

Figure 11.11

Promoters can integrate the input of multiple signaling pathways. In T cell signaling, expression of the cytokine IL-2 requires input from two pathways. The Ca^{2+}/calcineurin pathway leads to dephosphorylation and thus activation of the transcription factor NFAT. The Erk MAP kinase (MAPK) pathway leads to phosphorylation and thus activation of the transcription factor AP-1. Expression from the IL-2 promoter only occurs when NFAT and AP-1 interact synergistically. Thus, the IL-2 promoter functions like an AND gate.

Box 11.1 Signaling network architectures

At the simplest level, **nodes** in a signaling network represent individual signaling components, such as individual proteins, while **links** are the regulatory relationships between these components. For example, a node could represent an enzyme like a kinase or phosphatase, while the link from this node to another node (a substrate) represents the reaction in which the enzyme modifies the substrate (**Figure 11.12**).

Figure 11.12

Network representation of regulatory connections. (a) Links between nodes can be either positive or negative. (b) A series of two negative links can be represented by a single positive link.

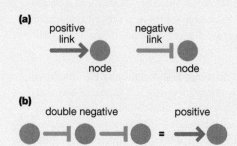

Links between nodes can be either positive or negative, depending on whether the action of the upstream node on the downstream node results in activation or repression. Often a link in a network diagram might actually represent a series of relationships that are mediated by a chain of one or more intermediate links. In these cases, a series of two negative links could also be represented by a single positive link (in other words, the sign of a series of links is the mathematical product of the signs of the individual links).

Aside from forming cascades (that is, a number of nodes linked in series), nodes can be linked in fan-in and fan-out configurations. A fan-in configuration is when one node is controlled by multiple upstream inputs. A fan-out configuration is when one node controls multiple output nodes (**Figure 11.13**).

Figure 11.13

Fan-in and fan-out network architectures.

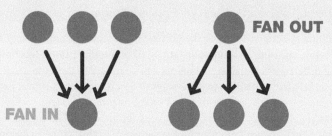

As described in the main text, *feedback* and *feed-forward* network architectures are responsible for many of the complex biological behaviors we see in signaling. **Feedback** is observed when the output from a given node follows a path of links that returns to regulate the node of origin. Feedback loops can be either negative or positive (**Figure 11.14**).

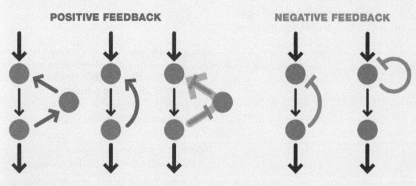

Figure 11.14

Examples of positive and negative feedback loops.

(Continued)

Box 11.1 Signaling network architectures (*Continued*)

Feed-forward is defined as a situation in which distinct paths that fan out from an upstream node reconverge on another downstream node (**Figure 11.15**). A feed-forward loop is considered **coherent** if the two divergent branches have the same overall sign, whereas it is considered **incoherent** if the two divergent branches have opposite overall signs (positive and negative).

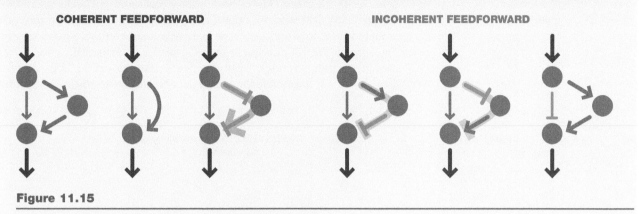

COHERENT FEEDFORWARD   **INCOHERENT FEEDFORWARD**

Figure 11.15

Examples of coherent and incoherent feed-forward loops.

parameters that define the network links. One can only state that a particular architecture type can perform a function or is often associated with that function. In **Table 11.1**, we list several commonly observed cellular network architectures and the range of behaviors with which they are often associated.

Signaling systems can respond to signal amplitude in a graded or a digital manner

Cells can respond to changes in input levels in different ways. In some cases, the situation may require what is referred to as a **linear** or **graded response**: a response in which the system detects a range of input levels and produces an output that is proportional to the input level. For example, in a stress-response pathway or a hormone-detection pathway, it may be useful for the cell to tune the degree of its response to match the level of the stress or hormone. Alternatively, in other physiological situations, a **switchlike** or **digital response** may be needed: the cell must ignore inputs below a certain threshold and only respond to strong signals above this threshold. Such systems are often described as having nonlinear or all-or-none behavior. An all-or-none response system might be advantageous in development, for example, where a cell only adopts a specific developmental fate when a morphogen input is present above a specific threshold.

Table 11.1

Network architectures and their associated behaviors

Network architecture module	Uses in cell signaling
Cascade	Amplification
	Switchlike activation
Negative feedback	Limit output level
	Increase precision of output level
	Adaptation
	Oscillation (nonrobust)
Positive feedback	Amplification
	Switchlike activation
	Bistability, memory
Interlinked positive/negative feedback	Robust, tunable frequency oscillations
Interlinked, multiple positive feedback	Increased synchronization, precision of switching
Feed-forward, coherent	Delay/filter for sustained input
Feed-forward, incoherent	Pulse response, adaptation

What determines whether a cell signaling system will respond in a graded manner or a digital manner? First, it is useful to consider the input/output behavior of an enzyme that is, for example, regulated by the binding of a single activating ligand (**Figure 11.16a**). Here, we consider the concentration of ligand as the input and the fractional activity of the enzyme as the output. In this case, activation will follow a hyperbolic curve (identical to the binding curve or *isotherm* for the binding reaction—see Figure 2.7). At low to medium input concentrations, output activity will be approximately linear with respect to input, although as the system approaches input ligand saturation, the output activity will level off to the maximal level. Thus, one can see that such simple signaling devices are intrinsically linear in their response, at least when significantly below their saturation point. In the next two sections, we will consider different ways of generating more switchlike responses;

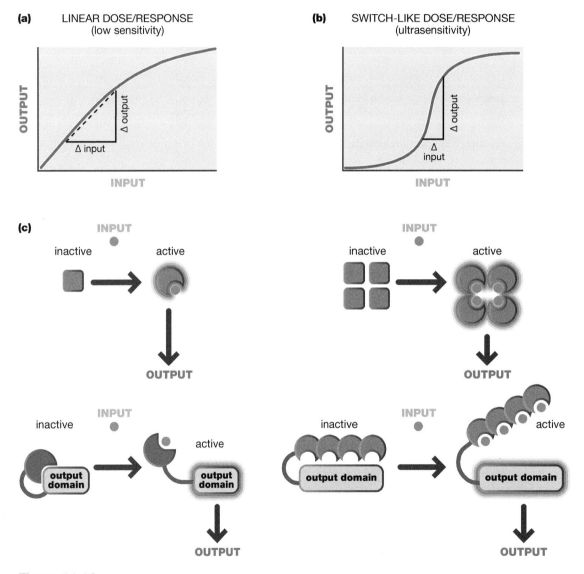

Figure 11.16

Linear versus switchlike activation curves. (a) A simple binding-driven activation process will yield a hyperbolic dose–response curve which, in the early part of the activation curve, is approximately linear. The sensitivity (change in output divided by change in input) of this type of response is relatively low. (b) More complex signaling systems can show a more switchlike response, in which the dose–response curve is sigmoidal. Within a narrow range of input values, switchlike devices have a high response sensitivity (ultrasensitivity). (c) Individual signaling molecules can show ultrasensitive responses if they are activated by input in a cooperative manner. This includes multisubunit allosteric enzymes (top) or enzymes regulated by multiple autoinhibitory domains (bottom). Cooperative switches are illustrated on the right; the corresponding noncooperative enzymes are illustrated on the left for comparison.

it is important to remember that there are often diverse molecular approaches for achieving a particular type of signaling behavior.

An enzyme can behave as a switch through cooperativity

It is possible for an enzyme to show switchlike behavior if it is activated in a highly cooperative manner (**Figure 11.16b** and **11.16c**). For example, if the enzyme (or enzyme complex) has multiple binding sites for the activating ligand, and if binding of ligand to one site allosterically increases the favorability of binding at the other sites (that is, ligand binding is *cooperative*), then this enzyme will be activated in a highly switchlike manner. The enzyme response curve (with increasing ligand) will be sigmoidal; that is, below a threshold point, adding input will lead to a sublinear increase in output. However, near and at the threshold point, adding input will lead to a very steep nonlinear increase in output level. Output will level off again, however, as ligand binding nears saturation and the enzyme approaches maximal activity. Such a response is said to be **ultrasensitive**: in some portion of the signal–response curve, a relatively small change in the level of input leads to a much larger than proportional change in the response (**Figure 11.17**).

 Cooperative binding is discussed in more detail in Chapter 2.

An example of a cooperative enzyme switch is protein kinase A (PKA), which was mentioned earlier in this chapter (see Figure 11.4c). The regulatory subunit of PKA (which binds and inhibits the catalytic subunit) has two cAMP-binding sites. Binding of cAMP to these sites is cooperative and leads to cooperative release and activation of the catalytic subunit. This mechanism results in a sigmoidal, switchlike activation of PKA in response to increasing cAMP concentration. This effect is amplified by the fact that inactive PKA is a heterotetramer (two catalytic and two regulatory subunits, with a total of four cAMP-binding sites), increasing the cooperativity of cAMP binding.

Networks can also yield switchlike activation

There are also network solutions to generating a switchlike detection system (**Figure 11.18**). For example, specific types of signaling cascades can lead to ultrasensitive, switchlike responses. A simple cascade of enzymes that successively activate one another will generally produce a linear input/output response (assuming all enzymes are operating below saturation). In some cases, however, cascades that involve multisite phosphorylation can show much more switchlike input/output responses (**Figure 11.18a**). Multisite phosphorylation may be *distributive* (one enzyme–substrate encounter leads only to one phosphorylation event, so two encounters are needed for multiple phosphorylation) or *processive* (one enzyme–substrate encounter leads to multiple phosphorylation events). Increasing the number of distributive events, as well as increasing the number of cascade steps, can lead to a sharper, more switchlike input/output transition of the cascade as a whole. For each step, the input is the concentration of the upstream activating kinase. In the case of a distributive multisite phosphorylation event, the effect of an increased amount of activating kinase is multiplied, because it contributes to each of the individual stepwise phosphorylation reactions (this assumes, as is the case in actual cells, that phosphorylation can be reversed at a constant rate by the action of phosphatases). This leads to a more sigmoidal activation curve. This is quite analogous to the way that increasing ligand concentration will contribute to each step of binding in an allosteric enzyme with multiple ligand-binding sites leading to cooperative activation.

 Processive and distributive phosphorylation are discussed in Chapter 4.

This type of distributive kinase cascade is observed to contribute to switchlike behavior of the MAPK cascade in *Xenopus* oocytes. However, it is also important to point out that not all MAPK cascades show switchlike activation, despite the fact that all such pathways involve three two-site phosphorylation steps. Other MAPK cascades, such as the yeast mating-response pathway, show a primarily

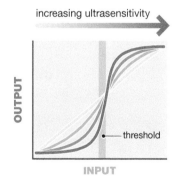

Figure 11.17

Graded (linear) versus ultrasensitive (nonlinear) responses. Increasing ultrasensitivity corresponds to increasingly sigmoidal activation curves. This sharper transition can approximate a threshold point—a level of input that is required for all-or-none activation.

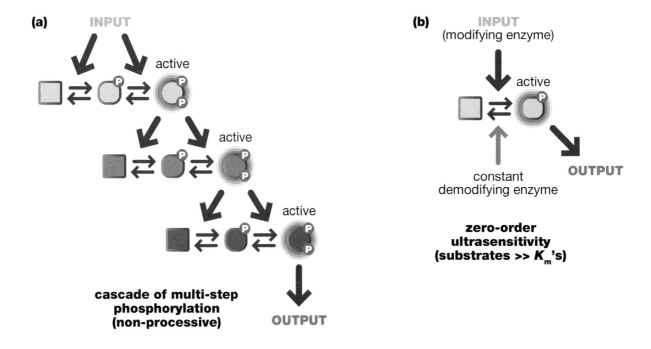

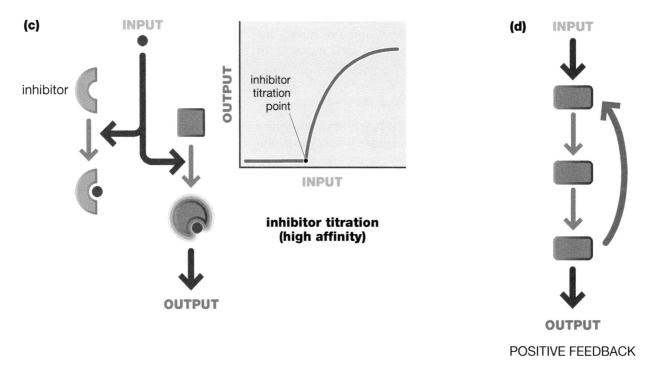

Figure 11.18

Network mechanisms that can yield ultrasensitive (nonlinear) input/output responses. A multistep activation reaction, and the organization of several of these reactions into a cascade, can lead to ultrasensitive input/output behavior for the intact cascade. (b) Zero-order ultrasensitivity. If a protein is subject to a reversible enzymatic modification (such as phosphorylation), then a change in the amount or activity of modifying enzyme can lead to an ultrasensitive change in the level of modified target protein. This behavior requires that the target protein concentrations be higher than the K_m of each of the two upstream enzymes (that is, both reactions are zero-order and thus insensitive to substrate concentration). (c) Inhibitor titration can also generate nonlinear activation. Increasing input will have no effect on output until the inhibitor is titrated out, whereupon there will be a sharp change in output. (d) A positive feedback loop can also increase ultrasensitivity.

linear response. In these cases, it is thought that factors such as scaffold proteins might increase the processivity of this process (making each step less distributive).

Switchlike activation can also occur with what is termed **zero-order ultrasensitivity** (**Figure 11.18b**). Reversible modifications typically use different enzymes for the forward and reverse reactions—in the case of phosphorylation, kinases and phosphatases. Under some situations, an increase in the amount or activity of the forward modifying enzyme will lead to an ultrasensitive increase in the amount of the modified target protein. For this to occur, the target protein concentrations need to be higher than the K_m of the two modifying enzymes (that is, they are operating at zero-order saturation kinetics—when the enzymes are fully saturated with substrate, so reaction velocity is at V_{max} and independent of substrate concentration). Over time, the slight increase in modifying-enzyme activity exerts a disproportionate effect on the level of the modified target because the demodifying enzyme is already operating at full capacity and therefore cannot counteract the increase.

Nonlinear activation can also be simply generated if there is a high-affinity inhibitor of the input. Increasing input will have virtually no effect on output until the inhibitor is titrated out, whereupon there will be a relatively sharp change in output (**Figure 11.18c**).

Another common and powerful solution for achieving switchlike activation is through a network that has strong **positive feedback**. Let us consider a cascade in which the output of the cascade acts to further activate the cascade itself (**Figure 11.18d** and Box 11.1). This is often the case when the system involves a protein kinase, where one activated kinase molecule is able to phosphorylate and activate multiple additional kinase molecules (remember that most kinases are activated by phosphorylation of activation-loop residues). Positive feedback networks of this type are often observed to display extremely sharp, switchlike input/output behavior. At low input levels, the pathway is poorly activated. However, near the threshold point, the pathway is sufficiently activated that its output begins to generate positive feedback, which in turn acts to further provide input to the pathway. In this way, the pathway becomes explosively activated in a sharp, all-or-none manner. Positive feedback of this type can also lead to other interesting behaviors, which we will discuss below. Note that a constant, low level of an opposing activity (for example, a phosphatase) is required to prevent the spontaneous, uncontrolled activation of such a system at low stimulus levels.

Signaling systems can distinguish between transient and sustained input

In some cases, it is very important for a cell signaling system to measure the duration of an input. For example, it may be advantageous to avoid turning on a costly response program in response to transient, noisy inputs, and instead only respond to sustained stimulation. This type of response is somewhat similar to the behavior of an automatic door that is programmed to stay open until a sustained period has passed during which no motion is detected, to avoid accidentally closing the door on a person. The ability to detect the duration or periodicity of an input signal (the input dynamics) provides an additional channel that can be used to convey information in cell signaling systems, a process termed **dynamic encoding**. How might a signaling system be programmed to distinguish between a sustained and a transient input?

One solution to this problem is a network with a coherent feed-forward architecture (Box 11.1 and **Figure 11.19a**). In such a network, the output from an upstream node fans out to two different pathway branches. In a coherent feed-forward architecture, these two branches reconverge to regulate a downstream node in the same direction or sign (thus the two branches have a "coherent" effect on the downstream node).

A detector for a sustained input can be built from a coherent feed-forward network with two features: first, the speeds at which the signal travels down the two divergent branches of the feed-forward network must differ; and second, the downstream converging node must act as an AND gate that is only activated when both branches of the feed-forward architecture are transmitting positive signals. The resulting system will generate output only when stimulated by an input whose duration is longer than the difference between the time scales of the two pathway branches. If stimulated by a more transient input, the fast and slow branches will be activated at different times, and the convergent AND-gate node will not be activated. Only with stimulation that is more sustained than the time difference between the two branches will there be dual input activation of the convergent AND-gate node (**Figure 11.19b**).

This type of coherent feed-forward architecture appears to be used to allow mammalian cells to detect the differences between sustained and transient activation of the Erk MAPK (**Figure 11.19c**). Some of the responses downstream of Erk are mediated by the transcription factor Fos, which serves the role of an AND gate. This transcription factor requires two inputs: first, its protein expression must be induced; and second, because it is inherently proteolytically unstable, its stability must be enhanced. Both of these inputs are mediated by activated Erk, but with different time scales. Direct phosphorylation of the Fos transcription factor by Erk leads to protection against proteolysis; this represents a fast, direct link. Expression of Fos, however, is mediated by a cascade of events downstream of Erk activation (Erk phosphorylates intermediate transcription factors that increase transcription of the *Fos* gene, leading to increased Fos protein production). This expression path represents a slow, indirect link between Erk and Fos. Thus, only when there is a sufficiently sustained pulse of Erk activation will the Fos transcription factor be both highly expressed and stabilized, ultimately resulting in expression of Fos-activated genes.

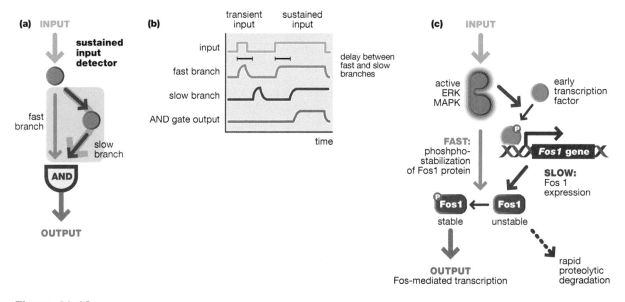

Figure 11.19

A network that responds only to sustained inputs: coherent feed-forward loop. (a) A network architecture that can distinguish sustained inputs from transient inputs is a coherent feed-forward loop. Critical features are a time delay between a fast branch (*blue*) and a slow branch (*purple*), and a fan-in node that acts as an AND gate to integrate the signals from the two branches. (b) Temporal profile of input, fast branch, slow branch, and output of this network. A transient input pulse (shorter than the delay between the two branches) results in no output, since the signals from the two branches do not arrive at the terminal AND gate at the same time. A sustained input (longer than the delay between branches) results in output. In this case, both signals arrive at the AND gate simultaneously. (c) A signaling network with this architecture is used to detect sustained Erk MAP kinase (MAPK) activity. Erk has two effects on the transcription factor Fos1. First, Erk induces expression of Fos1. This branch is slow, since it is dependent on transcription and translation. In the fast branch, Erk directly phosphorylates Fos1 and stabilizes the protein. Because of the time delay in the transcriptional branch, only sustained Erk activity will lead to stable accumulation of Fos1, and to Fos-mediated transcriptional output.

MODIFYING THE STRENGTH OR DURATION OF OUTPUT

It is critical for a signaling system to control the amplitude and duration of its output. Some responses in cells are transient, lasting only as long as the stimulus, or in some cases even less (such as in the case of *adaptation*, discussed below). Other signaling responses can last for very long time periods, even after the input stimulus is gone, thus representing a type of cellular memory (long-term changes in output after transient input). Here, we describe several common molecular and network mechanisms for tuning output and generating specific classes of behavior.

Signaling pathways often amplify signals as they are transmitted

A common challenge facing signaling systems is that an initiating stimulus might only be detected by a small number of receptor molecules, yet the signal needs to be propagated in a manner that leads to an output of altered cellular behavior involving huge numbers of downstream molecules. **Amplification** of signals as they are transmitted along a pathway yields high final output levels from relatively low input levels. One fundamental mechanism for signaling systems to amplify output is through the use of enzymes—one activated enzyme can act on many substrate molecules. For example, activation of one kinase molecule can lead to the phosphorylation of manyfold larger numbers of substrate molecules. Similarly, one activated molecule of adenylyl cyclase can generate manyfold higher equivalents of cAMP, which can then propagate signals to many more downstream effector molecules. Indeed, small signaling mediators such as cAMP often function to amplify the signal from a relatively small number of activated receptors to generate a massive and widespread response throughout the cell.

In principle, even higher levels of signal amplification will occur if multiple amplifying enzymes are linked into a cascade, such as a MAPK cascade, with each step potentially multiplying the degree of amplification (**Figure 11.20**). Proteases can also be linked in cascades, as in the activation of thrombin in blood clotting, or caspase activation in apoptosis. However, this type of exponential amplification is not always observed in endogenous cascades. First, significant amplification at every step of a cascade requires that there be a higher concentration of each successive enzyme in the cascade. This is not always the case, and the degree of amplification can be limited by saturating activation of the available downstream enzyme. Second, the organization of cascade components into multiprotein signaling complexes by scaffold proteins can cause some of the individual cascade steps to no longer serve as independent reactions, thereby dampening amplification. Thus, although most pathways amplify their signals at some point, the degree of amplification observed for each individual step can vary significantly.

Negative feedback allows fine-tuning of output

Signaling pathways are not always designed to maximally amplify their output; instead, it is often more important to control output levels to make them more precise. Relatively little is known about the ways in which the precision of pathway output levels is controlled, but evidence points to an important contribution of **negative feedback** loops. Negative feedback loops can dampen signaling through a pathway, leading to a steady-state output that is lower than would be expected without the negative feedback. Such negative feedback loops are also thought to increase the precision of output levels by counteracting fluctuations in input level.

For example, in pathways that utilize Ca^{2+} as a diffusible signaling mediator, upstream receptor activation often leads to an influx of Ca^{2+} into the cytosol. The maximal increase in cytosolic Ca^{2+} concentration, however, is limited by a negative

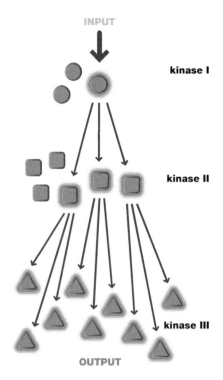

INPUT

kinase I

kinase II

kinase III

OUTPUT

Figure 11.20

Amplification in signaling cascades.
Input can be amplified by a cascade of enzymes, such as kinases or proteases, that activate one another. Amplification occurs because one activated enzyme molecule can activate many downstream substrate molecules. Cascades only result in amplification under conditions in which the enzymes are freely diffusible and each downstream enzyme is present at higher concentration than its upstream activator.

 Protease cascades are discussed in Chapter 9.

feedback loop: when cytosolic Ca^{2+} exceeds $0.6\,\mu M$, Ca^{2+} uptake into the mitochondria is induced. This negative feedback loop therefore acts to limit maximal output of the system (**Figure 11.21a**). Without such a mechanism, Ca^{2+} levels might easily rise to toxic levels.

An elegant and clear example of how negative feedback loops can tune the precision of output was observed in a synthetic gene expression network (**Figure 11.21b**). When an externally inducible promoter was used to express green fluorescent protein (GFP), cell-to-cell variability in GFP levels was high, due to random events such as differences in the copy number of the plasmid containing the GFP gene. However, the addition of a negative feedback loop, in which the inducible promoter not only expressed the GFP reporter but also a transcriptional repressor of that promoter, resulted in significant reduction of cell-to-cell variability. This is because the inhibitor sets up a kind of steady state or set point, where the activity of the inducer is balanced by the activity of the repressor. Whenever expression level rises beyond this set point, the negative feedback loop provides a self-correcting

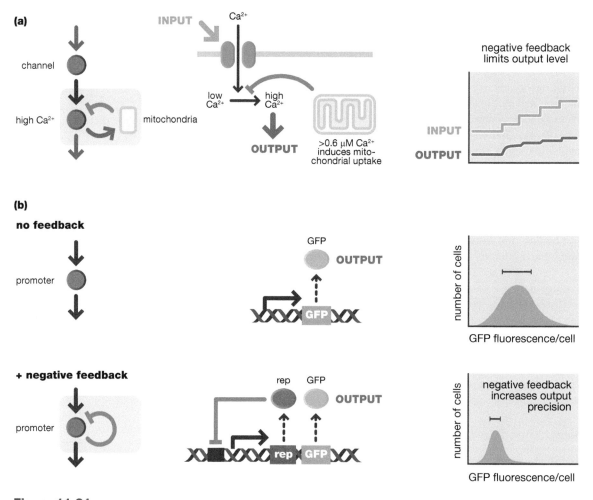

Figure 11.21

Negative feedback control of output level and precision. (a) Example of a negative feedback loop that limits output level. Stimuli that open Ca^{2+} channels result in increased concentrations of cytoplasmic Ca^{2+}. However, Ca^{2+} concentrations of $>0.6\,\mu M$ induce mitochondrial uptake of Ca^{2+}, thereby acting as a negative feedback loop to limit cytoplasmic Ca^{2+} despite increasing input stimulation. (b) Negative feedback can also increase output precision. One example involves two synthetic gene expression circuits, one with a simple promoter, and another with a repressor (rep)-controlled negative feedback loop (negative autoregulation). On the right, the distribution of fluorescence per cell is depicted. Expression of a green fluorescent protein (GFP) reporter is more precise in the presence of the negative feedback circuit—that is, there is less cell-to-cell variation in GFP fluorescence. Standard deviation from the average is represented by black bars.

mechanism that represses transcription in proportion to the deviation; similarly, when transcription goes down, repression will be relieved and transcription will tend to revert to the set point. Presumably, similar negative feedback architectures are used in natural signaling pathways to increase the precision of output and dampen the effects of random events, or "noise."

Adaptation allows cells to control output duration

It can also be critical for signaling systems to control the duration of their output. Many sensory systems therefore display a property known as **adaptation**—upon stimulation with a sustained input, the system initially responds with a burst of output, after which output levels return to their basal levels, even if the input stimulus remains at a high level (Figure 11.22). An adaptive response can be used to shut off output when it is harmful or toxic in excess, or costly in some other way for the organism. Perhaps more importantly, an adaptive system couples response to relative changes in the input, not to its absolute level, thus allowing the system to respond to a much wider range of input levels. By resetting output to its original basal level, adaptation allows the system to respond to further increases in input (as opposed to becoming quickly saturated, as occurs with nonadaptive sensing systems). The ability of adaptation to increase the dynamic range of input sensing is critical for many sensory systems, such as in our own vision, allowing our eyes to reset and operate under dramatically different ambient conditions, such as a dark room or bright daylight.

Theoretical analyses suggest that there are two basic ways to achieve adaptation (Figure 11.22b). The first is with a negative feedback loop mediated by a

(a)

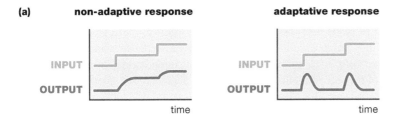

(b) negative feedback with buffering adaptive node

incoherent feedforward with proportioner adaptive node

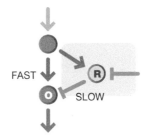

turns down upstream nodes by integrating system output

turns down output in proportion to system input

Figure 11.22

Adaptation circuits limit duration of output and increase dynamic range of sensory systems. (a) In a normal, nonadaptive response, output level increases with input and remains high during sustained input. The dynamic range of sensing is limited as the system reaches saturation. By contrast, in an adaptive system, input-induced increase in output level is transient—the system automatically resets to basal steady state, even with sustained input. (b) Two basic network architectures that can achieve adaptation. First, negative feedback loops (left) can allow a regulatory node (R) to monitor and integrate activity of the output node (O), and to down-regulate upstream or output nodes to return to basal steady state. The negative feedback loop must be slow enough that there is a transient delay before return to steady state. Second (right), a regulatory node (R) in an incoherent feed-forward loop can sense input levels and proportionally down-regulate output (O) to return to the basal steady state. Again, the negative branch of the feed-forward loop must be relatively slow to allow for a transient output response. Note that the specific architectures depicted are examples, and that other specific architectures are possible for each class of adaptation circuit.

regulatory node that acts to buffer output. In this case, the system sums over time (integrates) how much output is being transmitted through the pathway, then inhibits upstream signaling in a manner that returns it to a constant steady state. To achieve this, the two enzymes that act on the regulatory node (the output activity, and a basal enzyme that opposes it) must operate under saturating conditions (substrate concentration $>K_m$) such that regulatory-node activity only reflects the change in output enzyme activity and is independent of substrate concentration. As with the coherent feed-forward examples above, the forward signaling loop needs to be faster than the negative feedback loop to generate a transient response (deviation from steady state) followed by adaptation.

A second theoretical way to achieve adaptation is through an incoherent feed-forward loop (Figure 11.22b) in which a receptor node transmits a positive signal to both the output node and a regulatory node, while the regulatory node, in turn, acts to negatively regulate the output node in a feed-forward loop. Here, the regulatory node acts as a proportioner—it will negatively regulate the output node in proportion to the strength of the input signal that it receives, thus resulting in adjustment of the system back to a constant steady-state output. Once again, differences in the kinetics of the two branches of the feed-forward loop allow for transient output responses that deviate from steady state.

One of the best biochemically characterized examples of adaptation in cell signaling is found in bacterial **chemotaxis** (**Figure 11.23**). Bacteria can sense a gradient of chemorepellent molecules and swim away from the source. To do this, the bacteria perform a biased random walk—they switch between a straight swimming phase and a tumbling phase in which they reorient direction. When moving to a region of higher chemorepellent concentration, the bacteria show an increased probability of tumbling, allowing them to sample new directions; when moving away from chemorepellents, however, the probability of tumbling is lower, allowing them to swim with higher persistence. In essence, the cells sample in time whether they are moving to higher or lower chemorepellent concentrations, and adjust their behavior accordingly. Mechanically, tumbling is caused by clockwise rotation of the cell's flagella, which forces the flagella to act independently, while swimming is caused by counterclockwise rotation of the flagella, which then intertwine (because of the handedness of the helical flagella) and act in concert to propel the cell forward. In a controlled setting, if a bacterium is exposed to a step increase in chemorepellent, a transient increase in tumbling frequency is observed, followed by a rapid (within seconds) return to a basal tumbling frequency. This adaptation is the fundamental basis of the biased random walk.

How does the bacterial signaling system yield this critical adaptive behavior? The biochemical circuit for chemotaxis is shown in **Figure 11.23d** and **e**, and has the architecture of a negative feedback loop with adaptive node, as described above (see Figure 11.22b). Chemorepellent is detected by a chemotaxis receptor, which in turn activates the histidine kinase CheA, which can then activate the response regulator CheY via phosphorylation. Active (phosphorylated) CheY binds to the flagellar motor to cause clockwise rotation and tumbling. Thus, the increase in active CheY after stimulation explains the resulting increase in tumbling frequency. However, the bacteria's receptors quickly undergo adaptation, returning to basal output levels even at a constant high level of input. Adaptation is mediated by a second set of reactions—methylation of the receptor on several aspartate residues, which increases the output activity of the receptors. Receptor methylation acts as the "regulatory" node that buffers system output. Methylation is controlled by two opposing enzymes: the methylase, CheR, has constant basal activity, while the demethylase, CheB, requires activation by CheA. Thus, CheA, which is activated by the receptor, has two activities: it will activate CheY, leading to tumbling, and it will activate CheB, which leads to receptor demethylation and down-regulation. This system will return the level of active CheY (and tumbling frequency) to a constant steady state but only after a transient deviation caused by the slower kinetics of the CheB-mediated negative feedback loop.

Bacterial two-component signaling systems of this type are discussed in Chapter 4.

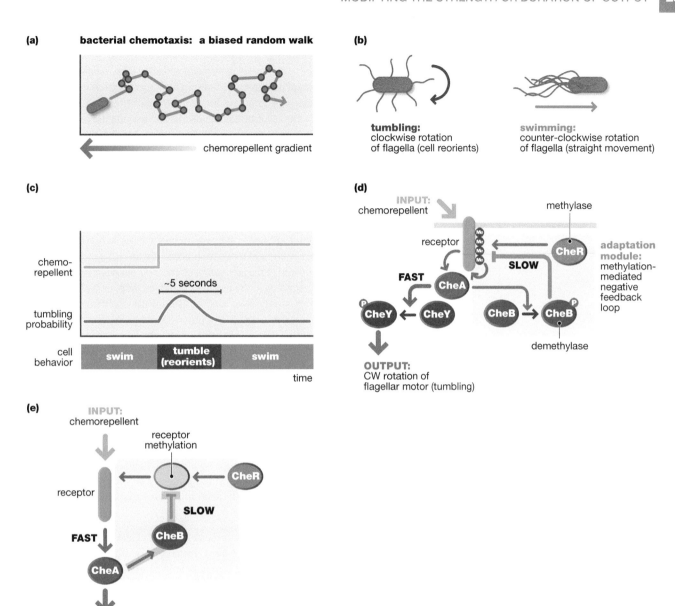

Figure 11.23

Bacterial chemotaxis is controlled by a negative feedback adaptation system. (a) Bacteria avoid chemorepellents by a biased random walk. The cell alternates between a straight swimming phase and a tumbling phase (*purple circles*) in which it reorients before swimming in a new direction. If the bacteria swim to a higher chemorepellent concentration, the probability of tumbling increases. (b) Tumbling occurs when the bacterial flagella rotate clockwise. Swimming occurs when the flagella rotate counterclockwise because of the natural helicity of the flagella, in this direction they intertwine and act in concert to propel the bacterium. (c) A stepwise increase in chemorepellent results in a transient increase in the probability of tumbling, followed by a return to the steady-state basal probability. Thus, increasing chemorepellent always induces a transient increase in tumbling, regardless of starting chemorepellent levels. (d) A negative feedback circuit underlies the adaptive behavior. Chemorepellent activates a chemotaxis receptor, which, in turn, activates the protein CheA. CheA activates two targets. First it rapidly phosphorylates CheY, which then binds to the flagellar motor to induce clockwise (CW) rotation (tumbling). However, CheA also activates CheB, a demethylase that inhibits the receptor. A counteracting enzyme, CheR, promotes receptor methylation and activation. Thus, the methylation system forms a negative feedback loop that can restore output to low levels after a delay. (e) The chemorepellent sensing network in (d) is presented schematically so the "negative feedback with buffering adaptive node" architecture can be more readily appreciated (compare with Figure 11.22b).

Feedback can cause output levels to oscillate between two stable states

Another important temporal response observed in biological regulatory systems is **oscillation**, when output levels fluctuate between states of high and low activity in a periodic manner. Examples of biological oscillation include the cell cycle and circadian rhythms (daily cycles, for example, in sleep or activity, that can persist even in the absence of external light/dark cues). Biological oscillators have been studied by numerous approaches, including the construction of synthetic oscillator systems. We will focus on several of these synthetic systems because of the key principles that they demonstrate.

Negative feedback is a core requirement for oscillations, but a simple negative feedback loop without sufficient steps or delays cannot achieve oscillation (**Figure 11.24a**). Such a system will move monotonically toward a reduced, single steady-state output (compared to the equivalent circuit without negative feedback). A circuit with at least three steps in the negative feedback loop, or with an explicit delay, can destabilize this steady state and yield minimal oscillators. However, such oscillators have been constructed in synthetic biology experiments and are found to depend on highly precise parameters, and often show damped oscillations. Biological systems, however, typically require oscillators that are highly **robust**—that is, they maintain their performance under a wide range of conditions and are relatively insensitive to perturbations of the system. More recent studies have shown that overlaying positive feedback on these simpler circuits can result in far more robust oscillators, with consistent amplitudes. Positive feedback, or any type of ultrasensitivity acting on the nodes that form the core negative feedback loop, acts to prevent an approach to a single steady state and makes the system more **bistable**—able to exist in two distinct stable output states, without a stable intermediate steady state (bistability is further discussed below). In principle, any of the mechanisms of ultrasensitivity outlined in Figures 11.16 and 11.18 can be used to increase oscillator performance.

A natural example of a robust oscillator is the cell cycle in cleavage-stage embryos of the frog *Xenopus laevis* (**Figure 11.24b**). Immediately after fertilization, the large amphibian egg undergoes repeated rounds of synchronous DNA replication and mitosis until ~1000 cells are generated; remarkably, this cycle can be reproduced *in vitro* using cell-free *Xenopus* egg extracts. At the core of this cycle is a negative feedback loop in which the active cyclin–CDK complex phosphorylates the *anaphase-promoting complex* (*APC*), promoting the binding of its Cdc20 subunit (see Figure 9.12). The APC^{Cdc20} complex then polyubiquitylates cyclin and targets it for destruction, thus turning the system off after a delay. However, overlaid on this system are two positive feedback loops that make the cyclin–CDK node act in a more ultrasensitive manner. Both of these act through a distinct inhibitory phosphorylation site on CDK (remember CDK is a complex multi-input node; see Figure 11.09). First, the Wee1 kinase that phosphorylates and inhibits CDK–cyclin is itself phosphorylated by CDK–cyclin and thereby inactivated and targeted for degradation (a double negative loop). Second, the Cdc25 phosphatase that dephosphorylates and activates CDK–cyclin is itself activated by phosphorylation by CDK–cyclin. These two positive feedback loops are thought to enhance the performance of this critical cell-cycle oscillator, resulting in consistent amplitude oscillations. Cell-cycle oscillations are most apparent in rapidly dividing embryos; in most other cells, additional signaling inputs are required before a new cell cycle is initiated.

Cell-cycle regulation is discussed in more detail in Chapters 9 and 14.

(a)

negative feedback

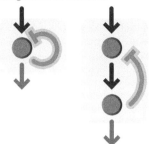

no oscillation
- system reaches lower
stable steady state

negative feedback + delay

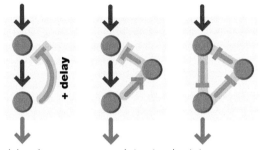

+ delay

delay dampens approach to steady state

poor, damped
oscillations

negative feedback + delay + positive feedback

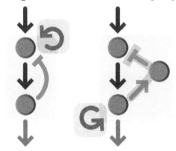

more robust oscillations
with consistent amplitude

non-linearity of postive feedback
destabilizes approach to steady state

(b)

example of natural robust oscillator:
Xenopus oocyte cell cycle

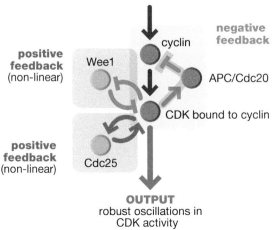

positive
feedback
(non-linear)

Wee1

cyclin

negative
feedback

APC/Cdc20

CDK bound to cyclin

positive
feedback
(non-linear)

Cdc25

OUTPUT
robust oscillations in
CDK activity

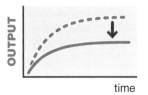

Figure 11.24

Robust oscillators require a core negative feedback loop with delay and nonlinear (ultrasensitive) nodes. (a) A negative feedback loop is required for oscillation, but on its own leads to monotonic approach to a reduced steady-state output, compared to a circuit with no negative feedback (dotted line). These minimal circuits cannot yield oscillation. A three-component negative feedback loop, or a two-component negative feedback loop with an intrinsic delay, can yield oscillations, but these are often damped and have inconsistent amplitudes. Addition of positive feedback can yield robust oscillations with consistent amplitudes. Positive feedback causes nodes to respond in a nonlinear (ultrasensitive) manner, which systematically destabilizes approach to a single steady state. (b) The *Xenopus* embryo cell cycle is an example of a robust oscillator constructed from interlinked positive and negative feedback loops. The core negative feedback loop involves activation of the APC^Cdc20 complex by CDK–cyclin, which leads to destruction of cyclin. There are two positive feedback loops, which revolve around the inhibitory phosphorylation site in CDK. APC, anaphase-promoting complex; CDK, cyclin-dependent kinase.

Bistable responses also underlie more permanent outputs

In simple signaling systems, changes in output are transient, persisting only as long as the input perturbation is maintained. When the input perturbation is removed, the system returns to its basal level—behavior akin to a buzzer that only stays on as long as the user pushes the input button. In some cases, however, signaling systems respond to inputs with a stable change in output that persists beyond the duration of a transient input. This behavior is analogous to a toggle switch, such as those commonly used to turn on a light—the light stays on after the switch is flipped, in the absence of any further input. This type of sustained change in the system after transient stimulation plays a critical role in complex behaviors such as development, learning, and the immune response. How can this form of **molecular memory**—the conversion of a transient input into a permanent (or semipermanent) change in output—be achieved?

A sufficiently strong positive feedback can lead to a system that shows bistability. In these cases, the state of the system depends on its history and starting conditions. These systems often show **hysteresis**; that is, the input level at which the system switches between the two output states will be different if one is moving up from low input to high input, or moving down from high input to low input (**Figure 11.25a**).

A hysteretic system could display memory by locking into a high-output state upon a transient increase in input (**Figure 11.25b**). If the basal input level is above the point required to reset the system, then the output will remain high even after a return to the basal input level. The memory can only be reset (reverting the system to the low-output state) by lowering input levels past the transition point for decreasing input. An extreme form of hysteresis is an irreversible system, in which no decrease in input level is sufficient to restore the system to the low-input state (**Figure 11.25c**).

The power of this positive feedback architecture has been demonstrated in several synthetic biology experiments, in which transcriptional or signaling circuits have been engineered to show this type of "lock-on" memory (**Figure 11.25d**). In these cases, activation of the system above an ON input threshold can stably toggle the system to a new activated state. A second distinct OFF input trigger must be induced to return the system to the inactive output state. Most permanent cellular changes, such as those that occur during development, are thought to involve this kind of strong positive feedback, lock-on circuitry.

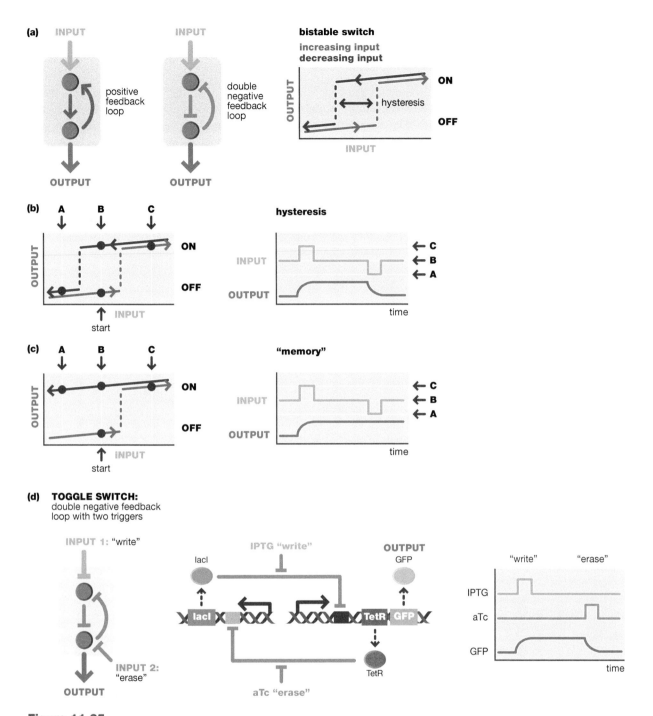

Figure 11.25

Circuits capable of memory: a sustained change in output level after a transient change in input. (a) Strong positive (or double negative) feedback systems can lead to bistability, in which the system can only exist in two distinct stable states (low and high output; no intermediate states). Bistable systems often show hysteresis—the transition point between low- and high-output states will differ with increasing versus decreasing input *(dark brown arrow)*. (b) A hysteretic system could display memory by locking into a high-output state upon a transient increase in input (labeled c). If the basal input level (B) is above the point required to reset the system, then the output will remain high even after a return to the basal input level. The system can be reset by lowering input levels past the decreasing input transition point (A). (c) In an irreversible system, no decrease in input level is sufficient to restore the system to the low-input state. (d) Synthetic toggle switches with memory can be constructed using this type of network design. One example involves a double negative feedback loop, in which the Tet repressor (TetR) inhibits expression of the lac repressor (lacI), and the lac repressor inhibits expression of both the Tet repressor and a green fluorescent protein (GFP) reporter gene. The system can exist in a high lac repressor (and GFP) state or a high Tet repressor state. The system can be induced to switch between states by adding small molecules that block the activity of one of the repressors (IPTG blocks lacI; aTc blocks TetR).

SUMMARY

Cell signaling systems must be able to take in information and adjust their output states accordingly. Individual signaling molecules can themselves function as complex gates and switches that process information. These proteins can also be hierarchically organized into networks that can perform higher-order information processing. Signaling molecules and pathways can integrate information from multiple inputs, and can exhibit different linear versus nonlinear input/output responses. These systems can also be organized to build networks that can monitor input amplitude and duration. Signaling networks can also be used to precisely control output amplitude and duration, resulting in complex dynamic behaviors such as adaptation, oscillation, and memory. Current work suggests that common network architectures are used to achieve particular functional signaling behaviors, even if the precise molecular implementation is different for each case.

Answers to these questions can be found online at www.routledge.com/9780367279370

QUESTIONS

1. Describe the difference between feedback and feed-forward loops. What different classes of feedback loops are possible? What different classes of feed-forward loops are possible?

2. When is it useful for signaling systems to behave as an OR rather than an AND gate?

3. Describe different strategies by which the modular arrangement of domains or motifs within signaling proteins can facilitate the integration of two separate inputs to make a coordinated decision.

4. When is it advantageous for a cell to respond to input in a graded fashion, as opposed to a switchlike (or ultrasensitive) fashion? What are the mechanisms by which an individual signaling molecule can yield an ultrasensitive response? What are the mechanisms by which a signaling pathway or network can yield an ultrasensitive response?

5. In what contexts is it useful for a cell to distinguish between transient and sustained input? What are the signaling network mechanisms that do this?

6. When is it useful for a cell to display adaptation after sensing an input? What are the general molecular strategies for achieving precise adaptation?

7. What is a bistable system? What kinds of physiological cellular responses require bistable behavior, and why? For what physiological functions might a bistable response be nonoptimal?

BIBLIOGRAPHY

SIGNALING SYSTEMS AS INFORMATION-PROCESSING DEVICES

Alon U (2019) *An Introduction to Systems Biology: Design Principles of Biological Circuits* (2nd ed.). Boca Raton, FL: Chapman & Hall/CRC.

Azeloglu EU & Iyengar R (2015) Signaling networks: Information flow, computation, and decision making. *Cold Spring Harb. Perspect. Biol.* 7(4), a005934. doi: 10.1101/cshperspect.a005934.

Ferrell J (2021) *Systems Biology of Cell Signaling: Recurring Themes and Quantitative Models*. Boca Raton, FL: CRC Press.

Lim WA, Lee CM & Tang C (2013) Design principles of regulatory networks: Searching for the molecular algorithms of the cell. *Mol. Cell* 49, 202–212.

INTEGRATING MULTIPLE SIGNALING INPUTS

Hirota T, Lipp JJ, Toh BH & Peters JM (2005) Histone H3 serine 10 phosphorylation by Aurora B causes HP1 dissociation from heterochromatin. *Nature* 438, 1176–1180.

Macián F, López-Rodriguez C & Rao A (2001) Partners in transcription: NFAT and AP-1. *Oncogene* 20, 2476–2489.

Prehoda KE & Lim WA (2002) How signaling proteins integrate multiple inputs: A comparison of N-WASP and Cdk2. *Curr. Opin. Cell Biol.* 14, 149–154.

Prehoda KE, Scott JA, Mullins RD & Lim WA (2000) Integration of multiple signals through cooperative regulation of the N-WASP-Arp2/3 complex. *Science* 290, 801–806.

RESPONDING TO THE STRENGTH OR DURATION OF AN INPUT

Alon U (2007) Network motifs: Theory and experimental approaches. *Nat. Rev. Genet.* 8, 450–461.

Brandman O & Meyer T (2008) Feedback loops shape cellular signals in space and time. *Science* 322, 390–395.

Dueber JE, Mirsky EA & Lim WA (2007) Engineering synthetic signaling proteins with ultrasensitive input/output control. *Nat. Biotechnol.* 25, 660–662.

Ferrell JE Jr (1996) Tripping the switch fantastic: How a protein kinase cascade can convert graded inputs into switch-like outputs. *Trends Biochem. Sci.* 21, 460–466.

Murphy LO, Smith S, Chen RH, et al. (2002) Molecular interpretation of ERK signal duration by immediate early gene products. *Nat. Cell Biol.* 4, 556–564.

Tyson JJ, Chen KC & Novak B (2003) Sniffers, buzzers, toggles and blinkers: Dynamics of regulatory and signaling pathways in the cell. *Curr. Opin. Cell Biol.* 15, 221–231.

Whitty A (2008) Cooperativity and biological complexity. *Nat. Chem. Biol.* 4, 435–439.

MODIFYING THE STRENGTH OR DURATION OF OUTPUT

Barkai N & Leibler S (1997) Robustness in simple biochemical networks. *Nature* 387, 913–917.

Becskei A & Serrano L (2000) Engineering stability in gene networks by autoregulation. *Nature* 405, 590–593.

Elowitz MB & Leibler S (2000) A synthetic oscillatory network of transcriptional regulators. *Nature* 403, 335–338.

Gardner TS, Cantor CR & Collins JJ (2000) Construction of a genetic toggle switch in *Escherichia coli. Nature* 403, 339–342.

Ma W, Trusina A, El-Samad H, et al. (2009) Defining network topologies that can achieve biochemical adaptation. *Cell* 138, 760–773.

Novák B & Tyson JJ (2008) Design principles of biochemical oscillators. *Nat. Rev. Mol. Cell Biol.* 9, 981–991.

Stricker J, Cookson S, Bennett MR, et al. (2008) A fast, robust and tunable synthetic gene oscillator. *Nature* 456, 516–519.

Tigges M, Marquez-Lago TT, Stelling J & Fussenegger M (2009) A tunable synthetic mammalian oscillator. *Nature* 457, 309–312.

Tsai TY, Choi YS, Ma W, et al. (2008) Robust, tunable biological oscillations from interlinked positive and negative feedback loops. *Science* 321, 126–129.

Yi TM, Huang Y, Simon MI & Doyle J (2000) Robust perfect adaptation in bacterial chemotaxis through integral feedback control. *Proc. Natl Acad. Sci. USA.* 97, 4649–4653.

Cell Signaling and Disease

INTRODUCTION

Cell signaling networks allow cells to recognize and respond to changes in their neighbors and surroundings, and are crucial for the normal homeostatic functioning of the body. Thus, not surprisingly, when signaling systems are mutated or malfunction, they can lead to many forms of disease.

Defects in signaling lie at the heart of diverse diseases (**Figure 12.1**). Many forms of *cancer* involve mutations within signaling pathways that control cell growth and proliferation, leading to cells that grow in an uncontrolled way. Many *infectious diseases*, especially those that involve intracellular bacterial or viral pathogens, actively block or hijack the signaling machinery of the cell, redirecting cell behavior in a way that benefits the pathogen. In some cases, pathogens use these strategies to more effectively invade or spread, and in other cases, they use such strategies to evade the host immune system. *Autoimmune disorders* often involve defects in the signaling systems of immune cells, leading to misrecognition and an overresponsive immune set point, in turn resulting in harmful damage to the body's own tissues. *Metabolic diseases*, such as *diabetes*, can involve defects in the extracellular (paracrine and autocrine) and intracellular homeostatic signaling networks that control nutrient and energy usage and storage. General tissue pathologies like *fibrosis*—the growth of excess fibrous connective tissue that can disrupt the function of an organ like the lung or the heart—involve normal healing responses (for example, wound healing and inflammation) gone awry.

It would be impossible to discuss all the ways in which signaling defects participate in disease. Thus, in this chapter we will focus on delving deeper into examples of how signaling defects play a central role in *cancer* and in *microbial pathogenesis*. We will also discuss how our deeper understanding of cell signaling has led to powerful new *therapeutic strategies to treat such diseases*.

DOI: 10.1201/9780429298844-12

Figure 12.1

Defects in cell signaling are associated with diverse diseases. (a) In cancer, cells that normally do not proliferate undergo changes that lead to uncontrolled proliferation. For example, the cells in an epithelial tissue are normally subject to contact inhibition, and limited growth, but cancer-causing mutations in signaling pathways that regulate growth control can lead these cells to dedifferentiate, escape the epithelium, and proliferate uncontrollably. These cancer cells can also metastasize to other sites. (b) Diverse microbes, including both viruses and pathogenic bacteria, interface with host signaling pathways, by producing virulence factors that block or send novel signals in ways that are advantageous to the pathogen. Usually, these virulence factors either make it easier for the pathogen to infect the host, or help it evade immune eradication. For example, viruses, once they enter the cell, can produce factors that modulate normal host signaling proteins. Bacteria can adhere to and send signals that stimulate uptake by the host cells, as well as use syringe-like assemblies to inject a cocktail of virulence factors into the cell. (c) Imbalanced regulation of immune system signaling can lead to autoimmune responses, where the immune system attacks normal host tissues. These responses could be mediated by autoreactive antibodies (produced by B cells) or autoreactive cytotoxic T cells. (d) Defects in metabolic signaling can lead to diseases like diabetes mellitus. Defects in glucose sensing can disrupt proper production of the hormone insulin, while defects in insulin sensing by cells in the body can disrupt glucose uptake from the bloodstream. (e) Tissue pathologies such as fibrosis involve overactive fibroblast signaling. Fibroblast overproliferation and deposition of an excess of extracellular matrix leads to mechanical malfunction of the tissue (for example, lung and heart). Here, pathways that normally contribute to wound healing contribute to pathology.

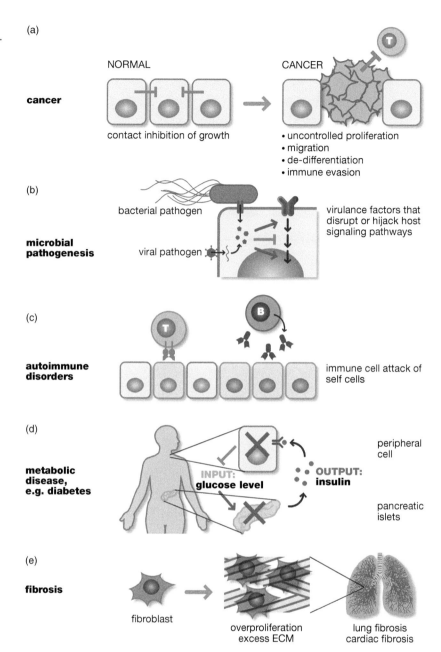

SIGNALING IN CANCER AND CANCER THERAPY

Cancer is perhaps the disease where the link to cell signaling is most obvious. At the most fundamental level, cancer results from failure of normal cell signaling mechanisms. In fact, one could argue that the field of cell signaling grew directly out of cancer research—by studying how signaling is broken in cancer, researchers uncovered the mechanisms and design principles that underlie normal development and maintain homeostasis in the different tissues and organs. In recent years, however, information has begun to flow the other way, as increased knowledge of the signaling mechanisms disrupted in cancer has spurred development of new, rationally designed therapies that specifically target the tumor cells while minimizing side effects.

Malfunctions in signaling contribute to the genesis of cancer

Cancer is a disease in which the cells of the body proliferate in an uncontrolled manner. A key compromise that was essential for the evolution of multicellular metazoan animals was that the individual cells of the body would exhibit highly regulated growth—the vast majority of cells do not proliferate extensively except during development, growth, and repair. Thus, metazoan cells have extensive signaling mechanisms that control cell proliferation in both a positive and negative manner. Proliferation is normally only triggered when explicit mitogenic signals are present. Conversely, there are opposing signaling pathways that induce cell death (apoptosis) or suppress its proliferation, in cases where a cell undergoes severe functional changes or escapes from a tissue.

Moreover, the immune system is programmed, under most circumstances, to be able to recognize and attack rogue mutated or malfunctioning cells. Together, these signaling pathways form a system of positive and negative checks and balances that normally controls proliferation and keeps unregulated cell proliferation in check. Disease states can arise, however, when multiple components of this regulatory network begin to malfunction.

> Mitogenic signaling is discussed in Chapter 14; apoptosis is discussed in Chapter 9.

As our understanding of cancer has matured, it has become clear that despite wide variation between specific cancer types, they share a number of common hallmarks, several of which involve defects in key cell signaling systems (**Figure 12.2**):

1. Cancers often show abnormal ***sustained proliferative signaling***, instead of proliferation that is dependent on the presence of mitogens. Constitutive proliferative signaling (gain of function) is often associated with the expression of *oncogenes*—genes that, when misregulated, can promote proliferation. Hyper-proliferation could be the result of overexpression of mitogenic signaling proteins, or the expression of mutant forms of such proteins with constitutive activity (that is, forms of the signaling proteins that constantly send an ON signal).

2. Cancers often show defects in the natural fail-safe signaling pathways that ***promote apoptosis or suppress proliferation*** in cells that show abnormal properties. When these *tumor suppression* pathways are mutated (loss of function), cells can grow unchecked.

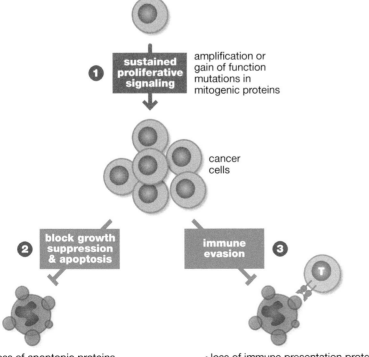

Figure 12.2

Signaling defects that contribute to cancer. A typical cancer cell must have three key features. First, cancer cells must have a gain-of-function mutation in proliferative signaling pathways, leading to sustained and unregulated proliferation. These often involve signaling pathways that mediate mitogenic signaling. Second, cancer cells must have loss-of-function mutations in tumor suppressor genes that normally cause apoptosis in malfunctioning cells. Third, cancer cells must have mechanisms to evade the immune system, such as loss of antigen presentation or expression of checkpoint proteins.

Table 12.1

Signaling protein defects that contribute to cancer

	Class of signaling molecule	Example proteins	Genetic alterations
Signaling proteins	Growth factor receptors (RTKs)	EGFR, ERBB2, MET, PDGFRA, FGFR, KIT, FLT3	*Amplification, constitutive mutations, or fusions*
	Cytoplasmic mitogenic proteins	RAS, BRAF, RAC, MAPK	*Constitutively active mutations*
	Signaling proteins that regulate proliferation	PI3K, AKT, MTOR	*Constitutively active mutations*
	Signaling proteins that suppress proliferation	PTEN	*Deletion or loss-of-function mutation*
Cell cycle regulators	Cell cycle activators	**CCND** (cyclin D1-G1/S cyclin) **E2F**—transcription factors G1/S transition **CDK2/4/6** (cyclin-dependent kinase)	*Amplification*
	Cell cycle inhibitors	**CDKN2**—CDK inhibitor 2 (p16ink4a) **RB**—binds E2F family members, blocks entry into S phase	*Deletion or loss-of-function mutation*
Apoptotic regulators	Apoptotic regulators	**P53**—triggers cell death in response to DNA damage	*Loss-of-function mutation*
	Anti-apoptotic factors	**Bcl2**—blocks apoptosis	*Amplification/overexpression*
Immune alterations	Immune presentation proteins	**MHC molecules** β2-microglobulin **TAP**—transporter of peptides for MHC presentation	*Deletion or loss-of-function mutation*
	Immune checkpoint regulators	**PD-L1**—ligand for PD-1 inhibitory receptor	*Overexpression*

Pink: cancer associated with amplification, gain-of-function mutations.

Blue: cancer associated with deletion, loss-of-function mutations.

bold = gene/protein names; italic = types of genetic alterations.

3. Cancers also often have mechanisms by which to **evade the immune system**. Cancer cells can use a wide range of evasive mechanisms. They can down-regulate presentation of antigen peptides by the major histocompatibility complex (MHC), recruit suppressor immune cells, or secrete suppressive factors.

When these three types of signaling changes occur in combination, this can lead to a cancer. It is only through successive genetic changes that cause changes in these multiple elements that stable tumors can form and become well-established. Mutated genes linked to all three of these cancer-causing mechanisms are summarized in **Table 12.1**.

Mitogenic and growth stimulating signaling proteins can drive cancer

As shown in **Figure 12.3**, many different mitogenic proteins are frequently mutated in cancer. These include *receptor tyrosine kinases (RTKs)* that normally mediate cell proliferation in response to stimulation by external growth factors. When bound to their ligand, RTKs propagate signals to downstream targets via activation of their intracellular kinase domains. Among the most commonly altered RTKs in cancers are members of the Epidermal Growth Factor Receptor (EGFR) family, including EGFR and related molecules such as ERBB2 (also referred to as HER2). It is common to find cancer-associated genetic rearrangements that amplify the expression of EGFR or HER2. In addition, several cancers are associated with mutations that lead to constitutive activation of these receptors. In general, amplification or mutation not only directly increases kinase activity, but can also saturate

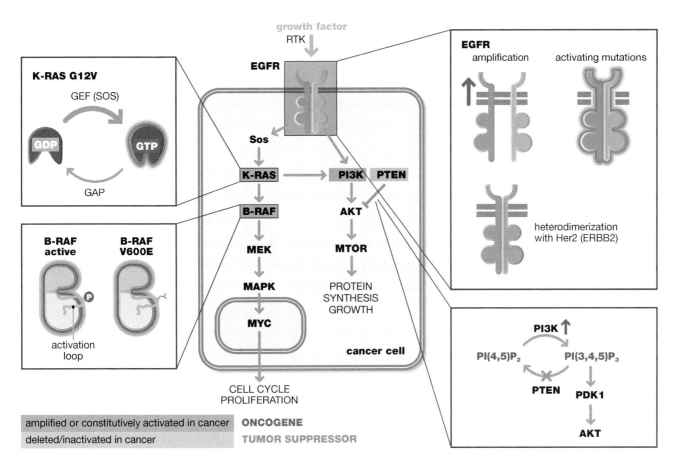

Figure 12.3

Mitogenic signaling proteins mutated in cancer. Many cancers contain mutations that lead to constitutive activation of proliferative signaling pathways. Commonly seen examples include receptor tyrosine kinases such as those of the EGFR family, which can be activated by overexpression or constitutively activating mutations, as well as downstream pathways that activate RAS/MAPK and PI3K/AKT signaling. RAS/MAPK signaling typically stimulates cell proliferation, while PI3K/AKT signaling typically promotes cell growth and survival.

the downregulatory systems that normally limit receptor accumulation (for example, ubiquitylation and degradation), thereby further amplifying growth signaling.

Hyperactive RTKs can also have secondary effects by stimulating the paracrine production of EGF or other mitogens, further strengthening proliferation.

 Receptor downregulation is discussed in Chapter 8.

Frequent activating mutations in the catalytic domain of EGFR include a point mutation in exon 21 (Leu858→Arg), and a deletion in exon 19. One of the most common extracellular domain alleles found in cancer is referred to as EGFRvIII, which is often seen in the brain cancer glioblastoma. EGFRvIII has a deletion of exons 2–7 of the *EGFR* gene; this mutation blocks the receptor from binding ligand, but the receptor now displays low-level constitutive signaling, which is further amplified by reduced internalization and downregulation.

Amplification of the EGFR family member ERBB2/HER2 is also found in many cancers, particularly breast cancer. ERBB2 is constitutively active (it does not require a ligand) and does not undergo efficient degradation, but it typically cannot homodimerize. ERBB2 can, however, strongly heterodimerize with other EGFR family members, potentiating their activity. Thus, ERBB2 is hypothesized to increase signaling amplitude and alter the activation/downregulation dynamics of other EGFR family members. In summary, receptor mutation or amplification can yield cells that proliferate even in the absence of the soluble mitogens normally needed to simulate these receptors.

RTKs activate a host of downstream cytoplasmic signaling proteins that regulate cell proliferation. Mutations of these downstream proteins, such as the small GTPase Ras and the MAPKKK Raf, can also lead to unregulated growth and are commonly seen cancer. The K-RAS gene, in particular, is mutated in a large number of cancers. Residue Glycine 12 is a specific hotspot for activating mutations, including the mutations G12V, G12S, and G12C. When Gly12 is mutated, it impairs the ability of Ras to hydrolyze GTP, even in the presence of a GAP, thus leading to sustained high levels of Ras activity (the GTP-bound state).

G protein signaling is discussed in Chapter 3

In the case of these mutant Ras molecules, the GAP catalytic residue (the "arginine finger" that promotes GTP hydrolysis) cannot sterically fit into the cavity surrounding the target phosphate group in GTP. This kind of steric occlusion of the catalytic Arg occurs when the minimally small glycine residue (which has no side chain) at position 12 is replaced with any larger alternative amino acid.

The most common mutation in the MAPKKK B-RAF associated with human cancers is V600E, a mutation in the kinase activation loop. This mutation in B-RAF is observed in papillary thyroid carcinoma, colorectal cancer, melanoma, and non-small-cell lung cancer. Current models suggest that the negatively charged glutamic acid residue (E) at position 600 acts like a phosphomimetic of the nearby T599 and S602 residues in the activation segment of B-RAF, which are normally phosphorylated upon activation in the wild-type protein. Thus, the mutant protein has higher constitutive kinase activity contributing to oncogenesis.

Regulation of kinase activity by the activation loop is discussed in Chapter 3

Another critical signaling protein involved in regulating cell growth is phosphatidylinositol 3-kinase (PI3K), which generates the phospholipid species phosphatidylinositol (3,4,5) trisphosphate (PIP_3). PIP_3 recruits the kinase AKT to the membrane, whereupon it is activated and participates in activation of the metabolic regulator mTOR.

Activation of AKT is discussed in Chapter 5

Mutations in the gene PIK3CA, which encodes the PI3K catalytic submit (also known as p110α), have been associated with human cancers. Most mutations are concentrated at a few hotspots, show increased kinase activity *in vitro*, and can induce **transformation** of mammalian cells in culture. Transformation is a term used to describe a malignant, cancer-like phenotype in cultured cells; generally this involves less dependence on serum and other mitogens for proliferation than normal cells; changes in cell shape; and the ability to grow under conditions where normal cells typically cannot, such as when they are crowded and touching other cells (contact inhibition), and when they are unattached to a solid substrate. Thus, constitutive activation of PI3K is also linked to cancer.

PTEN (Phosphatase and tensin homolog) is a lipid kinase that opposes the activity of PI3K—it acts as a PIP_3 phosphatase, converting PIP_3 to PIP_2 (phosphatidylinositol (4,5)-bisphosphate). Thus, PTEN counteracts the growth and survival signaling induced by PIP_3. PTEN thereby acts as a potent tumor suppressor, and loss-of-function mutation or deletion of the gene is associated with several cancers. Loss of PTEN likely allows a buildup of PIP_3 in the membrane, stimulating protein synthesis, growth, survival, and oncogenesis.

Cell cycle regulators can drive cancer

A critical step in cell proliferation is progression through the cell cycle, especially undergoing the transition between the G_1 and S phase, which marks the initiation of DNA replication.

Cell cycle regulation is described in Chapter 14.

Most cells in the body of metazoan organisms are not actively proliferating and are arrested in G_1, and mutations of regulators of the G_1/S transition can severely disrupt this normal tissue homeostasis. Key regulators of the G_1/S transition, both positive and negative, are shown in **Figure 12.4**.

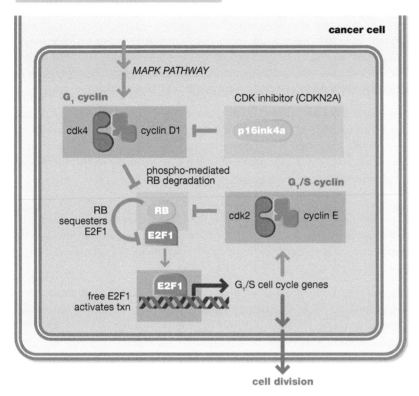

amplified or constitutively activated in cancer — **ONCOGENE**

deleted/inactivated in cancer — **TUMOR SUPPRESSOR**

Figure 12.4

Mutation of cell cycle regulators can drive cancer. Key positive and negative regulators of the G1 to S cell cycle transition are shown. Positive regulators, such as cyclins (D1 and E), CDKs, and the transcriptional regulator E2F1, are amplified or activated in cancers. Negative regulators, such as p16ink4a (CDK4/6 inhibitor) or RB (inhibitor of E2F1), are lost in many cancers.

Both increased expression and activity of positive activators of the cell cycle progression are linked to a number of cancers. Normally, growth factor stimulation leads to the accumulation of cyclins, which bind to and activate cyclin-dependent kinases (CDKs). For example, overexpression of cyclin D1, found in several cancers, can result in dysregulation of CDK activity, and bypass of normal cell cycle checkpoints. Activation of cyclinD/Cdk4 and cyclinD/Cdk6 (the G_1 cyclin/Cdks) is normally what commits the cell to exiting G_1 and initiating a new round of DNA replication. Similarly, the E2F family of transcription factors, which activate expression of multiple positive regulators of the cell cycle including G_1/S and S phase cyclins, are also often amplified in cancers.

Conversely, loss of cell cycle inhibitors is also linked to cancer. For example, a commonly found loss-of-function mutation in cancers affects the p16INK4 protein, a member of the INK4 family of cyclin-dependent kinase inhibitors (Inhibitors of CDK4; the gene encoding p16INK4 is CDKN2A, also known as cyclin-dependent kinase inhibitor 2A). The ubiquitously expressed INK4 proteins bind and inhibit both CDK4 and CDK6, preventing the cell from progressing through START (the point at which a cell is committed to exiting G_1) and initiating replication of the genomic DNA.

Cancers are also linked to loss of function of the Retinoblastoma protein (RB), a key negative regulator of the cell cycle. The RB protein normally binds to and sequesters the E2F transcription factors, such as E2F1, which as mentioned above, are crucial for initiating the G_1/S cell cycle transition. Among other targets, E2F1 stimulates expression of cyclins E and A. When conditions are right to commit the cell to replicate, the activated G_1 cyclin/CDKs phosphorylate RB, which blocks binding to E2F. RB phosphorylation thereby releases E2F1, allowing it to initiate its downstream cell cycle-associated transcriptional program. RB forms part of a larger positive feedback loop promoting the G_1/S transition, as E2F1 promotes further CDK activity and further RB phosphorylation.

Cancers are associated with loss of tumor suppressors—signaling molecules that induce apoptosis or suppress growth

Normally, cells have mechanisms to suppress tumorigenesis. For example, the p53 protein leads to cell cycle inhibition under various conditions of stress, such as DNA damage, and can lead to apoptosis if the damage cannot be repaired or the cause of stress is not alleviated.

 p53 and its regulation are discussed in Chapter 4.

Thus mutation or loss of p53 enables survival of cancers and allows them to accumulate further DNA damage and mutations. Mutations in p53 occur in anywhere from 10% to 100% of tumors, depending on cancer type. Many different P53 mutations are found, but most lead to loss of p53's ability to bind to DNA and activate transcription of its target genes (**Figure 12.5**).

Conversely, overexpression of anti-apoptotic factors, such as the protein Bcl2, can also play a role in cancer. Anti-apoptotic Bcl2 family members bind to and block the pro-apoptotic factors BAX and BAK, factors that act in the mitochondrial membrane to induce release of cytochrome *c* and reactive oxygen species (ROS), leading to apoptosis.

The intrinsic apoptosis signaling pathway is described in Chapter 9.

Bcl2 has long been linked to particular lymphomas, in which a chromosomal translocation placed Bcl2 under the control of the immunoglobulin heavy-chain promoter resulting in constitutively high expression. By blocking apoptosis, pro-survival factors such as Bcl2 increase tumor survival and resistance to chemotherapies and radiation.

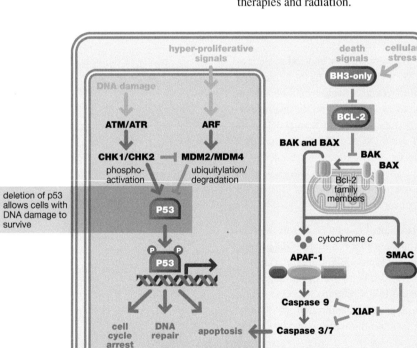

amplified/activated in cancer **ONCOGENE**
deleted/inactivated in cancer **TUMOR SUPPRESSOR**

Figure 12.5

Mutation of growth suppressors can contribute to cancer. Left: P53 is normally inactive, and is rapidly degraded by polyubiquitylation and proteasome-mediated degradation. It can be activated and stabilized by cell stresses such as DNA damage, or by overly strong or discordant proliferative signals (hyperproliferative signals). Once activated, it acts as a transcription factor to induce factors that block the cell cycle and lead to apoptosis. Right: Bcl2 blocks the activity of pro-apoptotic factors such as BAX/BAK in the mitochondrial membrane. Pro-apoptotic signals normally stimulate the intrinsic apoptotic pathway, causing release of cytochrome *c* and other factors from the mitochondria that activate caspases.

Immune evasion by cancers: downregulating antigen presentation

Normally, highly mutated cells such as cancers are recognized and eliminated by the host immune system, because such cells are in usually expected to express **neo-antigens** (immunogenic molecules not found on normal, unmutated cells) on their surface that would be recognized by host T cells as foreign. However, in many cases, cancers have mechanisms for evading immune recognition (**Figure 12.6**).

Some cancers have mutations that disrupt antigen presentation and surveillance. For example, in some cases, the *major histocompatibility complex (MHC)* molecules that present antigen peptides on the cell surface (or associated factors such as β2-microglobulin) are lost in cancers, thus effectively preventing presentation of any cancer neoantigens for T cell surveillance. Similar immune evasion can be achieved by mutations of the transporter associated with antigen processing (TAP1), an ABC transporter that plays a central role in the loading of class I MHC complexes with degraded cytosolic peptides inside the cell before they are presented on the cell surface.

Cancers can avoid immune surveillance by immune checkpoint activation

Another mechanism for cancer immune evasion involves expression of **immune checkpoint proteins**, natural "braking systems" used to dampen T cell immune responses. Activated T cells express the checkpoint receptor, programmed death 1 (PD-1), which provides a mechanism for downregulating T cell activation. More specifically, if a neighboring cell expresses the surface ligand PD-L1, this

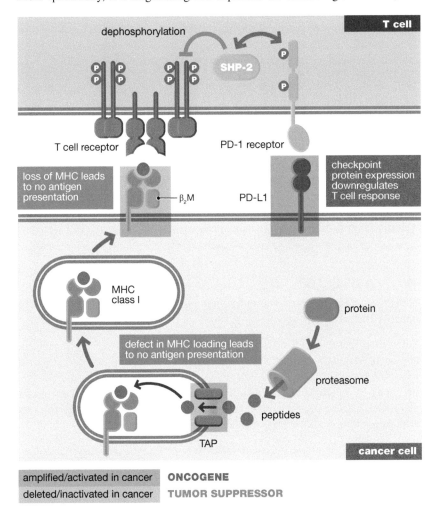

Figure 12.6

Mechanisms by which tumors can evade attack by host T cells. T cells recognize cancer neoepitopes presented on MHC molecules on the cancer cell surface. Some cancer cells evade recognition by downregulating epitope presentation. Mechanisms include mutation of molecules that form the MHC class I complex (such as loss of the β2 microglobulin protein) and mutation of the TAP protein, which helps to load the MHC complex with peptide antigens. A distinct mechanism for evasion is expression of checkpoint activators, like PD-L1, which is a ligand for the suppressive PD-1 receptor in T cells.

protein can bind to and activate PD-1 on the T cell, resulting in downregulation (see Figure 12.6). When activated by PD-L1 binding, PD-1 acts in a downregulatory manner because it contains intracellular immunoreceptor tyrosine-based inhibition motifs (ITIMs). ITIMs are a conserved tandem SH2 binding motif (containing two phosphotyrosine motifs) associated with immune inhibition.

When phosphorylated, the ITIMs bind and recruit the phosphotyrosine phosphatases, Shp1 and Shp2 (via their tandem SH2 motifs). These phosphatases, when recruited to PD-1 at the immune synapse, act to dephosphorylate and inactivate key signaling proteins downstream from the TCR complex (these phosphatases act in opposition to activating protein tyrosine kinases, such as Lck and ZAP70). Importantly, cancers often take advantage of this natural downregulatory mechanism by expressing PD-L1, thereby causing the checkpoint inactivation of attacking T cells.

See Chapter 14 for a discussion of the role of ITIMs in T cell signaling.

Targeted cancer therapies inhibit oncogenic signaling proteins

In the above sections, we have reviewed some of the many signaling proteins that can be mutated to promote cancer. Elucidating how the overexpression or mutation of oncogenic proteins can contribute to cancer has had major impact on the development of new anti-cancer drugs. Until recently, most therapies for cancer were relatively nonspecific, often targeting cells that were rapidly proliferating (through agents that disrupt DNA synthesis, for example). Such relatively nonspecific chemotherapies typically have severe side effects; this is because they kill normal proliferating cells as well, such as those in the hematopoietic system, hair follicles, and the gastrointestinal tract. In the past two decades, tremendous effort has gone into identifying small molecules and antibodies that are targeted to block the activities of nearly all the major classes of proliferative signaling molecules outlined above. **Figure 12.7** shows examples of such **targeted therapies** in clinical use, which target nearly all the major steps in mitogenic signaling.

Many of these inhibitors block protein or lipid kinase activity. One reason for this is that it is relatively straightforward for pharmaceutical companies to set up large-scale screens of libraries of small molecules to find kinase inhibitors. Often the "hits" from such screens are ATP analogues, which bind to the ATP binding site of the kinase and competitively inhibit the binding of its essential substrate, ATP. The first targeted kinase inhibitor, imatinib (commercial name: Gleevec®), was approved in the US by the Food and Drug Administration (FDA) in 2001. Imatinib blocks the Abelson (Abl) tyrosine kinase, as well as the breakpoint cluster region (BCR)-Abl fusion protein, which contains the same Abl kinase domain. The BCR-Abl fusion results in a deregulated oncogenic kinase that is present in all cases of chronic myelogenous leukemia (CML), a relatively common blood cancer.

Oncogenic activation of Abl is discussed in Chapter 10.

The approval and effectiveness of imatinib, and its relatively mild side effects, demonstrated for the first time that inhibition of an oncogenic kinase could be an effective therapeutic strategy.

Since then, many small molecule kinase inhibitors have been developed and approved, including those that target the kinase domains of oncogenic kinases including EGFR, B-RAF, MEK, and CDKs, as well as the lipid kinase PI3K (see Figure 12.7). In addition, monoclonal antibodies that target EGFR (cetuximab) and ERBB2/HER2 (transtuzumab) have been developed as effective therapies for cancers associated with overexpression or mutation of these RTKs.

More recently, small molecule inhibitors of the GTPase Ras have also been developed. In particular, small molecules have been identified that can bind to a cryptic binding pocket in the G12C oncogenic mutant form of K-Ras. This exciting

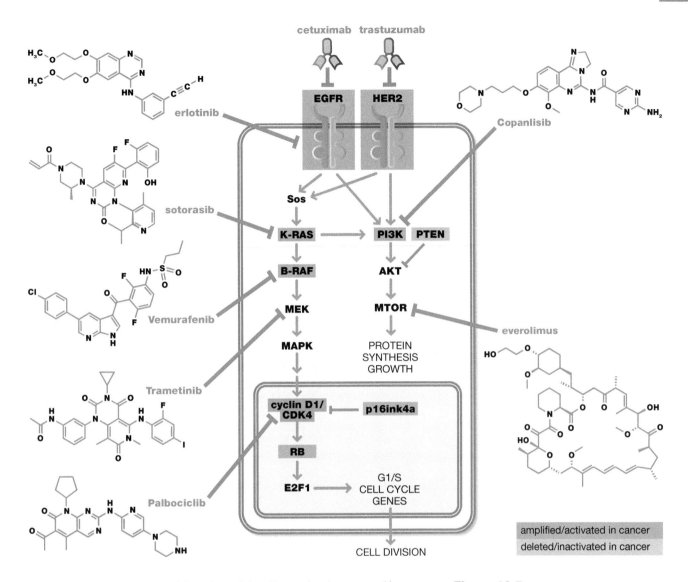

Figure 12.7

Mitogenic signaling proteins that are targets for therapy. Examples of targeted therapies now in the clinic include antibodies to extracellular domains of receptor tyrosine kinases, and small molecule inhibitors of cytosolic protein and lipid kinases and the small G protein K-Ras.

development shows how this critical class of signaling molecules, mutated in many different cancers, has the potential to be an effective drug target.

A major limitation of these targeted therapies, however, remains the emergence of drug resistance, often within months to a few years after the initiation of treatment. For example, treatment with EGFR inhibitors erlotinib and gefitinib classically leads to resistance through the emergence of the secondary mutation T790M in the EGFR kinase domain; more than 50% of the patients that show resistance have the T790M mutation. Structural studies have shown that this is an example of a class of mutations termed "gatekeeper mutations," which alter the shape of the ATP binding site such that binding of inhibitors is sterically blocked, while the binding of ATP is relatively unaffected. Second-generation inhibitors have now been developed, such as osimertinib, which can inhibit the T790 mutation. Resistance can also eventually develop to these drugs, however. Development of drug resistance remains a problem for long-term durability of these therapies, reflecting the intense Darwinian selection for growth and survival of tumor cells. Even when only a tiny fraction of cells in a tumor are resistant to a particular therapy, those resistant cells can rapidly grow and lead to recurrence of the disease.

It is interesting to consider that targeted therapies often inhibit key signaling proteins that are important for the normal functioning of cells. Yet in many cases, the side effects of such targeted therapies are relatively mild. Why might this be? It is thought that when a tumor is first developing from just a few cells, the signaling pathways in the tumor cells become increasingly dependent on the activity of the oncogenic changes that drive the tumor—that is, the tumor cells become "addicted" to the presence of high levels of activity from a constitutively active signaling protein. This re-wiring of signaling pathways is again likely due to very strong selection in the tumor for cells that can survive and proliferate better than their peers. Because they are so dependent on the activity of one overactive protein, when that protein is inhibited by targeted therapy, the tumor cells cannot survive as well as normal cells.

Immune checkpoint inhibitors can be effective cancer therapies

Another area of great advance in cancer therapies is the development of antibodies that block checkpoint proteins, known as checkpoint inhibitor therapeutics (**Figure 12.8**). Above we discussed the PD-1 checkpoint protein, which is found in activated T cells, and its ligand, PD-L1, which can often be found in tumor or tumor-associated cells. Tumors can suppress T cell-mediated responses if they express PD-L1, which acts through PD-1 in the T cells to inhibit T cell receptor (TCR) signaling. PD-L1/PD-1-mediated T cell suppression can be inhibited by treatment with antibodies that bind to and block PD-1, such as pembrolizumab and nivolumab, or antibodies that bind to and block PD-L1, such as atezolizumab. These so-called checkpoint inhibitor antibodies have proven to be remarkably effective therapies for certain types of cancers. In addition, the protein CTLA4 is another inhibitory checkpoint receptor in T cells that is activated by the B7 protein. Antibodies that block CTLA4, such as ipilimumab, are also strong anti-cancer agents.

These checkpoint inhibitors, particularly anti-PD1/PD-L1 drugs, can be highly effective for a subpopulation of patients with melanoma, lung cancer, and several other cancers, in some cases leading to durable remissions or what might be considered cures. There remain large subpopulations of patients, however, that are non-responders. In general, it is hypothesized that for a checkpoint inhibitor to work, there must be an existing anti-tumor T cell population that is held in check by the tumor; conversely, non-responder patients may lack an intrinsic T cell response (thus releasing checkpoint brakes would have no effect). For example, if a cancer has low mutational burden, then it may have few neoepitopes that

Figure 12.8

Checkpoint inhibitors. Immune checkpoint inhibitors function by blocking inhibitory signals sent from cancer cells to T cells. Signaling from the PD-1 and CTLA4 receptors in T cells recruits and activates the tyrosine phosphatase Shp2, which downregulates TCR signaling. Tumor cells often express ligands for these receptors, PD-L1 and B7. Antibodies targeting PD-L1, PD-1, and CTLA4 block these inhibitory signals, increasing the ability of T cells to recognize and kill tumor cells. A potential side effect of these inhibitors is autoimmunity due to disruption of normal controls on T cell activation.

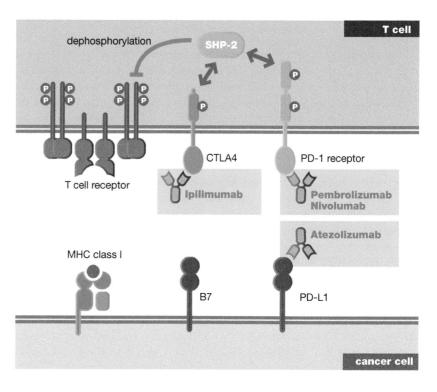

can drive T cell recognition. Thus, not surprisingly, some of the most responsive cancers are associated with *microsatellite instability (MSI)*; in such tumors, the ability of cells to repair DNA damage is compromised. Thus, MSI-high tumors have a high mutational load, presumably leading to a large number of neoepitopes with the potential to trigger strong T cell responses. Alternatively, tumors that fail to respond to checkpoint inhibitors may have other suppressive mechanisms unrelated to these checkpoint systems.

Notably, checkpoint inhibitors in some cases have the distinct but perhaps not surprising side effect of inducing or exacerbating autoimmune or inflammatory diseases in otherwise healthy normal tissues. These side effects are a consequence of altering the balance between immunity and tolerance, both in tumors and in normal tissues. Normally, many tendencies toward autoimmune diseases (where the immune system mistakenly recognizes one's own tissues as foreign, and attacks them) may be kept in check by tolerance mechanisms involving checkpoint receptors.

Engineered T cells can recognize and kill cancers

One of the most exciting recent advances in cancer therapy centers on the engineering of living T cells, altering their signaling behaviors in order to redirect them to kill cancer cells (**Figure 12.9**). These modified T cells utilize an engineered form of the TCR called a chimeric antigen receptor (CAR). When a CAR is expressed in a T cell, it can redirect the T cell to be activated by a novel surface antigen found on a cancer cell. Thus, when successful, the CAR T cells can be directed to kill cancer cells while sparing all normal cells not expressing the cancer-specific surface antigen.

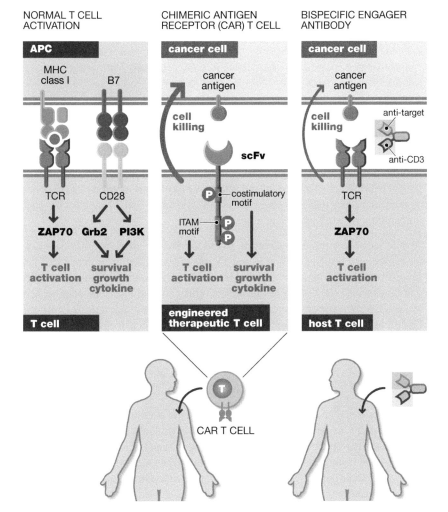

Figure 12.9

CAR T cells. Left: In normal T cells, strong activation requires stimulation of the T cell receptor by antigen–MHC complexes as well as costimulation via CD28. Middle: In CAR T cells, these activities are mimicked by a synthetic chimeric antigen receptor (CAR) expressed in recombinant T cells. The CAR has an extracellular single-chain antibody (ScFv) that recognizes a target on the tumor cell surface, and intracellular domains derived from the TCR and CD28. The CAR T cells are infused into patients, where they recognize and kill tumor cells. Right: A bispecific engager antibody recognizes both the T cell receptor and a tumor-specific surface antigen, leading to activation of host T cells. While this approach does not require *in vitro* modification of host T cells, it also does not provide co-stimulatory signals.

The CAR has a modular chimeric structure. The extracellular portion of the TCR is replaced with a single-chain antibody that is specific for a cancer-associated antigen—this region is responsible for retargeting the T cell toward a cancer cell. The intracellular domain contains ITAM signaling motifs from the CD3ς chain of the TCR. These are the core activation motifs that recruit the ZAP70 kinase (via binding by the ZAP70 tandem SH2 domains) and initiate the major T cell activation pathway. However, to create a more potent CAR, it is necessary to drive higher levels of T cell proliferation and cytokine secretion. Thus, a key advance in the development of modern CARs was the inclusion of co-stimulatory regions from immune co-activating receptors, such as CD28, which is stimulated by the co-ligand B7 when normal T cells encounter antigen-presenting cells. A CAR that contains both CD3ς ITAMs and a CD28 co-stimulatory motif shows much stronger proliferation and cytokine secretion than a CAR that only contains the CD3ς ITAMs. Strong proliferation is correlated with significantly improved clinical efficacy, likely because many T cells are required to generate a strong anti-tumor response. Current CARs fuse the CD3ς chain with co-stimulatory motifs derived from either CD28 or 4-1BB immune co-receptors.

See Chapter 14 for a description of normal TCR activation.

The first engineered CAR T cell therapy was approved in the US by the FDA in 2017. Currently, CAR T cells have been most successful in targeting blood cancers, including different forms of B cell leukemias and lymphomas. The first CARs targeted the antigen CD19, which is a lineage-specific surface marker unique to B cells. Thus, CARs directed against CD19 are able to kill all B cells, including both cancerous and normal B cells. While this activity is effective against cancer, it results in B cell aplasia, which is clinically tolerable.

Because the CAR directly targets a surface antigen protein on the cancer cell, this recognition bypasses the need for MHC-antigen presentation to the TCR. Thus, loss of MHC or antigen processing, a common mechanism of tumor immune evasion, does not hamper CAR recognition. CARs directed against CD19 have proven to be highly effective, and in many cases lead to highly durable cures lasting many years. In some cases, however, relapse is seen, especially with antigen loss or heterogeneity.

Treating a patient with a CAR T cell is an extremely complex process. Patient T cells must be isolated by leukopheresis, and then shipped to the manufacturer, where they are modified by infecting them with a recombinant virus vector expressing the CAR. The transduced T cells are then shipped back to be infused into the patient. Thus, current therapies use an autologous approach, in which a patient's own T cells are used for transplant. An active area of research is to develop allogeneic cells that can come from a universal donor (but which must be made immunologically tolerable when transplanted).

An alternative strategy for redirecting T cells to tumors is to use a molecular adaptor known as a bispecific engager (see Figure 12.9). This is an antibody that has at least two binding moieties, one that binds a cancer surface antigen, and another that binds to the CD3 chain of the TCR. In this case, the bispecific engager acts as an adaptor to tether together T cells and cancer cells. Remarkably, this can lead to relatively effective tumor killing. Current research focuses on engineering specific multi-valent engagers that both specifically target tumors and lead to high levels of T cell activation. A CD19-targeted bispecific engager is currently FDA-approved, and has the advantage of off-the-shelf convenience (that is, it should target CD19-positive cells in any individual).

CAR T cells and bispecific engagers have so far shown limited effectiveness against solid cancers. Solid cancers are more difficult to target specifically, because many tumor-associated antigens are also found in normal tissues. Thus, strategies are being developed to target antigen combinations or other features that are more tumor-specific. In addition, solid tumors often reside in immunosuppressive microenvironments, and thus many efforts focus on engineering CAR T cells that are resistant to or can modify these microenvironments. Nonetheless, CAR represents a remarkable advance, which illustrates clearly how our knowledge of cell signaling can be harnessed to create novel and therapeutically useful cellular responses.

PATHOGENIC MICROBES THAT HIJACK HOST CELL SIGNALING

Pathogenic microbes that infect humans and other mammalian hosts have evolved multiple ways to interact with host cell signaling proteins. Both bacterial and viral pathogens are thought to utilize these interactions to steer host responses that increase their ability to infect the host, to replicate and spread within the host more effectively, and to evade host immune or apoptotic responses. These interactions provide a way for relatively simple pathogens to plug into and take advantage of sophisticated host behaviors. Variations that alter these pathogen–host interactions can have dramatic effects on virulence and transmissibility, and thus can be linked to infectious disease pandemics. Furthermore, understanding the specific ways in which pathogens depend on signaling pathways will provide novel strategies for preventing and treating infectious diseases. Finally, the manipulation of cell signaling by pathogens has provided many tools and experimental systems that were enormously useful in deciphering normal cell signaling mechanisms.

It is worth briefly considering the logic that underlies the remarkable ability of microbial pathogens to modulate host signaling pathways. The success of such organisms is measured by their ability to replicate successfully over time—those that generate the most progeny over repeated rounds of infection will thrive and take over the population. Thus, avoiding the host's defenses (the adaptive and innate immune systems), replicating and spreading rapidly within the host, and ultimately being passed from one host to another are critical to their success. Regarding transmission, it is likely no accident that many human pathogens cause diarrhea or coughing/runny nose, which increase the likelihood that they can spread to others (through feces or respiratory secretions). Microbes are under strong selective pressure to minimize the size of their genomes, which makes it advantageous to take advantage of host cell machinery where possible. This is particularly true for viruses, which are almost entirely dependent on the host to provide the machinery and raw materials needed for their replication.

It would be impossible to describe comprehensively the many ways that human pathogens subvert and exploit host signaling pathways. Instead, here we will delve more deeply into several well characterized examples.

Yersinia pestis provides an example of the diverse ways in which pathogens can hijack host signaling

In this section, we will examine the many ways in which one pathogen rewires host cell signaling: *Yersinia pestis*, the causative agent of the plague. In the fourteenth century, an outbreak termed the Black Plague is thought to have killed up to one-third of the people in a broad area of the globe including Europe (**Figure 12.10**), and subsequent waves of infection struck for the next several centuries. The bacterium was transmitted through contact with infected fleas or other infected people, and most victims died within a few days of infection. Thankfully, although reservoirs of *Yersinia* still exist in the wild and lead to occasional human infections, modern antibiotics are usually effective in treating the disease.

One of the reasons why *Yersinia* is such an efficient pathogen is that it has a mechanism to inject host cells with a variety of proteins that rewire cell signaling pathways. The **Type III secretion system** (**T3SS**) is a plasmid-encoded mechanism

Figure 12.10

The Black Death. Contemporary depiction of plague victims being buried in 14th-century Belgium. (The Citizens of Tournai, Belgium, Burying the Dead during the Black Death of 1347–52. Detail of a miniature from *The Chronicles of Gilles Li Muisis (1272–1352), abbot of the monastery of St. Martin of the Righteous*, Bibliothèque royale de Belgique, MS 13076-77, f. 24v.)

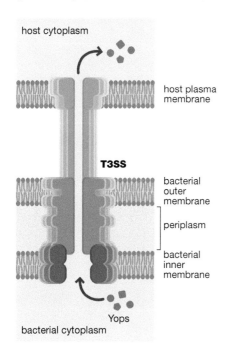

Figure 12.11

The Type III secretion system of *Yersinia pestis*. The Type III secretion system (T3SS) is assembled by the bacterium and spans the two bacterial membranes (inner and outer), as well as the plasma or endocytic vesicle membrane of the host cell. Bacterial effector proteins (Yops) contain an N-terminal targeting sequence that directs their transport through the hollow T3SS and into the host cell cytosol, where they exert their effects on host signaling.

used by some gram-negative bacteria that involves assembly of a syringe-like apparatus structurally related to the flagellum (**Figure 12.11**). This hollow channel spans the two bacterial membranes, as well as the plasma membrane or endocytic vesicle of the host cell. The *Yersinia* T3SS is used to inject a series of effector proteins known as *Yersinia* outer proteins (Yops). *Yersinia* use their T3SS to target critical host innate immune cells, including macrophages and neutrophils. Here we discuss several key Yops and their targets to illustrate the rich variety of ways in which pathogens can impact normal cell signaling (**Table 12.2**).

YopH is a constitutively active tyrosine phosphatase. Because tyrosine phosphorylation is not an important post-translational modification in bacteria, it has no effect in the pathogen. Once injected into the host cell, however, it targets a variety of tyrosine-phosphorylated substrates, most notably focal adhesion components such as p130Cas, paxillin, and focal adhesion kinase (FAK). Dephosphorylation of these and other targets is thought to block the ability of macrophages and other host phagocytic cells to engulf the bacterium, a critical defense against invading microorganisms. Other YopH targets include various components of the TCR; when *Yersinia* interacts with T cells, dephosphorylation by YopH blocks their activation, crippling the host's ability to mount an effective adaptive immune response.

Several other proteins secreted by *Yersinia* act in concert to interfere with normal innate immunity and inflammatory responses, which serve as a first line of defense against invading organisms. YopM, which consists of multiple leucine-rich repeats (LRR), assembles into an oligomeric scaffold in infected cells. It is thought to bind endogenous kinases such as PRK1 and PRK2 and promote the phosphorylation of pyrin, a component of the inflammasome. This in turn inhibits assembly of the host pyrin inflammasome, preventing activation of caspase 1 and the subsequent processing and secretion of inflammatory cytokines such as IL-1β by infected

Table 12.2

The Yop effectors secreted by *Yersinia pestis*

	Activity	Molecular targets	Cellular effect	Use to pathogen
YopE	RhoGAP	Rho family G proteins	Disrupt actin organization	Inhibit engulfment
YopH	Tyrosine phosphatase	Focal adhesion proteins	Inhibit adhesion, engulfment	Inhibit engulfment
		T cell receptor	Inhibit TCR activation	Inhibit adaptive immune response
YopJ	Acetyltransferase	MAPKKs, IKKβ	Inhibit NF-κB, MAPK signaling	Inhibit inflammation, stress response
YopM	Scaffold	PRKs, RSKs, pyrin	Inhibit inflammasome assembly	Inhibit inflammation
YopO	Protein kinase	WASP family, formins	Disrupt actin nucleation	Inhibit engulfment
	Rho-GDI	Rho family G proteins	Disrupt actin organization	
YopT	Protease	Rho family G proteins	Disrupt actin organization	Inhibit engulfment

cells. It is interesting to note that inhibition of Rho family GTPases (which is mediated by other Yops, as discussed below) would normally lead to activation of the pyrin inflammasome, but YopM prevents this during *Yersinia* infection. YopJ is an acetyltransferase that modifies specific serine and threonine residues in host kinases including MAPKKs and IKKβ, preventing their phosphorylation and thus their activation by upstream kinases. Inhibiting IKKβ blocks activation of the NF-κB pathway, thus muting inflammatory responses and promoting apoptosis during infection; inhibiting MAPK signaling likely blunts normal stress responses mediated by JNK, p38, and their relatives.

Three other Yops (YopE, YopT, and YopO) target the small G proteins that control organization of the actin cytoskeleton. YopE is a RhoGAP for RhoA, Cdc42, and Rac, serving to enhance their GTP hydrolyzing activity, and thus downregulate their activity. YopT is a protease that targets the same small G proteins, in this case by cleaving the C-terminal lipid modification site (which mediates membrane association) from the body of the G protein. Once liberated from the membrane, the small G proteins are functionally inactive. YopO is a more complex effector, encoding an N-terminal membrane targeting domain, a serine/threonine kinase domain, and a guanine nucleotide dissociation inhibitor (GDI) domain. Monomeric actin binds to the GDI and kinase domains, leading to activation of the kinase domain. This mechanism ensures that the kinase is only active once injected into the host, since eukaryotic actin is not present in the bacterium. Upon activation, YopO is thought to phosphorylate proteins that normally mediate actin assembly, such as WASP family members and formins (discussed below), while the GDI domain binds the inactive, GDP-bound forms of Rac and Rho, blocking nucleotide exchange and activation. These combined activities (phosphorylation of actin assembly factors and inactivation of Rho family GTPases) again impact actin cytoskeletal rearrangements, including those needed for phagocytosis of the microbes by macrophages and neutrophils.

Beyond *Yersinia*, manipulating G proteins (both small and heterotrimeric) is a common strategy used by many pathogenic bacteria. For example, cholera toxin (secreted by *Vibrio cholerae*, the bacterium that causes cholera) is an enzyme that ADP ribosylates the Gα$_s$ subunit of the heterotrimeric G protein. This locks it into the active conformation, leading to greatly elevated cAMP levels in the intestinal epithelium. This in turn leads to ion imbalances and the rapid efflux of water from the intestine, causing severe diarrhea. Because of their efficiency and specificity, bacterial toxins such as cholera toxin have been valuable tools to study cell signaling mechanisms.

 The role of the inflammasome in innate immunity is discussed in Chapter 9.

 G proteins and their regulation are discussed in Chapter 3.

Many pathogens manipulate the host actin cytoskeleton to facilitate adhesion, internalization, and cell-to-cell spread

We have already seen that *Yersinia* injects several proteins that regulate the activity of Rho family GTPases, thereby inhibiting engulfment and internalization of the pathogen by host cells. It turns out this is just one example of the many ways in which microbes and viruses subvert the host cell actin cytoskeleton. Here we will examine more closely two such mechanisms: how some pathogenic bacteria manipulate the structure of the intestinal epithelium, and how some bacteria and viruses use actin to propel themselves through the host cytosol and thus spread from cell to cell.

To understand how pathogens exploit actin during infection of their hosts, we must first briefly discuss how actin is normally regulated in cells (**Figure 12.12**). Monomeric actin (termed globular or G-actin), when bound to ATP, can assemble into long polarized filaments (filamentous or F-actin). These filaments can be arranged in various ways, including crosslinked bundles termed stress fibers (which typically

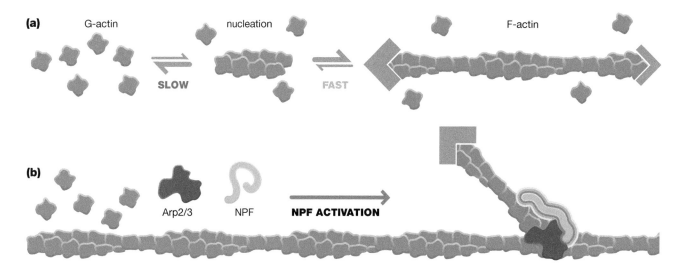

Figure 12.12

Control of actin polymerization by nucleation. (A) Monomeric actin (G-actin) bound to ATP can assemble into F-actin filaments. The spontaneous assembly of new filaments is very slow, because the oligomer is highly unstable until it reaches 3–4 actin subunits in size. Once nucleation occurs, polymerization is very rapid. Actin filaments are polarized and assemble much more rapidly at one end (termed the "barbed end") than the other (the "pointed end"). (B) Nucleation promoting factors (NPFs) bind the Arp2/3 complex and G-actin to promote the formation of new actin branches on existing filaments. Localized activation of NPFs leads to rapidly growing dendritic networks of actin filaments.

interact with focal adhesion complexes, myosin, and other proteins to generate contractile forces), and densely branched dendritic networks, which can push the membrane forward by adding new subunits, for example, at the leading edge of a lamellipodium. The rate-limiting step for actin polymerization is assembly of a nucleus consisting of three actin monomers; once this tiny filament forms, further assembly is very rapid, driven by the high concentration of G-actin in the cytosol (this is another example of how biology frequently takes advantage of reactions that are thermodynamically favorable, but kinetically very slow).

In the cell, the location where actin polymerization occurs is typically controlled by regulating the nucleation step. We have already encountered N-WASP, a member of a family of actin **nucleation promoting factors** (**NPFs**), which catalyze the initial addition of actin subunits to a specialized protein complex, the ARP2/3 complex, to nucleate new branches on existing actin filaments. Other nucleating mechanisms do not require ARP2/3; these include the formins, which nucleate long unbranched filaments, and members of the spire family.

The gram-negative bacterium *Escherichia coli* is the most abundant component of the human gut microbiome, and plays an important role in normal digestion. However, enteropathogenic *E. coli* (EPEC) and enterohemorrhagic *E. coli* (EHEC) are variants that have acquired additional features that make them dangerous human pathogens (for example, the EHEC *E. coli* O157:H7 is frequently linked to outbreaks of severe diarrhea associated with ground meat). These pathogenic *E. coli*, unlike their benign relatives, remodel the intestinal epithelium to attach themselves via "attaching and effacing" (A/E) lesions, where the surrounding intestinal villi are eliminated and the bacterium is closely attached to the membrane via an actin-rich "pedestal" structure (see **Figure 12.13**). While the precise function of the A/E lesions and the actin pedestal are not yet fully understood, many studies have shown that these structures are critical for pathogenicity.

Figure 12.13

Enterohemorrhagic *E. coli* (EHEC) actin pedestals. Immunofluorescence image of human HeLa cells fixed 3.5 h after infection with EHEC, showing actin pedestals assembled in the cytosol underneath bacteria adhering to the cell membrane. In this image, bacterial *Tir* protein is *red*, F-actin is *green*, and the nucleus is *blue*. Image courtesy of Katrina Velle and Ken Campellone.

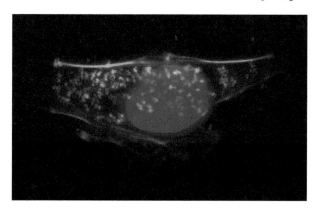

The initial step in generating A/E lesions is for the bacterium to firmly attach to the intestinal epithelium. Like *Y. pestis* and many other pathogenic bacteria, EPEC and EHEC use the T3SS to inject a variety of proteins and toxins into the host cell. One of these key effectors is the protein *Tir*, which inserts itself into the host cell plasma membrane and serves as a high-affinity binding partner for a bacterial outer membrane protein, intimin (**Figure 12.14a**). Thus, the bacterium injects its own receptor into the host cell, avoiding the need to use specific host cell proteins. Both intimin and *Tir* also self-associate into dimers and higher order structures, rapidly forming an extended lattice that tightly binds the bacterium to the intestinal epithelium. The *Tir* intracellular domains (exposed to the host cell cytosol) are critical for the ability of the bacterium to assemble the actin-rich pedestal structure in the underlying cell, though EPEC and EHEC do so through different but related mechanisms.

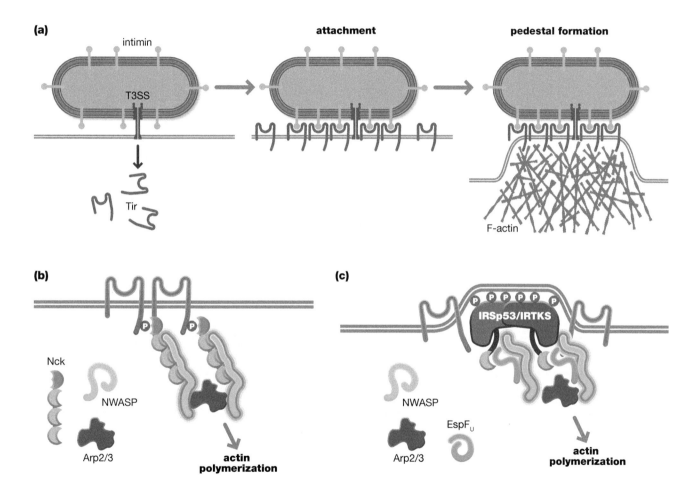

Figure 12.14

Attachment and pedestal formation by EHEC and EPEC. (A) When the bacterium encounters the intestinal epithelium, it uses its T3SS to inject *Tir* into the host cell, which inserts into the host plasma membrane and provides high-affinity binding sites for the bacterial outer membrane protein intimin. *Tir* also mediates localized assembly of actin under the bacterium to form a pedestal structure. (B) In EPEC infections, *Tir* is phosphorylated on Tyr474 by host Src or Abl family kinases, providing a binding site for the SH2 domain of the Nck adaptor. The Nck SH3 domains bind and activate N-WASP, leading to localized actin polymerization. WIP, which binds Nck SH3 domains and N-WASP and participates in N-WASP activation, is not shown here for clarity. (C) In EHEC infections, *Tir* binds to the I-BAR domain containing proteins IRSp53 and IRTKS via a NPY motif in *Tir*. The I-BAR proteins induce membrane protrusion and also bind to PIP_2 (*red* circles) on the membrane. The SH3 domain of I-BAR proteins binds the bacterial effector $EspF_U$, which recruits and activates N-WASP.

In the case of EPEC (**Figure 12.14b**), phosphorylation of tyrosine residues in the C-terminus of *Tir*, particularly Y474, is a key step. This is done by host cell non-receptor tyrosine kinases, particularly the Src and Abl families. Once phosphorylated, *Tir* serves as a binding site for the SH2 domain of the Nck SH2/SH3 adaptor. The SH3 domains of Nck recruit the NPF N-WASP, along with a second N-WASP binding protein, WIP. Together, WIP and Nck activate N-WASP, leading to formation of a rapidly polymerizing dendritic network of actin filaments under the bacterium.

 The role of N-WASP in integrating signals from signaling proteins such as Nck is discussed in Chapter 11.

For EHEC, a different C-terminal *Tir* motif is crucial, in this case an NPY-containing peptide (**Figure 12.14c**). This binds to two adaptor proteins, IRTKS and IRSp53, through a modular domain termed the I-BAR domain. The I-BAR domain has several interesting properties; in addition to binding NPY motifs, it also binds to the phosphoinositol lipid PIP_2, which itself binds many proteins involved in localized actin polymerization including N-WASP, and it induces negative membrane curvature (that is, causes a convex deformation or protrusion of the membrane).

 The related BAR domain, which induces positive membrane curvature, is illustrated in Chapter 10.

Furthermore, both IRTKS and IRSp53 also contain an SH3 domain, which binds yet another bacterial effector protein, $EspF_U$, that is injected into the host cell by the EHEC T3SS.

$EspF_U$ is a relatively simple protein, containing a C-terminal proline-rich region that binds with unusually high affinity to the IRTKS and IRSp53 SH3 domains, and an N-terminal region containing multiple helical segments that bind to N-WASP and relieve intramolecular interactions that normally hold N-WASP in an inactive conformation. Thus, though the specific mechanism is quite different, the *Tir* proteins of EHEC and EPEC both lead to activation of N-WASP to promote localized actin polymerization. Note that while EPEC does not encode $EspF_U$ and thus does not activate N-WASP through this mechanism (instead, it uses tyrosine phosphorylation and Nck to accomplish this), it is likely that the recruitment of I-BAR containing proteins through the NPY motif still plays an important role in inducing membrane curvature and the assembly of other factors needed to build the EPEC actin pedestal.

Unlike *E. coli* or *Yersinia* under most conditions, some pathogenic bacteria and all viruses replicate in the cytosol of infected host cells. For bacteria, this likely provides a protected niche in which they can hide from phagocytic cells and circulating antibodies while they replicate. Furthermore, viruses, which have very limited coding capacity in their tiny genomes, need access to host cell enzymes and raw materials in order to replicate their genetic material and generate the viral proteins needed to package it into new virion particles. Both viral and bacterial pathogens have evolved ways to exploit actin polymerization to propel themselves through the cytosol, and from one cell to another, allowing the pathogen to spread with minimal exposure to the immune system. A number of well-known human pathogens including the bacteria *Shigella*, *Listeria*, and *Rickettsia* and various poxviruses (a group that includes those that cause smallpox and chickenpox) use a variety of mechanisms to regulate actin polymerization; here, we will consider *Listeria* and vaccinia virus as examples.

Listeria monocytogenes is a gram positive bacterial pathogen. It is engulfed by host cells and internalized, whereupon the bacterium escapes from the endocytic vesicle into the cytosol. Once inside a cell, it is propelled through the cytoplasm at the tip of a long tail of polymerized actin (these structures are often termed "comet tails" for obvious reasons) (**Figure 12.15a**). Early studies showed that the actin comet tail remained stationary while the bacterium moved, implying that the force needed for propulsion is generated by assembly of new actin filaments near the surface of the bacterium. This actin polymerizing activity is provided by ActA, a bacterial protein that protrudes from the surface of the bacterium. ActA works as a bacterially-encoded NPF, binding G-actin and the Arp2/3 complex in a manner very similar to how N-WASP nucleates new actin branches in uninfected

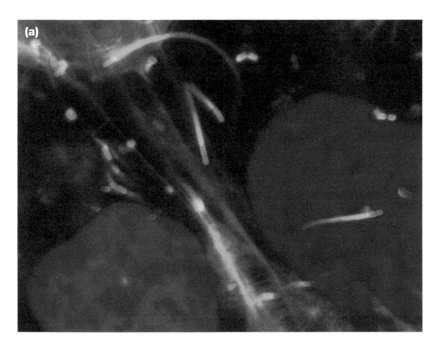

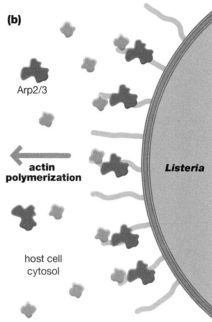

Figure 12.15

Listeria propels itself through the cytosol via actin comet tails. (A) Mammalian cells infected by *Listeria*: *Listeria* (*red*) causes infected cells to form comet-like tails from actin (*green*) that push them through their host cells (*blue* = cell nuclei). (© Pascale Cossart.) (B) ActA expressed on the bacterial cell membrane (*green* lines) binds the Arp2/3 complex and G-actin, acting as a constitutively active nucleation promoting factor (NPF). This nucleates new actin branches adjacent to the surface of the bacterium, which rapidly polymerize. ActA also binds Ena/VASP family proteins, which affect the directionality and speed of motility.

cells (**Figure 12.15b**). ActA is sufficient to induce rapid actin polymerization in the cytosol, or even *in vitro* in the presence of purified Arp2/3 complex and actin. This sort of molecular mimicry of host activities is commonly seen in pathogenic microbes.

Vaccinia virus replicates within cells, and when mature, the new membrane-covered viral particles bud from the plasma membrane, but remain associated with the host cell. These cell-associated virus particles can induce actin comet tail formation in the cytosol beneath the virion, allowing the virus to "surf" along the membrane of the infected cell and ultimately extend from the surface of the cell on long, rapidly growing protrusions (**Figure 12.16**). A transmembrane protein originating from the immature virion particle, A36, which is left behind on the host cell membrane once the virus buds off, was found to be the critical mediator of actin-based motility. A36 has a relatively short intracellular region with no obvious domain features, but it has a key tyrosine residue (Y112) that is phosphorylated by Src family kinases in infected cells. Once phosphorylated, this site binds tightly to the SH2 domain of the Nck adaptor; as we saw earlier in the case of EPEC, recruitment of Nck leads to activation of N-WASP via the Nck SH3 domains. This is a remarkable case of convergent evolution, in which a pathogenic bacterium and a virus have both engineered ways to recruit Nck and thereby induce localized actin polymerization. Indeed, aggregation of Nck SH3 domains on the plasma membrane of cells is sufficient to induce formation of actin comet tails that propel the aggregates in a manner very similar to what is seen in vaccinia-infected cells.

What selective advantages might actin-based motility confer to intracellular microbial pathogens? It is thought that one benefit might be in providing a way to infect nearby cells while minimizing exposure to the dangers of the outside environment (phagocytic cells, circulating antibodies, and T cells, for example) (**Figure 12.17**). In the case of intracellular bacteria, the force of actin polymerization can propel a bacterium into the cytosol of an adjacent cell (for example, in an epithelial sheet). After membrane scission and escape from the resulting vesicle, the bacterium can now replicate in the other cell without ever having exposed itself to the outside environment (**Figure 12.17a**). Similarly, actin comet tails can essentially shoot vaccinia virus particles to an adjacent cell relatively rapidly, with only brief exposure outside the cell (**Figure 12.17b**).

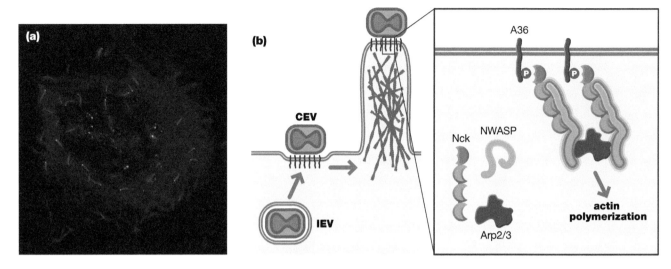

Figure 12.16

Vaccinia virus hijacks host cell actin machinery. (A) Comet tails in HeLa cell infected with vaccinia virus at 8 hours post-infection. In this image, actin is stained with phalloidin (*magenta*), and virus particles are labeled with a phospho-specific antibody against the viral protein A36 (*green*). (B) Viral particles mature within the cytosol, surrounded by two membranes (internal enveloped virus, or IEV). A virally encoded protein, A36, becomes embedded in the outer membrane, and when the outer membrane of the IEV fuses with the host cell plasma membrane, A36 remains in the plasma membrane under the cell-associated enveloped virus (CEV). A36 is phosphorylated on Y112 by host cell Src family kinases, and recruits Nck and activates N-WASP in a similar way as EPEC (see **Figure 12.14b**), leading to actin polymerization and formation of a comet tail. The virus particles can either move laterally along the plasma membrane or protrude from the membrane as shown here. a, courtesy of Angika Basant and Michael Way.

Figure 12.17

Actin propulsion allows infection of neighboring cells. (a) Intracellular bacteria such as *Listeria* can use the propulsive power of actin polymerization to protrude into a neighboring cell, for example, in an epithelial sheet. The bacterium then buds off, surrounded by a double plasma membrane derived from the host. It can then escape into the cytosol of the new cell. (b) In the case of vaccinia virus, the cell-associated virus can be projected from one cell to another via the actin comet tail generated in the original host cell. Once the virus particle encounters the plasma membrane of a neighboring cell, it can be internalized and replicate in the new cell.

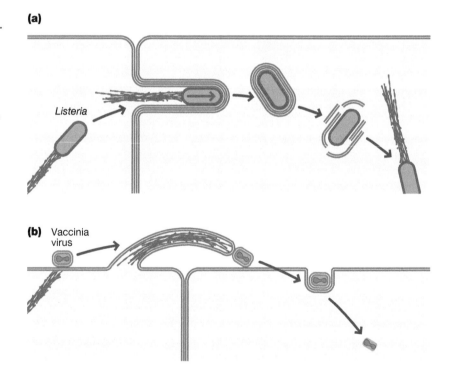

DNA tumor viruses manipulate the host cell to support their replication

While most human cancers are not caused by viruses, a few viruses can cause cancer in people under some circumstances; for example, human papilloma virus (HPV) is implicated in a large percentage of human cervical cancers worldwide.

Historically, tumor viruses (viruses that cause cancer) have been enormously important tools in cancer research, both in providing reliable, reproducible agents to induce cancer in experimental settings, and in highlighting the types of activities that can deregulate proliferation when hyperactivated or suppressed. However, from the perspective of the virus, tumorigenesis is more likely to be a bug than a feature, as it is typically the result of incomplete or aberrant viral replication. Here, we will briefly consider several **DNA tumor viruses**—viruses whose genomes consist of double-stranded DNA, and which depend on the host cell for most of the machinery required for their replication.

For such viruses, the fact that most cells in an organism are not actively proliferating is problematic. Cells that are not proliferating typically do not waste energy and resources making the machinery needed for DNA replication, such as DNA polymerases, or raw materials such as deoxynucleotides. Therefore, it is advantageous for the virus to be able to manipulate the host cell cycle—essentially, to trick the cell into believing the circumstances are appropriate for DNA replication.

Many viruses encode two sets of genes: "early" genes that are transcribed immediately after the virus enters the host cell, and which prepare the environment for replication of the genome, and "late" genes that encode major viral structural proteins needed for assembly of new virion particles. Once a virus has successfully replicated, making many new copies that can be transmitted to other cells, the infected cell typically dies. Occasionally this process can go awry, however. If some of the viral genes needed to make progeny virus are lost (for example, the late transcripts encoding structural genes), and if the defective viral genome can integrate into the host genome or otherwise be maintained, the resulting clone of cells may grow uncontrollably. The resulting tumor is a consequence of the virus's re-engineering of the cell cycle control machinery to facilitate its replication.

Two major targets of DNA tumor viruses are the key regulator of cell cycle progression, RB (the gene mutated in hereditary retinoblastoma), and the "guardian of the genome," p53 (**Figure 12.18**). These were both introduced earlier in this chapter; RB normally represses transcription of key genes needed for cell cycle progression, while p53 stops the cell cycle and can trigger apoptosis in cells when conditions are inappropriate for proliferation. Their central role in controlling cell cycle progression is highlighted by their targeting by multiple DNA tumor viruses, and by the fact that both are tumor suppressors frequently mutated in human cancers. Here we will consider the various ways in which these key regulators are inhibited by three DNA tumor viruses.

Simian Virus 40 (SV40) is a polyomavirus that normally infects non-human primates, but historically, it was often used as an experimental model of tumorigenesis in cultured cells. One of its early gene products, termed large T antigen, plays several roles in preparing the cell for viral replication, including acting as a DNA helicase for viral genomes during their replication. It adopts an oligomeric structure, consisting of ring-shaped hexamers that slide along the DNA. The outer arm

Figure 12.18

DNA tumor viruses manipulate cell cycle control mechanisms. RB and p53 are critical for normal host cell cycle control. RB normally represses transcription of genes needed for DNA replication, unless inactivated by mitogenic signals. P53 stops cell cycle when conditions are not right for proliferation, and induces cell death upon sustained activation. DNA tumor viruses SV40, adenovirus, and human papillomavirus (HPV) express early gene products (*green*) that inactivate RB and P53 to allow viral genome replication. In some cases, viral gene products enlist host E3 ubiquitin ligases (*orange*) to assist in degrading their targets.

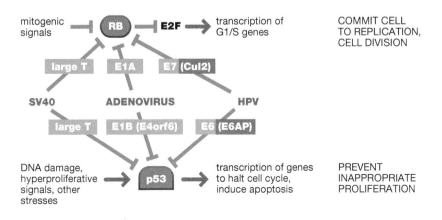

each subunit, however, binds tightly to p53 and to a number of other cell cycle regulators, including RB, and acts to sequester them and prevent them from performing their normal functions. In fact, p53 was first identified as a host cell protein that specifically bound to large T in cells oncogenically transformed by SV40. Subsequently, it was found that large T also bound and sequestered other proteins, including RB. This provides an example of the remarkable efficiency of viruses in making the most of their limited genomes—here a single viral protein plays a key role in genome replication, in preventing apoptosis, and in forcing quiescent cells into the cell cycle. Experimentally, large T has often been used as a tool to "immortalize" cells in the lab (allow them to proliferate indefinitely in culture).

Adenoviruses are common viral pathogens, typically causing mild infections that quickly resolve in humans. Adenoviruses can induce tumors in some animal models, however, and can also transform tissue culture cells. To date, adenoviruses have not been shown to cause cancer in people, presumably because the immune system is capable of successfully eradicating the infected cells. Cell transformation in cultured cells is due to the expression of two early viral genes, E1A and E1B (see Figure 12.18). E1A binds with high affinity to RB and its relatives (p107 and p130), via a conserved LxCxE sequence motif. E1A also binds a variety of other host cell proteins that could impact cell cycle regulation, including chromatin remodeling factors and ubiquitylation machinery. E1B, by contrast, binds p53 and targets it through a variety of mechanisms, including preventing it from activating transcription, relocalizing it, and perturbing its post-translational modification. Furthermore, in collaboration with another adenovirus early protein, E4orf6, E1B can also act as a ubiquitin E3 ligase, leading to the polyubiquitylation and degradation of p53.

In contrast to polyomaviruses and adenoviruses, human papillomaviruses are a significant cause of human cancers, primarily of the cervix. They are transmitted by intimate contact, and many young people become infected soon after they begin engaging in sexual activity, though only a small fraction of infected people go on to develop tumors later in life. Here again, tumorigenesis is due to the activity of early viral gene products, in this case the proteins E6 and E7. E7, like E1A, binds to and sequesters RB and its relatives via a similar LxCxE motif. E7 is also able to associate with the Cullin 2 ubiquitin E3 ligase complex, leading to polyubiquitylation and proteasome-mediated degradation of RB. By contrast, E6 binds p53 and, in complex with the cellular E3 ubiquitin ligase E6AP (also known as UBE3A), polyubiquitylates p53 and targets it for degradation. E6 is also able to target other apoptotic regulators for degradation, such as the pro-apoptotic Bcl2 family protein BAK.

Fortunately, effective vaccines are available to prevent HPV infection, which in principle could eliminate the major cause of cervical cancer. Research has also shown that downregulation or inhibition of the binding activity of E6 and/or E7 efficiently induces apoptosis in tumors caused by HPV, offering promising therapeutic strategies for those already infected. Such studies also provide clear evidence of the dependence of such tumors on the continuing activities of E6 and E7 in deregulating normal cell cycle and apoptotic mechanisms.

SUMMARY

Many human diseases result from dysregulation of normal cell signaling. This reflects the central importance of finely tuned signaling mechanisms in almost all aspects of normal life in multicellular organisms, including development, homeostasis, and the response to environmental stresses such as pathogen infection. Insight into the specific ways that signaling is disrupted in disease has been important in developing new treatment strategies and drugs that specifically target the affected pathways.

Cancer is a disease caused by the loss of normal control over cell proliferation and survival. Most tumor cells no longer require specific mitogenic signals to proliferate (that is, proliferation signals are constitutively active), and have also lost the ability to halt the cell cycle and induce apoptosis when conditions are inappropriate for proliferation. These changes are typically caused by mutations that lead to the overexpression or activation of oncogenes, and by the deletion or loss of activity of tumor suppressors. Tumors also often acquire the ability to evade surveillance by the host immune system. Increased understanding of the molecular basis of cancer has led to the recent development of targeted therapies to treat cancer. These include small molecules and antibodies that inhibit the activity of oncogenic proteins, and treatments that increase the efficiency of the immune system in detecting and killing tumor cells.

Microbial pathogens often subvert the host's signaling mechanisms to their advantage—enhancing their ability to infect host cells, to replicate and spread within the host, and ultimately to be transmitted to new hosts. Some pathogenic bacteria use specialized type 3 secretion systems to inject toxins and other effector proteins into host cells. These factors modify host signaling pathways, for example, to prevent engulfment of the bacteria by phagocytic cells, or to hamper the host's innate and adaptive immune defenses. Bacteria and viruses that replicate inside host cells often encode mechanisms to hijack the host's actin cytoskeleton, in order to propel themselves through the cytosol and infect neighboring cells while minimizing their exposure to the extracellular environment. Some viruses also manipulate the host's cell cycle and apoptosis pathways to enhance their own replication, in some cases leading to cancer.

QUESTIONS

1. The early stages of tumorigenesis are driven by somatic mutations to oncogenes and tumor suppressors. Targeted cancer therapies almost always target oncogenes and not tumor suppressors. Why do you think there are not more therapeutic strategies that target tumor suppressors such as p53, PTEN, and RB?

2. In some cases, simple overexpression of an oncogene is sufficient to induce constitutive proliferation of a cell, while in other cases, activating mutations in the coding sequence (either point mutations or larger structural rearrangements, such as deletions or gene fusions) are required. Discuss what types of signaling proteins are likely to require only overexpression, and what types would typically require structural mutation.

3. Some genes have the very unusual property of being able to act either as tumor suppressors or as oncogenes, depending on the type of mutation. What type of mutations might convert a protein that normally acts to suppress mitogenic signaling or cell survival into a protein that acts dominantly to stimulate proliferation and/or survival?

4. Despite the success of targeted therapies in treating cancer, the development of recurrent disease that has become resistant to the targeted therapy often occurs within months. One promising approach is to combine several targeted therapies at the same time (combination therapy). What do you think are the potential advantages and disadvantages of this approach?

5. In lung cancers, two of the most frequently mutated genes are the EGF receptor and K-Ras. However, it is very rare to see mutation of both genes in the same tumor. Why might this be?

Answers to these questions can be found online at www.routledge.com/9780367279370

6. Historically, bacterial toxins have provided valuable tools for investigators studying cell signaling. What properties of these toxins make them useful in the laboratory?

7. Viruses and some bacteria replicate inside host cells, while other bacteria remain extracellular. How might these two classes of pathogens differ in how they manipulate host cell signaling pathways?

8. Microbial pathogens that "jump" to a new host (for example, when influenza virus strains that normally infect birds or pigs acquire the ability to infect humans) can be particularly deadly in the new host. What does this tell us about the relationship between the pathogen's evolutionary fitness and its pathogenicity? What competing forces are at play here?

BIBLIOGRAPHY

MALFUNCTIONS IN SIGNALING CONTRIBUTE TO THE GENESIS OF CANCER

Cohen P, Cross D & Jänne PA (2021) Kinase drug discovery 20 years after imatinib: Progress and future directions. *Nat. Rev. Drug Discov.* 20(7), 551–569. doi: 10.1038/s41573-021-00195-4.

Fernández-Medarde A & Santos E (2011) Ras in cancer and developmental diseases. *Genes Cancer* 2(3), 344–358. doi: 10.1177/1947601911411084.

Hanahan D & Weinberg RA (2011) Hallmarks of cancer: The next generation. *Cell* 144(5), 646–674. doi: 10.1016/j.cell.2011.02.013.

Kent LN & Leone G (2019) The broken cycle: E2F dysfunction in cancer. *Nat. Rev. Cancer* 19(6), 326–338. doi: 10.1038/s41568-019-0143-7.

Lemmon MA, Schlessinger J & Ferguson KM (2014) The EGFR family: Not so prototypical receptor tyrosine kinases. *Cold Spring Harb. Perspect. Biol.* 6(4), a020768. doi: 10.1101/cshperspect.a020768.

Moore AR, Rosenberg SC, McCormick F & Malek S (2020) RAS-targeted therapies: Is the undruggable drugged? *Nat. Rev. Drug Discov.* 19(8), 533–552. doi: 10.1038/s41573-020-0068-6.

Prior IA, Lewis PD & Mattos C (2012) A comprehensive survey of Ras mutations in cancer. *Cancer Res.* 72(10), 2457–2467. doi: 10.1158/0008-5472.CAN-11-2612.

Sanchez-Vega F, Mina M, Armenia J, et al. (2018) Oncogenic signaling pathways in the cancer genome atlas. *Cell* 173(2), 321–337.e10. doi: 10.1016/j.cell.2018.03.035.

Sharma P, Hu-Lieskovan S, Wargo JA & Ribas A (2017) Primary, adaptive, and acquired resistance to cancer immunotherapy. *Cell* 168(4), 707–723. doi: 10.1016/j.cell.2017.01.017.

Sigismund S, Avanzato D & Lanzetti L (2018) Emerging functions of the EGFR in cancer. *Mol. Oncol.* 12(1), 3–20. doi: 10.1002/1878-0261.12155.

Simanshu DK, Nissley DV & McCormick F (2017) RAS proteins and their regulators in human disease. *Cell* 170(1), 17–33. doi: 10.1016/j.cell.2017.06.009.

Weinberg, RA. *The Biology of Cancer*, 3rd edition. New York: WW Norton & Co, 2023.

PATHOGENIC MICROBES THAT HIJACK HOST CELL SIGNALING

Alto NM & Orth K (2012) Subversion of cell signaling by pathogens. *Cold Spring Harb. Perspect. Biol.* 4(9), a006114. doi: 10.1101/cshperspect.a006114.

Du Z & Wang X (2016) Pathology and Pathogenesis of *Yersinia pestis*. In: R. Yang, A. Anisimov (eds.), *Yersinia pestis: Retrospective and Perspective, Advances in Experimental Medicine and Biology*, 918. doi: 10.1007/978-94-024-0890-4_7.

Lai Y, Rosenshine I, Leong JM & Frankel G (2013) Intimate host attachment: Enteropathogenic and enterohaemorrhagic Escherichia coli. *Cell Microbiol.* 15(11), 1796–1808. doi: 10.1111/cmi.12179.

Lamason RL & Welch MD (2017) Actin-based motility and cell-to-cell spread of bacterial pathogens. *Curr. Opin. Microbiol.* 35, 48–57. doi: 10.1016/j.mib.2016.11.007.

Pechous RD, Sivaraman V, Price PA, et al. (2013) Early host cell targets of *Yersinia pestis* during primary pneumonic plague. *PLoS Pathog.* 9(10), e1003679. doi: 10.1371/journal.ppat.1003679.

Pollard TD (2016) Actin and actin-binding proteins. *Cold Spring Harb. Perspect. Biol.* 8(8), a018226. doi: 10.1101/cshperspect. a018226.

Tessier TM, Dodge MJ, MacNeil KM, et al. (2021) Almost famous: Human adenoviruses (and what they have taught us about cancer). *Tumour Virus Res.* 12, 200225. doi: 10.1016/j.tvr.2021.200225.

Vats A, Trejo-Cerro O, Thomas M & Banks L (2021) Human papillomavirus E6 and E7: What remains? *Tumour Virus Res.* 11, 200213. doi: 10.1016/j.tvr.2021.200213.

Welch MD & Way M (2013) Arp2/3-mediated actin-based motility: A tail of pathogen abuse. *Cell Host Microbe.* 14(3), 242–255. doi: 10.1016/j.chom.2013.08.011.

Diversity of Signaling across Phylogeny

13

Living organisms show incredible diversity in their shape, size, lifestyle, and organizational complexity. Accordingly, different organisms have very different signaling needs and have evolved very different sets of signaling molecules and mechanisms to address these needs. In this book, we have mostly focused on the signaling machinery used in complex animals such as humans, and to a lesser extent those of simpler model organisms such as bacteria and yeast. In this chapter, we will take a much broader perspective, to examine the commonalities and differences in how various organisms solve signaling problems. We will also consider how the highly complex signaling machinery of multicellular organisms evolved in concert with the increasing complexity of multicellular life.

THE CONSTRAINTS ON SIGNALING IN DIFFERENT ORGANISMS

The process of natural selection works to increase the fitness of a population. In terms of cell signaling, this means that signaling mechanisms have been optimized to the circumstances unique to that group of organisms. These circumstances include physical constraints such as the size of an organism and its constituent cells, and the distance over which signals must travel; constraints due to the lifestyle of the organism, including the types of input signals that must be interpreted and the degree of precision needed in the response; and finally, the toolkit of components and mechanisms that was available to the organism as it evolved.

Differences in the physical properties of different cells constrain the signaling machinery

If we consider the extraordinary variety of cells across the kingdoms of life, it is not surprising that their signaling machinery is similarly varied. Contrast a free-living

DOI: 10.1201/9780429298844-13

bacterial cell just a micron or two long, a mammalian neuron whose axons can stretch over a meter in length, and the walled and immobile cells in the leaves of a tree, and it is obvious that the optimal solutions to each organism's signaling needs are likely to be quite different.

Size is one of the most consequential differences among the cells of different organisms, having a major effect on the speed and fidelity of signal transmission. Perhaps the most fundamental task performed by signaling systems is to transmit information from one location to another in the cell—often from the outer membrane, which is exposed to signals from the environment, to the interior of the cell and in particular to the genomic DNA. In most cases, this means that a molecule must passively diffuse the distance from the membrane to the DNA. The distances involved, of course, depend entirely on the size and shape of the cell.

As we briefly discussed in Chapter 6, the average time needed for a particle to diffuse from one place to another is a function of both the distance and the size of the diffusing particle. More specifically, the time is proportional to the mean squared displacement of a diffusing particle and inversely proportional to its diffusion constant (which itself is proportional to the radius of the diffusing molecule). Thus, a typical protein would take on average 16 ms to diffuse 1 μm (approximately the diameter of bacterium), while the same protein would take almost 3 minutes to diffuse 100 μm, the approximate diameter of many eukaryotic cells. And for it to diffuse a meter, the length of some neuronal processes, would take over a thousand years!

How do such differences impact signaling? It depends on the timescale of the response and the distance that must be covered. As can be seen in **Figure 13.1**, both of these parameters vary over many orders of magnitude in biological systems. In one example that we will discuss in more detail below, bacteria convert many signals from outside the cell into changes in gene expression or other outputs through *two-component systems*, which involve the diffusion from the membrane of a response regulator protein that is post-translationally modified by phosphorylation of aspartic acid. The small dimensions of the bacterium mean that the

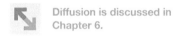

Diffusion is discussed in Chapter 6.

Figure 13.1

The length and timescales of signaling vary enormously. Left: Distances relevant to signaling range from the diameter of a typical protein (5–10 nm) up to the size of the largest organisms. Right: timescales relevant to signaling range from the time it takes an enzyme to catalyze a reaction (less than a microsecond for very efficient enzymes), to days or even years for long-term systemic responses. Where speed is of the essence (for example, the time it takes for visual information to be received by the retina, processed by the brain, and to evoke a response in the muscles), signaling can be very fast, although in many other cases the response time is much slower (minutes to hours). Any signaling response that depends on new transcription and translation must necessarily be slower than the time it takes to transcribe and process mRNA and translate it into protein.

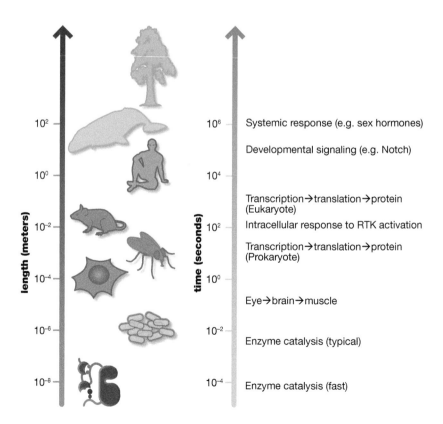

phosphorylated response regulator can rapidly find its binding sites on the bacterial genome, and the highly labile acyl phosphate group of phosphoaspartate is unlikely to be hydrolyzed before the response regulator reaches its destination. The much larger distances involved in the typical eukaryotic cell, however, mean that the trip from membrane to DNA takes much longer on average. While a response time of over a minute may be adequate for many situations, it may be too long for phosphoaspartate to survive before spontaneous hydrolysis. It is thought that this is one reason why eukaryotes are more reliant on phosphoserine and phosphothreonine, which are much more stable to spontaneous hydrolysis.

Relative stability of different phosphoamino acids is discussed in Chapter 4.

Where distances are even larger, for example, in the case of a neuron or muscle cell, it is clear that protein diffusion is nowhere near fast enough to allow timely communication between the cell body and distant cell processes. Here, other mechanisms must be used. In vertebrates, at the organismal level, signals such as hormones can be transmitted throughout the organism relatively quickly through the bloodstream, which is actively pumped by the heart. In the nervous system, electrical waves (action potentials) are used to transmit information almost instantaneously across the vast distances spanned by axons and dendrites. At the cellular level, cells take advantage of ions and small molecules, which can diffuse much more rapidly than a protein by virtue of their much smaller diameter. Signaling systems that involve regulated opening and/or closing of ion channels on the membrane, and/or generation of small molecule second messengers such as cAMP, are used to rapidly transmit information from one side of the cell to another. Finally, in some cells, proteins and nucleic acids can be actively transported to their destination along cytoskeletal tracks via motor proteins, a process driven by ATP hydrolysis.

The lifestyle of an organism dictates what type of signaling mechanisms are needed

Prokaryotes, such as bacteria, lead relatively simple lives, often requiring only hospitable conditions and a variety of small organic and inorganic molecules to grow and reproduce. For the most part, their activities do not need to be coordinated with those of their neighbors (though as we discuss below, such coordination is possible in some circumstances). Accordingly, their needs in terms of signaling are relatively basic, focused on adapting to the presence or absence of particular nutrients, and on moving toward beneficial stimuli and away from noxious ones.

The evolutionary transition from prokaryotes to eukaryotes brought a number of additional complexities, even for those that maintained a free-living, single-celled lifestyle. Not only did eukaryotic cells become bigger, but the cytoplasm became much more diverse, sub-divided into a number of membrane-bound compartments. Therefore, more elaborate mechanisms to transfer information between compartments in response to signals needed to evolve. Many eukaryotic cells lost the rigid cell walls of their prokaryotic ancestors and developed highly complex cytoskeletal systems that allowed them to regulate cell shape, and thus adapt to and migrate in response to extracellular cues. For some, the need to actively hunt food sources such as bacteria necessitated mechanisms to sense, track, and capture prey, and then to engulf it. Sexual reproduction introduced a need to recognize mating partners, and added additional complexity to the regulation of the cell cycle and of DNA replication.

The advent of multicellularity imposed even greater demands on cell signaling systems compared to single-celled organisms (this topic is further discussed later in this chapter). Development of a complex body plan consisting of different specialized cell types, organized into distinct tissues, required exquisite coordination among cells on a variety of scales. Contacts between adjacent cells needed to be controlled by adhesive and repulsive cues, and information on the status of the surrounding cells and extracellular matrix communicated to the interior of the cell. Distant tissues needed ways to communicate with each other via secreted molecules such as hormones. Ways to sense other organisms and respond to them were strongly selected for. In plants, sophisticated mechanisms were needed to recognize microbial pathogens or the damage from insects or larger herbivores, and to mount defenses against such attacks. In multicellular animals, which unlike plants have the ability to roam their environment, sensory systems such as vision and hearing developed to better enable the organism to eat and avoid being eaten. As we will see, a host of new cell signaling systems first appeared around the transition to multicellularity to make such advances possible.

Comparative genomics provides insight into the evolution of signaling mechanisms

The advent of rapid whole-genome sequencing has allowed us to analyze the relatedness of different organisms at the molecular level by comparing their DNA sequences. This has revolutionized our understanding of the evolution of signaling mechanisms, by allowing us to identify where in evolution a particular family of signaling proteins first emerged, what other signaling proteins emerged around the same time (that is, co-evolved), and how the arrangement of specific modular signaling domains (their number, what other domains are found in the same protein) differs in various lineages.

Such analyses might seem conceptually straightforward, but because we cannot actually go back and examine the genomes of the ancestral forms that gave rise to present-day organisms, we can only infer their properties indirectly. Analysis is further complicated by issues such as gene loss and horizontal transmission. Gene loss describes genes that were present in a progenitor, but were lost once they were no longer useful for its descendants (we have already encountered this with the bacterial two-component systems, which are not found in animals but have been retained in plants and fungi). The horizontal transfer of genetic material between divergent organisms, for example, between bacteria and eukaryotic cells, can also complicate establishing where a particular signaling protein first evolved. Furthermore, current genomic databases are far from comprehensive, and we are unlikely to have the genomic sequences of the majority of species for the foreseeable future. Despite these challenges, however, we are rapidly gaining insight into the differences in signaling machinery across phylogeny, and how such differences might have arisen under evolutionary pressure.

Considering signaling systems as they now exist across different lineages, one might assume that evolution has produced the optimal solution for each signaling problem. In reality, however, evolution works with the material it has (that is, the existing protein domains and architecture, and the existing signaling networks), and changes to the existing machinery occur incrementally. Once a system evolves that works reasonably well for one purpose, it is more likely to be adapted to new purposes than it is for entirely new systems to evolve from scratch. For example, many vertebrates share the same overall body plan (typically four limbs, a head and a tail), but obviously this is not the only viable solution to the problem—it is what evolution had to work with in this lineage. Animals in other lineages solved the same problem very differently (consider insects and cephalopods, for example). The same principle applies to signaling: all vertebrates share very similar signaling systems, despite hundreds of millions of years since their last common ancestor.

Conversely, when similar structures or functions evolve independently in different lineages, their signaling mechanisms can be quite different. The eye, for example, has evolved a number of different times in the animal lineage. Even simple unicellular organisms have the ability to detect light and adjust their behavior based on its intensity. In simple multicellular organisms, it is likely photoreceptor cells that could detect light began to organize over time into localized patches, the beginning of dedicated visual organs. In diverse lineages from arthropods to mollusks to vertebrates, such structures independently developed the ability to focus light, and to discriminate shapes and colors.

The mechanism of visual signaling in vertebrates is discussed in Chapter 14.

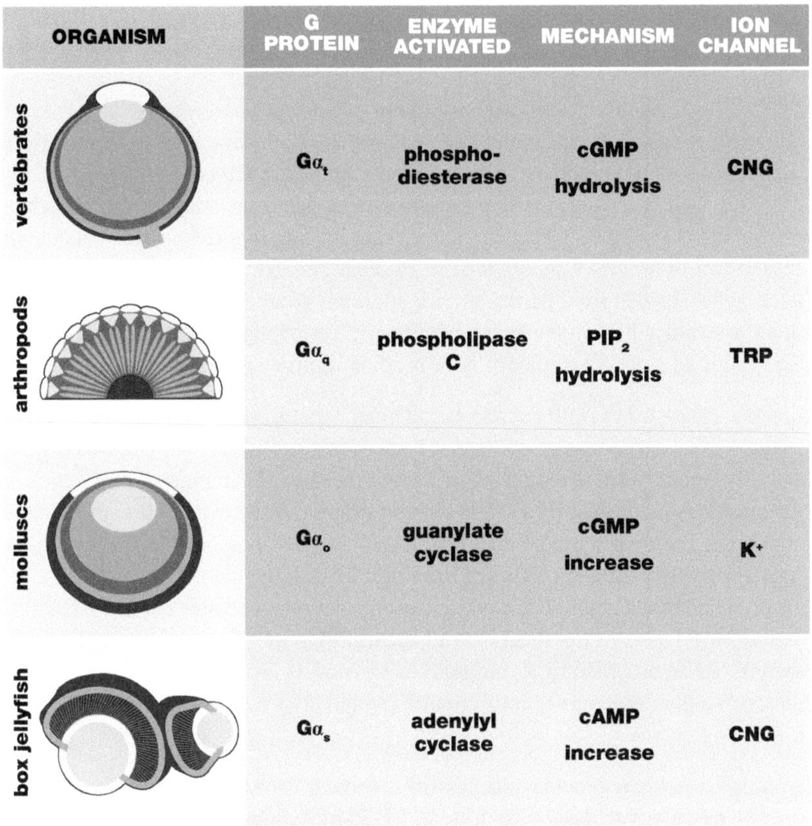

Figure 13.2

Signaling mechanism of vision in different lineages. The eye has independently evolved a number of times. In all cases, a photon striking rhodopsin activates a heterotrimeric G protein, which in turn activates a downstream enzyme. The product in turn regulates the opening and closing of gated ion channels, ultimately resulting in an electrical signal. Each of the four examples here uses a different G protein, G protein effector, and channel, illustrating how different signaling machinery can be harnessed to perform similar tasks in different organisms. (Adapted from Oakley and Speiser, *Annu. Rev. Ecol. Evol. Syst.* 2015; 46:237–260. With permission from Annual Reviews.)

ORGANISM	G PROTEIN	ENZYME ACTIVATED	MECHANISM	ION CHANNEL
vertebrates	$G\alpha_t$	phospho-diesterase	cGMP hydrolysis	CNG
arthropods	$G\alpha_q$	phospholipase C	PIP_2 hydrolysis	TRP
molluscs	$G\alpha_o$	guanylate cyclase	cGMP increase	K⁺
box jellyfish	$G\alpha_s$	adenylyl cyclase	cAMP increase	CNG

Once a few organisms possessed a primitive eye, the evolutionary pressure for both prey and predators to develop better eyes must have been enormous. But if we compare the eyes of different lineages, despite structural similarities and their dependence on photosensitive proteins of the opsin family to detect photons, the signaling pathways downstream of the opsin can vary enormously (**Figure 13.2**). Apparently, a variety of mechanisms can fulfill the same basic signaling task: greatly amplifying the signal from light-activated opsins as quickly as possible.

SIGNALING THEMES IN PROKARYOTIC ORGANISMS

Bacteria and archaea are likely to resemble to the earliest forms of life, where the ability to detect and respond to changes in environmental conditions first evolved. Prokaryotic signaling can therefore provide insight into the origins of more elaborate mechanisms seen in eukaryotes. While we are apt to think of prokaryotes as relatively simple or primitive organisms, their signaling systems are extremely well suited to their tasks. Because prokaryotes have been around for billions of years, their machinery has been subject to selection for optimal growth and survival over an enormous number of generations (perhaps trillions of generations, given the rapid doubling times of many bacteria). Their bare-bones signaling mechanisms are thus much more highly optimized than those of multicellular organisms, which arose more recently and tend to have much longer generation times. We cannot do justice to the extraordinary diversity of prokaryotes and their signaling in this brief section, but instead we will consider just a few examples of the relatively simple but exquisitely fine-tuned mechanisms used by bacteria.

Bacterial operons couple transcriptional regulation with metabolic activity

The earliest signaling mechanisms likely regulated *homeostasis*, allowing an organism to adapt its metabolic and biosynthetic pathways to the availability of nutrients and other raw materials in the environment. Such adaptation is typically accomplished by changes in the transcription of genes encoding proteins such as membrane transporters and catabolic or anabolic enzymes (those that mediate the breakdown or biosynthesis of nutrients or other metabolites, respectively). In bacteria, genes needed for a particular task are often grouped together in the genome in an **operon**, where a single promoter and a set of regulatory elements control expression of a polycistronic mRNA encoding multiple proteins.

In some cases, transcriptional regulation in bacteria is quite simple and direct. For example, the *ara* operon of the enteric bacterium *Escherichia coli* controls expression of proteins needed to take advantage of the sugar L-arabinose. Expression of the operon is controlled by a DNA-binding protein, AraC, which directly binds to arabinose. In the absence of arabinose, AraC forms a homodimer that binds two sites upstream of the *ara* promoter, forming a DNA loop that inhibits transcription. When AraC binds arabinose, however, conformational changes convert AraC into a transcriptional activator that binds adjacent sites in the *ara* promoter, and thereby recruits RNA polymerase to promote expression (**Figure 13.3**). These types of direct, "single-component" transcriptional regulatory systems are quite common in bacteria.

A more complex regulatory mechanism is seen in the *lac* operon of *E. coli*. This operon has a particularly rich history, as many fundamental concepts of transcriptional regulation were first demonstrated in this system more than 60 years ago. The *lac* operon consists of three genes involved in lactose metabolism: *lacZ*, encoding a β-galactosidase that cleaves lactose into glucose and galactose; *lacY*, encoding a permease that transports lactose into the cell; and *lacA*, encoding a transacetylase that modifies lactose. Because glucose is a more efficient energy source for the cell than lactose, when it is plentiful the transport and metabolism of

Figure 13.3

AraC mediates transcription of the Ara operon in the presence of arabinose.
(A) The domain organization of AraC.
(B) In the absence of arabinose, the AraC dimer binds to two distant sites, O_2 and I_1, which results in DNA looping and prevents the binding of RNA polymerase. The affinity for AraC is higher for I_1 than it is for I_2. (c) In the presence of arabinose, conformational changes in AraC allow it to bind cooperatively to two closely adjacent sites, I_1 and I_2, leading to recruitment of RNA polymerase and transcription of the operon.

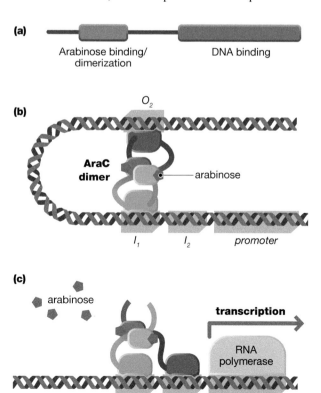

lactose are suppressed to save energy. This is accomplished by several interlocking mechanisms.

First, the lac repressor (LacI) binds to the "operator" region of the promoter and prevents binding of RNA polymerase when lactose is absent from the cell (**Figure 13.4**). As in the case of AraC (discussed above), the repressor binds cooperatively to multiple operator sites, forming a stable DNA loop in the repressed conformation. When lactose is present, however, LacZ converts some of it to allolactose (the rest is cleaved into lactose and glucose), which binds to LacI and induces a conformational change, lowering its affinity for the operator and allowing RNA polymerase to bind and transcribe the genes in the operon. This sets up positive feedback, as the LacY and LacZ produced by new transcription further increase the transport and processing of lactose, and further inhibit the repressor. This also allows a modest amount of expression when both glucose and lactose are present.

The second control element is a transcriptional activator, the catabolite activator protein or CAP (also termed cAMP receptor protein or CRP). CAP binds near the promoter and activates transcription only when it is bound to cAMP, which is present when cytosolic glucose levels are low. Glucose levels are sensed in the cell by a phosphotransfer system coupled to glucose transport across the membrane, termed the phosphoenolpyruvate-carbohydrate phosphotransferase system (PTS). This system allows operon activity to be dynamically coupled both to membrane transport of nutrients, and to the overall metabolic activity of the cell.

The PTS system consists of three major components: Enzyme I (EI), which uses phosphoenolpyruvate (PEP) as a donor to autophosphorylate itself on histidine; this phosphate is then transferred to a histidine residue on the so-called Histidine Protein (HPr); and finally, the phosphate is transferred to a multicomponent EII complex on the plasma membrane (**Figure 13.5**). The EII complex for glucose

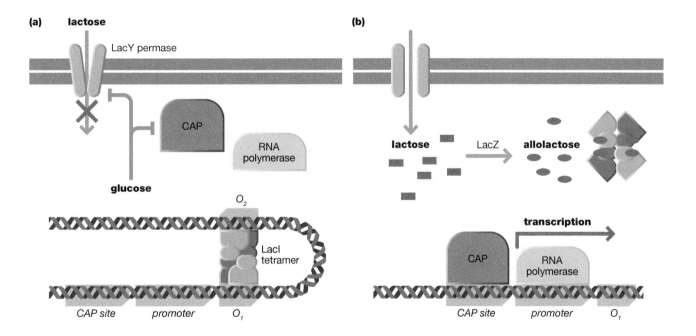

Figure 13.4

Regulation of Lac operon by levels of glucose and lactose. (A) Under conditions where glucose concentrations are high in *E. coli*, transcription of the Lac operon is inhibited by three interlocking mechanisms. The tetrameric LacI inhibitor binds to two operator sites, looping the DNA and preventing binding of RNA polymerase. High glucose levels also lead to inhibition of the CAP transcriptional activator and to inhibition of the Lac permease (LacY). (B) When glucose levels are low, LacY and CAP are activated, and LacZ converts a fraction of lactose to allolactose, which binds to and inhibits LacI. This leads to recruitment of RNA polymerase and robust transcription of the Lac operon.

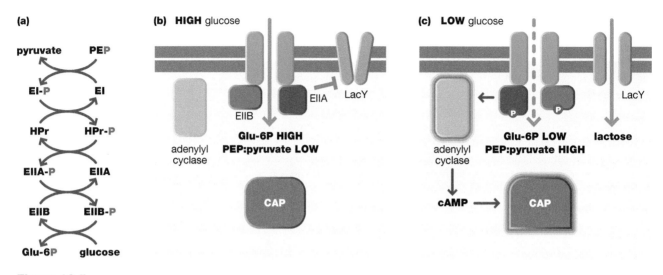

Figure 13.5

The PTS system couples Lac operon activity to metabolic state. (A) The PTS system transfers phosphate from phosphoenolpyruvate (PEP) to glucose and other sugars through a number of intermediate steps. (B) When glucose transport is high, phosphate is rapidly transferred to glucose, and the levels of phosphorylation of the EIIA and EIIB subunits of the glucose transporter complex are low. Unphosphorylated EIIA inhibits LacY, the lactose permease. (C) When glucose levels are low and the cell needs to utilize other sugars such as lactose, transport of glucose is low and levels of EIIA and EIIB phosphorylation are high. In this case, phosphorylated EIIA activates adenylyl cyclase, and the resulting elevated levels of cAMP activates CAP, which can then bind to the Lac operon and stimulate its transcription.

consists of three components (EIIA, EIIB, and EIIC), which couple the transport of glucose into the cell with the transfer of the phosphate from EII to glucose, yielding glucose-6 phosphate. Thus, when flux of glucose is high, phosphate is rapidly transferred from PEP through the proteins of the PTS system to glucose, and levels of phosphorylated PTS components are low. Phosphorylation of PTS components also depends on the overall metabolic activity in the cell via the ratio of PEP to pyruvate; when cells are starved, the PEP:pyruvate ratio is high, and PTS components will be more highly phosphorylated.

Regulation of CAP activity is coupled to *Lac* transcription through the phosphorylation state of one of the PTS proteins, EIIA^{Glc}. When phosphorylated, EIIA^{Glc} can bind to and activate adenyl cyclase, the enzyme that generates cAMP.

Adenyl cyclase and cAMP are discussed in Chapter 6.

Thus, in conditions where flux of glucose is low, and/or overall metabolic rate is low (conditions where the cell needs to use alternative energy sources), cAMP levels go up, and CAP binds cAMP and activates transcription from the *Lac* operon.

Finally, Lac transcription is regulated by yet another mechanism, termed *inducer exclusion*. As noted above, the Lac operon can only be expressed if there is some lactose present: its metabolite allolactose must bind to LacI to inactivate it and relieve transcriptional repression. The lactose permease, LacY, is itself regulated by glucose levels and the PTS system through EIIA^{Glc}. As in the case of adenyl cyclase, this depends on the phosphorylation state of EIIA^{Glc}, which when unphosphorylated binds to and inactivates LacY. In these conditions (when glucose is abundant), lactose cannot enter the cell, and LacI is active and represses Lac transcription. When glucose transport and/or metabolic levels are low, however, EIIA^{Glc} is phosphorylated and cannot inhibit LacY, lactose can enter the cell, and LacI can be inactivated leading to Lac transcription.

These three interlocking regulatory mechanisms lead to very tight control of lac operator in response to multiple metabolic inputs. Similar systems of transcriptional control linked to PTS systems regulate other operons in *E. coli* that control the import and catabolism of other nutrients. In other bacteria, some of the specific regulatory details differ, but the broad themes of regulation of transcription via activity of the PTS system are similar. Importantly, the phosphorylation of PTS components is highly transient due to the very high-energy phosphohistidine (or in some case phosphocysteine) bond, so the response to changing metabolic conditions can be nearly instantaneous.

The instability of phosphohistidine is discussed in Chapter 4.

Bacterial populations can exchange information and coordinate their activities

While almost all prokaryotes are unicellular, under some conditions their cells can work together to enhance the overall survival of the population. Such microbial communities may consist entirely of the same species of organism, or of a mixture of species that make complementary contributions. An example of such communities are **biofilms**, which form on surfaces and consist of microbes embedded in a secreted extracellular polymer matrix. Biofilms can impact human health, as they can provide a sheltered environment for potentially harmful microbes to thrive, as in dental plaque, endocarditis (a persistent infection of the heart valves), or the colonization of indwelling medical devices.

In microbial communities such as biofilms, the behavior of individual cells must be coordinated, so from a signaling perspective these communities have some of the properties of multicellular organisms. In both instances, communication among individual cells is essential for cooperative behavior. One way in which bacteria can exchange information is through the process of **quorum sensing**, in which individual bacteria secrete signaling molecules called autoinducers (AI) that are bound by receptors and can induce transcriptional responses, both in the secreting cell and in other cells in the community. These quorum sensing mechanisms typically regulate responses that would be costly for an individual cell, but can be beneficial to a larger population when it responds collectively. In addition to biofilm formation, quorum sensing regulates other collective responses such as bioluminescence, the induction of virulence factors during infection, and the production of "public goods" such as secreted enzymes that can liberate essential nutrients by digesting surrounding extracellular materials.

Among different types of bacteria, the secreted AI molecules vary greatly in structure. In gram-negative species, the AI are typically membrane-permeable small molecules derived from S-adenosyl methionine, such as acyl-homoserine lactones (AHLs). By contrast, in gram-positive bacteria, the AI are usually derived by modification of small peptides (**Figure 13.6**). In all systems, once the AI levels have reached a threshold (indicating that the local density of bacteria of the same species is sufficiently high), the transcriptional program switches to promote the transcription of genes needed for biofilm formation, virulence, or other communal activities.

The AI receptors are critical to the quorum sensing response. For both gram-negative and gram-positive organisms, receptors can be either typical transmembrane two-component receptors or cytoplasmic "one-component" receptors (discussed above). In the two-component systems, AI binding regulates the intrinsic histidine kinase/phosphatase activity of the receptor, which in turn couples to response regulator (RR) proteins that can directly bind DNA and modify transcription. The one-component receptors contain both DNA- and AI-binding domains; AI binding induces conformation changes that affect DNA binding and thus transcriptional regulation.

Two-component signaling systems are described in Chapter 4.

Typically, a cell will express AI receptors that bind with high specificity only to the AI generated by the cell and others of the same species, and thus can read out the local density of closely related organisms. Some receptors have wider specificity, however, and are able to respond to AI generated by other species as well. Furthermore, some bacteria express multiple AI and/or receptors. As in the much more intricate combinatorial signaling of multicellular organisms, complex gating and

Figure 13.6

Quorum sensing in bacterial populations. (A) Structure of a few representative autoinducers (AI) is shown. In gram-negative bacteria, membrane-permeable acyl-homoserine lactones (AHLs) such as AI-2 and OHHL are common, while in gram-positive bacteria AI are typically cyclized or otherwise modified peptides, such as AIP-1 from *Staphylococcus aureus* (*orange* circles represent standard amino acids, using single letter code). (B) Transcription of target genes depends on the local concentration of AI secreted by bacteria; when population density is low, AI concentrations (*red* shading) are below the threshold for inducing transcription of target genes.

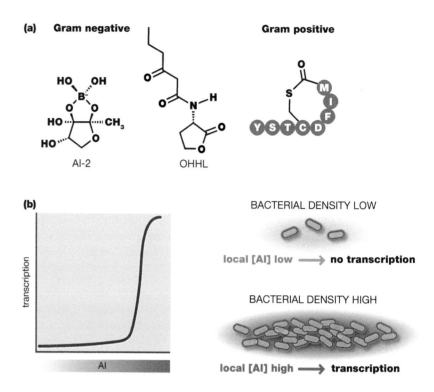

regulatory responses can thus be generated. In this way, each cell has the potential to read out changes in who is in the neighborhood, whether they are potentially helpful or harmful, and to adjust its transcriptional program accordingly to maximize its chances of replication and survival.

SIGNALING MECHANISMS IN MULTICELLULAR PLANTS

Plants are thought to have evolved from an early eukaryotic cell that engulfed a free-living cyanobacterium, entering into an endosymbiotic relationship (a similar, earlier event is thought to have led to the eukaryotic mitochondrion). The ability to capture carbon and store solar energy by photosynthesis gave the new cell and its descendants an enormous advantage. Eventually these simple single-celled aquatic organisms underwent further momentous transitions, evolving complex multicellular forms and then emerging from the oceans to colonize the land. Now, by some measures, plants are the most successful and abundant multicellular organisms. Because land plants evolved multicellularity independently of the animal (metazoan) lineage, it is not surprising that the signaling mechanisms they use can be quite different from those seen in animal cells. In the following sections, we will explore a few of these mechanisms, highlighting distinctions from animal cells and the ways these mechanisms are suited to the particular evolutionary constraints faced by plants.

Signaling needs of plants are affected by their structure

Plants have a number of obvious differences from animals, which helped shape the type of signaling mechanisms that evolved. One obvious difference is that plants are literally rooted to their environment, and thus cannot move to avoid unfavorable conditions, microbial pathogens, or herbivores of various sorts. As we will see below, however, plants have evolved highly elaborate innate immune systems that allow the organism to mount a coordinated defense against a variety of threats. Plants also share a distributed body plan in which no single part is essential, unlike the more specialized division of labor seen in metazoans. Another obvious difference is the central importance of light as an environmental cue. Although specific mechanisms won't be discussed here, plants have evolved a number of light-responsive signaling proteins, some of which have been exploited by molecular biologists to develop light-regulated (**optogenetic**) tools for research (see Chapter 15).

One major difference from animals is that the cells of multicellular plants are surrounded by a rigid cell wall, consisting primarily of cellulose. An obvious consequence of this is that individual cells cannot freely migrate within the plant. In terms of cell signaling, the cell wall physically prevents the plasma membranes of adjacent cells from interacting directly with each other, and also limits the size of macromolecules that can be transported between cells across the cell wall. Perhaps to compensate for the challenges that this presents for cell-to-cell communication and tissue homeostasis, in most plant cells the cytosol is connected to that of their neighbors by membranous tubules called **plasmodesmata**. These structures, consisting of an outer membrane continuous with the plasma membrane of adjacent cells and an inner core continuous with their endoplasmic reticulum, allow proteins, nucleic acids, and other molecules up to ~70 kDa in size to move freely between cells (**Figure 13.7**). While most cells of the plant are interconnected in this way, some can be permanently or inducibly sealed off from their neighbors by a substance called callose, which physically plugs the plasmodesmata. When a cell is stressed or infected by pathogens, for example, callose deposition prevents the damage from spreading beyond the original cell. The regulation of transport through plasmodesmata, and how this transport can be modulated in development and in response to various signals, is an area of active research in plant biology.

Note that cell–cell communication through plasmodesmata is relatively slow, as it depends on diffusion, and certainly would not be sufficient to allow timely communication throughout the entire plant. For more rapid transport of nutrients and other signaling molecules, terrestrial plants have an extensive vascular system that actively moves such compounds by bulk fluid flow.

Another unique challenge faced by plants is the need to coordinate the activity of the chloroplast, which has a reduced but still functional genome derived from the ancestral cyanobacterium, with that of the nucleus. Because many of the proteins needed in the chloroplast are encoded by nuclear genes, to maintain **proteostasis**, it is essential that the nucleus adjust transcription of these genes in concert with the needs in the chloroplast. It is also important that the level of photosynthetic activity in the chloroplast be communicated to the nuclear genome, not only to prevent the accumulation of damaging **reactive oxygen species** (**ROS**) generated by photosynthesis under high-light conditions, but also to coordinate cell- and organ-wide responses (for example, the opening and closing of stomata in leaves to regulate water loss and gas exchange) with the amount of photosynthesis (**Figure 13.8**). It has been suggested that tight control of this so-called *retrograde signaling* from

An example of a light-responsive protein is provided in Chapter 10.

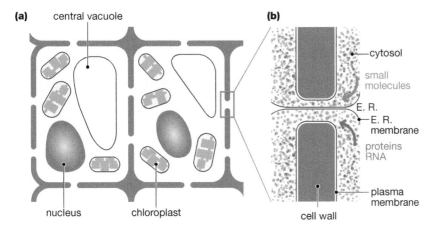

Figure 13.7

Plant cells are connected by plasmodesmata. (A) Overall structure of typical plant cells, which are separated from each other by cell walls consisting of cellulose. (B) Magnified depiction of plasmodesmata. The plasma membranes of adjacent cells are continuous, spanning a gap in the cell wall. A thin tube of membrane continuous with the endoplasmic reticulum (ER) runs through the middle. Cytosolic components, including small molecules, proteins, and nucleic acids, can pass between cells in the sheath between the plasma membrane and ER membranes.

Figure 13.8

Retrograde signaling from chloroplast to nucleus. Many proteins needed for chloroplast function are derived from nuclear transcripts. These proteins (*orange*) are imported into the chloroplast, where they function in complex with proteins derived from the chloroplast genome (*brown*). Nuclear transcription must be adjusted depending on chloroplast activity and stress (for example, high-light conditions). Chloroplast state is signaled via reactive oxygen species (ROS) such as H_2O_2 and singlet oxygen, small molecule intermediates in pigment biosynthesis such as tetrapyrroles, and other compounds which regulate transcription of nuclear genes.

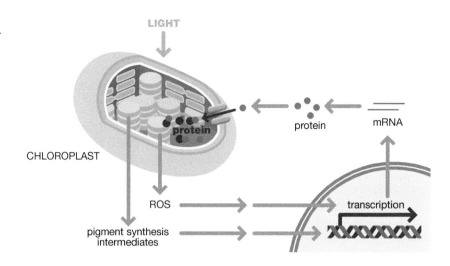

plastid to nucleus was essential for the emergence of land plants, because stresses due to excessive sunlight and desiccation are much more likely on land than in aquatic environments.

While the mechanisms of retrograde signaling are still being elucidated, some of the known signals originating in the chloroplast are small molecules such as tetrapyrroles, which are intermediates in the biosynthesis of pigments used in the photosynthetic machinery. In other cases, ROS generated as a by-product of photosynthesis is thought to directly oxidize key signaling proteins, leading to their degradation or inactivation, which in turn generates downstream signals that modulate nuclear transcription.

Plant hormones (phytohormones) use a variety of mechanisms to modify transcriptional programs

Plants need to coordinate the activities of various tissues (such as roots, stems, and leaves) in development and in response to changes in environmental conditions. This presents particular challenges given the sheer size of some plants (extending many hundreds of feet in some cases) and their decentralized organization. Plants use a variety of hormones, termed **phytohormones**, to communicate between cells both locally and throughout the entire organism. Unlike in animals, where most hormones are synthesized and secreted by specialized endocrine organs, phytohormones can be synthesized throughout the plant; the response, however, often varies considerably depending on the responding cell. Another difference is that the receptors for many phytohormones are located inside the cell. Many phytohormones can pass from the cytosol of one cell to another via plasmodesmata; alternatively, they may passively cross the plasma membrane by diffusion, as is the case for gases like ethylene, or they are small molecules that can be actively transported through membranes. In this section, we will consider three representative phytohormones—cytokinin, ethylene, and auxin—and highlight the distinct mechanisms they use to signal to the nucleus.

Cytokinins regulate a number of aspects of plant development, including cell proliferation. Cytokinins are modified forms of adenine, and while they are found in most organisms, they seem to have signaling activity only in plants. They signal via a two-component phospho-relay signaling system similar to those found in bacteria.

Two-component systems are discussed in Chapter 4.

The cytokinin receptor contains both a histidine kinase domain and a receiver domain, to which phosphate is transferred upon cytokinin binding (**Figure 13.9**). Phosphate is then transferred yet again via cytosolic histidine phosphotransferases to aspartic acid residues on the receiver domains of RR proteins. Some of these (termed Class B RR proteins), once activated by phosphorylation, bind

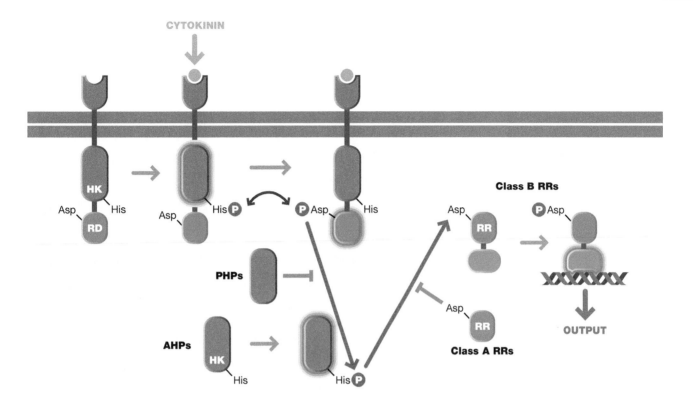

Figure 13.9

Cytokinin signaling. Cytokinin receptors contain both a histidine kinase (HK) and a receiver domain (RD). Upon ligand binding, a phospho-relay system transfers phosphate to authentic histidine phosphotransferases (AHPs) and ultimately to Class B response regulators (RRs), which bind DNA and promote transcription when phosphorylated. Pseudo-histidine phosphotransferases (PHPs) and Class A response regulators serve as negative regulators that tune output levels.

to cytokinin-dependent promoters in the nucleus and stimulate transcription directly. Negative regulation can occur at several levels: pseudo-histidine phosphotransferases (PHPs), which lack the histidine phosphorylation site, can inhibit the phospho-relay by competing with the authentic histidine phosphotransferases (AHPs); in addition, class A RR proteins, which lack DNA-binding activity, can competitively inhibit the activation of Class B RR proteins. Such inhibitors can either shape the initial response to cytokinins or be induced by cytokinin pathway activity, and thus serve as feedback inhibitors.

We had earlier suggested that two-component systems were better suited for smaller organisms such as bacteria, as the short half-life of the high-energy phosphohistidine and phosphoaspartate bonds suggests they might hydrolyze before being able to diffuse the relatively long distances needed to find their DNA-binding sites in eukaryotic cells. How might plant cells have avoided this problem? First, most of the cytokinin receptors are found on the endoplasmic reticulum, which typically is continuous with the nuclear membrane and thus closer to its DNA target than the plasma membrane. In addition, the fact that the phospho-relay has multiple steps may help make long-distance transmission more efficient, as the histidine-phosphorylated AHPs can diffuse from the cytosol to the nucleus before phosphate is transferred to aspartate in the RR proteins. Phosphohistidine is somewhat more stable than phosphoaspartate under physiological conditions.

Ethylene also uses a signaling mechanism based on two-component signaling, but with an interesting twist—it dispenses with the typical RRs that directly regulate transcription in other systems. Ethylene gas synthesized by plants regulates a variety of behaviors; it is perhaps best known for its role in hastening the ripening of many kinds of fruit. Ethylene binds to a family of receptors related to two-component receptors. Each has a membrane-spanning ethylene binding domain linked to an intracellular kinase domain, which can have either histidine or serine/threonine kinase activity. In the model plant *Arabidopsis*, three of the five ethylene receptors also have a receiver domain. The functional unit for signaling is likely to be a dimer or higher-order oligomer of receptors. Surprisingly, neither histidine nor serine/threonine kinase catalytic activity is absolutely required for signaling, suggesting that the main consequence of ethylene binding is conformational change.

It is thought that ethylene primarily regulates the interaction of its receptors with two downstream proteins: the serine/threonine kinase CTR1 and a scaffolding protein EIN2 (**Figure 13.10**). In the absence of ethylene, the receptor facilitates the phosphorylation of EIN2 by CTR1, presumably by acting as a scaffold. This leads to the polyubiquitylation of EIN2 and its degradation; the absence of EIN2, in turn, leads to the ubiquitylation and degradation of a class of transcription factors including EIN3 that mediate ethylene responses. In the presence of ethylene, however, EIN2 phosphorylation decreases, leading not only to more EIN2, but also to a specific proteolytic cleavage that leaves the N-terminal half of EIN2 embedded in the endoplasmic reticulum (ER) membrane and freeing its C-terminus to diffuse away. The C-terminus of EIN2 has two distinct activities: first, it binds to and induces the degradation of mRNAs encoding the ubiquitin E3 ligases that target the transcription factor EIN3 and its relatives, thus increasing levels of those transcription factors; and second, it can localize to the nucleus itself and participate directly in transcription of ethylene-responsive genes.

While a number of details remain to be worked out, we see that the evolutionarily ancient two-component signaling system has been adapted to new uses for ethylene signaling in plants. Here, it is coupled to the activity of ubiquitin ligases that directly and indirectly regulate the levels of the transcription factors that mediate

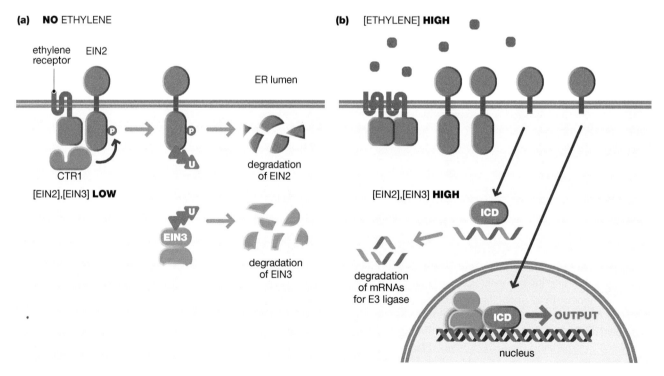

(a) NO ETHYLENE

ethylene receptor   EIN2

ER lumen

CTR1

[EIN2],[EIN3] **LOW**

degradation of EIN2

degradation of EIN3

(b) [ETHYLENE] HIGH

[EIN2],[EIN3] **HIGH**

ICD

degradation of mRNAs for E3 ligase

ICD OUTPUT

nucleus

Figure 13.10

Ethylene signaling. (A) In the absence of ethylene, the ethylene receptor facilitates phosphorylation of EIN2 by CTR1. Phosphorylated EIN2 is polyubiquitylated and degraded. EIN3 and related transcriptional activators are also polyubiquitylated and degraded. (B) In the presence of ethylene, the receptor dimerizes and no longer promotes EIN2 phosphorylation. Unphosphorylated EIN2 accumulates and is cleaved, liberating its intracellular domain (ICD). The ICD binds to and promotes the degradation of mRNAs encoding E3 ubiquitin ligases that target EIN3; in addition, the ICD relocalizes to the nucleus and promotes transcription by EIN3.

the ethylene response. Interestingly, while the kinase activity of the ethylene receptor is not absolutely required for signaling, it does modulate receptor output. One way this is thought to occur is through phosphotransfer from the receptor to the downstream components of the cytokinin pathway described above (that is, AHPs and RR proteins). This illustrates a general theme that one phytohormone often modulates the cellular responses to others. This provides a broad range of possible responses, depending on the other phytohormones and downstream components that may be present in the cell.

The final example, *auxin* signaling, also harnesses the proteasomal degradation machinery to regulate transcriptional activity. Auxins, which are small molecules related to indole-3-acetic acid (IAA), are perhaps the most celebrated and longest-studied phytohormones; they are regarded as master regulators of plant development, eliciting a wide range of specific responses depending on the cell environment. In the cytosol, auxin serves as a heterodimerizer to bring together two distinct proteins (**Figure 13.11**): TIR/AFB family proteins, which are F-box containing, substrate-binding components of a SCF-like ubiquitin E3 ligase complex, and transcriptional repressors of the AUX/IAA family. In the absence of auxin, these AUX/IAA proteins inhibit the transcription of auxin-responsive genes by binding to auxin response factors (ARFs), which are bound to their promoters. In the presence of auxin, however, AUX/IAA proteins are recruited to the SCF complex, polyubiquitylated, and degraded by the proteasome. Once enough AUX/IAA proteins are proteolyzed, repression is relieved and transcription of auxin-responsive genes can occur.

The SCF complex is described in Chapter 9.

This system lends itself to a broad range of quantitative transcriptional responses. First, there are many TIR/AFB and AUX/IAA family members, each with different affinities for each other and for auxin, and thus different AUX/IAA proteins have a broad spectrum of half-lives across a wide range of auxin concentrations. There is also diversity in the ARFs—not only do they have different affinities for various AUX/IAA proteins, but some ARFs also lack transcriptional activation domains and can themselves act as repressors in the absence of AUX/IAA proteins. Finally, both ARFs and AUX/IAA proteins can dimerize and form higher-order oligomers. Thus, cells can differ in their repertoire of AUX/IAA proteins, in how the concentrations of these proteins respond to different concentrations of auxin, and in which ARF proteins are bound to each promoter and which AUX/IAA proteins will bind

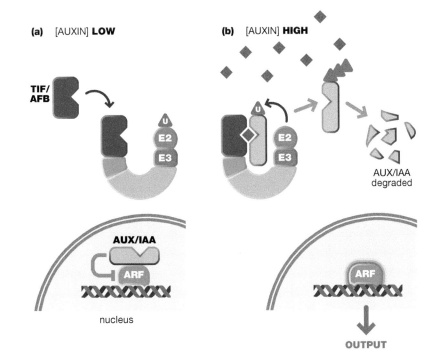

Figure 13.11

Auxin signaling. (A) In the absence of auxin, AUX/IAA proteins associate with ARF proteins in the nucleus and repress transcription of auxin-responsive genes. (B) When auxin is present, however, it acts to heterodimerize AUX/IAA with TIR/AFB proteins. This recruits AUX/IAA to a SCF-like ubiquitin ligase complex, leading to polyubiquitylation and degradation of AUX/IAA. ARF can then act as a transcriptional activator to promote transcription of auxin-responsive genes. Relative levels of different ARF, AUX/IAA, and TIR/AFB proteins in each cell tune the response to auxin.

them. As a result, both the dynamics and nature of the transcriptional response can vary widely from cell to cell, consistent with the role of auxin as a global coordinator of developmental signaling throughout the plant.

Plants have extensive stress response and innate immunity systems to respond to pathogens and predation

Like animals, plants are susceptible to infection by a host of pathogens, including viruses, bacteria, fungi, and parasites. They are also eaten by herbivores ranging from insects to large mammals. For plants, defense against these many hazards is complicated by their inability to move and by the cell walls that preclude the motile immune cells that are central to the adaptive immune system in animals.

Adaptive immunity is discussed in Chapter 12.

However, plants have developed a series of highly sophisticated innate defenses that serve to protect their cells from successful infection by pathogens and to physically limit the extent of damage to protect the larger organism. In this section, we will look at the signaling mechanisms that form the basis for these defenses, and highlight some surprising similarities to innate immunity in animals.

The defensive mechanisms used by plants consist of two overlapping and complementary layers. In the first line of defense, plasma membrane receptors that recognize common pathogen features, termed *pattern-recognition receptors (PRRs)*, initiate intracellular responses, while the second layer consists of intracellular **nucleotide-binding and leucine-rich repeat receptors (NLRs)**, which respond primarily to so-called *effectors* that are secreted by pathogens in an effort to overcome the plant's initial defensive responses (**Figure 13.12**).

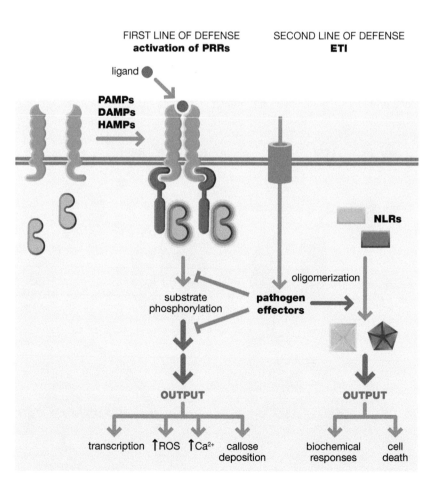

Figure 13.12

Plant innate immunity. A series of pattern-recognition receptors (PRRs) with intrinsic or associated serine/threonine kinase domains are activated by pathogen-, stress-, or herbivore-associated molecular patterns. Upon activation, signals are initiated by substrate phosphorylation, ultimately leading to cellular responses. Microbial pathogens often secrete effectors that interfere with host immune responses mediated by PRRs. In response, plants mount an effector-triggered immunity (ETI) response, which recognizes the effectors themselves or their modified substrates. Target recognition by nucleotide-binding and leucine-rich repeat receptors (NLRs) leads to their oligomerization and to host defensive responses, which can include cell death.

PRRs either have their own intracellular serine/threonine kinase domains, or associate with co-receptors and/or cytosolic kinases that provide kinase activity. In most plants, there are several hundred different PRRs that respond to different pathogen-, damage-, or herbivore-associated molecular patterns (sometimes termed PAMPs, DAMPs, and HAMPs, respectively). For example, one such receptor recognizes a peptide from the bacterial flagellar protein, flagellin, while another recognizes peptides found in caterpillar salivary secretions. When PRRs engage their cognate ligands, their intrinsic or associated kinase domains are activated through auto- or transphosphorylation. This in turn leads to phosphorylation of intracellular substrates, including cytosolic kinases (RLCKs), and ultimately to the activation of downstream signaling cascades, such as the MAP kinase cascade. Cellular consequences include calcium influx, deposition of callose to seal off the plasmodesmata connections to other cells, production of reactive oxygen species (ROS), and transcriptional activation of genes such as those needed for biosynthesis of antimicrobial or insect-repelling compounds. These responses are designed to either kill or repel invaders, or localize damage and prevent its spread. Echoes of these evolutionarily ancient innate immune mechanisms are seen in metazoans in the Toll-like receptors, which lead to activation of NF-κB and other downstream signals.

Toll-like receptors and their roles in innate immunity are discussed in Chapter 8.

To counter these cellular defenses, many pathogens secrete molecules termed effectors that interfere with them in various ways. In the plant, this in turn elicits *effector-triggered immunity* (*ETI*) to counteract the effects of the pathogen effectors. For example, one effector secreted by pathogenic bacteria, AvrAC, uridylates and inactivates key kinases that work downstream of PRRs (uridylation is the enzymatic addition of uridine monophosphate [UMP] to serine or threonine residues of a target protein). The key players in the second line of defense in plants, the NLRs, recognize either the effectors themselves or their modified substrates (in the AvrAC example above, the ZAR1 NLR recognizes and binds to uridylated kinases). Once bound to their targets, NLRs typically undergo conformational change and oligomerization (see Figure 13.12). Downstream consequences of NLR activation are similar to those outlined above for PRRs, and strong activation can also lead to the *hypersensitive response* (*HR*)—the death of the affected cell.

Plants contain a wide variety of NLRs (often numbering in the hundreds), some of which work alone and some of which bind to "helper" NLRs. They typically consist of three domains: a leucine-rich repeat (LRR) region that recognizes specific substrates, a nucleotide-binding domain, and a more variable N-terminal domain, which can include coiled-coil or TIR domains.

TIR domains were introduced in Chapter 8.

Recent structural studies by cryo-electron microscopy (cryo-EM) have shed light on the response of NLRs when they encounter their targets (**Figure 13.13**). Conformational changes in monomeric NLRs upon target binding lead to exchange of bound ADP for ATP, inducing further conformational changes leading to assembly of an oligomeric wheel-like structure, termed a *resistosome*. In the case of ZAR1, interaction with uridylated kinases induces assembly of a pentameric structure. Coiled-coil regions at the center of the resistosome form a pore-like structure with the potential to span the membrane, which is likely to serve as a plasma membrane calcium channel that mediates Ca^{2+} influx and ultimately induces cell death. A second type of NLR that contains TIR domains, exemplified by RPP1, assembles into a tetrameric resistosome upon encountering its target effector. In this case, the TIR domains of the NLR assemble at the hub to generate NAD^+-cleaving enzymatic activity; this produces ADP-ribose (ADPR) and its variant cyclized versions (v-cADPRs), which induce cell death by a poorly understood mechanism.

The NLR system in plants has many parallels to mechanisms used in mammals to generate the apoptosome ("wheel of death") and the inflammasome, though the downstream effects are different—in mammals, the primary output is the activation of caspases, while in plants the molecular consequences are more varied.

The apoptosome and inflammasome are discussed in Chapter 9.

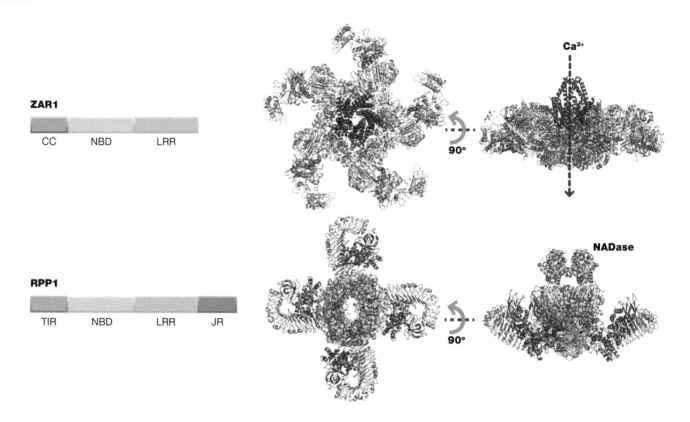

Figure 13.13

NLR structures. On the left, two types of NLRs are depicted schematically. ZAR1 consists of an N-terminal coiled-coil domain (CC), a nucleotide-binding domain (NBD), and a series of leucine-rich repeats (LRR). RPP1 has an N-terminal TIR domain, followed by NBD and LRR domains, and a C-terminal Jelly Roll (JR) domain. On the right, the oligomerized resistosome structures determined by cryo-EM are shown; colors correspond to those on the left. Also included in the structure are target proteins recognized by the LRR domains; for ZAR1, RKS (a protein kinase) and uridylated PBL2 are shown in *teal* and *gray*, while for RPP1, ATR1 is shown in *orange*. Right panels are adapted from Maruta, et al., *Immunogenetics* 74:5–26, 2022. With permission from Springer Nature.

THE TOOLKIT NEEDED FOR THE TRANSITION TO MULTICELLULAR LIFE

The transition from free-living single-celled organisms to multicellular organisms must have increased cell signaling requirements enormously; therefore, it is not surprising that the emergence of new signaling machinery and mechanisms would be associated with such a transition. Although multicellularity has evolved independently a number of times (most prominently in the case of multicellular plants, fungi, and animals), we will focus here on the emergence of the metazoans or multicellular animals. Not only there is more molecular data available for the metazoan transition than in the case of other lineages, but this transition is also of particular interest to us because it was essential for the emergence of our own species.

Metazoans are now thought to have evolved from colonial, single-celled relatives of the sponges around 600 million years ago. The exact properties of the "ur-metazoan," or last common ancestor of the metazoan lineage, are of course lost to history. However, recent work based on the genome sequences of various modern representatives of metazoans and closely related lineages has provided insight into what signaling machinery was present in that organism, and by extension what signaling capabilities were important for and perhaps essential to the evolution of complex, multicellular animals.

New signaling pathways emerge through a combination of novel elements and the repurposing of existing elements

Compared to unicellular life, the development of tissues and organs required vast increases in the ability of cells to communicate with each other, and to adjust their behavior in concert with their neighbors. Where did this new signaling capacity come from? In most cases, signaling mechanisms do not depend only on a single gene product, but instead on many different gene products working together to transmit the signal and regulate its output.

Signaling pathways and networks are discussed in Chapter 11.

It seems unlikely that multiple essential elements of the machinery could independently emerge *de novo* at the same time. Instead, it is thought that existing proteins are adapted to new uses when combined with other newly evolving components.

This principle is clearly illustrated by the emergence of tyrosine kinase signaling close to the time when the metazoan lineage first emerged. Phosphotyrosine signaling is based on a typical writer-eraser-reader system, in which protein tyrosine kinases (PTKs) post-translationally mark substrates by phosphorylating them on tyrosine; protein tyrosine phosphatases (PTPs) remove these post-translational marks; and the marked proteins are "read" by the binding of phosphotyrosine-specific modular binding domains such as SH2 domains. Proper signaling function in this system requires all three components.

Writer-eraser-reader systems are discussed in Chapter 4.

Genomic analysis shows that individual components of the pTyr signaling system were present in earlier lineages pre-dating the emergence of metazoans. For example, at least one SH2 domain is found in a fairly wide range of current eukaryotes, including fungi and protists such as amoebas and ciliates. Tyrosine-specific phosphatases have a similarly widespread distribution. Modern tyrosine-specific kinases, by contrast, first appear in the holozoan lineage, which contains the metazoans as well as choanoflagellates (about which we will hear more later), and a few lesser-known lineages including filastereans and ichthyosporeans.

Based on this analysis, the three components needed for modern pTyr-dependent signaling first came together in the holozoan lineage. The SH2 and PTP domains, which apparently evolved much earlier than the dedicated PTK domain, must have performed other useful roles in ancestral single-celled organisms. We can speculate that the PTPs were selected for their ability to efficiently reverse tyrosine phosphorylation generated by existing dual-specificity protein kinases. Early SH2 domains may have served a general scaffolding or phosphate-binding role in pre-holozoan organisms; consistent with this, the lone SH2 domain found in yeast does not bind to phosphotyrosine, but instead it plays a role in transcription by binding the serine/threonine-phosphorylated tail of RNA polymerase II.

Once the writer, eraser, and reader elements finally came together in the same cell in early holozoans, however, they underwent rapid diversification as they were enlisted to regulate a host of new activities. This concept was highlighted when the first choanoflagellate genomes were sequenced. Choanoflagellates are free-living or transiently colonial unicellular organisms closely related to sponges, and are thought to be the modern lineage most similar to the direct precursors of multicellular animals. One striking feature of choanoflagellate genomes is that the architecture of the proteins containing PTK, PTP, and SH2 domains is in many cases wildly different from those seen in all present-day metazoans. It was as if the new-found signaling capabilities enabled by the presence of these three domains in the same cell unleashed a flood of evolutionary experimentation in the metazoan and choanoflagellate lineages. More distantly related holozoans, such as filastereans and ichthyosporeans, show similar diversity in protein architectures. Key nonreceptor tyrosine kinase families, however, including the Src, Abl, Csk, and Tec families, are deeply conserved among all holozoans and may have taken on important core functions (in adhesion, for example) relatively early in the evolution of holozoans. Somewhat surprisingly, in modern metazoans, the domain arrangement of proteins containing PTK, PTP, and SH2 domains is very similar. This suggests that after the initial burst of experimentation at the dawn of multicellular life, a set of basic domain architectures emerged that was sufficient for the signaling needs of all animal species and has not changed significantly for hundreds of millions of years.

Eight signaling pathways are closely associated with the emergence of animals

As we have already noted, it is challenging to identify precisely when in evolution a particular signaling pathway became fully functional, as some elements of a pathway may have existed long before they were combined with newer components to form the modern pathways seen in humans and other vertebrates. Furthermore, non-metazoan organisms may encode proteins with significant sequence similarity to metazoan signaling proteins, but they lack key functional domains involved in signaling in animal cells. These complexities are compounded by uncertainties in the precise relationship between various lineages on the tree of life; for example, there is still some uncertainty about whether sponges or ctenophores (comb jellies) branched more recently from the main metazoan lineage. There is good evidence, however, that eight core signaling pathways emerged around the time of the last common ancestor of all metazoans, and thus are likely to have played important roles in enabling the emergence of complex multicellular animals. These are four pathways important for establishing the multicellular body plan during development (TGF-β, Wnt, Notch, and Hedgehog); three pathways important for long-distance communication via hormones and cytokines (nuclear receptors, and the receptor tyrosine kinase and JAK/STAT pathways); and the integrin-containing complexes that allow communication between the extracellular matrix and the cytoskeleton.

 These core signaling pathways are discussed in Chapter 8.

Each of these is very briefly outlined below, and their emergence is illustrated in **Figure 13.14**.

TGF-β: Core components of the TFG-β pathway include the transmembrane receptors, which have serine/threonine kinase activity; peptide ligands such as TGF-β, bone morphogenic proteins (BMPs), and activin; and signal transducers of the SMAD family, which relocalize to the nucleus and mediate transcriptional responses upon receptor engagement. While receptor serine/threonine kinases are common in other eukaryotes, including plants and fungi, direct homologs of the TGF-β receptor and the other core pathway components first emerged simultaneously in the metazoan lineage, and are not present in choanoflagellates or other metazoan relatives. In complex animals, this pathway is often involved in specifying cell fate and the axes of the developing embryo.

Wnt: The Wnt pathway also participates in key developmental events such as specifying cell fate, the site of gastrulation, and the main body axes in embryogenesis. Core components include peptide ligands such as Wnt, receptors and co-receptors of the Frizzled and LRP families, downstream transducers such as the serine/threonine kinase GSK3 and β-catenin, and scaffolding proteins such as disheveled, APC, and axin. With the exception of GSK3, which is present in many eukaryotes, most of the core components emerged in the first metazoans.

Notch: The Notch/delta pathway mediates juxtacrine signaling between adjacent cells, one of which expresses the ligand (Delta) and the other expresses the receptor (Notch). While the proteases involved in processing Notch upon ligand binding and in the release of its intracellular, DNA-binding domain are present in a number of primitive single-celled eukaryotes, Notch and Delta arose later, around the time of the first metazoans. The complete canonical Notch signaling pathway, which

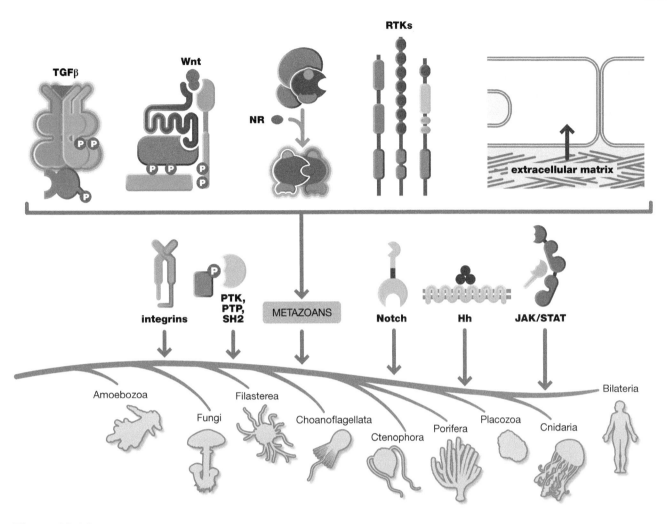

Figure 13.14

Key signaling pathways that emerged around the metazoan transition. Evolutionary tree of eukaryotes leading to the emergence of metazoans (multicellular animals) is depicted. The approximate position where key functional signaling pathways first emerged is indicated. In most cases, some elements of the signaling pathway emerged earlier that the complete, functional pathways seen in modern metazoans. See the text for details. (Adapted from Babonis and Martindale, *Philos. Trans. R Soc. Lond. B Biol. Sci.* 372:20150477, 2017. With permission from The Royal Society.)

typically functions to determine cell fate at tissue boundaries, seems to have only been fully realized in more complex metazoans, after they branched from the simpler sponges and ctenophores.

Hedgehog: Hedgehog is another key signaling pathway involved in tissue specification during development. In this case, all of the components of the signaling pathway (including the ligand hedgehog (Hh), its receptors patched (Ptc) and smoothened (Smo), and its nuclear effector (Gli)) are only found together in cnidarians (sea anemones, corals, and jellyfish) and bilaterians (organisms with bilateral symmetry at some point in development, including most familiar animals). However, many pathway components are found in more primitive metazoans such as sponges, which lack only the complete Hh ligand (the individual "hedge" and "hog" domains of Hh are present, but on different proteins). Therefore, it is likely that the full Hh pathway was first functional in early metazoans, but that key components were lost in some lineages.

Nuclear receptors: These intracellular receptors bind hydrophobic hormones such as steroids, as well as some lipids; ligand binding induces conformational changes in the receptors that enable DNA binding and transcriptional regulation. The key defining components are the DNA-binding domain and the hydrophobic ligand-binding pocket. They likely evolved from similar "one-component" receptors seen in prokaryotes and other eukaryotic lineages, but direct homologs to modern nuclear receptors are only seen in the metazoans, including sponges and ctenophores. Their role as receptors for steroid hormones (including the major sex hormones, testosterone and estrogen) is important for long-distance signaling in metazoans.

Receptor tyrosine kinases: As discussed above, tyrosine kinase, tyrosine phosphatase, and pTyr binding domains first came together just before the emergence of metazoans. Non-metazoans, such as choanoflagellates and filastereans, have transmembrane receptor tyrosine kinases (RTKs), but these all have different domain architectures (particularly in the extracellular ligand-binding domains) than seen in modern metazoans. Direct homologs of modern RTKs are first seen in sponges and ctenophores, while the compete complement of modern RTK families is first seen in cnidarians and bilaterians.

JAK/STAT: The cytosolic tyrosine kinase JAK and its target, the SH2 domain-containing transcription factor STAT, are critical in transducing signals from the receptors for cytokines and other hormones such as interferon and growth hormone. Elements of the pathway are present in pre-metazoans (for example, apparent homologs of STAT are found in the amoebozoan *Dictyostelium* and in filastereans), but the full complement of pathway components is first seen in the bilaterians.

Integrin: The ability to regulate a cell's adhesion to the extracellular matrix (ECM) and to other cells is essential to many aspects of multicellular life, including organogenesis and cell motility. The integrin pathway is particularly important for communication between intracellular structures such as focal adhesions and the actin cytoskeleton, and ECM components such as fibronectin. The ECM itself plays important roles in organizing tissues, and in establishing concentration gradients of ligands for various cell surface receptors. The intracellular components of focal adhesions, such as talin, paxillin, and vinculin, appeared quite early, and some are present in amoebozoans and fungi. Homologs of the key transmembrane receptors, the integrins, are found in filastereans. By contrast, most ECM components, such as collagens and laminins, are first seen in metazoans (though a homolog of collagen IV is present in filastereans). Presumably, their emergence marked the transition from the need to organize and regulate the actin cytoskeleton in unicellular organisms, to the need for complex extracellular structures which could provide the organizational framework for multicellular tissues.

SUMMARY

The signaling needs of different organisms differ widely, and the signaling machinery that has evolved for each is optimized to fit those needs. Specific signaling constraints on organisms include their physical properties, such as their size, as well as their lifestyles and environment. The recent explosion of genomic information for organisms across the tree of life has allowed the direct comparison of their signaling machinery, shedding light on how different organisms meet their signaling needs and how this machinery evolved over time. Prokaryotic signaling is largely focused on optimizing the utilization of resources from the environment; elaborate mechanisms have evolved to couple transcription to the availability of specific nutrients. Bacteria also have evolved the ability to communicate with each other to optimize their growth and survival in larger communities. Multicellular plants differ in many respects from animals in their signaling machinery, consistent with their decentralized organization, immobility, and ability to use light to generate energy and fix carbon. A number of pathways evolved in plants to enable long-distance communication and coordination, both within tissues and throughout the plant. In multicellular animals, new signaling capabilities were needed to regulate embryogenesis and tissue development, and to facilitate local and long-distance communication. A number of new signaling pathways first evolved around the time of the emergence of metazoans, and these are likely to have been important for the transition to multicellularity.

QUESTIONS

1. How do the signaling needs of multicellular animals and multicellular plants differ? What kinds of input stimuli must they respond to, and over what timescales and distance?

2. Two phylogenetically very distant species contain signaling proteins that have the same molecular function, and which share modest sequence similarity. How might you decide whether the presence of these signaling proteins is more likely due to evolution from a common ancestor with that protein, convergent evolution, or horizontal transmission?

3. Many bacterial transcriptional regulators, such as AraC and LacI, function as dimers or multimers. How might this affect their response to changes in the environment? How might this be useful to the cell?

4. Compare signaling among bacteria in biofilms (quorum sensing) with hormone signaling between the cells of multicellular animals. In what important ways are the two processes similar, and how are they different?

5. Most receptors for plant phytohormones are cytosolic. By contrast, most receptors in animal cells are found on the plasma membrane. What might explain this difference?

6. The phytohormone auxin plays many different roles in multicellular plants. More specifically, different cells in the plant can respond very differently to the same concentration of auxin. What explains this diversity of responses to the same input signal?

7. Superficially, there is great similarity between the resistosomes generated by plant effector-triggered immunity, and the apoptosome and inflammasome structures seen in metazoans (see Chapter 9). What do these structures have in common, and what are some key differences between them?

8. Why is it so difficult to determine precisely where in evolution a particular signaling mechanism evolved?

Answers to these questions can be found online at www.routledge.com/9780367279370

BIBLIOGRAPHY

THE CONSTRAINTS ON SIGNALING IN DIFFERENT ORGANISMS

Oakley TH & Speiser DI (2015) How complexity originates: The evolution of animal eyes. *Ann. Rev. Ecol. Evol. Syst.* 46(1), 237–260. doi: 10.1146/annurev-ecolsys-110512-135907.

Rubin GM, Yandell MD, Wortman JR, et al. (2000) Comparative genomics of the eukaryotes. *Science* 287(5461), 2204–2215. doi: 10.1126/science.287.5461.2204.

SIGNALING THEMES IN PROKARYOTIC ORGANISMS

Deutscher J, Aké FM, Derkaoui M, et al. (2014) The bacterial phosphoenolpyruvate:carbohydrate phosphotransferase system: Regulation by protein phosphorylation and phosphorylation-dependent protein-protein interactions. *Microbiol. Mol. Biol. Rev.* 78(2), 231–256. doi: 10.1128/MMBR.00001-14.

Görke B & Stülke J (2008) Carbon catabolite repression in bacteria: Many ways to make the most out of nutrients. *Nat. Rev. Microbiol.* 6(8), 613–624. doi: 10.1038/nrmicro1932.

Hawver LA, Jung SA & Ng WL (2016) Specificity and complexity in bacterial quorum-sensing systems. *FEMS Microbiol. Rev.* 40(5), 738–752. doi: 10.1093/femsre/fuw014.

Lewis M (2011) A tale of two repressors. *J. Mol. Biol.* 409(1), 14–27. doi: 10.1016/j.jmb.2011.02.023.

Papenfort K & Bassler BL (2016) Quorum sensing signal-response systems in Gram-negative bacteria. *Nat. Rev. Microbiol.* 14(9), 576–588. doi: 10.1038/nrmicro.2016.89.

Reznikoff WS (1992) The lactose operon-controlling elements: A complex paradigm. *Mol. Microbiol.* 6(17), 2419–2422. doi: 10.1111/j.1365-2958.1992.tb01416.x.

Schleif R (2003) AraC protein: A love-hate relationship. *Bioessays* 25(3), 274–282. doi: 10.1002/bies.10237.

SIGNALING MECHANISMS IN MULTICELLULAR PLANTS

Binder BM (2020) Ethylene signaling in plants. *J. Biol. Chem.* 295(22), 7710–7725. doi: 10.1074/jbc.REV120.010854.

Brunkard JO & Zambryski PC (2017) Plasmodesmata enable multicellularity: New insights into their evolution, biogenesis, and functions in development and immunity. *Curr. Opin. Plant Biol.* 35, 76–83. doi: 10.1016/j.pbi.2016.11.007.

Calderon RH & Strand Å (2021) How retrograde signaling is intertwined with the evolution of photosynthetic eukaryotes. *Curr. Opin. Plant Biol.* 63, 102093. doi: 10.1016/j.pbi.2021.102093..

Kieber JJ & Schaller GE (2018) Cytokinin signaling in plant development. *Development* 145(4), dev149344. doi: 10.1242/dev.149344.

Leyser O (2018) Auxin signaling. *Plant Physiol.* 176(1), 465–479. doi: 10.1104/pp.17.00765.

Maruta N, Burdett H, Lim BYJ, et al. (2022) Structural basis of NLR activation and innate immune signalling in plants. *Immunogenetics* 74(1), 5–26. doi: 10.1007/s00251-021-01242-5.

Ngou BPM, Ding P & Jones JDG (2022) Thirty years of resistance: Zig-zag through the plant immune system. *Plant Cell* 34(5), 1447–1478. doi: 10.1093/plcell/koac041.

Wu GZ & Bock R (2021) GUN control in retrograde signaling: How GENOMES UNCOUPLED proteins adjust nuclear gene expression to plastid biogenesis. *Plant Cell* 33(3), 457–474. doi: 10.1093/plcell/koaa048.

THE TOOLKIT NEEDED FOR THE TRANSITION TO MULTICELLULAR LIFE

Babonis LS & Martindale MQ (2017) Phylogenetic evidence for the modular evolution of metazoan signalling pathways. *Philos. Trans. R Soc. Lond. B Biol. Sci.* 372(1713), 20150477. doi: 10.1098/rstb.2015.0477.

Lim WA & Pawson T (2010) Phosphotyrosine signaling: Evolving a new cellular communication system. *Cell* 142(5), 661–667. doi: 10.1016/j.cell.2010.08.023.

Richter DJ, Fozouni P, Eisen MB & King N (2018) Gene family innovation, conservation and loss on the animal stem lineage. *Elife* 7, e34226. doi: 10.7554/eLife.34226.

Richter DJ & King N (2013) The genomic and cellular foundations of animal origins. *Annu. Rev. Genet.* 47, 509–537. doi: 10.1146/annurev-genet-111212-133456.

Ros-Rocher N, Pérez-Posada A, Leger MM & Ruiz-Trillo I (2021) The origin of animals: An ancestral reconstruction of the unicellular-to-multicellular transition. *Open Biol* 11(2), 200359. doi: 10.1098/rsob.200359.

Suga H, Torruella G, Burger G, Brown MW & Ruiz-Trillo I (2014) Earliest Holozoan expansion of phosphotyrosine signaling. *Mol. Biol. Evol.* 31(3), 517–528. doi: 10.1093/molbev/mst241.

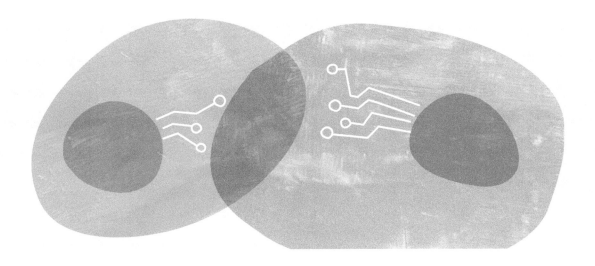

How Cells Make Decisions

In the previous chapters of this book, we introduced the molecular components of cell signaling systems and examined how these components link together to build more complex devices and networks. In this chapter, we turn to a more classical physiological view and examine several model signaling pathways in an effort to synthesize earlier concepts and to provide a broader context of how cells employ signaling machinery to make important physiological decisions.

We focus on four physiological processes:

Section 1: VERTEBRATE VISION—*how photoreceptor cells sense and amplify light inputs*

Section 2: PDGF SIGNALING—*how fibroblast cell proliferation is triggered as part of wound healing*

Section 3: THE CELL CYCLE—*how cells control their replication and division*

Section 4: T LYMPHOCYTE ACTIVATION—*how key cells in our immune system are mobilized to fight infection*

While these examples represent only a few of the myriad complex decisions that the cells in our body must make, they epitomize common problems that cells face, such as how signal integration is used to make very specific decisions, how signals are propagated and amplified through cascades, and how feedback networks control the amplitude and timing of the induced response.

This chapter is organized in a visual-interrogative style, as a series of panels with questions. The first panels of each section introduce the physiological signaling system at different scales—from the organismal, to the cellular, to the molecular levels. Subsequent panels in each section address questions that focus on key functional problems that cells need to address to perform specific tasks. Within these panels, we discuss the molecular- and network-level mechanisms that cells have

DOI: 10.1201/9780429298844-14

evolved to solve these problems. It is important to keep in mind that these systems are extremely complex and still the subject of considerable ongoing research. Thus, the models shown here represent simplified versions of our current understanding, and there is little doubt that future research will lead to modifications of these models. Nevertheless, the types of mechanisms and solutions shown in these pathways are also employed by many other cells in diverse physiological processes, and exemplify fundamental principles of cell signaling.

SECTION 14.1 VERTEBRATE VISION— how photoreceptor cells sense and amplify light inputs

As organisms evolved and became more complex, they also developed increasingly sophisticated sensory systems which improved their ability to interact with the surrounding environment. One of the most sophisticated and best understood sensory systems is vision. Light that we sense enters the eye, which is an extremely complex organ, and this light hits specific photoreceptor cells in the retina. There are two basic types of photoreceptor cells, named for their overall cell shape: rods, which have the highest sensitivity to low levels of light, and cones, which are used primarily for color vision. Activation of these photoreceptor cells by light ultimately leads to propagation of a neuronal signal through the optic nerve to the visual cortex in the brain, where these signals are processed to yield the overall perceived image that we "see." Here, we focus specifically on the question of how the molecular signaling machinery in the photoreceptor cells of the eye functions to detect light and convert it to a signal that can be efficiently transmitted to the optic nerve.

Our everyday experience reveals many aspects of vision that make it a particularly useful sensory system. We can see even when it is quite dark, because rod photoreceptor cells are very sensitive to low levels of light. We can also see when it is much brighter, although it can take a few seconds to adjust to dramatic changes in brightness, because the visual system has a wide dynamic range and can adapt to prolonged stimulation by ambient light levels. At the same time, the visual system has high temporal resolution—when we close our eyes, it immediately appears dark because signals are turned off rapidly. This fast resolution allows us to respond to rapidly changing cues, such as an approaching baseball, which can take less than a second to travel from the pitcher's hand to the batter.

Here, we describe how a rod photoreceptor cell functions, focusing on the cellular and molecular signaling events that occur in the cell upon light stimulation. In particular, we will focus on the following questions:

1. **Question 1: How does the photoreceptor cell convert light into a biochemical signal that can be transmitted to the brain? (see page 350)**

2. **Question 2: How does the photoreceptor cell detect low levels of light, even a single photon? (page 352)**

3. **Question 3: How can the response be so rapid? (page 353)**

4. **Question 4: How does the photoreceptor cell reset itself quickly to allow for detection of further increases in light? (page 354)**

THE SYSTEM

ORGAN | vertebrate eye

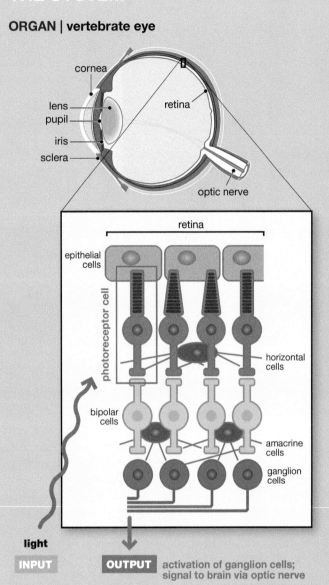

THE VERTEBRATE RETINA | light input detected by the retina leads to an action potential in ganglion cells that transmits information to the brain.

Light from the environment enters the eye and passes through the outer layers of the retina until it reaches the photoreceptor cells. Light stimulation of photoreceptor cells initiates a cascade of cell–cell communication that leads to the activation of ganglion cells, which transmit the visual signal to the brain via the optic nerve. Note that light must first pass through the ganglion cells to reach the photoreceptor cells, and then the signal propagates back to the ganglion cells.

The cell-signaling processes begin when light induces hyperpolarization of the photoreceptor cell. Rod cells are the class of photoreceptor cells most sensitive to low levels of light.

CELL | photoreceptor cell

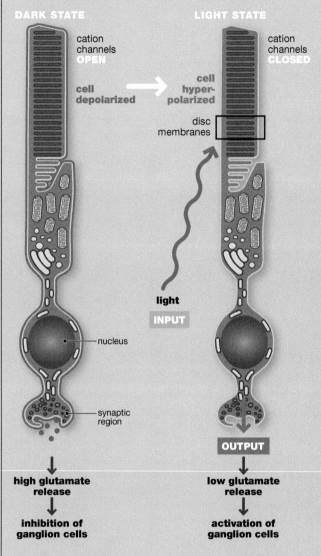

PHOTORECEPTOR CELL (ROD CELL) | light causes hyperpolarization of photoreceptor cells, inhibiting glutamate release.

In the dark state, cation channels in the photoreceptor cell membrane are open, and the cell is depolarized. In this depolarized state, the cell releases high amounts of the neurotransmitter glutamate, which activates the bipolar cells, which in turn inhibit the ganglion cells.

Light leads to hyperpolarization of the photoreceptor cells by closing cation channels in the plasma membrane. Hyperpolarization in turn inhibits glutamate release from the cell, ultimately activating the downstream ganglion cells.

This light-induced response is initiated through a molecular signaling network in the photoreceptor cells, as shown in the next panel.

MOLECULAR NETWORK | visual transduction cascade within photoreceptor cell

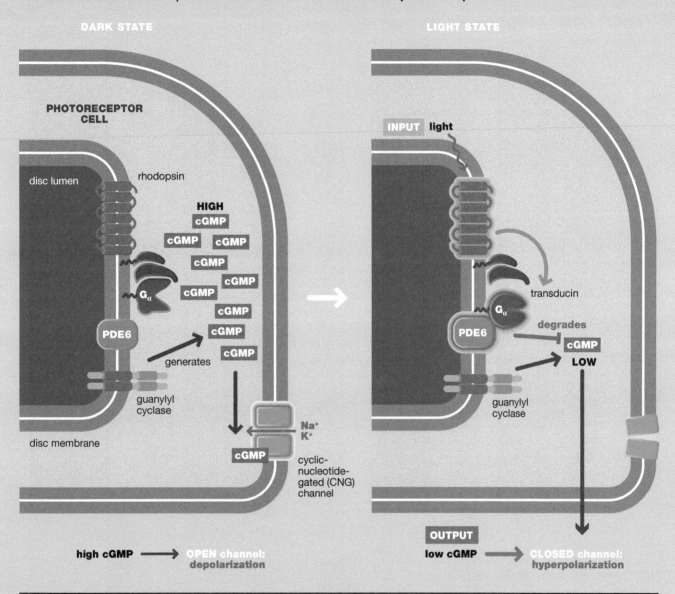

SIGNALING NETWORK IN THE PHOTORECEPTOR CELL DISC MEMBRANE | light activates a signal transduction cascade in the photoreceptor cell that leads to closing of cation channels and hyperpolarization of the cell. Hyperpolarization leads to activation of ganglion cells (optic nerve).

In the dark state, the enzyme guanylyl cyclase is constitutively active and produces a high level of the second messenger cyclic GMP (cGMP) from GTP. cGMP binds to and allosterically opens the cyclic-nucleotide-gated (CNG) channel, which allows cations such as sodium and potassium into the cell, leading to a constitutively depolarized state. The photoreceptor cell is thus constantly utilizing energy, in the form of GTP, to maintain itself in a highly responsive state.

When light input strikes the photoreceptor cell, it activates the G-protein-coupled receptor (GPCR) rhodopsin, which is highly concentrated in the disc membranes on the outer segment of the cell. Activated rhodopsin in turn activates the heterotrimeric G protein alpha subunit (Gα) named transducin. One of the major downstream targets of activated transducin is the enzyme phosphodiesterase 6 (PDE6). PDE6 hydrolyzes cGMP (to GMP) leading to a rapid decrease in cGMP concentration. Under these low-cGMP conditions, the CNG channel closes, leading to the transient light-induced hyperpolarization of the cell.

Question 1: How does the photoreceptor cell convert light into a biochemical signal that can be transmitted to the brain?

LIGHT-INDUCED CONFORMATIONAL CHANGE

The photoreceptor cell contains signaling proteins that can convert light into a series of protein conformational changes. These conformational changes trigger changes in enzymatic function.

THE RECEPTOR RHODOPSIN CONVERTS LIGHT INTO A CHANGE IN CONFORMATION AND A CHANGE IN ENZYMATIC ACTIVITY

Incoming light is sensed by rhodopsin, a membrane-spanning GPCR. Rhodopsin is composed of retinal, a light-sensitive cofactor, covalently linked to the protein opsin. When retinal absorbs a photon, it isomerizes from the 11-*cis* to the all-*trans* conformation. This results in a conformational change in the opsin protein which activates rhodopsin for downstream signaling by reorganizing the transducin-binding site. Active rhodopsin acts as a GEF enzyme that activates the GTPase transducin.

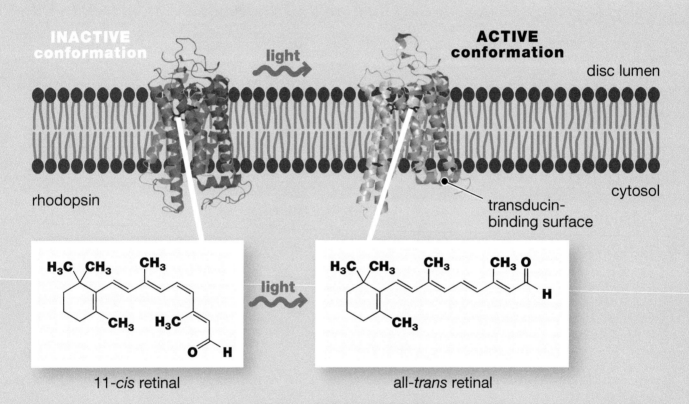

11-*cis* retinal   all-*trans* retinal

Rhodopsin binds the cofactor retinal, which isomerizes upon stimulation with light. The change in retinal structure forces a major change in the protein conformation.

A G PROTEIN AND SECOND MESSENGER CASCADE LEADS TO CHANNEL CLOSING

The activated state of rhodopsin starts a biochemical cascade that leads to CNG channel closing and hyperpolarization of the photoreceptor cell.

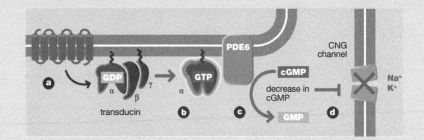

a. Like other GPCRs, activated rhodopsin acts as a GEF to activate a heterotrimeric G protein, in this case the protein transducin, by causing it to exchange bound GDP for GTP.

b. The activated alpha subunit of transducin (GTP-bound) dissociates from its beta and gamma subunits, and binds to the effector protein phosphodiesterase 6 (PDE6), allosterically activating it.

c. Active PDE6 hydrolyzes the second messenger cyclic GMP (cGMP) into GMP.

d. Reduction in the concentration of cGMP leads to closing of the CNG channel, blocking entry of Na^+ and K^+ ions, thus leading to hyperpolarization of the cell.

G proteins are discussed in Chapter 3
Second messengers like cGMP are discussed in Chapter 6

GUANYLYL CYCLASE AND PHOSPHODIESTERASE ACT AS OPPOSING "WRITER" AND "ERASER" ENZYMES CONTROLLING cGMP LEVELS

cGMP level in the photoreceptor cell ultimately determines if the cell propagates a signal to the brain.

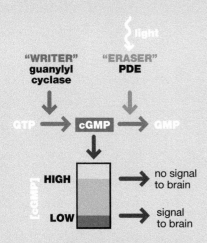

The second messenger molecule, cyclic guanine monophosphate (cGMP), is one of the key regulatory nodes in photoreceptor signaling. The concentration of cGMP ultimately determines whether the photoreceptor cell is depolarized and sends a signal to the brain.

The intracellular concentration of cGMP is controlled by the balance between two opposing enzymes, the "writer" enzyme guanylyl cyclase which synthesizes cGMP from GTP, and the "eraser" enzyme PDE which degrades cGMP to GMP. In the dark state, the writer is more active, leading to high cGMP. With light stimulation, the eraser also becomes highly active, thus resulting in a transient decrease in the level of cGMP. Note that although not explicitly shown, the cell is constantly paying energy for this highly responsive signaling system, in the form of a constant resynthesis of GTP.

Second messengers like cGMP are discussed in Chapter 6

SUMMARY

The GPCR rhodopsin senses light, which triggers an intracellular signaling cascade leading to hyperpolarization of photoreceptor cells. Hyperpolarized photoreceptor cells then activate the ganglion neurons that transmit signals to the brain via the optic nerve.

Question 2: How does the photoreceptor cell detect low levels of light, even a single photon?

ENZYMATIC AMPLIFICATION
Several steps in the signaling pathway help greatly amplify the output from a single photon.

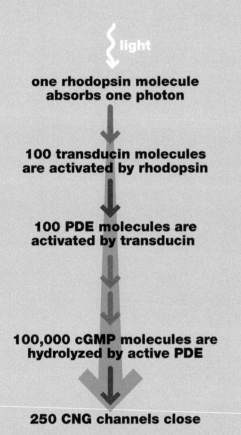

light

one rhodopsin molecule absorbs one photon

100 transducin molecules are activated by rhodopsin

100 PDE molecules are activated by transducin

100,000 cGMP molecules are hydrolyzed by active PDE

250 CNG channels close

The visual transduction cascade relies on enzymes and small signaling mediators, which can convert a small number of input molecules into many more active output molecules. This allows the system to respond in a robust manner even when stimulated by very few photons in dim light. In fact, even a single photon can lead to a measurable response from the photoreceptor cell. Amplification occurs primarily at two steps. First, when a rhodopsin molecule is activated by a single photon, it can activate over a hundred molecules of transducin per second, each of which can go on to activate a PDE6 molecule. Second, each PDE6 molecule, in turn, hydrolyzes about 1000 molecules of cGMP per second. The resulting change in cytosolic cGMP concentration results in the closing of a few hundred cation channels. This is sufficient to hyper-polarize the membrane and suppress glutamate neurotransmitter release by the photoreceptor cell.

 Amplification is discussed in Chapters 3 and 11.

SUMMARY

The visual system is very sensitive, and it uses enzymatic amplification cascades to convert small numbers of photons into a robust response.

Question 3: How can the response be so rapid?

SPATIAL ORGANIZATION

The organization of the signaling proteins and their properties lead to efficient and rapid communication.

ONE OF THE MOST REMARKABLE ASPECTS OF VISUAL SIGNALING IS ITS SPEED. SEVERAL FACTORS ACCOUNT FOR THIS RAPIDITY

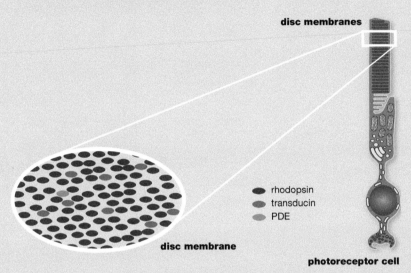

disc membranes

disc membrane

rhodopsin
transducin
PDE

photoreceptor cell

• Signaling proteins are very densely packed on the surface of the rod disc membrane (rhodopsin, transducin, and PDE6 cover 25%, 10%, and 1% of the membrane surface, respectively). This ensures that activated rhodopsin almost immediately encounters transducin, and that activated transducin almost immediately encounters PDE6.

• Two-dimensional diffusion on the membrane increases the likelihood of productive interactions compared to three-dimensional diffusion in solution.

• PDE6 is almost a "perfect enzyme": once activated, it operates nearly at the diffusion-limited rate, so virtually every cGMP molecule it encounters is converted to GMP.

• Signal output depends on cGMP and cations, which diffuse very rapidly due to their small size.

 Diffusion and its role in reaction rates is discussed in Chapters 6 and 7.

SUMMARY

Photoreceptor cell signaling proteins are co-localization at high density to optimize reaction rates and yield rapid response.

Question 4: How does the photoreceptor cell reset itself quickly to allow for detection of further increases in light?

ADAPTATION

The photoreceptor cell signaling network contains several negative feedback loops that mediate adaptation.

NEGATIVE FEEDBACK LOOPS

The activation of the photoreceptor cell (hyperpolarization) is only transient, even in constant light. The cell is able to automatically reset itself to its original baseline output (depolarized state). This sensory adaptation is critical to allow the cell to respond to further increases in light stimulus, thus giving the cell a much higher dynamic range of light detection. These adaptive mechanisms are part of what allow the visual system to function in a wider range of ambient light conditions.

The adaptation of the photoreceptor cell involves at least three characterized negative feedback loops, outlined below.

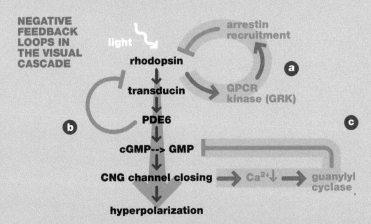

Downregulation of GPCR signaling is discussed in Chapters 3 and 8; Second messengers like cGMP are discussed in Chapter 6.

a. PHOSPHORYLATION FEEDBACK

Activated rhodopsin binds and allosterically activates a GPCR kinase (GRK), which then phosphorylates rhodopsin on multiple sites. These phosphorylation sites recruit the protein arrestin, which prevents rhodopsin from activating transducin. This usually occurs within 200 ms of activation. Arrestin also acts as an adaptor to couple rhodopsin to the endocytosis machinery, resulting in internalization into endosomes. From the endosomes, rhodopsin may be recycled to the rod disc membranes, or it can be degraded in lysosomes. Thus, arrestin serves to desensitize rhodopsin after activation.

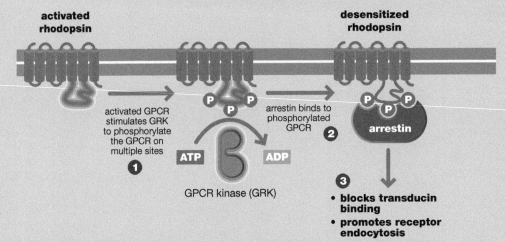

b. GAP FEEDBACK

Activated PDE6, in addition to hydrolyzing cGMP, has some GTPase-activator protein (GAP) activity that reciprocally cata-lyzes inactivation of transducin.

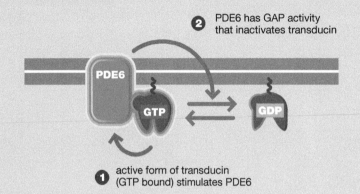

❷ PDE6 has GAP activity that inactivates transducin

❶ active form of transducin (GTP bound) stimulates PDE6

c. CALCIUM-GUANYLYL CYCLASE FEEDBACK

Closing of the CNG channels after pho-toreceptor cell activation also leads to a decrease in Ca^{2+} concentration in the cell. This decrease in Ca^{2+} also acts to increase guanylyl cyclase activity. Guanylyl cyclase activity counteracts PDE6 activity to restore a high level of cGMP, thus opening the CNG channels and restoring the photore-ceptor cell to a depolarized state.

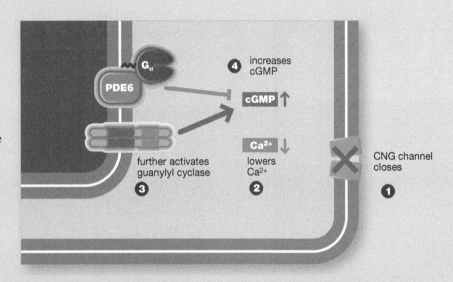

❹ increases cGMP

❸ further activates guanylyl cyclase

❷ lowers Ca^{2+}

❶ CNG channel closes

In summary, the negative feedback loops rapidly restore the cell to a high cGMP/depolarized state that is ready to respond to further input stimuli.

 Negative feedback and adaptation are discussed in Chapter 11.

SUMMARY

Multiple negative feedback loops rapidly down-regulate the phototransduction cascade, allowing changes in light to be detected with high temporal resolution.

SECTION 14.2 PDGF SIGNALING—triggering controlled cell proliferation during wound healing

While a developing and growing organism has many proliferating cells, in an adult body, most cells are quiescent (arrested in the G_1 phase of the cell cycle), and rapid proliferation often is a hallmark of diseases like cancer. However, there are specific situations in which cells in the adult must be able to rapidly proliferate. Here we examine how fibroblasts, cells which most of the time are quiescent, are able to initiate rapid proliferation when the body is wounded. We explore how these cells are able to initiate this response only when they receive precise signaling instructions, and how this response is transmitted and controlled.

After an injury, a wound in the skin and underlying connective tissue must be rapidly repaired to restore integrity of the tissue and to reconstruct a barrier to microorganisms, which will otherwise infect and colonize the wounded site. The process of wound healing requires that numerous signals be transmitted between different cell types to elicit a coherent response. Upon wounding, blood spills into the site and the platelets it contains are exposed to components of the extracellular matrix. Integrin-mediated signaling then triggers platelets to release clotting factors, leading to the formation of the initial clot. The platelets also release mitogens and other bioactive molecules, such as platelet-derived growth factor (PDGF) and transforming growth factor β (TGF β), which have several effects. One outcome is that they precipitate an inflammatory phase by recruiting neutrophils and macrophages to the wound. These cells engulf and kill invading bacteria and, in the case of macrophages, produce more PDGF.

But how is new tissue made to repair the wound? Wound repair is mediated largely by a class of cells called fibroblasts, which normally lie dormant in a quiescent state. However, upon injury, the fibroblasts near the site of the wound detect the release of PDGF and awaken. They migrate to the area of the wound, begin to proliferate, and secrete extracellular matrix proteins such as collagen that are needed to repair the damaged tissue.

Thus, fibroblasts rapidly sense a chemical signal and initiate a diverse program of behaviors, which includes directed migration toward the wound, proliferation to produce more fibroblasts, and repair and remodeling of extracellular matrix. In this section, we will focus on the proliferation response, which must be exquisitely regulated because aberrant proliferation could lead to cancer. This is one example of many analogous mitogen response pathways that trigger similar sets of potent but tightly controlled proliferative responses.

Here, we will focus on the following questions:

1. Question 1: How do fibroblasts detect and respond to the extracellular PDGF signal? (page 360)
2. Question 2: How is this signal propagated within the fibroblast to trigger cell proliferation? (page 361)
3. Question 3: How is misactivation of the proliferation response prevented? (page 362)
4. Question 4: How is the proliferative response terminated? (page 363)

THE SYSTEM

TISSUE | process of wound healing

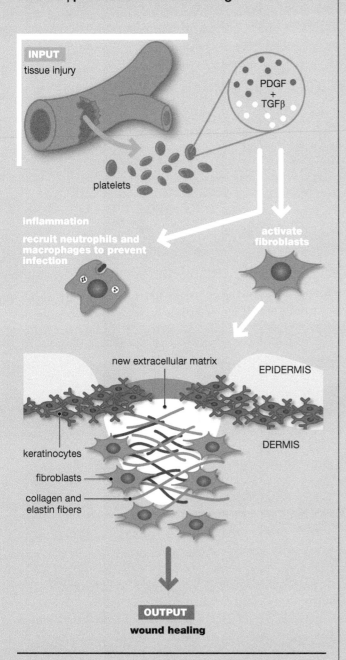

INPUT
tissue injury

PDGF
+
TGFβ

platelets

**inflammation
recruit neutrophils and
macroblasts to prevent
infection**

**activate
fibroblasts**

new extracellular matrix

EPIDERMIS

DERMIS

keratinocytes

fibroblasts

collagen and
elastin fibers

OUTPUT
wound healing

CELL | fibroblast response to wounding and platelet activation

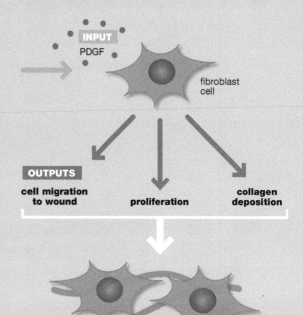

INPUT
PDGF

fibroblast
cell

OUTPUTS

**cell migration
to wound**

proliferation

**collagen
deposition**

wound healing

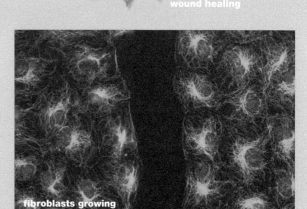

**fibroblasts growing
across a wound**

Courtesy of Jan Schmoranzer

WOUND HEALING | platelets mobilize fibroblasts.

At the site of injury, damaged tissue activates blood platelets, which initiate clotting and also secrete factors such as PDGF that act upon fibroblasts in the connective tissue (dermis). Collagen secreted by the fibroblasts is then cross-linked to repair and strengthen the extracellular matrix, as well as to support the regrowth of epithelial cells of the overlying epidermis.

FIBROBLAST RESPONSES | migration, proliferation, and collagen deposition.

The gradient of PDGF released by platelets is sensed by fibroblasts, and these cells respond in several ways. First, they migrate to the site of the wound. Second, they begin to rapidly proliferate. Third, they begin to deposit new collagen and other extracellular matrix proteins.

MOLECULAR NETWORK | control of fibroblast proliferation

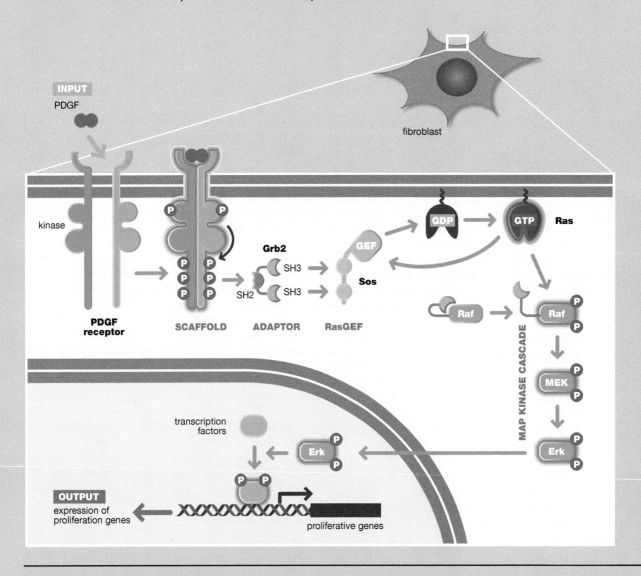

FIBROBLAST PROLIFERATION PATHWAY | how PDGF stimulation leads to expression of proliferation control genes such as Myc.

Human platelets contain two closely related chains of PDGF, A and B, which are synthesized as precursors, processed by proteolytic cleavage, and are then linked through disulfide bonds to form homo- or heterodimers. The mature PDGF dimers exert their effects on target cells by binding to the extracellular regions of the PDGF receptor (PDGFR), which has intrinsic tyrosine kinase activity. PDGF binding induces dimerization and autophosphorylation, producing at least nine phosphotyrosine sites that serve as docking sites for signaling proteins with SH2 and PTB domains. One such adaptor protein is Grb2, which recognizes phosphorylated PDGFR through an SH2 domain and

contains two SH3 domains that recruit Sos, a guanine nucleotide exchange factor (GEF) for Ras. Localization to the plasma membrane allows Sos to activate Ras, which itself can then activate the Erk MAP kinase pathway. Ultimately, activated Erk translocates to the nucleus, where it induces the transcription of proliferative genes, including cyclins (which drive the cell cycle) and Myc. Myc is a transcription factor that orchestrates a complex gene expression program required for cell growth and proliferation. Among other effects, Myc induces transcription of genes involved in glycolysis and metabolism, ribosome biogenesis, mitochondrial biogenesis, DNA replication, and the cell cycle.

Question 1: How do fibroblasts detect and respond to the extracellular PDGF signal?

A TRANSMEMBRANE RECEPTOR IN FIBROBLASTS SENSES PDGF AND TRANSMITS THE SIGNAL ACROSS THE MEMBRANE INTO CELLS

The platelet-derived growth factor receptor (PDGFR) is a receptor tyrosine kinase.

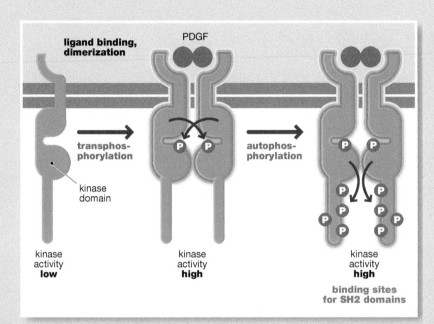

PDGFR ACTIVATION REQUIRES LIGAND BINDING AND DIMERIZATION

PDGF binds to PDGFR with very high affinity ($K_d \sim 10^{-10}$ M), so binding is favored even at low PDGF concentrations. This is important so that fibroblasts can respond robustly to relatively weak input signals. Binding to PDGF induces dimerization of PDGFR monomers, and juxtaposition of the two cytoplasmic kinase domains stimulates their catalytic activity. Prior to activation, the kinase domain is inhibited through intramolecular interactions, with the net result that the activation segment of the kinase domain adopts a nonproductive conformation so that residues important for formation of the active site are not properly positioned, and the substrate-binding site is blocked. This multilayered inhibitory device is important because the inappropriate activation of PDGFR could have pathologic effects, as indeed is seen in diseases such as fibrosis, scleroderma, and cancer, in which these constraints are overridden. Catalytic activation is achieved because the two adjacent kinase domains of the ligand-bound dimer are able to cross-phosphorylate one another on tyrosine residues that are essential for maintaining the autoinhibited state, and whose inhibitory effects are negated once they are phosphorylated. Thus, the proximity effect, initially induced by binding of PDGF to the PDGF receptor, is thereby converted by a phosphorylation-driven allosteric switch into catalytic activation of the receptor's kinase domain.

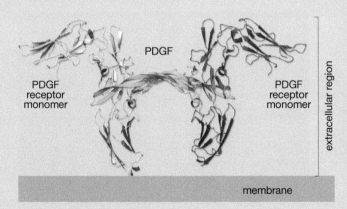

 Kinase activation is discussed in greater detail in Chapter 3; Information transfer across a membrane is discussed in greater detail in Chapter 8.

SUMMARY

In fibroblasts, PDGF is detected by the PDGFR, a receptor tyrosine kinase. Upon activation by PDGF, the PDGFR autophosphorylates, translating an extracellular input into intracellular signals.

Question 2: How is this signal propagated within the fibroblast to trigger cell proliferation?

> **MULTIPLE PHOSPHORYLATION SITES ON PDGFR ALLOW MEMBRANE RECRUITMENT OF ADAPTORS/EFFECTORS, LEADING TO CO-LOCALIZATION OF SIGNALING MACHINERY**
>
> Receptor phosphorylation creates docking sites for SH2-domain-containing proteins. SH2-domain-mediated recruitment assembles key signaling complexes that propagate the signal intracellularly.

PDGFR AUTOPHOSPHORYLATION CREATES BINDING SITES FOR MULTIPLE EFFECTORS, ULTIMATELY LEADING TO MULTIPLE OUTPUTS

PDGFR is usually the most abundantly tyrosine-phosphorylated protein in PDGF-stimulated cells. It has at least nine autophosphorylation sites that are not directly involved in regulating kinase activity, but rather serve as docking sites for cytoplasmic signaling proteins with one or more phosphotyrosine-binding domains (mostly SH2 domains). The autophosphorylated receptor is therefore converted into a scaffold that recruits a range of proximal targets based on selective phosphopeptide-SH2 domain interactions. Phosphoinositide 3-kinase (PI3K) catalyzes the conversion of PIP_2 to PIP_3, an early step in the migration response. PI3K binds PDGFR through SH2 domains on its p85 regulatory subunit which, in turn, binds to the p110 catalytic subunit. Grb2 is an adaptor protein that recruits Sos to the membrane, leading to the proliferation response.

MEMBRANE LOCALIZATION OF SOS DRIVES THE PROLIFERATION RESPONSE

In addition to its SH2 domain, Grb2 also possesses N- and C-terminal SH3 domains that engage proline-rich sequences in the C-terminal tail of Sos, which acts as a GEF for the Ras GTPase. These Grb2-mediated interactions concentrate Sos at the membrane, where it has access to its substrate Ras, which is itself anchored in the plasma membrane by its C-terminal isoprenyl modifications. Upon recruitment, Sos catalyzes the exchange of Ras-bound GDP to GTP, thereby converting Ras from the inactive form to the active form.

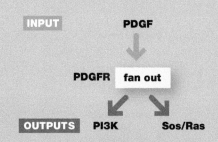

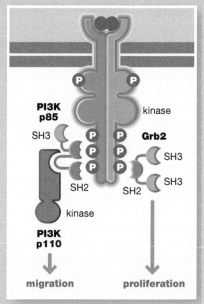

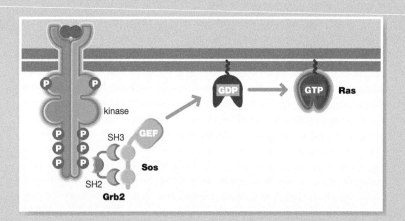

GTPases and their activation by GEFs are discussed in greater detail in Chapter 3. Phosphorylation-dependent interactions are discussed in greater detail in Chapter 4. The role of subcellular localization in signaling is discussed in greater detail in Chapter 5.

SUMMARY

Autophosphorylation of PDGFR creates binding sites for several effectors containing SH2 domains, allowing one signal to fan out into multiple responses.

The proliferative response to PDGF requires membrane localization of the GEF Sos, which leads to activation of the G protein Ras and subsequently to activation of the Erk MAP kinase signaling pathway. MAPK phosphorylation of key transcription factors leads to expression of proliferation genes.

Question 3: How is misactivation of the proliferation response prevented?

KEY SIGNALING MOLECULES LIKE SOS AND RAF FUNCTION AS SWITCHES THAT REQUIRE MULTIPLE INPUTS TO BECOME ACTIVATED

Ensuring that Ras is activated only in the correct time and place is critical. Indeed, hyperactive Ras mutations are frequently observed in cancer. Combinatorial gating of signaling molecules is a common theme for tightly controlling cellular response.

THE GEF PROTEIN SOS IS REGULATED BY COMBINATORIAL INPUTS AND FORMS A POSITIVE FEEDBACK LOOP WITH RAS

Sos is a multidomain protein with an N-terminal histone-like domain, followed by a Dbl homology (DH) domain, a pleckstrin homology (PH) domain, a helical linker, the catalytic REM-Cdc25 domain, and the Grb2-binding region at the C-terminus. In addition to its interaction with Grb2, Sos is also recruited to the membrane and activated by binding to the phospholipid phosphatidylinositol 4,5-bisphosphate (PIP_2) through its PH domain. In addition to the catalytic site, Sos also binds Ras-GTP at a second site: an allosteric regulatory site in the REM-Cdc25 domain. Binding at this allosteric site is enhanced by binding of PIP_2, and it increases Sos catalytic activity. This constitutes a positive feedback loop, as the product of Sos activity, Ras-GTP, serves to further activate Sos, leading to greater Ras-GTP production. PIP_2 and Ras-GTP at the plasma membrane cooperatively stimulate Sos, and these interactions may be initiated or stabilized by Grb2. The dependence of Sos on multiple inputs may ensure that it is buffered against erroneous activation by weak upstream signals, while the positive feedback afforded by allosteric Ras provides a mechanism to turn on Sos rapidly and maintain it in the active state when a signal such as PDGF exceeds a threshold level.

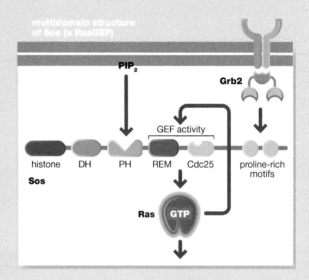

MULTIPLE INPUTS CONTROL ACTIVATION OF RAF

The primary downstream targets for Ras-GTP are the Raf serine/threonine kinases, which possess an N-terminal domain that selectively interacts with Ras in the GTP-bound form. In the absence of a signal, the N-terminal region of Raf interacts with the kinase domain to repress catalytic activity; binding to Ras-GTP coordinately releases this autoinhibition and relocalizes Raf to the plasma membrane. Although binding to Ras is the primary signal for activation, Raf is also subject to several other regulatory controls that prevent its inappropriate activation, allow for additional upstream inputs, and co-localize it with downstream substrates. For example, the inactive Raf conformation is maintained by a 14-3-3 dimer that simultaneously binds two phosphorylated threonine sites that flank the kinase domain

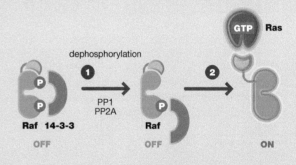

and clamps the kinase in the inactive state. For Raf activation, the N-terminal site must be dephosphorylated by the phosphatases PP1 or PP2A, relieving the 14-3-3 clamp. 14-3-3 binding to the remaining site (of two Raf molecules) promotes Raf dimerization and stabilizes the active conformation (see Figure 10–26).

GTPases and kinases are discussed in greater detail in Chapter 3. The modular domain architecture of signaling proteins is discussed in greater detail in Chapter 10. Positive feedback loops are discussed in greater detail in Chapter 11.

SUMMARY

Combinatorial regulation and positive feedback loops prevent misactivation by inappropriate or weak inputs, but produce strong activation when appropriate.

Question 4: How is the proliferative response terminated?

NEGATIVE FEEDBACK LOOPS ALLOW SIGNALS TO BE TERMINATED, SO THAT PROLIFERATION ONLY OCCURS FOR A LIMITED TIME

In the long term (days to weeks), proliferation of fibroblasts is limited by the degradation of PDGF and other mitogens that were initially released by platelets or inflammatory cells at the wound site. In the short term, however, cell-signaling mechanisms also blunt the proliferative response over time. Both the active receptor and the downstream MAPK Erk trigger several negative feedback loops.

PDGFR ACTIVATION ALSO RECRUITS PATHWAY DOWN-REGULATORS

In addition to recruiting positively acting effectors such as PI3K and Sos (via Grb2), PDGFR phosphotyrosine sites also recruit SH2-domain-containing enzymes that down-regulate the pathway, including a Ras GTPase-activator protein (which inactivates Ras by promoting the hydrolysis of GTP to GDP) and tyrosine phosphatases Shp1 and Shp2 (which remove phosphotyrosine sites on PDGFR). Sustained activation of PDGFR also ultimately leads to its internalization into endosomes and possible degradation. This is, in part, mediated by recruitment of the ubiquitin E3 ligase Cbl.

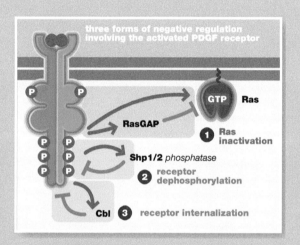

ERK PARTICIPATES IN MULTIPLE NEGATIVE FEEDBACK LOOPS

The Erk MAP kinase is also involved in several negative feedback loops. First, Erk suppresses signaling near the beginning of the pathway by phosphorylating the C-terminal tail of Sos and interfering with its binding to the Grb2 SH3 domains. Second, in addition to inducing proliferative genes, Erk induces the transcription of genes encoding dual-specificity phosphatases (MAP kinase phosphatases, or MKPs), which dephosphorylate the activation loop of Erk and thus inhibit its activity. These relatively slow negative feedback loops allow attenuation of the pathway after activation.

 Negative feedback loops are discussed in greater detail in Chapter 11.

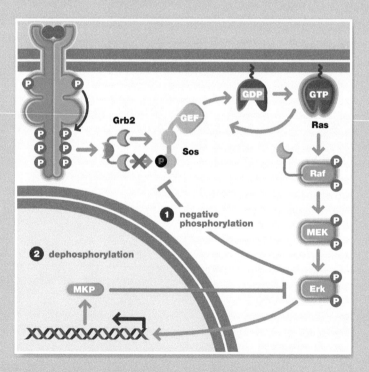

SUMMARY

Multiple negative feedback loops coordinate attenuation of signaling to ensure proliferation is not erroneously maintained.

SECTION 14.3 THE CELL CYCLE—how cells control their replication and division

Cell division is a hallmark of living organisms. For a single-celled organism, cell division is a means of reproduction. For multicellular organisms, cell division is required for development from an embryo, and it is also crucial for maintaining the health of the organism. Thus, it is critical that cell division occurs in a precise and regulated manner.

The eukaryotic cell cycle is a series of committed steps. For example, once a cell initiates DNA replication, the entire process must be completed; having two copies of some genes and only one copy of others for a prolonged period could lead to misregulation of many important pathways. Thus many cell-cycle steps occur in a switchlike (ultrasensitive) and irreversible manner. In this section, we will discuss some signaling mechanisms that promote sharp and decisive transitions between key phases in the cell cycle.

It is also critical that each step in the cell cycle only occurs after the previous steps are successfully completed. Failing to follow this strict order could result in severe consequences for the progeny of the division. In particular, the cell must ensure that genomic DNA is accurately replicated and segregated so that each daughter cell receives the appropriate complement of genes. Therefore, the cell cycle incorporates several checkpoints; each process cannot proceed until some prerequisites have been completed. Here, we will review two such checkpoints: the spindle assembly checkpoint and the DNA damage checkpoint.

The cell cycle is a beautiful example of tightly choreographed and precisely timed signaling events. In this section, we will review how cyclin-dependent kinases (CDKs) act as a central switch to control the cell cycle, and how their interactions with key regulatory factors drive progression through the distinct phases of the cell cycle.

We will use the very well-characterized budding yeast cell cycle as a model in this section. Specifically, we will address the following questions:

1. Question 1: What drives the distinct phases of the cell cycle? (page 367)

2. Question 2: How does the circuitry of the cell cycle drive sharp and irreversible transitions between its different phases? (page 369)

3. Question 3: How does the cell arrest progression through the cell cycle when it detects critical problems? (page 373)

THE SYSTEM

CELL | distinct phases of the cell cycle

KEY CELLULAR EVENTS AND TRANSITIONS IN THE CELL CYCLE

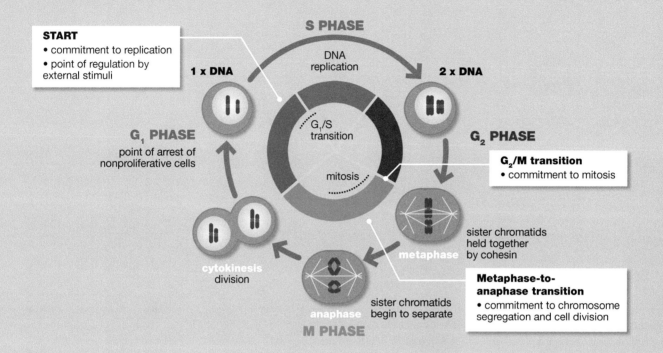

START
- commitment to replication
- point of regulation by external stimuli

S PHASE

DNA replication

1 x DNA

2 x DNA

G_1/S transition

G₁ PHASE
point of arrest of nonproliferative cells

G₂ PHASE

G₂/M transition
- commitment to mitosis

mitosis

sister chromatids held together by cohesin

metaphase

cytokinesis division

Metaphase-to-anaphase transition
- commitment to chromosome segregation and cell division

sister chromatids begin to separate

anaphase

M PHASE

The four phases of the cell cycle are as follows: (1) the G_1 (gap 1) phase; (2) S phase, when the DNA is replicated; (3) G_2 (gap 2) phase; and (4) M phase, when the replicated chromosomes are segregated into daughter cells (mitosis) and cell division takes place. Nondividing cells usually are paused in the G_1 phase. Mitosis can be further divided into several subphases. The two most central subphases are metaphase, when the paired sister chromatids align, and anaphase, when the sister chromatids are pulled apart by the mitotic spindle. Mitosis is followed by cytokinesis—when the cell constricts to become two new daughter cells.

Key commitment transition points in the cell cycle are highlighted and include START (just prior to the G_1/S transition; also termed the *restriction point*), which is a commitment to DNA replication; the G_2/M transition, which is a commitment to start mitosis; and the metaphase-to-anaphase transition, which represents a commitment to chromosome segregation and division. We will be focusing on what drives these major transitions, how they occur in a sharp and irreversible manner, and how they are blocked if something goes wrong.

Question 1: What drives the distinct phases of the cell cycle?

KEY TRANSITIONS CORRESPOND TO THE PERIODIC RISE AND FALL OF CYCLIN PROTEINS

Changes in the phases of the cell cycle are driven by the rise and fall in the abundance of a related family of key regulatory proteins known as the cyclins. The levels of cyclins change periodically, with each one associated with a distinct phase of the cell cycle. The plot to the left shows the periodic fluctuations in major classes of cyclins: the G_1/S cyclins, which spike during START; the S cyclins, which rise during S phase; and the M-phase cyclins, which rise during the beginning of mitosis, but disappear when anaphase begins. Note that these cyclins have different names in different organisms, and some organisms have multiple cyclins of the same class.

Functionally, these cyclins play a crucial role in activating the CDKs—the central switch regulating the cell cycle—which are described in more detail in the following panel. Cyclins bind to the CDK and have two effects: first, the cyclins act as allosteric activators to shift the CDK into an active conformation (although full activation of CDK also requires phosphorylation on the CDK activation loop); second, the cyclins can act as adaptors that bind to specific substrates, thus, in part, determining what specific substrates CDK phosphorylates for a given phase of the cell cycle.

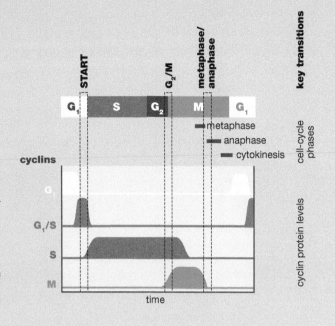

CYCLINS ARE EXCHANGEABLE SUBUNITS THAT ACTIVATE AND DIRECT CDK ACTIVITY IN A PHASE-SPECIFIC MANNER

The central molecular switch controlling the cell cycle is the CDK. This family of closely related kinases is defined by low activity unless associated with a cyclin subunit. In addition to activating the bound CDK, each cyclin recognizes distinct docking motifs on preferred substrates. Thus, each cyclin directs CDK to phosphorylate a distinct set of targets. These characteristic sets of substrates form the physiological program of that particular phase of the cell cycle: the G1/S program

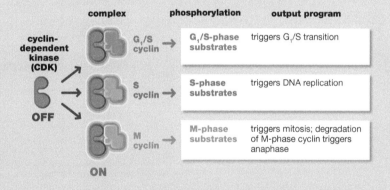

triggers the start of the cell cycle; the S-phase program phosphorylates targets that trigger DNA replication; the M-phase program triggers the initiation of mitosis, while the destruction of the M-phase cyclin then allows progression to anaphase and cell division.

CDK allosteric regulation is discussed in Chapter 3.

OTHER FACTORS THAT REGULATE CDK ACTIVITY

Cyclins are not the only way to regulate CDK activity. There are several other modes of regulation:

- CDK inhibitors: CDK–cyclin complexes can be inhibited by the binding of specific CDK inhibitor subunits. Thus, destruction of inhibitors can be critical for activating CDK.

- Activating phosphorylation site: activation of CDK also requires phosphorylation of Thr160 in the activation loop of the kinase (both pThr160 and cyclin binding are required for CDK activity). Regulatory kinases and phosphatases can modulate this site.

- Inhibitory phosphorylation site: CDK activity can be inhibited by phosphorylation on specific Tyr and Thr residues. Regulatory kinases and phosphatases can also modulate these sites.

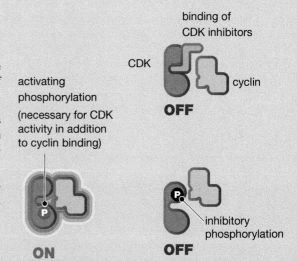

MAKING A CYCLE: FEEDBACK LOOPS DRIVE THE RISE AND FALL OF CYCLINS, LEADING TO PHASE TRANSITIONS

Cyclins and other regulatory factors control the CDK and direct it to execute the current cell-cycle phase program. How then does the cell transition to the next phase of the cell cycle? This system can progress through multiple phases because in addition to executing the current phase program, CDK phosphorylation also leads to feedback regulation that can help end the current cycle phase and drive entry into the next cycle phase. For example, active CDK complexes can phosphorylate and

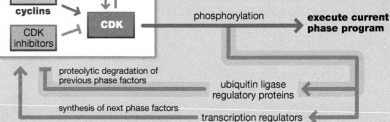

activate ubiquitin ligase regulatory proteins, which can in turn lead to targeted ubiquitylation and degradation of factors associated with maintaining the prior cell-cycle phase. Destruction of these factors leads to exit from the prior phase. Active CDK complexes can also phosphorylate and activate transcriptional regulators, thus driving the synthesis of factors that are associated with the next phase of the cell cycle. Expression of new factors thus helps drive entry into the next phase.

SUMMARY

Stages of the cell cycle are controlled by the CDKs which associate with a series of phase-specific factors known as the cyclins.

Cyclin levels rise and fall through the cell cycle, with distinct cyclins associated with each phase of the cycle.

Specific cyclin–CDK complexes phosphorylate phase-specific targets, leading to the execution of that phase's program. These specific CDK complexes also control the timing of the eventual transition to the next phase in the cell cycle.

Question 2: How does the circuitry of the cell cycle drive sharp and irreversible transitions between the phases of the cell cycle?

POSITIVE FEEDBACK LOOPS AND PROTEOLYTIC DEGRADATION MAKE KEY CELL-CYCLE TRANSITIONS SHARP AND IRREVERSIBLE

Progression through many of the key transitions in the cell cycle is driven by positive feedback loops, which help make the transitions sharper and more switchlike (all-or-nothing). Although the exact proteins involved are different at each stage of the cell cycle or in different organisms, many of the network-level circuit architectures are well conserved. In addition, several key steps in the cell cycle are driven by decisive proteolytic events that are inherently irreversible. Together, these mechanisms make transitions in the cell cycle reliable and committed.

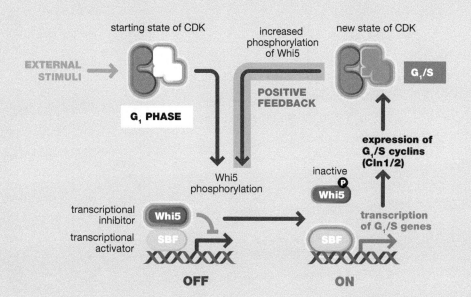

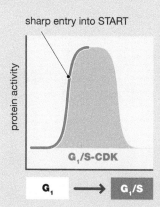

POSITIVE FEEDBACK REGULATION OF G_1/S TRANSCRIPTION PRODUCES SHARP COMMITMENT TO START

Entering START (leaving G_1 and beginning the G_1/S transition) represents the commitment to replicate the cell's genome. This needs to be a sharp and decisive transition. An important transcriptional positive feedback loop contributes to making this a sharp transition. When the cell is in G_1 phase, genes associated with the G_1/S program are not expressed because of the presence of a transcriptional inhibitor—in yeast, this inhibitor is the protein Whi5 (in vertebrates, the tumor suppressor RB and its relatives fulfill this role). Whi5 interacts with and represses the function of the transcriptional activator SBF (E2F in vertebrates).

At the initiation of START, the G_1-CDK enzyme is stimulated to initiate phosphorylation of Whi5. Phosphorylation acts to

inactivate Whi5. Since Whi5 is a transcriptional inhibitor, this results in the initiation of gene expression by the positive transcription factor SBF. Genes associated with the G_1/S program are expressed, beginning entry into START.

Among these newly activated genes are those expressing the G_1/S cyclins (in yeast, Cln1 and Cln2). Expression of these new cyclins leads to formation of the new G_1/S cyclin–CDK complex. This new form of CDK efficiently phosphorylates Whi5 (which leads to further SBF-driven expression of G_1/S genes). This positive feedback loop further drives the transition, such that even a small initial threshold amount of Whi5 phosphorylation will lead to an all-or-nothing commitment. The ultrasensitive increase in the activity of the G_1/S cyclin–CDK complex leads to a sharp entry into the START phase of the cell cycle.

Ultrasensitive responses and positive feedback circuits are discussed in Chapter 11.

POSITIVE-FEEDBACK-DRIVEN DEGRADATION OF CDK INHIBITOR PRODUCES SHARP ENTRY INTO S PHASE

An analogous positive feedback loop occurs during the transition to S phase. In yeast, S-cyclin–CDK complex begins accumulating during the late G_1, but is maintained in an inactive state by a specific CDK inhibitor protein called Sic1 (p27^{Kip1} in vertebrates). Sufficient G_1/S cyclin–CDK, however, can initiate phosphorylation of Sic1 on several key sites. Phosphorylated Sic1 is recognized by the ubiquitin ligase complex SCF-Cdc4 and targeted for proteolytic degradation. As more Sic1 is degraded, the active form of S cyclin-CDK is released. This newly activated CDK complex provides positive feedback by efficiently phosphorylating more Sic1, leading to further Sic1 degradation. Thus, an initiating threshold level of Sic1 phosphorylation will trigger this positive feedback and a sharp, ultrasensitive entry into S phase.

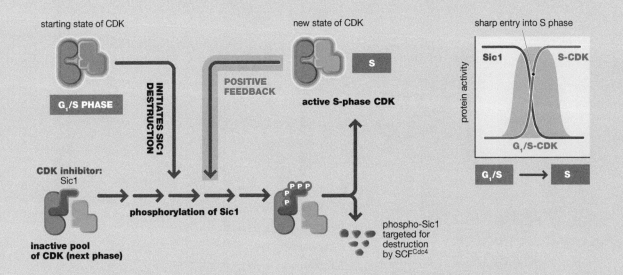

 Ubiquitin-directed proteolysis is discussed in Chapter 9.

POSITIVE FEEDBACK OR DOUBLE-NEGATIVE FEEDBACK LOOPS ARE A COMMON THEME IN DRIVING SHARP CELL-CYCLE TRANSITIONS

The two examples of positive feedback loops described above, involved in two successive cell-cycle transitions, are shown in circuit form here. These circuits represent common themes in sharp cell-cycle transitions. Both are positive feedback loops based on double-negative feedback (where activation occurs through inhibition of an inhibitor). In the case of the G_1-to-G_1/S transition, the transcriptional repressor Whi5 is inactivated by CDK phosphorylation. In the case of the G_1/S-to-S transition, the CDK inhibitor Sic1 is degraded in response to CDK phosphorylation. In addition, a negative feedback loop in which S cyclin–CDK phosphorylates and inactivates the SBF transcriptional activator provides another key link that drives exit from the G_1/S program. Here, we have focused on the yeast circuits for these transitions, although analogous feedback circuits are observed for the same transitions in the mammalian cell cycle.

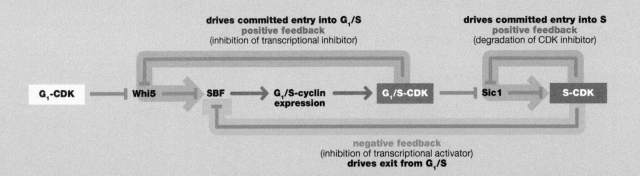

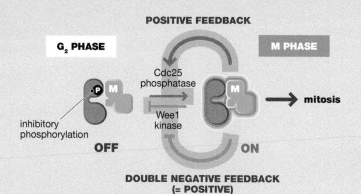

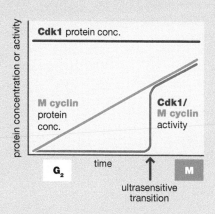

DIRECT POSITIVE FEEDBACK: CDK CAN POSITIVELY REGULATE ITS OWN ACTIVATION TO DRIVE SHARP ENTRY INTO M PHASE

A complex comprising the M-phase cyclin (cyclin B) and Cdk1 is critical for initiating mitotic spindle assembly and other key mitotic processes. Prior to the onset of mitosis, Cdk1 is maintained in an inactive state by phosphorylation catalyzed by the inhibitory kinase, Wee1. When mitosis is set to begin, the inhibitory phosphorylation is removed by the activating phosphatase, Cdc25. The activated Cdk1 molecules then participate in two positive feedback loops: Cdk1 phosphorylates and activates more Cdc25 molecules (activator of Cdk1 activity), while it also phosphorylates and inhibits Wee1 (inhibitor of Cdk1 activity). As a result of these two positive feedback loops, entry into mitosis is an ultrasensitive event (see the graph). The exact nature of the trigger that initiates the positive feedback loop is not fully understood.

Feedback loops are discussed in greater detail in Chapter 11.

PROTEOLYSIS PROVIDES AN IRREVERSIBLE SWITCH: M CYCLINS AND SECURIN ARE KEY TARGETS FOR DEGRADATION IN THE METAPHASE-TO-ANAPHASE TRANSITION

Unlike most regulatory post-translational modifications, proteolytic degradation is essentially irreversible because regenerating the intact protein requires further protein synthesis. Ubiquitin ligases, which mediate polyubiquitylation of target substrates and their subsequent degradation by the proteasome, are key cell-cycle regulators. Degradation-based regulation ensures that the cell cycle progresses to the next stage and cannot go backward.

A key cell-cycle regulator is the anaphase-promoting complex (APC), which ubiquitylates substrates and targets them for degradation during the metaphase-to-anaphase transition. The APC is a large, multi-subunit complex, whose specificity is controlled by two primary activators: Cdc20 and Cdh1. When chromosomes are properly aligned in mitosis, the APC is activated by phosphorylation and Cdc20 binding, marking the metaphase-to-anaphase transition.

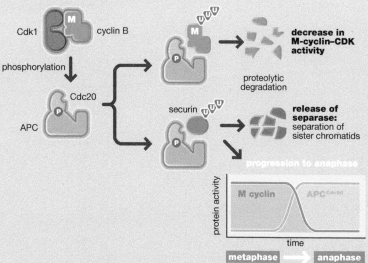

Two primary substrates of APC^{Cdc20} are securin and cyclin B. Degradation of securin leads to activation of separase, a protease that degrades the linkage that holds sister chromatids together (allowing their separation in anaphase). Degradation of cyclin B allows exit from mitosis.

Regulated protein degradation is discussed in Chapter 9.

SUMMARY

CDK complexes initiate many positive feedback loops (or double-negative feedback loops), which play a central role in generating sharp and committed transitions between distinct phases of the cell cycle. These committed transitions are required for ratchet-like forward progression of the cell cycle.

Some key steps in the cell cycle are regulated by ubiquitin-mediated proteolysis, which makes these transitions irreversible and committed.

Question 3: How does the cell arrest progression through the cell cycle when it detects critical problems?

THE CELL HAS KEY CHECKPOINTS THAT LEAD TO CELL-CYCLE ARREST WHEN TRIGGERED

The cell has sensor proteins that detect if critical problems arise during the cell cycle. If triggered, these sensor proteins initiate checkpoint programs that lead to cell-cycle arrest. Below, we discuss checkpoints for DNA damage (which results in blocking entry into or progression through S phase) and spindle misassembly (which results in blocking progression to anaphase and chromosome segregation).

These checkpoint blockades give the cell time to correct problems that would otherwise lead to serious consequences, such as genomic instability or chromosome mis-segregation.

DNA DAMAGE CHECKPOINT BLOCKS INITIATION OF S PHASE BY ACTIVATING KINASES THAT SENSE AND AMPLIFY DNA LESIONS

One of the most important functions of the cell cycle is to replicate the genome accurately and segregate the chromosomes equally between the two daughter cells. Failure to detect damage to genomic DNA could result in the transmission of mutations to a cell's progeny. Indeed, cells have several proteins that link DNA damage to regulation of the cell cycle. In G_1, DNA damage pauses the cell cycle before S phase, thereby preventing any errors in DNA replication. Detection of DNA damage in G_2 results in a similar pause before initiating chromosome segregation.

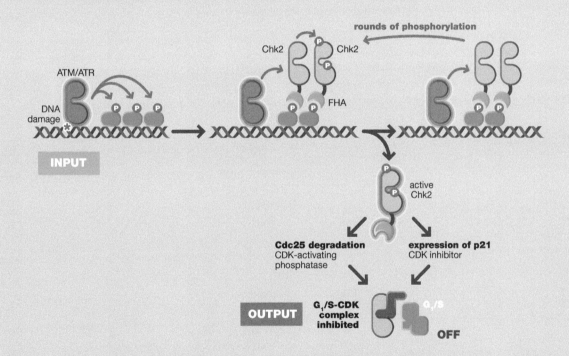

DNA DAMAGE INDUCES FORMATION OF KINASE COMPLEXES THROUGH PHOSPHORYLATION-DEPENDENT INTERACTIONS

The response to DNA damage requires the action of ATM/ATR kinases. ATR and ATM phosphorylate oligomeric adaptor proteins, including Rad9 and 53BP1, which also localize to sites of DNA damage. These phosphorylated adaptors are recognized by the Chk2 kinase through an FHA domain. Chk2 is then activated by phosphorylation, both by ATR and through autophosphorylation, and released, thereby allowing additional molecules of Chk2 to localize to the signaling complex. This allows amplification of the response: a single site of DNA damage can activate several molecules of Chk2. Activated Chk2 phosphorylates Cdc25, which targets Cdc25 for degradation. In the absence of Cdc25, phosphorylation by Wee1 inhibits CDK2, the G_1/S-CDK. Chk2 also indirectly induces the transcription of p21, a CDK inhibitor, further ensuring that S phase is not initiated.

Kinase activation and phosphorylation-dependent interactions are discussed in greater detail in Chapters 3 and 4.

THE SPINDLE ASSEMBLY CHECKPOINT BLOCKS THE METAPHASE-TO-ANAPHASE TRANSITION USING A SENSOR PROTEIN THAT CONFORMATIONALLY DETECTS UNATTACHED KINETOCHORES

Segregation of the chromosomes during mitosis is performed by the mitotic spindle, which consists of microtubules. Before sister chromatids are separated, they must be properly aligned and attached to both spindle poles. Each chromatid is attached to the microtubules of the mitotic spindle by a specialized structure called the kinetochore. If the kinetochores are not properly secured to spindle microtubules, sister chromatids could be mis-segregated, resulting in one daughter cell receiving two copies, and the other daughter receiving none. The spindle checkpoint acts to ensure that chromosome segregation occurs only after all the kinetochores are properly attached.

Mad2, a critical component of the spindle assembly checkpoint, has two primary binding partners: Mad1, which binds to unattached kinetochores, and Cdc20, which is also a key activator of the APC. Cdc20 cannot activate the APC when it is bound to Mad2.

In the absence of a binding partner, Mad2 exists in an open conformation (O-Mad2), where a "safety belt" region (pink, in the figure

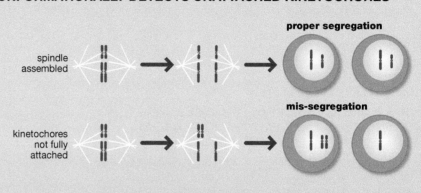

Initiation of anaphase before proper spindle assembly can result in mis-segregation of chromosomes.

below) is bound tightly to the Mad2 core. In order to bind Mad1 or Cdc20, Mad2 must undergo a conformational change where the safety belt loosens so that it can wrap around the binding partner. In the ligand-bound closed conformation (C-Mad2), the safety belt interacts with another region of Mad2.

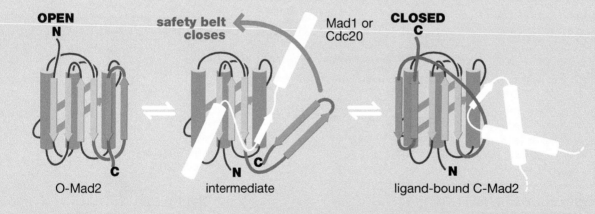

Spindle assembly checkpoint protein Mad2 can exist in two conformations, OPEN and CLOSED.

Kinetochores that are not attached to the spindle bind to a stable complex of Mad1 and C-Mad2. This copy of C-Mad2 can dimerize with soluble O-Mad2, and this binding loosens the safety belt, which allows binding to Cdc20 and converts the O-Mad2 molecule into a C-Mad2 molecule. In a similar fashion, C-Mad2 bound to Cdc20 will also potentiate the binding of O-Mad2 to Cdc20.

Thus, C-Mad2 acts as a catalyst: a single unattached kinetochore can help convert many unbound O-Mad2 molecules to complexes of C-Mad2 bound to Cdc20. The binding of Cdc20 to Mad2 acts to block the metaphase-to-anaphase transition by sequestering Cdc20 and preventing it from activating the APC. Thus, anaphase cannot be induced until all the kinetochores are attached.

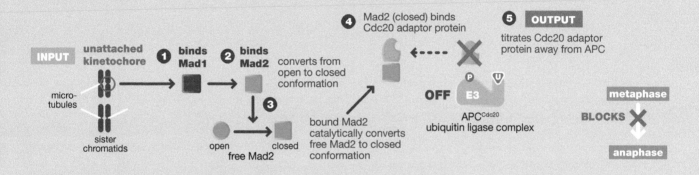

Unattached kinetochores potentiate Mad2 binding to Cdc20 resulting in inhibition of the APC.

 Protein interactions are discussed in greater detail in Chapter 2. Conformational changes are discussed in greater detail in Chapter 3.

SUMMARY

Cell-cycle progression can be arrested by checkpoint programs that detect key problems such as DNA damage or unattached kinetochores. These checkpoints provide time for the cell to fix problems that would be exacerbated if the cell cycle had continued to progress.

SECTION 14.4 T LYMPHOCYTE ACTIVATION— how key cells in our immune system are mobilized to fight infection

Some of the most remarkable examples of complex cellular signal processing are displayed by the lymphocytes of the mammalian adaptive immune system. The adaptive immune system recognizes molecules unique to invading pathogens, and subsequently activates sophisticated downstream response programs that eliminate specific pathogen types, kill infected cells, and provide long-term immunity against future infection with that pathogen.

In the course of an infection, pathogens initially encounter chemical mechanisms and specialized cells, such as neutrophils and macrophages, which try to kill and eliminate the invading microbes and infected cells. This first, more generalized response is known as the innate immune response. In vertebrates, pathogens also initiate a second response, known as the adaptive immune response, which is much more specifically aimed at the particular invading pathogen.

An adaptive immune response typically begins when antigen-presenting cells, such as dendritic cells, engulf pathogens or pathogen-derived molecular products. Inside dendritic cells, the pathogens or their derivative proteins are broken down into small peptide fragments. As a part of normal protein turnover in the cell, these foreign peptides bind to major histocompatibility complex (MHC) receptors in the endoplasmic reticulum and are then transported to the cell surface, where they are displayed on the outside of the cell as MHC–peptide complexes. MHC receptors, in effect, display the intracellular protein diversity of the cell on its extracellular surface. The majority of MHC complexes expressed on the cell contain self-derived peptides; pathogen-derived peptides are referred to as antigens because of their ability to be recognized as foreign, and to eventually generate antibodies and trigger other aspects of the adaptive immune response.

Dendritic cells displaying a pathogen-derived antigen will migrate from sites of infection to lymph nodes where they interact with T lymphocytes, also known as T cells. T cells have specialized receptors on their surfaces, called T cell receptors (TCRs), that can bind with great specificity to foreign peptides displayed in MHC–peptide complexes on dendritic cells. Each individual T cell expresses a unique TCR variant that can recognize and bind only a very specific set of antigenic peptide sequences. Therefore, each T cell recognizes only one or a few antigens. However, since vertebrates typically produce over a million different variant TCRs within an organism, the repertoire of T cells is capable of identifying nearly all foreign antigens. (T cells that recognize self-antigens are eliminated during development, as such self-reactive cells would give rise to autoimmunity.)

A T cell that has not yet been activated through recognition of a specific foreign peptide is said to be "naïve." A T cell becomes activated when it recognizes the specific, cognate antigen for its TCR, in addition to other co-stimulatory signals provided by the antigen-presenting cell. Once a naïve T cell is activated, it begins to proliferate in order to create more T cells with receptors specific to that particular antigen. These cells also differentiate into more specialized types of T cell, called killer (or cytotoxic) T cells and helper T cells, which perform specific functions that enable the adaptive immune system to eliminate pathogens. Helper T cells stimulate antibody production through B cells and also stimulate neutrophils. As the name suggests, killer T cells attack and kill infected cells. This multipronged response stimulated by antigen recognition and T cell activation is extremely powerful and can eliminate many infections.

In this section, we focus on the signaling events that allow T cells to detect foreign antigens presented in MHC–peptide complexes. Specifically, we will address the following questions:

1. **Question 1: How does the TCR detect and propagate signals in the T cell? (page 382)**

2. **Question 2: How does the T cell launch a robust response when stimulated by as few as ten antigenic peptide complexes? (page 383)**

3. **Question 3: How does the T cell signaling system recognize only foreign antigens and filter against activation by weak or transient self-peptide signals? (page 386)**

THE SYSTEM

ORGANISM | launching the adaptive immune response

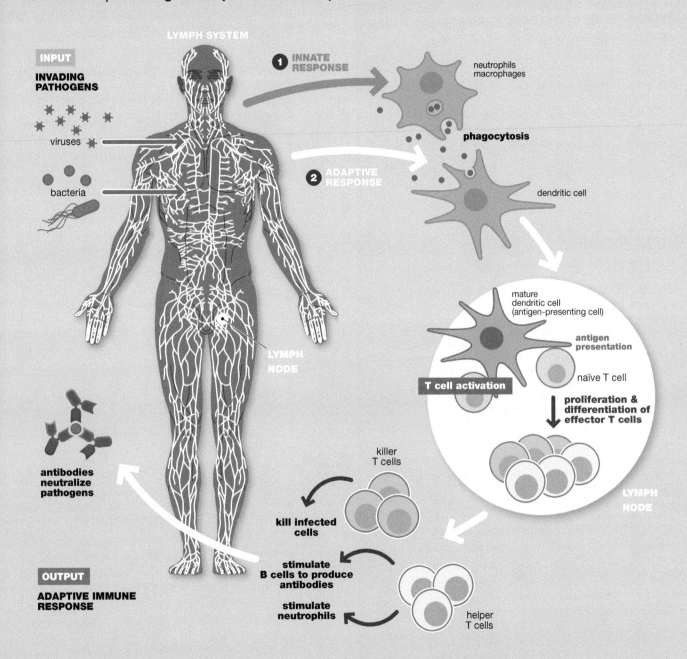

THE ADAPTIVE IMMUNE RESPONSE IS ACTIVATED DURING AN INFECTION.

The adaptive immune response begins when specialized antigen-presenting cells, such as dendritic cells, take up pathogen-derived material in the periphery through phagocytosis. They then traffic to the lymph nodes, where they interact with naïve T cells. The body has an incredible diversity of naïve T cells, each with a distinct TCRs that can recognize potential antigenic peptides. If the antigen-presenting cell encounters a cognate T cell that recognizes one of the molecules that it presents on its surface as an antigen, then that T cell becomes activated. Through a cytokine-stimulated autocrine loop, stimulated by the cytokine IL-2, that T cell will clonally proliferate and differentiate into killer and helper T cells (effector T cells). Together, these cells will launch a series of complex responses, including the killing of infected cells presenting the cognate antigen, and stimulation of B cells to produce cognate antibodies. These responses help neutralize the infection.

CELL | engagement of T cell and antigen-presenting cell

ANTIGEN PRESENTATION TO THE T CELL.
Multiple inputs from the dendritic cell are required for T cell activation.

In the lymph node, dendritic cells and T cells will directly contact each other, literally crawling over one another, scanning for proper molecular interaction partners (panel a). Once the cells recognize each other, they form extensive contacts, and full activation of a cognate T cell can require hours of sustained interaction with the antigen-presenting cell (here a dendritic cell).

At the heart of the recognition process is the TCR, a multiprotein cell-surface complex on the T cell which directly binds and recognizes the peptide–MHC complex on the dendritic cell (upper right). Most peptides displayed on the dendritic cell will be self-peptide–MHC complexes, which TCRs do not recognize. If, however, the dendritic cell displays an antigen peptide that is recognized by the TCR with sufficient affinity and duration, then that T cell will become activated, leading to clonal proliferation and differentiation.

Proper interaction between the TCR and MHC complex, however, is not sufficient for activation of the T cell (lower right). Multiple other cell–cell interactions are required, including the interaction of cell adhesion receptors (LFA1 with ICAM), co-receptor molecules such as CD4 (in helper T cells) and CD8 (in cytotoxic T cells), and other co-stimulatory receptors such as CD28. The cell adhesion molecules are necessary to form a tight and extensive area of contact between the dendritic cell and the T cell. The CD4 and CD8 co-receptors participate directly in the formation of the TCR-MHC complex and are necessary for TCR activation. The CD28 co-receptor recognizes the ligand B7, which is a partner cell-surface

protein in the dendritic cell. The activation process then leads to major cytoskeletal rearrangements in the T cell (leading to more extensive interactions between the cells), changes in transcription (including expression of key secreted cytokines), and cell proliferation and survival. All of the receptor systems cooperate, providing tight control over whether the T cell actually undergoes sustained activation. The T cell thus acts like an AND gate that requires multiple inputs for activation.

Human dendritic cell (*blue*) interacting with T cell (*yellow*). Olivier Schwartz/ Science Photo Library

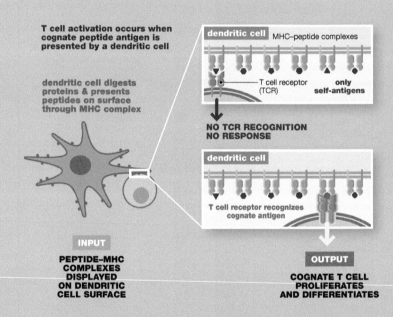

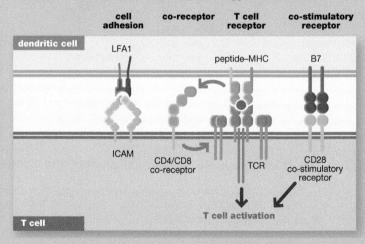

MOLECULAR NETWORK | T Cell Receptor (TCR) Signaling Network

T CELL ACTIVATION
Signal propagation involves four main molecular complex modules

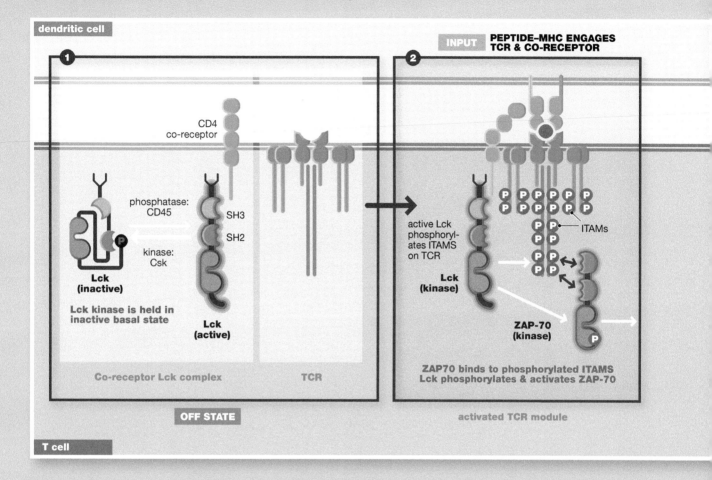

1. OFF STATE. Lck is a Src family tyrosine kinase that is associated with the co-receptor (in this case, for a helper T cell, the co-receptor is CD4). Lck is the key kinase that phosphorylates and activates the TCR upon activation. Lck's activity is basally regulated by opposing enzymes—the tyrosine kinase Csk which phosphorylates and inhibits Lck, and the membrane-associated phosphotyrosine phosphatase CD45 which "erases" this phosphorylation and activates Lck. In the basal state, Lck is inactive. Exactly how peptide–MHC engagement with the TCR triggers Lck activity is unclear, but it is thought to involve the engagement of the co-receptor CD4 with the TCR–peptide-MHC complex, leading to transphosphorylation by other Lck molecules in the complex. These changes ultimately shift the balance in the amount of activated Lck. In the next panel, we discuss how Lck then activates the TCR.

2. Activated TCR complex. Within minutes of stimulation by engagement with the antigenic peptide–MHC complex, Lck phosphorylates key tyrosine motifs in the TCR known as immunoreceptor tyrosine-based activation motifs (ITAMs). Phosphorylation of the ITAMs converts them into binding sites that are recognized by the tandem SH2 domains from the cytoplasmic

tyrosine kinase ZAP-70. Once recruited to the TCR, ZAP-70 is also phosphorylated and allosterically activated by the co-localized Lck kinase. Thus, the Lck kinase is responsible for both recruitment and activation of ZAP-70. In the next panel, we describe how ZAP-70 leads to assembly of a major signaling complex.

3. Assembly of LAT/SLP-76 scaffold complex. The activated and localized ZAP-70 kinase is now in a position to phosphorylate two of the most important substrates, the scaffolding proteins LAT and SLP-76. LAT is a membrane-associated protein with five tyrosine phosphorylation sites, while SLP-76 is a soluble protein with four tyrosine phosphorylation sites. Phosphorylation of the key tyrosines in these scaffold proteins creates a set of SH2-domain-binding sites, and thus allows the assembly of a large multiprotein signaling assembly (a *biomolecular condensate*) that is essential for T cell activation. The SH2- and SH3-domain-containing protein GADS acts as an adaptor to mediate this assembly. The enzyme phospholipase Cγ (PLCγ), which also has SH2 and SH3 domains, forms a part of this complex and is activated as a result (PLCγ is illustrated here using a simplified domain representation; see Figure 10.21 for the complete domain structure). This complex is also responsible for activating the enzyme Sos

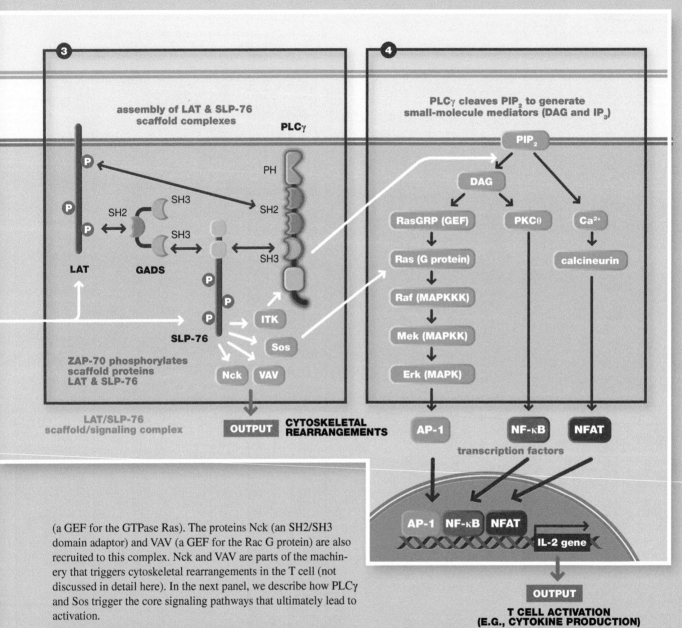

(a GEF for the GTPase Ras). The proteins Nck (an SH2/SH3 domain adaptor) and VAV (a GEF for the Rac G protein) are also recruited to this complex. Nck and VAV are parts of the machinery that triggers cytoskeletal rearrangements in the T cell (not discussed in detail here). In the next panel, we describe how PLCγ and Sos trigger the core signaling pathways that ultimately lead to activation.

4. Core signaling pathways lead to activation of key T cell genes. The activated LAT/SLP-76 complex leads to activation of a number of core signaling pathways. One of the key effectors is the enzyme PLCγ, which when activated in this complex, generates the signaling mediators diacylglycerol (DAG) and inositol trisphosphate (IP$_3$) by hydrolyzing the lipid signaling molecule phosphatidylinositol (4,5) bisphosphate (PIP$_2$). DAG leads to the activation of the Ras-MAP kinase pathway (through recruitment of the GEF RasGRP; activated Sos also helps activate Ras) and the protein kinase Cθ pathway. IP$_3$ stimulates the release of calcium from the ER, leading to activation of the calcium-dependent

phosphatase calcineurin. The net result is the activation of a set of transcription factors, AP-1, NF-κB, and NFAT, that cooperatively mediate a number of critical transcriptional responses needed for T cell activation (co-stimulatory signals are also important for NFAT activation, but the details are not discussed here). Here we illustrate one key transcriptional output, which is the increased expression of the cytokine IL-2.

Question 1: How does the TCR detect and propagate signals in the T cell?

T CELL RECEPTOR PHOSPHORYLATION INITIATES ACTIVATION

The initial stages of signal transmission upon T cell activation are controlled by tyrosine phosphorylation. Engagement of the TCR with the peptide–MHC complex leads to phosphorylation of the TCR by the Lck kinase. Phosphorylation of the TCRs occurs on peptide motifs known as ITAMs. Upon phosphorylation, the ITAMs act as binding sites for the SH2 domains from the kinase ZAP-70. This recruitment of ZAP-70 to the TCR complex then triggers the next set of phosphorylation events in the T-cell-activation cascade.

The TCR is composed of eight subunits: the α and β chains contain the extracellular domains that interact with the MHC–peptide complex, while the CD3 γ, δ, ε, and ζ chains contain intracellular regions that communicate with downstream signaling proteins. The intracellular segments of the TCR collectively contain 10 ITAMs—a specific dual tyrosine motif that is phosphorylated by Lck kinase upon activation. The phosphorylated ITAMs become docking sites for the tandem SH2 domains from the ZAP-70 kinase. Once bound to the TCR, ZAP-70 is itself phosphorylated and activated by Lck. In turn, activated and localized ZAP-70 phosphorylates a number of critical downstream targets that initiate the T cell response.

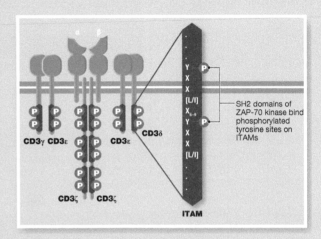

The ITAMs act as recruitment sites for the tandem SH2 units from the kinase ZAP-70. The tandem recognition unit has higher affinity and specificity than many other SH2–peptide interactions due to avidity effects. Because the interaction is dependent on phosphorylation of the tyrosine residues in the ITAMs, the interaction only occurs after the receptor has been activated. Binding of the ITAMs also relieves autoinhibition of the ZAP-70 kinase. The ZAP-70 kinase is now in an activated conformation and localized to the activated receptor complex, where it is in a position to phosphorylate a number of key downstream targets.

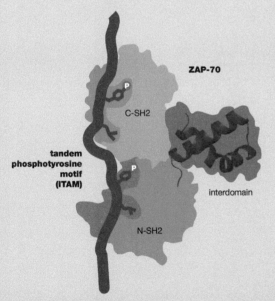

Receptor activation is discussed in Chapter 8. SH2-mediated phosphopeptide recognition is discussed in Chapter 10. Multivalent protein interactions are discussed in Chapter 2.

SUMMARY

The T cell scans for antigenic peptide–MHC complexes and co-stimulatory signals on the surface of an antigen-presenting cell. Activation of the naïve T cell only occurs with the proper combination of signals.

The TCR recognizes the peptide–MHC complex and propagates signals via tyrosine phosphorylation on ITAM peptides. The phosphorylated ITAMs then serve as recruitment sites for the ZAP-70 kinase (via its SH2 domains), thereby triggering further downstream signaling.

Question 2: How does the T cell launch a robust response when stimulated by as few as ten antigenic peptide complexes?

> ## POSITIVE FEEDBACK LOOPS IN THE T-CELL-ACTIVATION NETWORK CAN AMPLIFY A SMALL ANTIGENIC INPUT SIGNAL TO TRIGGER A ROBUST RESPONSE
>
> T cells are remarkable in that as few as 10 antigenic peptide–MHC complexes presented by an antigen-presenting cell can lead to robust activation of a cognate T cell. Positive feedback loops in the T-cell-activation pathway play an important role in amplifying this small stimulus to result in full activation. Here we review four examples of positive feedback loops involved in T cell activation. These examples illustrate how feedback regulation of signaling processes can operate at many different levels, including intracellular molecular interactions, cellular reorganization and localization, and intercellular (paracrine) communication.

POSITIVE FEEDBACK THROUGH CELLULAR REORGANIZATION: FORMATION OF THE IMMUNE SYNAPSE LEADS TO RECEPTOR AND SIGNALING-PROTEIN CLUSTERING

Upon antigen presentation, the antigen-presenting cell and the T cell undergo major structural reorganization, forming an extensive cell–cell adhesion junction known as the immune synapse or the supramolecular activation complex (SMAC). The immune synapse structure forms within a few minutes of the encounter, but can remain stable for well over an hour. Within this new cellular junction structure, cell adhesion molecules like ICAM segregate to the periphery (pSMAC), while signaling

receptors, including the TCR, CD4 and CD8 co-receptors, and CD28 co-stimulatory receptors, are corralled and clustered in the center (cSMAC). This clustering of key signaling receptors, especially during a sustained interaction between T cell and antigen-presenting cell, is thought to enhance and amplify their communication with one another, allowing them to accurately assess the presence of even a few molecules of the cognate antigenic peptide–MHC complex. The immune synapse is one of the best characterized examples of a biomolecular condensate, stabilized by a multivalent interactions among the various components.

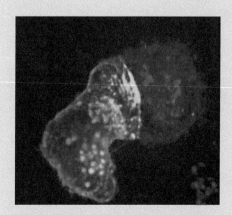

T cell (*blue*) interacting with antigen-presenting cell (*red*). The zone of contact between the cells is the immunological synapse (*green*). Courtesy of Tomasz Zal, M. Anna Zal, and Nicholas R.J. Gascoigne.

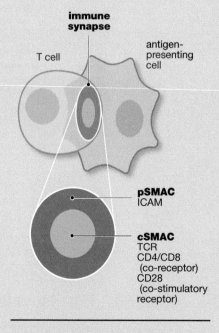

Supramolecular activation complex (SMAC).

POSITIVE FEEDBACK THROUGH RECEPTOR–RECEPTOR SPREADING

Although an antigen-presenting cell–T cell interface may only contain a few agonist (non-self) peptide complexes within a sea of non-agonist (self) peptide complexes, recent evidence suggests that agonist–receptor complexes can interact with non-agonist–receptor complexes. Thus, activated Lck associated with an agonist–receptor complex may be able to phosphorylate other nearby receptors, thus leading to amplified signal transmission.

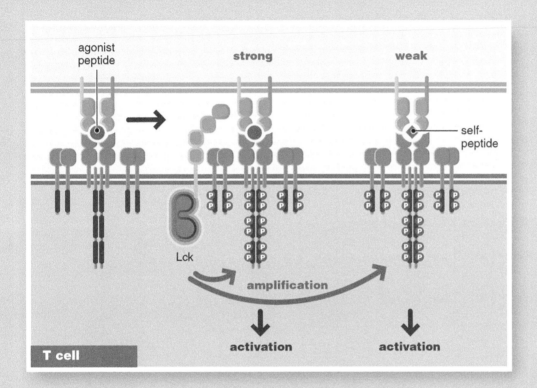

 Biomolecular condensates are discussed in Chapter 5.

POSITIVE FEEDBACK THROUGH IL-2 CYTOKINE AUTOCRINE AND PARACRINE SIGNALING

One of the key outputs of naïve T cell activation is the secretion of the cytokine IL-2. IL-2 is a potent mitogen and survival factor for T cells. Thus, IL-2 secreted by an activated T cell can then form an autocrine loop (stimulates the same cell) and a paracrine loop (stimulates its neighboring T cells). Self-activation by this paracrine loop is detected by the IL-2 receptor in T cells. IL-2 activation itself leads to increased expression of both IL-2 and IL-2 receptor in the signal-receiving cell, thus forming an even more potent positive feedback mechanism for amplifying activation. Ultimately this process leads to expansion of a single cognate T cell into a clonal army of activated T cells.

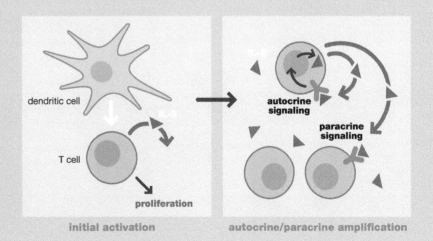

POSITIVE FEEDBACK THROUGH A RAS-SOS ALLOSTERIC FEEDBACK LOOP LEADS TO SWITCHLIKE ACTIVATION OF THE MAPK ERK

Activation of the Ras-MAP kinase pathway is a central part of T cell activation. Initial activation of Ras is thought to occur via activation of the GEF RasGRP in response to diacylglycerol (DAG) production (see the molecular network, panel 4). RasGRP leads to a linear increase in Ras activation. However, a second RasGEF, Sos, is known to be allosterically activated by its own product

(GTP-bound Ras). Thus, it is postulated that the initial amount of Ras that is activated by RasGRP is sufficient to allosterically activate Sos GEF activity, thus triggering a positive feedback loop that can strongly amplify output once past a threshold of activation. Consistent with this model, increasing stimulation of T cells is observed to lead to a bistable (all-or-none) response, with respect to phosphorylation of the MAPK Erk (right). Disruption of the Ras-Sos positive feedback loop disrupts this all-or-none response.

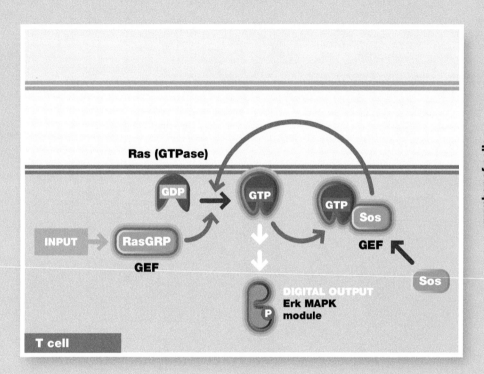

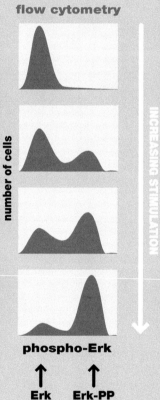

SUMMARY

Positive feedback loops in T cell signaling amplify weak but correct inputs to yield robust activation, even by as few as 10 non-self-peptide–MHC complexes on an antigen-presenting cell.

Question 3: How does the T cell-signaling system recognize only foreign antigens and filter against activation by weak or transient self-peptide signals?

NEGATIVE FEEDBACK LOOPS HELP PREVENT AND LIMIT T CELL ACTIVATION

It is critical that T cells are not activated by improper inputs. We now know that there are a number of negative feedback loops that may help keep the system relatively quiet and prevent misactivation. These negative feedback loops can be used as a way to abort misactivation of a T cell by transient or partial input signals (for example, absent co-stimulation). Full activation of a T cell requires several hours of sustained interaction with the antigen-presenting cell. Here, we describe two such negative feedback systems.

NEGATIVE FEEDBACK: ACTIVATION OF CO-INHIBITORY RECEPTORS THAT BLOCK CO-STIMULATORY SIGNALS

Activation of T cells requires co-stimulatory signals from the dendritic cell, in addition to the MHC–peptide signal detected directly by the TCR. For example, a co-stimulatory receptor in the T cell, CD28, is activated by the co-stimulatory cell-surface molecule, B7, expressed on the antigen-presenting cell. Dual activation of the TCR and the CD28 co-receptor is necessary for T cell activation.

CD28 co-receptor signaling appears to be an important target for negative feedback regulation to prevent misactivation of the T cell. When the TCR is stimulated, in addition to the better-characterized positive effects, this also leads to increased cell-surface expression of molecules known as co-inhibitory receptors, such as CTLA-4. The CTLA-4 co-inhibitory receptor binds to the same ligand (B7) in the antigen-presenting cell as that recognized by the CD28 co-receptor, but with much higher affinity. Thus, CTLA-4 competitively blocks co-receptor stimulation of the T cell. In addition, the CTLA-4 receptor has intracellular motifs known as immune tyrosine inhibitory motifs (ITIMs), which, upon phosphorylation, recruit inhibitory factors such as inhibitory phosphatases, which turn down T cell activation (ITIMs have an opposing function to ITAMs).

Monoclonal antibodies that block inhibitory co-receptors like CTLA-4 have provided a major breakthrough in potential cancer therapies, because they can effectively stimulate the host's own immune response against tumor cells. These antibodies, however, are also associated with autoimmune-like side effects, as might be expected to occur if T cells lack this negative feedback mechanism to prevent misactivation.

Activation of the TCR leads to its tyrosine phosphorylation. While these phosphorylation events lead primarily to recruitment of activating downstream factors, they can also lead to recruitment of SH2-domain-containing ubiquitin ligases of the Cbl family. Recruitment of Cbl leads to ubiquitylation and ultimately internalization and degradation of the TCR, creating a negative regulatory loop.

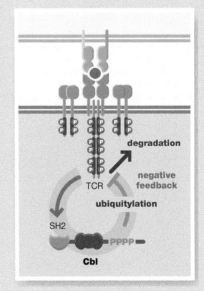

Negative feedback loops are discussed in Chapter 11. CTLA-4 as a target of cancer therapy is discussed in Chapter 12.

COORDINATED FAST NEGATIVE FEEDBACK AND SLOW POSITIVE FEEDBACK COULD YIELD A DISCRIMINATORY FILTER FOR T CELL ACTIVATION

As a T cell encounters and contacts an antigen-presenting cell, it will engage millions of self-peptide–MHC complexes, scanning for a potential antigenic peptide that is a match for its specific TCR. The affinities of antigenic versus non-antigenic peptide complexes for the TCR are not significantly different (for example, they may only differ by a few fold in K_d), so receptor occupancy alone cannot account for the remarkable specificity of T cell activation—over a million non-antigenic peptide complexes are unable to activate the T cell, while as few as ten antigenic peptide complexes are sufficient to fully activate the T cell. We do not have a complete answer to this complex problem, but some data suggest that a critical way in which non-self-peptides differ from self-peptides is the lifetime of the complex that they form with the cognate TCR—non-self-peptides bind with lifetimes at least tenfold longer than self-peptides. Thus, the TCR may use lifetime of receptor engagement to discriminate between self and non-self signals. Below, we discuss models for how dynamically coordinated positive and negative feedback loops might act in concert to achieve both effective filtering against misactivation by self-peptides, yet still allow robust activation by antigenic (longer-lifetime) peptides.

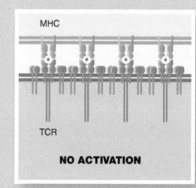

excess of self-peptides presented on MHC

- low affinity
- short complex lifetime (<2 sec)

MHC

TCR

NO ACTIVATION

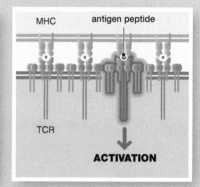

<10 agonist peptide–MHC complexes are sufficient for strong activation

- low to moderate affinity
- longer complex lifetime (2–10 sec)

MHC   antigen peptide

TCR

ACTIVATION

FAST NEGATIVE FEEDBACK MEDIATED BY SHP1 CAN BLOCK ACTIVATION BY SHORT STIMULATION TIME INPUTS.

We have described how negative feedback loops in T cell signaling may abort and filter against improper activation, while positive feedback loops may amplify a sustained signal to yield full activation. Yet it remains unclear how such positive and negative feedback loops can yield discrimination between self and antigenic peptide inputs. It is postulated that one way in which positive and negative feedback loops can act in a discriminatory way is by having tightly coordinated fast negative feedback loops and slow positive feedback loops. Such a network could act as a sharp time filter in which only signals that are sustained enough (that is, longer-lifetime antigenic peptide binding) will switch the system on. Here, we describe a set of coordinated positive and negative feedback loops involving the Lck tyrosine kinase and Shp1 phosphotyrosine phosphatase that have been postulated to contribute to T cell temporal discrimination.

The Lck tyrosine kinase is activated upon TCR engagement and generates the first positive signal of TCR activation, ITAM phosphorylation. Nevertheless, in addition to this positive signal, Lck also generates a negative signal—it phosphorylates the Shp1 phosphotyrosine phosphatase. Phosphorylated Shp1 is recruited to the Lck SH2 domain, where it can exert negative effects by catalyzing the dephosphorylation of the TCR, ZAP-70, and Lck itself. This negative feedback loop may act to filter out transient stimulation.

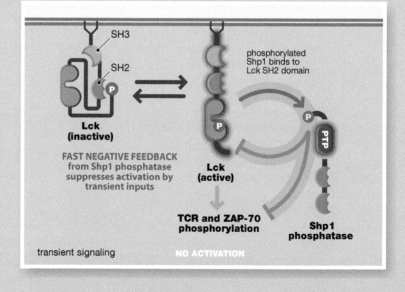

SH3

SH2

P

Lck (inactive)

FAST NEGATIVE FEEDBACK from Shp1 phosphatase suppresses activation by transient inputs

transient signaling

phosphorylated Shp1 binds to Lck SH2 domain

P

P

PTP

Lck (active)

TCR and ZAP-70 phosphorylation

Shp1 phosphatase

NO ACTIVATION

SLOW POSITIVE FEEDBACK IN WHICH ACTIVATED ERK OVERRIDES SHP1 NEGATIVE REGULATION MAY ALLOW FOR FULL ACTIVATION WITH SUSTAINED INPUT STIMULATION.

If activating signals received by the T cell are sustained enough (and there are repeated rounds of receptor engagement), then this should lead to the gradual accumulation of the activated form of the MAPK Erk. Erk has been found to phosphorylate Lck at residue Ser59, which disrupts the interaction of Shp1 with Lck. Thus, activated Erk can disrupt the Lck→Shpl negative feedback loop, acting as a double-negative (or positive) feedback loop. Thus, Erk provides a slower, delayed positive feedback loop that overrides Shp1 fast negative feedback

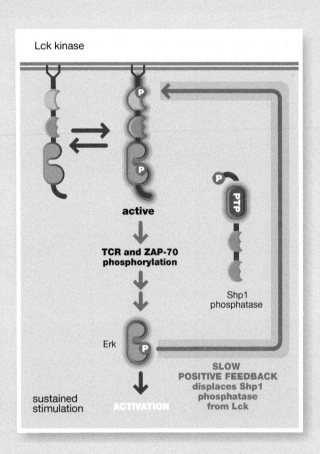

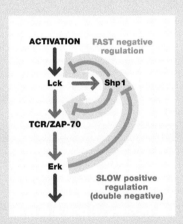

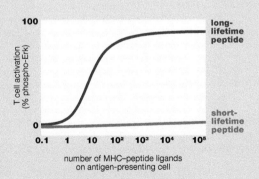

Theoretical plot of how interlocked fast negative and slow positive feedback loops could lead to a high level of T cell discrimination between long-lifetime and short-lifetime peptide–MHC ligands.

COORDINATED FAST NEGATIVE FEEDBACK AND SLOW POSITIVE FEEDBACK CAN LEAD TO AN ALL-OR-NONE TIME FILTER.

These negative and positive feedback loops differ in time scale, and are mutually exclusive (they both determine whether Shp1 is or is not recruited to Lck). Quantitative modeling indicates that such a circuit could yield activation that is highly dependent on the lifetime of the stimulating input (peptide–MHC engagement with the TCR). Because Erk activation in T cells is such a switchlike response, one expects to see a sharp threshold for this Erk-mediated time filter. Such a dual-timescale feedback circuit could explain how a system can be activated fully by a very small number of long-lifetime peptide–MHC ligands, but can ignore stimulation by large numbers of short-lifetime peptide–MHC ligands. This circuit design is analogous to a *kinetic proofreading* mechanism, where only sustained input can reliably result in downstream output.

 Positive and negative feedback loops are discussed in Chapter 11. Kinetic proofreading is discussed in Chapter 5.

SUMMARY

Negative feedback loops in T cell signaling prevent misactivation by transient or partial signals, such as those presented by the large excess of self-antigen peptide–MHC complexes (millions of which are presented on any antigen-presenting cell).

Coordination between fast negative and slow positive feedback loops may allow the T cell to discriminate between self and non-self (antigenic) peptides, only yielding a full response for non-self-peptides that form a longer-lived complex with the TCR.

BIBLIOGRAPHY

SECTION 1 (VISION)

Burns ME & Pugh EN Jr (2010) Lessons from photoreceptors: turning off G-protein signalling in living cells. *Physiology (Bethesda)* 25, 72–84.

Calvert PD, Govardovskii VI, Krasnoperova N et al. (2001) Membrane protein diffusion sets the speed of rod phototransduction. *Nature* 411, 90–94.

Fu Y & Yau KW (2007) Phototransduction in mouse rods and cones. *Pflugers Arch.* 454, 805–819.

Leskov IB, Klenchin VA, Handy JW, et al. (2000) The gain of rod phototransduction: reconciliation of biochemical and electrophysiological measurements. *Neuron* 27, 525–537.

Palczewski K (2012) Chemistry and biology of vision. *J. Biol. Chem.* 287, 1612–1619.

SECTION 2 (MITOGENIC SIGNALING)

Andrae J, Gallini R & Betsholtz C (2008) Role of platelet-derived growth factors in physiology and medicine. *Genes Dev.* 22, 1276–1312.

Boykevisch S, Zhao C, Sondermann H et al. (2006) Regulation of ras signaling dynamics by Sos-mediated positive feedback. *Curr. Biol.* 16, 2173–2179.

Demoulin JB & Essaghir A (2014) PDGF receptor signaling networks in normal and cancer cells. *Cytokine Growth Factor Rev.* 25, 273–283.

Lemmon MA & Schlessinger J (2010) Cell signaling by receptor tyrosine kinases. *Cell* 141, 1117–1134.

SECTION 3 (THE CELL CYCLE)

Bertoli C, Skotheim JM & de Bruin RA (2013) Control of cell cycle transcription during G1 and S phases. *Nat. Rev. Mol. Cell Biol.* 14, 518–528.

Craney A & Rape M (2013) Dynamic regulation of ubiquitin-dependent cell cycle control. *Curr. Opin. Cell Biol.* 25, 704–710.

Ferrell JE Jr (2013) Feedback loops and reciprocal regulation: recurring motifs in the systems biology of the cell cycle. *Curr. Opin. Cell Biol.* 25, 676–686.

Fisher D, Krasinska L, Coudreuse D & Nov\aacutek B (2012) Phosphorylation network dynamics in the control of cell cycle transitions. *J. Cell Sci.* 125, 4703–4711.

Johnson A & Skotheim JM (2013) Start and the restriction point. *Curr Opin. Cell Biol.* 25, 717–723.

Morgan DO (2007) *The Cell Cycle: Principles of Control*, New Science Press Ltd.

Musacchio A & Salmon ED (2007) The spindle-assembly checkpoint in space and time. *Nat. Rev. Mol. Cell Biol.* 8, 379–393.

Reinhardt HC & Yaffe MB (2009) Kinases that control the cell cycle in response to DNA damage: Chk1, Chk2, and MK2. *Curr. Opin. Cell Biol.* 21, 245–255.

Teixeira LK & Reed SI (2013) Ubiquitin ligases and cell cycle control. *Annu. Rev. Biochem.* 82, 387–414.

SECTION 4 (T LYMPHOCYTE ACTIVATION)

Altan-Bonnet G & Germain RN (2005) Modeling T cell antigen discrimination based on feedback control of digital ERK responses. *PLoS Biol.* 3, e356.

Das J, Ho M, Zikherman J, et al. (2009) Digital signaling and hysteresis characterize ras activation in lymphoid cells. *Cell* 136, 337–351.

Dustin ML & Groves JT (2012) Receptor signaling clusters in the immune synapse. *Annu. Rev. Biophys.* 41, 543–556.

Kortum RL, Rouquette-Jazdanian AK & Samelson LE (2013) Ras and extracellular signal regulated kinase signaling in thymocytes and T cells. *Trends Immunol.* 34, 259–268.

Lin JJY, Low-Nam ST, Alfieri KN, et al. Mapping the stochastic sequence of individual ligand-receptor binding events to cellular activation: T cells act on the rare events. *Sci Signal.* 2019 Jan 15;12(564):eaat8715.

Morris GP & Allen PM (2012) How the TCR balances sensitivity and specificity for the recognition of self and pathogens. *Nat. Immunol.* 13, 121–128.

Sherman E, Barr V & Samelson LE (2013) Super-resolution characterization of TCR-dependent signaling clusters. *Immunol. Rev.* 251, 21–35.

Sykulev Y (2010) T cell receptor signaling kinetics takes the stage. *Sci. Signal.* 3, pe50.

Tkach K & Altan-Bonnet G (2013) T cell responses to antigen: hasty proposals resolved through long engagements. *Curr. Opin. Immunol.* 25, 120–125.

Zarnitsyna V & Zhu C (2012) T cell triggering: insights from 2D kinetics analysis of molecular interactions. *Phys. Biol.* 9, 045005.

Zehn D, King C, Bevan MJ & Palmer E (2012) TCR signaling requirements for activating T cells and for generating memory. *Cell. Mol. Life Sci.* 69, 1565–1575.

Methods for Studying Signaling Proteins and Networks

15

Our current understanding of cell signaling is the result of decades of experimental research in a variety of fields. In this chapter, we describe some of the most commonly used methods and approaches for studying cellular signaling. These range from tools for the analysis of individual signaling proteins to those that can characterize entire networks within living cells.

BIOCHEMICAL AND BIOPHYSICAL ANALYSIS OF PROTEINS

Changes in the physical state of proteins are central to cell signaling, so methods to probe the properties of proteins are essential for understanding signaling mechanisms. In this section, we will look at methods used to study the properties of purified proteins, including the quantitative analysis of their binding and enzymatic activities, and determination of their three-dimensional structure.

Analytical methods can determine quantitative binding parameters

The dissociation constant, on-rate, and off-rate for a binding reaction determine both the likelihood that the interaction will occur in the cell and its dynamic behavior. Measuring these parameters accurately becomes ever more important as biologists strive to understand the dynamic properties of signal transduction networks and to develop quantitative computational models to describe their behavior. Here, we discuss a few of the most widely used methods for determining such parameters experimentally.

The *dissociation constant* (K_d) is often calculated from binding experiments in which a small amount of one protein (A) is incubated with varying concentrations of a binding partner (B) and the binding reaction is allowed to achieve equilibrium.

 Quantitative measures of binding are discussed in Chapter 2.

DOI: 10.1201/9780429298844-15

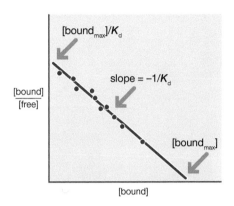

Figure 15.1

Scatchard analysis. The results of binding studies can be presented in the form of a Scatchard plot. When [AB]/[B] (bound/free) is plotted versus [AB] (bound), in simple cases a straight line is obtained where the slope is $-1/K_d$ and the intercept on the horizontal axis is the total number of binding sites for B.

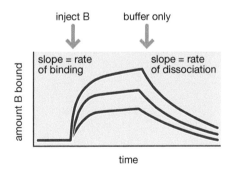

Figure 15.2

Determining binding parameters by surface plasmon resonance (SPR). Protein A is fixed to a chip in the instrument; a solution containing its binding partner (protein B) is injected at the time indicated by the arrow, and the association of A and B is monitored. After a time, buffer alone (lacking protein B) is injected to allow the dissociation of B to be monitored. Plots of binding versus time for three different concentrations of B are shown. Initial rates of association or dissociation can be obtained from the slopes of the curves, and K_d can be calculated.

From the amount of A bound to B (its *fractional occupancy*) at different concentrations of B, a binding curve can be fitted to the data, and from this, the dissociation constant can be calculated (see Figure 2.7). Such methods depend on being able to quantify precisely the amount of A that is bound to B. For binding in solution, this can often be determined by monitoring differences in the spectral properties of A upon binding (for example, changes in fluorescence polarization when one of the species is fluorescently labeled).

In another experimental set-up, A is fixed to a solid surface and B is tagged with a fluorescent or radioactive label. Binding reactions are then performed at different concentrations of B, and for each concentration, the amount of B bound to A is measured directly after washing away unbound B. As above, such binding data can be used to determine the K_d. Although this approach is technically straightforward, one must keep in mind that one of the proteins is immobilized at relatively high concentrations on a solid surface, which can distort binding behavior.

Computational fitting of binding data directly to the binding equation is used to calculate binding parameters with high precision. A convenient way to visualize binding data, however, is provided by the Scatchard plot (**Figure 15.1**). The basis for **Scatchard analysis** is the algebraic rearrangement of the equation for equilibrium binding (see Equation 15.1) into the form of a straight line (y = mx + b), to obtain:

$$\frac{[AB]}{[B]} = -\frac{[AB]}{K_d} + \frac{[A_{total}]}{K_d} \qquad (15.1)$$

If one plots [AB]/[B] (the amount of B that is bound to A, divided by the amount that remains unbound—that is, bound over free) on the vertical axis and [AB] (the bound fraction) on the horizontal axis for many different concentrations of B, in simple cases a straight line is obtained for which the slope is $-1/K_d$, and the horizontal intercept is $[AB]_{max}$, or the total number of binding sites. The latter number can be quite useful, for example, when analyzing the binding of a labeled hormone to cells, where the total number of binding sites corresponds to the total number of receptors for the hormone in those cells. In some cases, the Scatchard plot may not yield a straight line, suggesting the existence of more than one class of binding site for B, each with a different affinity. In such situations, the Scatchard plot can provide an estimate of the number of sites and K_d for each class of binding site. In practice, computational fitting of multiple datasets covering a wide range of concentrations of components is required for accurate determination of both K_d and the number of binding sites.

To measure the on-rate and off-rate for a binding reaction, binding and dissociation must be monitored in real time. This can be done using a **surface plasmon resonance** (**SPR**) instrument, which measures the change in the angle of reflection of light bouncing off a metallic surface as the mass on the surface is increased. In SPR, one binding partner is immobilized on a sensor chip in the apparatus, and solutions containing the other partner (or buffer alone) are injected and allowed to flow over the chip. Binding is then monitored over time, as a function of changes in the protein mass bound to the surface. By measuring initial rates of binding and dissociation at different concentrations of injected binding partner, on-rates and off-rates can be calculated (**Figure 15.2**) as well as the overall dissociation constant for the interaction. One advantage of this approach is that none of the binding partners need to be labeled. SPR binding data must be interpreted with care, however, as one of the binding partners is immobilized on a surface, thus potentially distorting rates of binding and dissociation.

It is also possible to determine the thermodynamic parameters [the changes in *enthalpy* (H) and *entropy* (S)] underlying a binding interaction. For example, rates of binding and dissociation obtained from SPR experiments at different temperatures can be used to derive the change in enthalpy (ΔH) and in entropy (ΔS) associated with binding. **Isothermal calorimetry (ITC)** is another analytical method that provides direct information about thermodynamic parameters, in this case when both binding partners are in solution. In ITC, small amounts of one binding partner are sequentially injected into a highly sensitive calorimeter containing a solution of the other binding partner. After each injection, the heat released or absorbed by the solution is measured precisely (**Figure 15.3**). The amount of heat absorbed or released, how this amount changes as more of the components are bound, the temperature of the system, and the concentrations of the components are used together to calculate K_d, ΔH, and ΔS.

Michaelis–Menten analysis provides a way to measure the catalytic power of enzymes

The properties of enzymes and their roles in signaling are described in Chapter 3.

Many of the key proteins involved in signal transduction are enzymes, such as kinases and phosphatases, and their function in signaling often revolves around regulated changes in their activity. Thus, it is essential to be able to measure enzyme function quantitatively in standard ways. The catalytic power of an enzyme is characterized by the degree to which it enhances the reaction rate over that of the uncatalyzed reaction. However, assessing this activity can be relatively complex, as reaction rates are highly dependent on the concentrations of the enzyme and the substrates. For the simple case of an enzyme binding to a single substrate (S) and reacting to form product(s) (P), the reaction rate can be described by the **Michaelis–Menten equation** (Equation 15.2), which defines the reaction velocity (V) as a function of enzyme (E) and substrate (S) concentrations. The derivation of the Michaelis–Menten equation can be found in any elementary biochemistry textbook.

Experimental biochemical analysis of an enzyme is typically carried out by measuring how the reaction rate varies as a function of substrate concentration. This analysis is performed at several different enzyme concentrations, and the data can be fitted to the Michaelis–Menten equation to determine the key kinetic parameters k_{cat} and K_m.

For the reaction scheme:

$$E + S \underset{k_{-1}}{\overset{k_1}{\rightleftharpoons}} ES \xrightarrow{k_{cat}} E + S$$

$$V_{obs} = \frac{d[P]}{dt} = \frac{k_{cat}[E]_0[S]}{K_m + [S]}$$ (15.2)

$$K_m = \frac{K_{-1} + K_{cat}}{K_1}$$

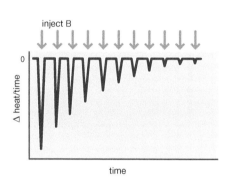

inject B

Δ heat/time

0

time

Figure 15.3

Determining binding parameters by isothermal calorimetry (ITC). Small amounts of a solution of B are injected into a calorimeter containing a solution of A. Changes in the heat of the solution are monitored after every injection. As more of A is bound by B, each subsequent injection results in a smaller change in heat because less free A is available for binding. K_d, ΔH, and ΔS can be derived from these data.

Figure 15.4a shows a schematic of typical experimental data used to determine k_{cat} and K_m. Each line on the graph represents an experiment conducted at a different initial substrate concentration, with the concentration of the reaction product plotted as a function of time for each experiment. The initial slopes ($d[P]/dt$: change in product concentration as a function of time) of these lines equal the initial reaction velocities, V, which can then be plotted as a function of initial substrate concentration, [S], to yield the Michaelis–Menten plot (**Figure 15.4b**). A typical enzyme will show a hyperbolic plot, in which velocity initially increases linearly with increasing substrate concentration, but then asymptotically approaches a maximal velocity, V_{max}, when substrate approaches saturation.

Intuitively, the two terms k_{cat} and K_m can be used to understand basic functional parameters of an enzyme: k_{cat} reflects the maximum rate achievable in the presence of saturating substrate concentration, and K_m is the substrate concentration at which half-maximal velocity is achieved (see Figure 15.4b). It is important to note that many signaling enzymes have multiple substrates—for example, a protein

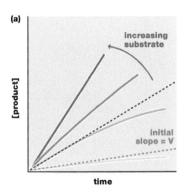

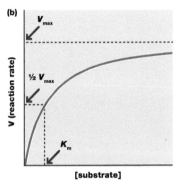

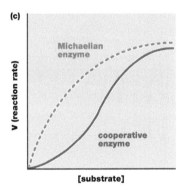

Figure 15.4

Quantifying enzyme catalysis by Michaelis–Menten analysis. (a) In a typical series of experiments, the formation of product over time is measured for different starting substrate concentrations and with a substoichiometric concentration of enzyme, [E]. Each line represents an experiment performed at a different substrate concentration. (b) The initial rates (V) from part (a) are then plotted as a function of substrate concentration [S] to give the classical hyperbolic Michaelis–Menten plot. The maximal rate achieved at saturating substrate concentrations is defined as the V_{max}. The parameter K_m reflects the substrate concentration at which the rate is half-maximal. The catalytic rate constant, k_{cat}, is equal to $V_{max}/[E]$. (c) Cooperative enzymes, where the substrate acts to allosterically activate the enzyme, deviate from Michaelis–Menten behavior and show a sigmoidal dependence of reaction velocity on substrate concentration. These types of enzymes are well suited for switchlike control.

kinase uses both a peptide and ATP as substrates. In these cases, each substrate has its own distinctive K_m that reflects the concentration at which half-maximal velocity is reached. This K_m is typically measured and reported for conditions in which the other substrate is already saturating.

Overall, the Michaelis–Menten equation tells us how a typical enzyme will behave at different substrate concentrations. If substrate concentration is high (that is, much greater than K_m), then the enzyme active site will be fully saturated with substrate and the rate equation simplifies to a form (Equation 15.3) that depends only on the total enzyme concentration $[E]_0$:

$$V_{obs} = V_{max} = k_{cat}[E]_0 \qquad (15.3)$$

If the substrate concentration is low (that is, considerably lower than K_m), then the enzyme will not be operating near saturation and the rate equation simplifies to Equation 15.4:

$$V_{obs} = \frac{k_{cat}}{K_m}[E]_0[S] \qquad (15.4)$$

k_{cat}/K_m is the apparent bimolecular rate constant for the enzymatic reaction. Because concentrations of substrates in a cell are often low, k_{cat}/K_m provides a good estimate for how an enzyme performs *in vivo*. k_{cat}/K_m is often referred to as the **specificity constant** because the relative values of k_{cat}/K_m for two different substrates reflect the degree of specificity of the enzyme for the two substrates. It is important to note that k_{cat}/K_m determines specificity regardless of whether the substrate concentration is saturating or not; readers are referred to specialized biochemistry textbooks (see the reference section at the end of the chapter) for a rigorous derivation.

Not all signaling enzymes show simple Michaelian (hyperbolic) behavior. In particular, a key function of many signaling enzymes is to act as *allosteric* regulatory nodes. Briefly, allostery is the property of being able to exist in two or more structural states of differing activity. Many signaling enzymes may show a dramatic change in k_{cat} and/or K_m upon stimulation by an upstream signal, such as covalent modification or ligand binding. These changes in enzyme kinetic parameters are often caused by allosteric conformation changes induced by the upstream input. In cases where the substrate itself is an allosteric activator, cooperative activation of the enzyme is observed, which leads to significant deviations from simple Michaelian behavior and a sigmoidal plot of V versus [S] (**Figure 15.4c**). Cooperative activation is often seen for oligomeric enzymes, where binding of substrate to one subunit affects binding to the other subunits. Such cooperative enzymes have an important function—they make the enzyme output less like an analog signal and more like a digital signal (all or none). For a cooperative enzyme, there is a threshold substrate concentration below which the rate of catalysis is very low, and above which the rate is near maximal; unlike a standard Michaelis–Menten enzyme, there is only a small range of substrate concentrations that leads to intermediate reaction velocities.

Methods to determine and analyze protein conformation are central to the study of signaling

The three-dimensional structure or *conformation* of signaling proteins is what determines their interactions and their catalytic activity. In addition, how their conformations change based on input signals, and how this alters their output functions, is at the heart of how they function as allosteric signal transmission devices. Thus, methods to determine and analyze protein conformation are central to the study of signaling. Recent years have seen enormous progress in the development of computational methods, such as AlphaFold and similar platforms, that can predict the three-dimensional structure of a protein with impressive accuracy from its amino acid sequence alone. For the time being, however, experimental structure determination is still important to validate such predictions. Here, after first describing the fundamental elements of protein structure, we outline the three major experimental approaches used to determine protein structure: X-ray crystallography, nuclear magnetic resonance (NMR), and cryo-electron microscopy (cryo-EM).

The structure of a protein can be described on a number of levels. Here, these different levels are illustrated using the serine/threonine kinase Pak1 (which is activated by Rho family G proteins).

The **primary structure** of the protein is the simple linear sequence of amino acids in the polypeptide chain (**Figure 15.5**). The **secondary structure** consists of local structural elements, predominantly α helices and β strands (described further below). In Figure 15.5, the α helices are depicted as cylinders and the β strands as block arrows. Each helix and strand is numbered by its order of appearance in the protein, starting at the N-terminus. The **tertiary structure** of the protein is how these α helices and β strands, and the loops that connect them, are folded together. In most proteins, this fold generates one or more globular domains (**Figure 15.6**). Finally, the **quaternary structure** refers to how multiple subunits are arranged in a multiprotein complex. In this case, Pak1 exists as a dimer.

In an **α helix**, the N–H group of every peptide bond is hydrogen-bonded to the C=O of a neighboring peptide bond located four peptide bonds away in the same chain (**Figure 15.7a–c**), thus forming a regular right-handed helix with 0.54 nm between turns. The side chains (R) of each amino acid project out from the axis of the helix. The individual polypeptide chains (strands) in a **β sheet** are held together by hydrogen-bonding between peptide bonds in different (adjacent) strands, and the amino acid side chains in each strand alternately project above and below the plane of the sheet (**Figure 15.7d–f**). In the example of a β sheet shown in Figure 15.7, adjacent peptide chains run in opposite (antiparallel) directions. Sheets with strands oriented in the same (parallel) direction are also possible.

Allosteric enzymes are described in Chapter 3.

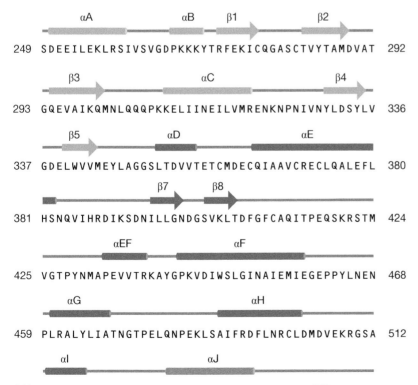

249 SDEEILEKLRSIVSVGDPKKKYTRFEKICQGASCTVYTAMDVAT 292

293 GQEVAIKQMNLQQQPKKELIINEILVMRENKNPNIVNYLDSYLV 336

337 GDELWVVMEYLAGGSLTDVVTETCMDECQIAAVCRECLQALEFL 380

381 HSNQVIHRDIKSDNILLGNDGSVKLTDFGFCAQITPEQSKRSTM 424

425 VGTPYNMAPEVVTRKAYGPKVDIWSLGINAIEMIEGEPPYLNEN 468

459 PLRALYLIATNGTPELQNPEKLSAIFRDFLNRCLDMDVEKRGSA 512

513 KELLQHQFLKIAKPLESLTPLIAAAEEEATKNNH 545

Figure 15.5

Primary and secondary structure.
The amino acid sequence of the catalytic domain of the serine/threonine kinase Pak1 (human, amino acids 249 to the end) in single-letter code. Positions of major secondary structure elements (α helices and β strands) are depicted above the sequence. (Adapted from M. Lei, et al., *Cell* 102:387–397, 2000. With permission from Elsevier.)

Figure 15.6

Tertiary and quaternary structure. The Pak1 structure is depicted here as a ribbon diagram, with α helices and β strands indicated schematically. The colors in the structure correspond to the colors on the secondary structure diagram above the sequence in Figure 15.5. An N-terminal portion of Pak1 (from residues 78 to 147) is also included here (*green*); residues 1–77 and 148–248 are not shown. The second subunit of the Pak1 homodimer is shown to the right and is colored *yellow*. (Adapted from M. Lei, et al., *Cell* 102:387–397, 2000. With permission from Elsevier.)

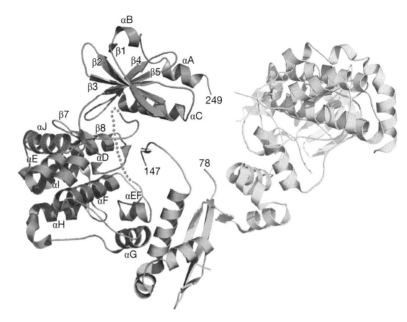

Figure 15.7

Common secondary structure elements: α helix and β sheet. (a, d) All the atoms in the polypeptide backbone are shown, but the amino acid side chains are truncated and denoted by R. (b, e) The backbone atoms only are shown. (c, f) The shorthand symbols that are used to represent the α helix and the β strand in ribbon drawings of proteins. (Adapted from B. Alberts, et al., *Molecular Biology of the Cell*, 5th ed. Garland Science, 2008.)

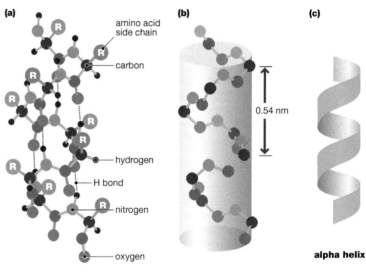

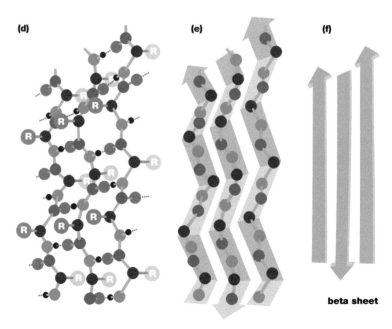

It is important to note that not all protein regions adopt a regular secondary structure. In humans, up to 30% of the proteome is estimated to consist of *intrinsically disordered regions* (*IDRs*). Such regions do not fold into stable globular domains but instead sample a wide variety of possible conformations. IDRs, which are often rich in peptide binding motifs and sites for post-translational modification, may not be well resolved in 3D structures due to their conformational flexibility.

X-ray crystallography provides high-resolution protein structures

Among methods used to analyze the structure of proteins and to detect changes in conformation, the highest resolution is provided by **X-ray crystallography**. In this method, highly concentrated solutions of purified protein are induced to form crystals, in which the protein is arrayed in a regular lattice. These crystals are then exposed to an X-ray beam, and the resulting diffraction pattern can be analyzed to calculate a three-dimensional electron density map of the crystallized protein. The known amino acid sequence of the protein can be fitted into this map to provide a high-resolution structural model (often with a resolution of approximately 2Å). At this high resolution, one can identify the position of most individual atoms in the protein (as long as that region of the protein is well ordered in the crystal). A comparison of crystal structures of a protein in two different conformational states can reveal precisely which residues undergo movements (**Figure 15.8**).

A potential disadvantage of X-ray crystallography is that it provides a static picture of the protein structure, because the protein must be immobilized in the crystal lattice for analysis. Even when a protein is capable of adopting multiple conformations, only a single conformation is likely to be observed in one crystal lattice. Furthermore, not all proteins can be crystallized successfully. Crystallization is particularly challenging in the case of proteins with transmembrane segments, such as channels and receptors, as these hydrophobic regions tend to aggregate in solution. Nonetheless, in quite a few cases, researchers have been able to crystallize

(a)

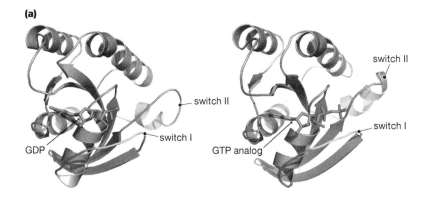

(b)

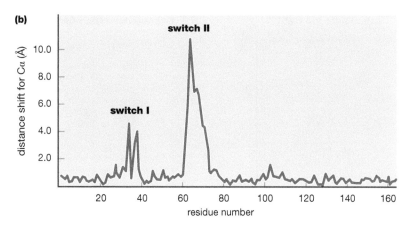

Figure 15.8

X-ray crystal structures illustrate conformational changes. (a) X-ray crystal structures of human H-Ras in the GDP-bound (left) and GTP-bound (right) conformations; switch I and II regions are shown in *green*, and bound nucleotide is shown in *blue*. (b) Graph shows differences in distances between Cα atoms in the structures of GDP-bound and GTP-bound conformations, plotted as a function of residue number. The largest differences occur in the switch I and II regions. (b, Adapted from M.V. Milburn, et al., *Science* 247:939–945, 1990. With permission from AAAS.)

and determine the structure of a signaling protein in several distinct states (that is, bound to different allosteric ligands, or in distinct phosphorylation states). In such cases, one can determine with very high resolution precisely how the protein structure changes in each state and how this is linked to protein function.

Nuclear magnetic resonance (NMR) can reveal the dynamic structure of small proteins

The structure of small proteins can also be determined by **nuclear magnetic resonance (NMR) spectroscopy**. In this approach, a concentrated solution of the protein sample is placed in a strong magnetic field, which leads to a difference in the energy of the two oppositely oriented spin states of atomic nuclei. Radiofrequency energy is then used to flip the spin states of individual atomic nuclei within the protein (this occurs at the resonance frequency). If an atom of a particular type is in a unique chemical environment, it will show a unique and altered resonance energy—also known as a "chemical shift." An NMR spectrum of the protein will reveal all the unique nuclei of a certain type in the molecule (most often protons, though the stable isotopes ^{13}C and ^{15}N can be used to monitor carbon and nitrogen nuclei, respectively). Because nuclei that are close in space or bonded to one another can perturb each other's chemical environment, the interactions between resonance peaks can be analyzed to reveal spatial relationships such as interatomic distances and bond angles. These pieces of structural information can be combined to generate a three-dimensional model of the protein structure. Because there is no complete structural map, as in the case of X-ray crystallography, structures determined by NMR typically have somewhat lower resolution.

An advantage of NMR is that it can also reveal dynamic motions involved in conformational changes, because the protein is studied in aqueous solution rather than crystallized form. In some cases, NMR can also monitor multiple conformations in the same sample and provide information on the fraction of the total in each population, even if the structure is not absolutely known. For example, one can monitor the change in chemical shifts corresponding to atoms that undergo a dramatic change in chemical environment in one conformation versus another, or before and after ligand binding (**Figure 15.9**). Technical issues currently limit NMR analysis to relatively small proteins or domains.

Electron microscopy can map the shape of large proteins and complexes

Although proteins are in general too small to observe directly by microscopy, it is possible to map their overall shape using **electron microscopy (EM)**. Because the wavelength of electrons is much shorter than that of light, the potential resolution of EM is much higher than for light microscopy. To obtain images, purified protein samples are immobilized in a very thin layer of vitreous ice at extremely

Figure 15.9

NMR spectroscopy reveals structural changes upon ligand binding. Calmodulin, which has four Ca^{2+}-binding sites, was analyzed by two-dimensional NMR in the absence and presence of Ca^{2+}. Resonance peaks that undergo shifts due to changes in local environment are highlighted by dotted arrows, from their positions in the absence of Ca^{2+} (*light blue*) to their positions in high Ca^{2+} (*pink*). In this experiment, calmodulin was selectively labeled so only glycine (G) residues appear in the NMR spectra. Knowledge of which peaks correspond to which residues (resonance assignments) is required to interpret the structural changes in an NMR spectrum. In NMR, chemical shifts are typically expressed in PPM (parts per million), based on differences in measured values compared to a reference value. (Adapted from H. Ouyang & H.J. Vogel, *Biometals* 11:213–222, 1998. With permission from Springer Science and Business Media.)

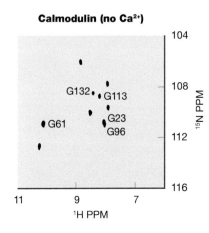

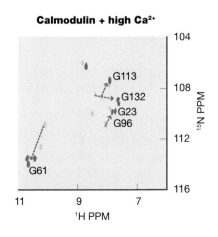

low temperatures, in a method called **cryo-electron microscopy (cryo-EM)**. Individual images from the electron microscope are of relatively low resolution and rather noisy, so images of many individual particles are collected and analyzed by computational methods (single particle analysis) to generate an average image, which has much higher resolution than each of the individual images. Images of a large number of particles in many different orientations provide enough information to generate three-dimensional density maps, which can then be fitted to the protein sequence. Recent improvements in instrumentation and computational methods have increased the resolution of cryo-EM to near that of X-ray crystallography and NMR.

This approach provides a way to visualize large-scale structures and conformational changes under relatively native conditions. Some examples of early cryo-EM-derived structures can be seen in Chapter 9, for the proteasome (Figure 9.9), the anaphase-promoting complex (APC) (Figure 9.10), and the apoptosome (Figure 9.22). More recent high-resolution structures can be seen in Chapter 8, including a number of G-protein-coupled receptor complexes as seen in Figures 8.7 and 8.30. One advantage of cryo-EM over other methods is that it is possible to analyze relatively crude or heterogeneous samples; in some cases, it is even possible to visualize the conformation of protein complexes *in situ* within an intact cell.

Specialized spectroscopic methods can be used to study protein dynamics

Other spectroscopic methods, such as circular dichroism or intrinsic fluorescence, can be used to monitor physical properties of proteins, such as the amount of α helix structure or the extent that aromatic residues are buried in the hydrophobic core. Because such approaches monitor structural properties of proteins in solution, they can also be used to dynamically track changes in conformation. These methods, however, do not provide detailed atomic information about structural changes, only global shifts.

Specifically modified proteins can be used to track conformational changes also. For example, if a protein is thought to undergo some type of hinge-bending motion in which the distance between two points on the protein changes dramatically, then one can chemically attach spectroscopic probes to these points and monitor properties that change with distance. For example, fluorescent probes (fluorophores) can be attached to these points and the distance between them can be monitored by *fluorescence resonance energy transfer* (*FRET*), which is discussed in more detail below. Essentially, FRET measures how the excitation of one fluorophore modifies the fluorescence properties of a second nearby fluorophore. FRET is exquisitely sensitive to the distance between the two fluorophores, so it can detect when they move further than a few angstroms apart. If such FRET experiments are performed using fusions of genetically encoded fluorescent proteins, such as yellow fluorescent protein (YFP) and cyan fluorescent protein (CFP), then these molecules can be used as *biosensors* to report on conformational changes (for example, as a result of binding or post-translational modification) in living cells (**Figure 15.10**).

Figure 15.10

Detecting ligand-induced conformational changes by fluorescence resonance energy transfer (FRET).
(a) Diagrammatic depiction of the structure of a ligand-binding protein, showing positions of the FRET donor (*green*) and FRET acceptor (*pink*). Binding of ligand (*orange circle*) reduces the distance between the FRET donor and acceptor, increasing FRET efficiency. (b) FRET efficiency plotted as a function of ligand concentration.

(a)

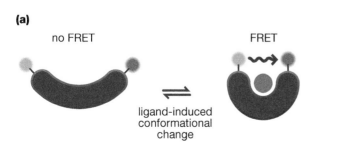

(b)

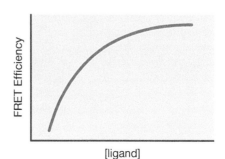

MAPPING PROTEIN INTERACTIONS AND LOCALIZATION

Given the importance of protein–protein interactions in signal transduction, it is no surprise that extraordinary effort has been expended to identify binding partners for signaling proteins. Through characterizing such physical interactions, investigators have been able to work their way up and down signaling pathways and to forge connections between different pathways. In this section, we describe some of the methods used to identify specific physical interactions between proteins, and discuss methods for analyzing protein interactions and subcellular localization in living cells.

Interacting proteins can be identified by isolating protein complexes from cell extracts

One approach to identifying binding partners for a protein of interest is to isolate complexes containing that protein from **cell lysates** made by solubilizing cells with detergent. The most usual method of obtaining such complexes is by **co-immuno-precipitation** (**co-IP**). A specific antibody that binds the protein of interest is used to separate it and its associated proteins from all the other proteins in the extract. The antibody-bound proteins are then analyzed, usually by **gel electrophoresis**, which separates proteins on the basis of their size and/or charge. The separated proteins can then be visualized and identified by various methods (**Figure 15.11**). Binding partners are often identified by **immunoblotting** (also termed **Western blotting**), in which the proteins separated on the gel are transferred (blotted) onto a membrane which is then probed with labeled antibodies specific for possible candidate proteins (see also Figure 15.21c, later in this chapter). Although co-IP provides good evidence for the *in vivo* association of two proteins, false-negative and false-positive results can occur. For example, two proteins that interact *in vivo* might not be detected by co-IP if the off-rate of the interaction is high, as the complex will dissociate during the multiple incubation and washing steps. On the other hand, the process of dissolving cells in detergent may allow two proteins to interact that might never encounter each other in the intact cell.

A common variation on the co-IP assay involves the expression in cells of the protein of interest modified by an **epitope tag**. This is a small peptide sequence that is specifically recognized by a well-characterized (usually monoclonal) antibody; the advantage is that the same antibody can be used to precipitate any protein tagged with that epitope, so that a different antibody is not needed for each protein. Various epitope tags and other types of so-called affinity tags have been developed for purifying proteins. Multiple distinct affinity tags can also be combined into one large tag, allowing sequential purification steps that greatly increase the specificity of the overall purification. In another variation on co-IP, one potential binding partner is tagged with an epitope tag and the other with an enzyme such as luciferase. This enables sensitive, high-throughput, automated detection of the enzyme-tagged binding partner after the immune precipitation of the partner bearing the epitope tag.

A **pull-down assay** is frequently used as an alternative to co-IP. In this method, large amounts of a pure protein or a protein fragment are first produced, usually by expression in bacteria. The purified protein is coupled to tiny beads, which are then incubated with the cell extract. The beads, along with any proteins that have bound during the incubation, are then separated from the extract, washed, and analyzed to identify binding partners. This approach has the advantage over co-IP in that the physical state of the protein of interest can be controlled experimentally, but pull-down assays may detect interactions that do not normally occur under physiological conditions.

Identifying the proteins that associate with the target protein in co-IP or pull-down experiments can be a challenge. Where a particular interaction is suspected, the

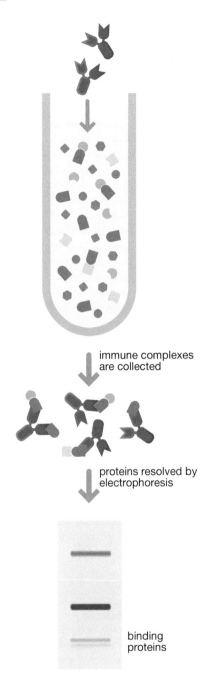

immune complexes
are collected

proteins resolved by
electrophoresis

binding
proteins

Figure 15.11

Detecting binding partners by co-immunoprecipitation. A cell extract is incubated with an antibody (*purple*) that specifically binds a protein of interest (*orange*). The immune complexes are collected and washed, thus separating the protein of interest and any proteins that associate with it (*green*) from the mixture. The individual proteins in the isolated complexes are then resolved by gel electrophoresis and identified.

suspicion can be confirmed by immunoblotting; but often, little information is available to guide the investigator. In such cases, *mass spectrometry (MS)* can be used to identify the binding partners directly. MS is an analytic method that provides extremely accurate information on the atomic mass of molecules such as peptides (described more fully below; see Figure 15.23a). For a protein, the atomic masses of its proteolytic fragments often provide sufficient information to identify it. Other MS approaches allow a selected peptide to be further fragmented during analysis, providing direct information on the amino acid sequence of the peptide. MS is now the basis for many large-scale efforts to characterize protein interactions.

Binding partners can be identified by screening large libraries of genes

The **yeast two-hybrid (Y2H) assay** is one example of various methods that have been developed to identify the binding partners of a protein by screening **cDNA expression libraries**—large collections of cDNAs reverse-transcribed from total cell mRNA. Such methods are particularly valuable because they can identify specific partners from literally millions of candidates without any prior information on their identity. In the Y2H assay, the cDNA for a protein or domain of interest is fused to a sequence encoding a DNA-binding domain, and the fusion protein is expressed in yeast. An expression library consisting of random fragments of cDNA fused to a transcriptional activation domain is then introduced into the yeast strain expressing the first fusion protein, such that an individual yeast cell expresses the first fusion along with one of the second fusions encoding a possible binding partner. If the two fusion proteins bind to each other in the cell, a functional transcriptional activator is assembled and transcription of a marker gene is induced (**Figure 15.12**). The marker gene is one that allows growth or induces a color change in yeast colonies grown on selective media. The cDNA of the binding partner can then be recovered and its identity determined by DNA sequencing. One important advantage of such an approach over *in vitro* methods is that binding must occur in the nucleus of a living cell to be detected, increasing the likelihood that those interactions are relevant under normal physiological conditions.

Direct protein–protein interactions can be detected by solid-phase screening

The apparent association of two proteins in co-IP and pull-down assays may, in some cases, be the result of indirect interaction—an unknown molecule may act as a bridge between the two. This can even be the case for interactions obtained by Y2H screening. **Far-Western blotting** is designed to visualize direct interactions only. In this approach, the proteins in a cell extract are separated by gel electrophoresis and transferred to a solid membrane, as in immunoblotting. The membrane is then incubated with a purified protein or protein domain that is labeled or tagged in some way so it can be visualized. If the labeled protein binds to any of the proteins in the extract, a labeled band (or bands) will be visible on the membrane after washing (**Figure 15.13**). Because proteins in the cell extract are usually completely denatured by boiling in detergent before gel electrophoresis, this approach is most useful for detecting interactions, such as protein–peptide interactions, that do not require the native, folded structure of the partner in the extract.

Other types of solid-phase binding assays take advantage of **microarray** technology, which increases the ability to score many possible interactions at once in a single experiment. For instance, many recombinant proteins, synthetic peptides, or protein lysates can be arrayed as tiny spots on a solid support, which can then be probed with labeled purified proteins or cell lysates. Such experimental approaches allow interactions to be probed on a proteome-wide scale, but can be subject to artifacts because one binding partner is immobilized at high concentration on a surface.

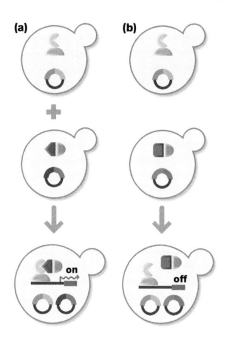

Figure 15.12

The yeast two-hybrid assay. A protein of interest (*green*) is expressed in yeast as a fusion with a DNA-binding domain (*blue*). Other proteins are expressed as fusions with a transcription activation domain (*orange*). (a) When the two fusion proteins bind to each other, transcription of a selectable gene is induced (*pink*). (b) When the two fusion proteins do not bind to each other, no transcription is seen.

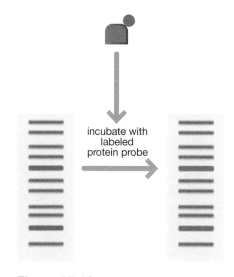

Figure 15.13

Detecting binding partners by far-Western blotting. Proteins in a cell lysate are separated by gel electrophoresis and transferred to a membrane, which is then incubated with a labeled protein. Proteins on the membrane that directly bind the labeled protein can be visualized (*pink*).

Fluorescent protein tags are used to locate and track proteins in living cells

Proteins are not uniformly distributed in the cell, and signal-induced changes in protein localization provide an important signaling mechanism. Furthermore, binding interactions in signaling are highly dynamic and can vary dramatically at different locations and times in the same cell. Although the experimental methods outlined above provide useful information on protein–protein interactions that might occur, it is obviously much more useful to know precisely when and where in the cell such interactions actually do occur or, more fundamentally, whether they occur at all during normal signaling. For this, we need to detect and quantify changes in localization and in specific protein–protein interactions in living cells. Advances in live-cell imaging have begun to make this possible.

For many years, biologists have used antibodies labeled with fluorescent dyes to probe the subcellular localization of proteins (a method termed **immunofluorescence**). However, this approach requires cells to be killed by chemical fixation before analysis, and thus provides only a static picture of protein localization. The discovery and exploitation of small fluorescent proteins have revolutionized the imaging of proteins in living cells. When a fluorescent protein is expressed in a cell as a genetically encoded fusion with a protein of interest, the subcellular localization and local concentration of the fusion protein can be followed over time by fluorescence microscopy. The first such fluorescent protein to be widely used was **green fluorescent protein** (**GFP**), isolated from a jellyfish and named for the color of fluorescent light emitted upon excitation. The isolation of fluorescent proteins from other organisms and the creation of new variants by mutation have broadened the spectrum of available colors, which now include cyan (CFP), yellow (YFP), and red (RFP), among others, and have improved the brightness, stability, and other desirable properties of the proteins. Using different fluorescent proteins, it is possible to track the location of multiple fusion proteins in the same cell (**Figure 15.14**).

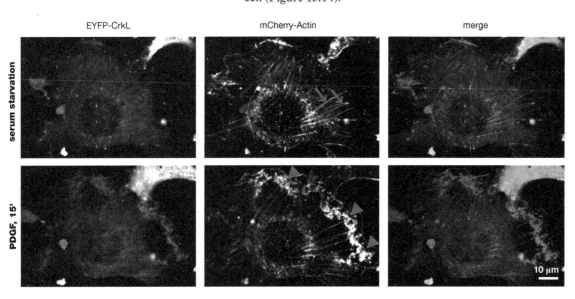

Figure 15.14

Monitoring subcellular localization of fluorescently tagged proteins. Mouse fibroblasts were engineered to express actin tagged with the red fluorescent protein mCherry, and the CRKL adaptor tagged with the yellow fluorescent protein EYFP. Serum-starved cells were photographed before (top) and 15 min after (bottom) treatment with the mitogen platelet-derived growth factor (PDGF), which causes actin cytoskeletal rearrangements including loss of actin cables (*pink arrows*) and formation of transient structures termed dorsal actin ruffles (*pink arrowheads*). In starved cells, CRKL is localized to focal adhesions, structures at the ends of actin cables that attach the cell to the underlying extracellular matrix. Upon PDGF stimulation, CRKL is released from focal adhesions and relocalizes to dorsal actin ruffles. CRKL and actin can each be visualized individually using different wavelengths of ultraviolet light; in the right panels, the two images have been merged to better visualize areas of co-localization (EYFP is colored *green* and mCherry is colored *red* in the merged images). (Images courtesy of Susumu Antoku).

This means of visualizing proteins in a living cell provides a wealth of information that can be used to evaluate whether or not a specific protein–protein interaction is likely to occur. For example, if two fluorescent proteins are highly enriched in the same subcellular compartment, their local concentrations at that site will be high and they are much more likely to bind than if they were evenly distributed throughout the cell. The fact that each molecule emits a fixed amount of light makes it possible to calculate the absolute number of molecules and approximate local concentration of each fluorescent species present. A caveat in this type of analysis is that, except in situations where the gene for the normal endogenous protein can be replaced precisely with the fluorescent fusion version, it is likely that the expression level of the fluorescent protein will be different from that of its normal counterpart. Also, care must be taken to ensure that fusion to the fluorescent moiety does not alter the properties of the protein of interest. But at minimum, such experiments can provide valuable clues about whether a particular interaction is likely.

Fluorescent proteins have also been exploited as a means of visualizing the intracellular locations of binding sites for proteins of interest. Small, modular protein-binding or lipid-binding domains can be fused to GFP or other fluorescent proteins, and changes in their subcellular localization, and thus in the local concentrations of their binding sites, are tracked after stimulation of the cells or some other manipulation of signaling pathways. Such fluorescent fusions exemplify a class of molecules known as *biosensors*, which are designed to detect and monitor changes in the signaling status of living cells (biosensors are described in more detail below). Such probes can provide information on the dynamics and subcellular localization of signaling events that would not be apparent from biochemical analysis of whole-cell lysates.

Protein–protein interactions can be visualized directly in living cells

Fluorescently tagged signaling proteins provide hints about whether a specific protein–protein interaction may occur in the course of signaling, but typically cannot demonstrate conclusively that the interaction actually occurs. One reason is that the resolution of light microscopy is generally limited by the wavelength of light to ~200 nm, many times the diameter of a typical protein (~5 nm). Thus, even if two proteins are shown to co-localize precisely by fluorescence microscopy, they may not physically interact. More specialized imaging methods have been developed, however, that do allow the detection of direct, physical interactions of proteins in living cells.

One such approach is based on **fluorescence resonance energy transfer (FRET)**. FRET depends on the ability of one fluorescent molecule to excite a second fluorescent molecule with different excitation and emission spectra when the two are in close proximity (**Figure 15.15**). The strength of FRET diminishes with the sixth power of the distance between the two fluorescent molecules, so in practice it only occurs when the two molecules are separated by less than 80 Å—that is, when they are physically associated. In a typical FRET experiment, one protein of interest will be expressed in cells as a CFP fusion and a second will be expressed as an YFP fusion. In the absence of FRET, excitation of CFP with short-wavelength laser light leads to emission of cyan fluorescence only (**Figure 15.15a**). Wherever the CFP fusion and YFP fusion proteins associate in the cell, however, some of the energy from CFP will be transferred to YFP, resulting in a decrease in cyan fluorescence and an increase in yellow fluorescence that can be detected through quantitative image analysis (**Figure 15.16**). Thus, where and when the two proteins associate can be observed in living cells. Three-color FRET experiments are also possible, opening up the prospect of monitoring the assembly of multiprotein complexes in living cells. FRET is technically rather challenging, however, and requires careful analysis to exclude possible artifacts.

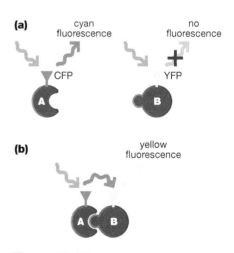

Figure 15.15

Detecting protein interactions by fluorescence resonance energy transfer (FRET). (a) Protein A is expressed in the cell as a fusion with cyan fluorescent protein (CFP), and protein B is expressed as a fusion with yellow fluorescent protein (YFP). Excitation of CFP with short-wavelength light (*light pink arrow*) gives cyan fluorescence, whereas YFP is not excited at this wavelength. (b) When the two fusion proteins are physically associated, some of the energy is transferred from CFP to YFP, resulting in quenching of cyan fluorescence and a corresponding increase in yellow fluorescence.

Figure 15.16

Visualizing protein binding in a living cell with FRET. The association between MEK (a MAP kinase kinase) and its substrate Erk (a MAP kinase) is visualized by fluorescence resonance energy transfer (FRET). Cultured cells (HeLa cells) expressing CFP–Erk as the FRET donor and MEK–YFP as the FRET acceptor were stimulated with epidermal growth factor (EGF), which activates the MAP kinase pathway, and fluorescence images were taken at different times (left panel). FRET was calculated from these images and displayed with colors indicating the amount of association (right panel). The cFRET (corrected FRET) images show the net amount of Erk–MEK complex at each pixel. By contrast, the cFRET/CFP images show the proportion of MEK-bound Erk versus free Erk at each pixel. Initially, most Erk is bound to MEK in the cytoplasm (*red*). After stimulation with EGF at time 0, a large proportion of Erk dissociates from MEK in the cytoplasm and migrates to the nucleus. CFP, cyan fluorescent protein; YFP, yellow fluorescent protein. (Images courtesy of Michiyuki Matsuda.)

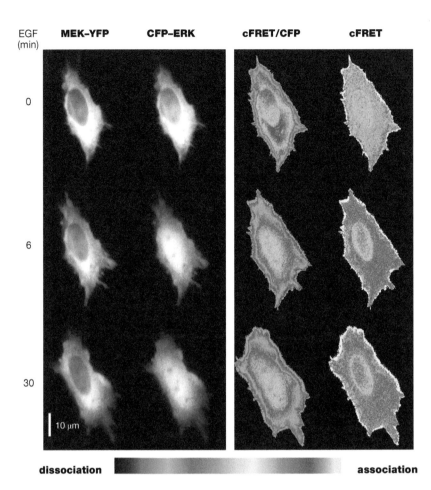

(a)

(b)

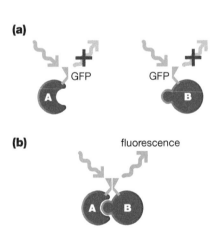

Figure 15.17

Detecting protein–protein interactions with the protein-fragment complementation assay (PCA). (a) Protein A is expressed in the cell as a fusion with a nonfunctional fragment of green fluorescent protein (GFP). Protein B is expressed as a fusion with a complementary nonfunctional fragment of GFP. Neither fusion protein is fluorescent. (b) When proteins A and B bind to each other, the two halves of GFP fold together and reconstitute a functional, fluorescent molecule that can be observed in the cell.

Another approach that is conceptually similar to FRET is called the **protein-fragment complementation assay** (**PCA**). In PCA, two proteins are expressed as fusions with complementary fragments of a reporter enzyme or fluorescent protein; only if the two proteins bind to each other can the two reporter fragments fold together correctly and assemble into a functional reporter. In a system based on fragments of GFP, for example, the two proteins by themselves are not fluorescent, but upon their association a functional, fluorescent GFP molecule is assembled (**Figure 15.17**). In principle, this is a highly sensitive method of detecting protein interactions, because the background of fluorescence in the absence of binding is essentially zero, and thus the dynamic range of the assay is much higher than for FRET. On the other hand, folding and activation of the fluorescent protein are quite slow and are essentially irreversible, so the method is not ideal for detecting rapid and/or transient changes in binding.

METHODS TO PERTURB CELL SIGNALING NETWORKS AND MONITOR CELLULAR RESPONSES

So far in this chapter, we have focused on methods to analyze the properties of individual signaling proteins and their interactions with substrates or binding partners. However, we are often interested in the behavior of larger signaling networks, which may involve complex and dynamic interactions between many different components. Ultimately, analyzing cellular signaling networks requires methods both to stimulate or perturb cells (to manipulate signaling inputs) and to analyze how such perturbations change the cells' internal states and behaviors (to monitor signaling outputs). In this section, we first discuss the range of methods available to perturb signaling networks; we then go on to discuss methods that have been developed to monitor various signaling readouts within the cell and to track their changes dynamically within populations as well as in individual cells.

Genetic and pharmacological methods can be used to perturb networks

Signaling is often studied in cells stimulated by a physiologically relevant input, such as the addition of a mitogen or hormone, or by subjecting cells to environmental stresses, such as shifting to a new temperature or to medium with altered solute concentration (osmolarity). However, in addition to these physiological inputs, there are other approaches to perturb and interrogate the underlying network (**Figure 15.18**). One can use genetic deletion mutations (knockouts) or RNA interference (RNAi)-mediated knockdown to eliminate or decrease the expression of key protein nodes within the signaling network, in order to identify those nodes that are critical for signal transduction (**Figure 15.18b**). Another traditional approach is to use pharmacological small-molecule inhibitors to map the role of a particular inhibitor target in the function of the pathway, as well as to analyze pathway relationships (**Figure 15.18c**). In some cases, natural products or synthetic chemicals can inhibit a particular signaling protein with high specificity, allowing for a very precise analysis. In other cases, inhibitors block a large class of signaling molecules in a relatively nonspecific way; for example, okadaic acid blocks the activity of several major classes of Ser/Thr protein phosphatases. Although less specific, the use of such molecules can still be very informative.

More recently, it has been possible to use hybrid chemical genetic methods to generate systems that can be conditionally inhibited in a very specific way (**Figure 15.18d**). For example, a key conserved "gatekeeper" residue in the ATP-binding pocket of many protein kinases can be mutated, without altering its natural function. However, this mutation allows the modified kinase to bind to

Figure 15.18

Methods to perturb cell signaling networks. (a) A simple cell signaling network is illustrated, in which binding of input ligand to a cell-surface receptor leads to signaling output. This network can be perturbed in a variety of ways. (b) It can be perturbed by decreasing or eliminating the expression of one of the components. (c) It can be perturbed by adding a small-molecule inhibitor of one of the components. (d) It can be perturbed using chemical genetics: a component of the network, such as a kinase, can be replaced with an engineered version that is uniquely inhibited by an ATP analog. (e) Strategy for design of a mutant kinase (*green*) that is engineered to be inhibited by an ATP analog that cannot bind other ATP-binding proteins, including the corresponding wild-type kinase (*gray*), because of steric incompatibility. Here, the kinase is mutated to create a "hole" that can accommodate the extra "bump" on the ATP analog.

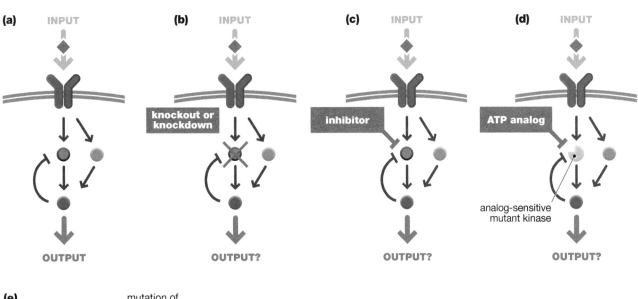

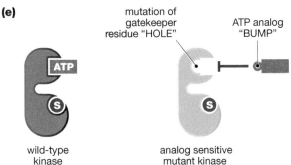

and be blocked by a bulkier ATP analog inhibitor that cannot fit into any endogenous kinase—the ATP analog has a "bump" which can only be accommodated by a kinase with an engineered complementary "hole" (**Figure 15.18e**). When this type of chemically sensitive allele can be genetically introduced into a researcher's system of interest (replacing the wild-type allele), one can create a system in which a targeted kinase can be uniquely and selectively inhibited, in the absence of off-target effects on other proteins.

Chemical dimerizers and optogenetic proteins provide a dynamic way to artificially activate pathways

In addition to methods to inhibit specific nodes within a pathway, there are also methods to artificially activate signaling nodes (**Figure 15.19**). Most simply, a protein of interest can be overexpressed, or a constitutively active mutant can be introduced, in order to probe the consequences of increased activity. In the case of molecules that are activated by localization or by binding to a specific partner (**Figure 15.19a**), it is possible to link these molecules to small-molecule-dependent heterodimerization domains (**Figure 15.19b**). In these cases, formation of an active complex or recruited state of a molecule can be induced by addition of a small-molecule artificial dimerizer. The advantage of this system is that a single, well-defined molecular event (the binding of two proteins of choice) can be rapidly induced in the absence of other confounding changes. The most commonly used examples of these artificial dimerization domains are derived from proteins that bind to the rapamycin family of immunosuppressant drugs.

⬉ An example of such a light-sensitive conformational change is illustrated in Figure 10.27.

Another way to artificially activate signaling nodes is through **optogenetic** methods (**Figure 15.19c**). These take advantage of light-sensitive proteins and domains, originally found in plants or other light-responsive organisms. When illuminated with light of a specific wavelength, these domains undergo conformational changes that result in functional changes. In some cases, light-sensitive domains show light-induced

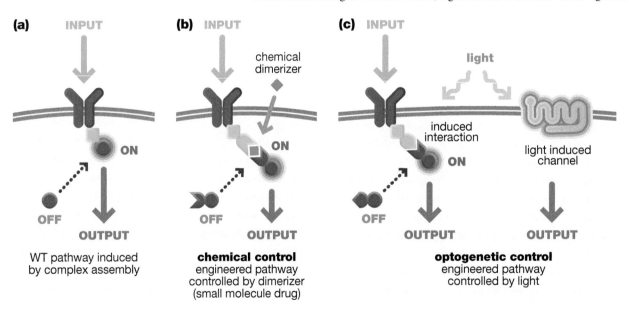

Figure 15.19

Non-natural ways to activate a cell signaling pathway. (a) In this example, normal wild-type (WT) signal output depends on the binding of a cytosolic protein (*purple*) to another protein localized on the membrane (*green*). (b) This binding interaction can be experimentally induced if the two components are expressed as fusion proteins that bind to a small-molecule chemical heterodimerizer (*blue diamond*). (c) In optogenetic systems, the association of the two components can be controlled by fusion to light-induced protein–protein interaction domains (left). Light can also be used to activate specific membrane ion channels (right). Such light-sensitive proteins have been isolated from plants and algae, but can be expressed in other cell types.

heterodimerization. By fusing the two optogenetically controlled partner domains to signaling proteins that are controlled by co-recruitment or localization, and expressing these in cells, one can create a system that is specifically activated by light. In some cases, this may allow localized activation only in the small region of the cell that is illuminated. Another type of optogenetic control involves light-sensitive channel proteins, such as channel rhodopsins, which can be used to activate excitable cells such as neurons. This kind of optogenetically controlled system can be used to specifically activate signaling systems with complex temporal or spatial patterns.

High-throughput sequencing is used to monitor the transcriptional state of a cell

Many signaling pathways ultimately lead to changes in transcription. Changes in the levels of specific mRNAs can be monitored by a number of methods, for example, by quantitative polymerase chain reaction (qPCR). The most powerful insights, however, come from high-throughput methods that provide a global overview of a cell's transcriptional response (**Figure 15.20**). **Microarray** analysis was the first method that could quantify global changes in transcription by measuring mRNA levels in the cell before and after stimulation. This method depends on hybridization of labeled cDNA from the cells to be analyzed to DNA or oligonucleotide microarrays printed on chips. Today, such information is typically provided by direct high-throughput **cDNA sequencing analysis** (**RNA-seq**). In this approach, millions of independent cDNA fragments are sequenced for each sample, allowing the relative abundance of different messages to be determined. While fast, powerful, and comprehensive, these methods have two major limitations. First, they only provide information about transcriptional changes in the cell, and not about post-transcriptional changes, such as changes in protein abundance or post-translational modifications. Second, they typically require material from a large number of cells to be pooled for analysis, and thus information about what is happening in individual cells is lost. However, recent innovations have made the amplification and sequencing of the mRNA from single cells or nuclei (**single-cell RNA-seq** or **scRNA-seq**) a standard approach to analyze cell-to-cell variation in gene expression within a population.

Modification-specific antibodies provide a method to track post-translational changes

Antibodies that specifically recognize particular post-translational modifications are powerful tools for characterizing the state of these modifications in a cell that has been exposed to a particular history of inputs. For example, antibodies are available that recognize tyrosine-phosphorylated proteins, without particularly strong discrimination between distinct sites. But highly specific antibodies have also been generated against individual phosphorylated sites, for example, the activation-loop phosphorylation sites in a particular kinase. These antibodies recognize phosphate in the context of a specific amino acid sequence (**Figure 15.21**). Such antibodies can be extremely useful in measuring the activity of particular signaling pathways through methods such as immunoblotting or *flow cytometry* (see below). For example, antibodies recognizing the activating phosphorylation sites or known substrates of Erk or Akt kinases can be used as readouts of Ras/MAPK or PI3K–Akt pathway activity, respectively. Similar antibodies have been generated recognizing other specific post-translationally modified sites, for example, histones that have been methylated or acetylated at specific sites.

Not only do such antibodies allow the quantification of changes in modification over time in different cell samples, but they can also be used to purify proteins and other components associated with the modified proteins. For example, in the **chromatin immunoprecipitation (ChIP)** method, antibodies to specific modified histone sites are used to purify other proteins and genomic DNA associated with the modified

Figure 15.20

Analysis of changes in cellular gene expression. (a) In microarray analysis, mRNA from two different cell populations is isolated and reverse-transcribed into fluorescently tagged cDNA. The two cDNA samples are then mixed and hybridized to a panel of DNA probes arrayed on a glass slide. Data about the absolute amount of each cDNA, and the relative amounts of each cDNA in the two samples, are obtained. (b) In RNA-seq, cDNA from different cell samples is directly sequenced in a high-throughput DNA sequencer. The abundance of different mRNAs can be inferred from the abundance of sequencing reads that correspond to each mRNA. Cluster analysis is a bioinformatic method that groups or "clusters" cell samples and cDNAs together based on their similarity in a series of quantitative values, such as the expression levels of different cDNAs. For example, two genes with very similar expression patterns in different samples will cluster closely together.
(Image of Illumina HiSeq® 2500 System courtesy of Illumina, Inc. ©2012. All rights reserved.)

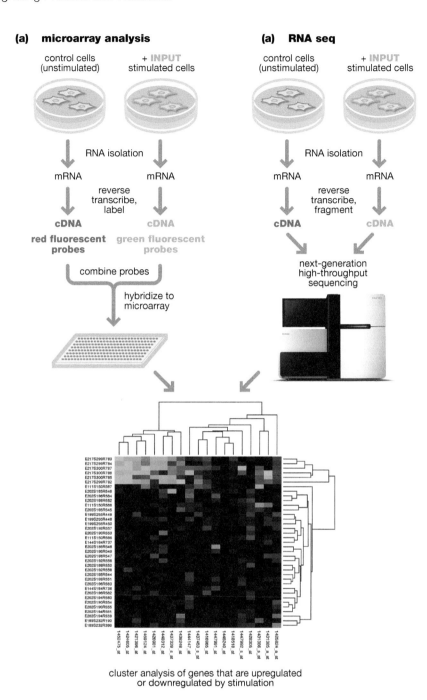

cluster analysis of genes that are upregulated or downregulated by stimulation

histones. Analysis of the associated DNA provides a measure of which genes and sequences are associated with particular chromatin modifications (**Figure 15.22**).

Protein-binding domains that recognize specific modified sites (reader domains) can also provide the basis for purifying their binding partners or quantifying changes in their abundance. For example, SH2 domains can be used in a manner similar to phosphospecific antibodies to detect and quantify their tyrosine-phosphorylated binding sites in cell lysates or fixed cells. When expressed in cells, fused to a fluorescent marker, modification-specific protein-binding domains (such as PH domains that can bind to specific phosphoinositol lipids) can also be used to probe changes in the amount or subcellular localization of their binding sites in living cells (discussed in more detail below).

(a)

(c)

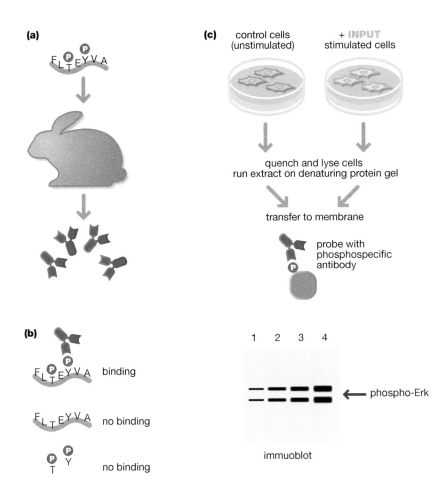

Figure 15.21

Analysis of changes in phosphorylation using phosphospecific antibodies.
(a) A phosphorylated peptide (in this case, the phosphorylated activation-loop sequence of the MAP kinase Erk1) is used to immunize rabbits, and antibodies are prepared. (b) Antibodies that bind to the phosphorylated peptide, but not to unphosphorylated peptide or to other phosphorylated sites, are isolated. (c) A hypothetical experiment in which lysates of control and stimulated cells are subjected to immunoblotting with the phosphospecific Erk antibody. Phospho-Erk is revealed as two bands on the immunoblot (arrow), and the extent of phosphorylation is indicated by the intensity of the bands.

Figure 15.22

Chromatin immunoprecipitation (ChIP).
DNA in chromatin is cross-linked to associated proteins and fragmented into small pieces. The fragmented chromatin is then incubated with antibodies to a specific protein or post-translational modification. In this example, an antibody to acetylated histone is used. Antibody-bound chromatin is isolated, and protein is removed. The DNA in this fraction is enriched in DNA that was associated with chromatin that contained acetylated histones (*pink*).

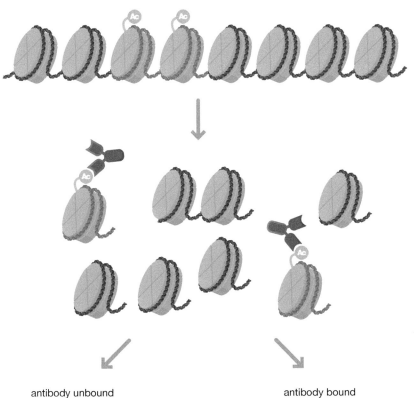

Mass spectrometry is the workhorse for identification of proteins and their modifications

Advances in **mass spectrometry (MS)** methods and instrumentation have dramatically improved our ability to identify proteins in complex mixtures. These approaches can be used to determine not only which proteins are post-translationally modified, but also the specific amino acids where the modification takes place. In a typical experiment (**Figure 15.23**), a mixture of proteins is digested into peptides using a protease such as trypsin, which cleaves C-terminal to lysine or arginine. These peptides are separated by liquid chromatography and converted to ions in the gas phase, and the intensity and mass-to-charge (m/z) ratio of every ion are measured as they pass through the mass spectrometer. Individual peptides can then be excited such that chemical bonds are broken, producing a series of fragment ions (referred to as the MS/MS or MS^2 spectrum), usually corresponding to the sequential loss of amino acids from the initial peptide. Thus, the fragmentation

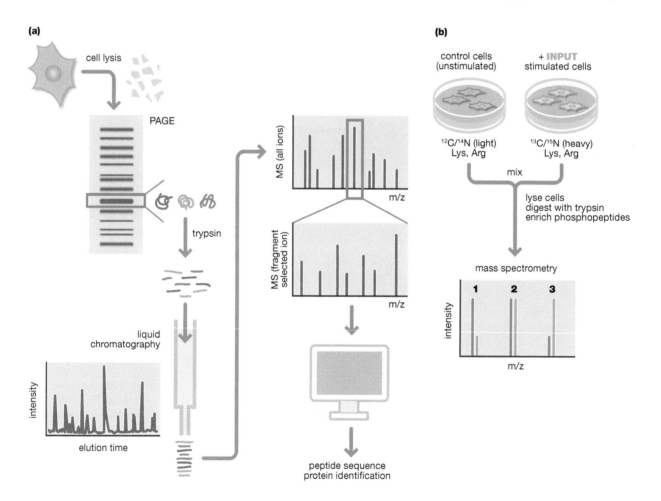

Figure 15.23

Analysis of changes in protein abundance and post-translational modification using mass spectrometry (MS). (a) In a typical experiment, cells are lysed and proteins fractionated by size using polyacrylamide-gel electrophoresis (PAGE). Gel slices containing multiple proteins are digested using a protease such as trypsin, producing a set of peptides for each protein. The peptides are separated by liquid chromatography and ionized. In the initial MS spectrum, the intensity and mass-to-charge (m/z) ratio of each ion is measured. Individual ions are then fragmented, resulting in an MS/MS spectrum that can be used to deduce the sequence of the peptide. These sequences are then compared to proteome databases, thus allowing the identification of proteins in the initial sample. (b) In SILAC, cells are grown in the presence of essential amino acids labeled with distinct isotopes. In this example, unstimulated cells are grown in $^{12}C/^{14}N$ (light) lysine and arginine, while stimulated cells are grown in $^{13}C/^{15}N$ (heavy) lysine and arginine. The samples are mixed and processed together, and the relative intensity of the light and heavy peaks in the MS spectrum can reveal changes in abundance. In this example, stimulation reduces the abundance of peptide 1, while peptide 2 is unchanged and peptide 3 increases upon stimulation.

pattern of an ion essentially allows the peptide to be sequenced, and the corresponding protein can be identified by comparing these sequences to proteome databases.

Post-translational modifications can be identified by MS because they change the mass of the peptide to which they are attached, and also because they produce characteristic fragment ions. For example, phosphorylated peptides produce a fragment ion at m/z 79 (HPO_3^-). Phosphates attached to tyrosine are more stable to fragmentation, but produce a phosphotyrosine immonium ion with m/z 216. Ubiquitylated lysines are protected from digestion by trypsin, so the complement of peptides detected from a given protein is changed. In addition, treatment with trypsin leaves the C-terminal Gly-Gly motif from ubiquitin attached to the modified lysine.

Post-translational modifications are often substoichiometric and transient, so the fraction of peptides that are modified is often fairly low. Therefore, it is usually beneficial to enrich for peptides that carry the modification of interest, particularly in studies seeking to identify large numbers of sites. Several methods are used (alone or in combination) to enrich for phosphorylated peptides. Immobilized metal affinity chromatography (IMAC) exploits the affinity of metal ions such as Fe^{3+} for the negatively charged phosphate group. Similarly, TiO_2 or other metal oxides are often used. Another common enrichment strategy employs anti-phosphotyrosine antibodies, which can be used on intact proteins or on peptide mixtures. Pan-specific phosphoserine/phosphothreonine antibodies provide poor enrichment for MS, but some motif-specific antibodies that recognize sequences phosphorylated by specific kinases have been successfully used. Analogous antibodies can be employed to enrich for other post-translational modifications.

In addition to identifying sites of post-translational modification, it is useful to quantify changes that occur under different conditions. For example, this can help distinguish background phosphorylation (which remains stable) from pathway-specific phosphorylation (which increases or decreases upon stimulation). A number of techniques have been developed to quantify the amount of protein or post-translational modification in different samples. In one method, referred to as stable-isotope labeling of amino acids in cell culture (SILAC), cells are grown in the presence of either natural $^{12}C/^{14}N$- (light) or $^{13}C/^{15}N$-labeled (heavy) amino acids such as lysine and arginine. After labeling, the two cell cultures are mixed and processed together. Isotopic variants of a given peptide have the same chemical characteristics, so they co-fractionate, but they can be distinguished because of their slight difference in mass. As a result, every peptide produces a pair of peaks in the mass spectrum. The intensities of the two peaks can then be used to determine the relative abundance of the protein or modification in the two samples (**Figure 15.23b**). A related technique is iTRAQ (isobaric tag for relative and absolute quantification), in which peptides are chemically modified with tags that have the same mass but produce distinct fragmentation patterns. One advantage of iTRAQ is that it can simultaneously quantify up to eight different samples in a single experiment. Label-free techniques that allow absolute quantification (AQUA) have also been developed. For each peptide of interest, a variant labeled with ^{13}C and/or ^{15}N isotopes is chemically synthesized, and a defined quantity is added as an internal standard to the mixture of peptides. The stoichiometry of post-translational modifications can be determined by measuring the absolute concentrations of both unmodified and modified versions of the same peptide.

The specific MS approach taken depends on the purpose of the experiment. In some cases, the objective is to discover new sites of post-translational modification by identifying large numbers of peptides. These experiments are inherently irreproducible because not all peptides are selected for fragmentation. In experiments that focus on specific signaling pathways, it is useful to focus on pre-selected peptides in a technique referred to as multiple reaction monitoring (MRM).

Live-cell time-lapse microscopy provides a way to track the dynamics of single-cell responses

A shortcoming of the methods described above for analyzing cell state is that they all require the lysis and pooling of large numbers of cells to provide sufficient material for analysis, and thus these methods have limited ability to yield single-cell information. Furthermore, limits on the throughput and speed of these analyses can prevent the researcher from obtaining a high-resolution dynamic picture of the timing of various responses. Single-cell information is often very important, since there can be a great deal of stochastic cell-to-cell variability in a response, due to, for example, slight variations in the number of signaling molecules or ribosomes in each cell or its precise cell-cycle stage at any given time. Thus, when one looks at a large population of cells, important information about the nature of a signaling response at the cellular level could be lost.

Time-lapse microscopy of live cells provides one approach to overcome these limitations (**Figure 15.24**). Imaging can be done in standard culture vessels, or cells may be loaded into a microfluidic device that traps cells within a viewing window for microscopy (**Figure 15.24a**). Such devices, depending on how they are constructed, allow the researcher to use various in-flow ports to change the media of the trapped cells in a highly controlled, dynamic way. For example, as shown in **Figure 15.24b**, the input signal molecule concentration can be rapidly shifted from low to high in a step function, allowing real-time monitoring of cellular responses after this sudden change in input. To monitor responses, one requires a live-cell reporter or biosensor: for transcriptional responses, this could be expression of a fluorescent reporter protein from a promoter; for post-translational modifications,

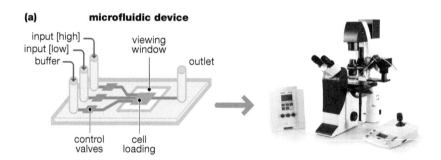

(a) **microfluidic device**

input [high]
input [low]
buffer
viewing window
outlet
control valves
cell loading

Figure 15.24

Live-cell time-lapse microscopy.
(a) Cells can be imaged in a microfluidic chamber. Cells are loaded into the device, and specific inlet valves are used to control media flow over the cells, allowing precise spatiotemporal control over the input stimulation that the cells experience. The field of cells in the device can be tracked by microscopy. (b) For each individual cell, multiple outputs can be monitored over time using different fluorescent reporters (*green* and *pink* in this example). The response dynamics of each individual cell in the field of observation can be tracked. In this example, the individual cell traces suggest that there is cell-to-cell variation in the time of reporter 1 activation, but reporter 2 is consistently activated after reporter 1. Statistical analysis of many cells would reveal underlying patterns in the response. (Microscope image courtesy of Leica Microsystems.)

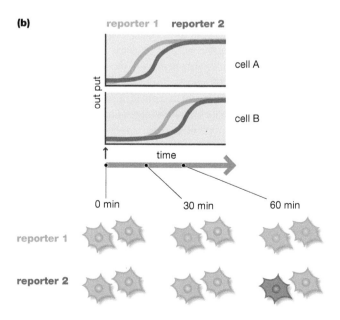

(b)

reporter 1   reporter 2

output

cell A

cell B

time

0 min   30 min   60 min

reporter 1

reporter 2

this could be a FRET reporter of a specific phosphorylation event (described further below). With such reporters, one can track how each individual cell in the field of view responds to the input change, and can monitor variation in response within the population. In the example illustrated in **Figure 15.24b**, when two individual cells are tracked, there appears to be stochastic variability in the timing of when the reporter 1 response is initiated after the change in input. By contrast, there seems to be a tighter correlation of the timing of the reporter 2 response after the initiation of the reporter 1 response. Thus, through analysis of multiple individual cells, one can see how different aspects of a cellular response show high or low variability.

Overall, time-lapse microscopy provides a powerful way to follow single-cell responses and the best way to follow the dynamics of these responses (**Table 15.1**). However, this approach is limited by the need for a live-cell reporter (for example, not all specific phosphorylation events have a suitable reporter). It is also limited by throughput—although microfluidic devices help tremendously, usually experiments are limited to the analysis of tens to hundreds of cells, which is several orders of magnitude lower than what can be analyzed by methods like flow cytometry (described below).

Biosensors allow signaling activity to be monitored in living cells

Monitoring the activity of signaling pathways and networks at the single-cell level requires some way to convert changes in a biochemical activity of interest into changes that can be visualized in the microscope. To accomplish this, a variety of **biosensors** have been developed, typically based on fluorescent molecules that can be directly visualized by fluorescence microscopy. A very simple type of biosensor can be made by fusing a modification-specific modular binding domain to a fluorescent protein.

For example, GFP can be fused to a PH domain that recognizes a specific phosphoinositide species such as phosphatidylinositol 3,4,5-trisphosphate (PIP_3). When such a biosensor is expressed in cells, changes in the level of PIP_3 can be visualized as changes in the amount of GFP fluorescence on the membrane. Not only can such a biosensor monitor overall changes in the level of PIP_3 over time, but it can also reveal specific subcellular locations where PIP_3 levels are particularly high or low. An example of the use of such PH domain-based sensors is provided in Figure 7.11. Other modular domains that recognize modified targets, such as reader domains that bind phosphorylated motifs (SH2 domains, for example), can similarly be used to monitor changes in the abundance or location of their binding sites in living cells.

Another simple type of biosensor consists of cell-permeable fluorescent dyes whose spectral properties change based on their chemical environment. For example, Fura-2 is a dye that binds calcium. The amount of fluorescence emitted when it is excited at particular wavelengths depends on whether calcium is bound or not. Thus, it is relatively straightforward to monitor the absolute concentration of calcium in cells, as well as the spatiotemporal dynamics of changes in intracellular calcium, in the presence of this dye. Such a calcium-sensitive dye was used to visualize calcium waves in Figure 6.12. Other specific dyes can be used to monitor changes in different properties, such as the potential difference (voltage) across a membrane, or the local pH.

Live-cell imaging	Flow cytometry
Single-cell dynamics	No single-cell dynamics
Medium throughput (hundreds of cells)	High throughput (10^3–10^5 cells)
Requires live-cell reporter	Live-cell reporter not required
Information on subcellular localization	Little information on localization

Table 15.1

Comparison of single-cell analysis methods

More sophisticated biosensors can monitor changes in properties such as the activity of specific G proteins or protein kinases, or the concentration of small molecules such as cAMP. Such biosensors are typically based on FRET; changes in the conformation of the biosensor due to pathway activity (G protein binding, phosphorylation, cAMP binding) are converted into changes in the FRET ratio, which can in turn be monitored by microscopy. For example, to monitor the activity of a protein kinase, a biosensor can be constructed that contains a specific substrate peptide sequence favored by that kinase, along with a modular protein-binding domain that binds to the substrate site only when it is phosphorylated. Thus, the conformation of the biosensor will be quite different in the unphosphorylated state versus when it is phosphorylated; this change in conformation can be coupled to changes in FRET when the biosensor contains a FRET donor and FRET acceptor. Biosensors to monitor G protein activity often contain the G protein itself fused to an effector domain that binds the G protein only when it is active (in the GTP-bound state) (**Figure 15.25**). Thus, an intramolecular interaction between the two occurs only when the G protein has been activated by endogenous guanine nucleotide exchange factors (GEFs). Again, the conformational change can be monitored by changes in FRET. Expression of such FRET-based biosensors in cells allows the magnitude and location of the activity of interest to be monitored by microscopy (**Figure 15.25b and c**).

Biosensors such as those described above are very powerful tools for interrogating pathway activity in individual cells in real time. Experimental results must be interpreted carefully, however, particularly in cases where the biosensor itself may perturb signaling. Often, relatively high levels of biosensor are needed for visualization, and this may result in inhibition of the pathway being monitored. For example, a GFP-PH domain fusion may prevent the binding of the endogenous PH domain-containing effectors to the membrane, or a kinase sensor may prevent normal levels of phosphorylation of the endogenous targets of the kinase. Furthermore, development of a useful biosensor may require a great deal of protein engineering to optimize for low background, high sensitivity, and appropriate specificity.

Figure 15.25

A fluorescence resonance energy transfer (FRET) biosensor for activation of Rho GTPases. (a) The nucleotide state of Rho GTPases is regulated by GEFs (guanine nucleotide exchange factors) and GAPs (GTPase-activator proteins), with the GTP-bound form being active and capable of interacting with downstream effectors. An example of a RhoA biosensor consists of full-length RhoA protein, a RhoA-binding domain derived from its effector PKN (*brown*), and a FRET pair of fluorescent proteins [cyan fluorescent protein (CFP) and yellow fluorescent protein (YFP)]. These components are connected with linkers in such a way that the activation of RhoA in the sensor causes an intramolecular conformational change that increases FRET efficiency. (b) Representative ratio (R) images (FRET/CFP) of an MCF-7 breast cancer cell are shown before and after addition of lysophosphatidic acid (LPA), a known activator of RhoA. (c) By conducting time-lapse imaging of multiple cells, the temporal kinetics of RhoA activation can be estimated based on changes of ratios (data shown as mean of whole cells ± standard error of the mean). Scale bar 5 μm. (Images courtesy of Taofei Yin and Yi Wu.)

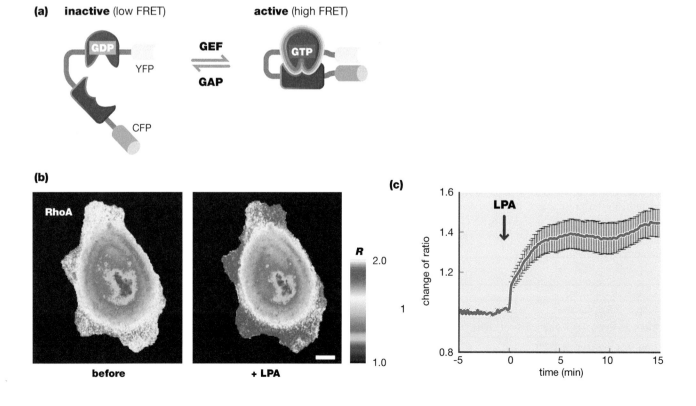

Flow cytometry provides a method to analyze single-cell responses rapidly

One of the most powerful and rapid methods for high-throughput analysis of single-cell responses is **flow cytometry** (**Figure 15.26**). Typically, a large population of cells are stimulated and the response is monitored using fluorescent reporters. In the case of a transcriptional response, this can be expression of a fluorescent protein, like GFP. In the case of a phosphorylation-based response, for example, one can fix, permeabilize, and stain cells with a fluorescently labeled phosphospecific antibody (or other output-specific labeling reagent). In either case, cells are fixed at various times after stimulation and each sample is analyzed in the flow cytometer. Hydrodynamic focusing allows the cells to pass through the instrument in single file; lasers are used to excite each cell, and the amount of fluorescence is measured by a detector. A large number of different reporters can be followed within each cell, if the instrument is equipped with multiple lasers and detectors (the most modern instruments can now quantify more than 20 different fluorescent labels in one experiment). The data can then be processed to give a histogram of the distribution of cells with a given level of fluorescence within the population.

An example of flow cytometry analysis is shown in **Figure 15.26b**, where a population of cells has been stimulated for 0, 30, 60, or 90 minutes, and the output of a fluorescent reporter measured (here, for example, it might be the amount of phospho-Erk after mitogen stimulation). In the example on the left, the cells respond in a *graded* fashion—the cell population gradually shifts with stimulation time to higher fluorescence values, and the cells pass through a stage of intermediate fluorescence. By contrast, in the example on the right, the cells respond in a *bistable* fashion—the cells exist only in a low or high fluorescence state, and it is the distribution of cells in each of these states that changes with the time of stimulation. Although these two classes of responses (graded versus bistable/all or none) are very different, and can lead to very different biological behaviors, it is important to note that these two responses could not be distinguished by a population-based

Figure 15.26

Flow cytometry. (a) Cell populations (either unstimulated or stimulated) are labeled with fluorescent antibodies or other fluorescent reporters of signal output; in this example, a phosphospecific antibody is used. The flow cytometer rapidly quantifies fluorescence for each cell; forward and side scatter are used to monitor the size and granularity of cells. (b) The number of cells and fluorescence level are plotted for each cell population. A graded response (left), in which individual cells can have intermediate levels of output, can be distinguished from a bistable response (right), where cells only can exhibit either low or high output.

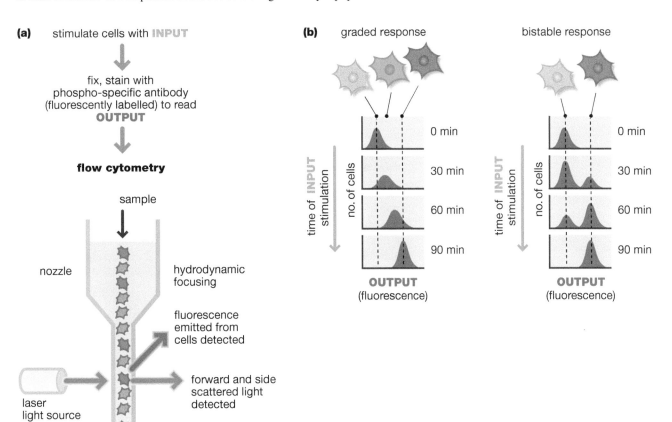

assay such as an immunoblot using a phosphospecific antibody. The pooling of cells required for obtaining enough material for detection on the immunoblot would result in the loss of all single-cell information, and even the cells showing a bistable response would appear graded because of variation in the timing of when each cell shifted from the low to high fluorescence state.

Overall, flow cytometry is a very powerful way to obtain single-cell information because of the very large number of cells (10^3–10^5) that can be analyzed in this way, allowing one to obtain excellent statistics from large cell populations (see Table 15.1). Another advantage of flow cytometry is that it does not require a live-cell reporter, and thus can be used, for example, to track simultaneously many different phosphorylation events using phosphospecific antibodies labeled with different fluorophores. A limitation of flow cytometry is that one cannot obtain single-cell dynamic data—that is, how one specific cell is responding at time A and then at time B. This is because in flow cytometry, one does not track the same cells over time, but follows a large population of cells by sampling their response distributions at each time point.

QUESTIONS

Answers to these questions can be found online at www.routledge.com/9780367279370

1. You are studying the development of a particular region of the mouse brain, and have genetic evidence that protein X is involved. From your search of the literature, you find that protein X has several identified domains and a fairly long region with no known mapped function. You suspect this region may be a new protein-binding domain, and in searching databases you think you find weak sequence similarity between this region and other proteins. Describe an experimental strategy to test your hypothesis and to find which proteins (or other biological molecules) bind to the putative domain.

2. Based on theoretical considerations, you believe that the modification of protein Y by enzyme X is operating at saturation in the cell (that is, the reaction is zero order). What information is needed to evaluate whether this is the case in the cells that you are studying? Provide a set of experiments to test your hypothesis.

3. You find that stimulation of cells with a cytokine leads to rapid changes in the actin cytoskeleton. The receptor for the cytokine binds to protein X upon activation, and you develop an optogenetic tool that allows you to recruit protein X to the plasma membrane. How might this tool help you dissect the mechanism of actin rearrangement? What are the advantages and disadvantages of studying the response to the normal stimulus (the cytokine) versus responses to the optogenetic tool?

4. You are interested in cellular responses to low pH and perform a SILAC (stable-isotope labeling of amino acids in cell culture) experiment to look for changes in phosphorylation upon a change in pH. You find a phosphorylated peptide from protein Y that is rapidly phosphorylated on serine when the pH of the culture medium is lowered. Design a series of experiments to determine what fraction of the total protein Y in the cell is phosphorylated at this site, and to test whether phosphorylation of Y is important for inducing the transcription of pH-responsive genes.

5. You have developed an inhibitory antibody that binds to a mitogen receptor and are in the early stages of clinical testing to see whether this antibody might provide a new way to stop the growth of certain tumors that have abnormally high receptor activity. You find that the antibody decreases growth of a tumor cell line by ~10%. Discuss how you might determine whether this represents a weak response by all cells or a strong response by a small fraction of the cells. If the latter, how would you address why some cells respond and not others?

BIBLIOGRAPHY

BIOCHEMICAL AND BIOPHYSICAL ANALYSIS OF PROTEINS

Branden C & Tooze J (1998) *Introduction to Protein Structure*, 2nd ed. New York: Garland Science.

Hammes GG (2000) *Thermodynamics and Kinetics for the Biological Sciences*. New York: Wiley-Interscience.

Jumper J, Evans R, Pritzel A, et al. (2021) Highly accurate protein structure prediction with AlphaFold. *Nature* 596(7873), 583–589.

Klotz IM (1997) *Ligand-Receptor Energetics: A Guide for the Perplexed*. New York: John Wiley & Sons.

Kuriyan J, Konforti B & Wemmer D (2012) *The Molecules of Life: Physical and Chemical Principles*. New York: Garland Science.

Lei M, Lu W, Meng W, et al. (2000) Structure of PAK1 in an autoinhibited conformation reveals a multi-stage activation switch. *Cell* 102, 387–397.

Menten L & Michaelis MI (1913) Die Kinetik der Invertinwirkung. *Biochem Z* 49, 333–369.

Michaelis L, Menten ML, Johnson KA, & Goody RS (2011) The original Michaelis constant: Translation of the 1913 Michaelis-Menten paper. *Biochemistry* 50(39), 8264–8269. doi: 10.1021/bi201284u.

Rich RL & Myszka DG (2000) Advances in surface plasmon resonance biosensor analysis. *Curr. Opin. Biotechnol.* 11, 54–61.

Saibil HR (2022) Cryo-EM in molecular and cellular biology. *Mol. Cell* 82(2), 274–284.

Voet D, Voet JG & Pratt CW (2013) *Principles of Biochemistry*, 4th ed. New York: Wiley.

Winzor DJ & Sawyer WH (1995) *Quantitative Characterization of Ligand Binding*. New York: Wiley-Liss.

Wodak SJ, Vajda S, Lensink MF, et al. (2023) Critical assessment of methods for predicting the 3D structure of proteins and protein complexes. *Annu. Rev. Biophys.* 52, 183–206. doi: 10.1146/annurev-biophys-102622-084607.

MAPPING PROTEIN INTERACTIONS AND LOCALIZATION

Choudhary C & Mann M (2010) Decoding signalling networks by mass spectrometry-based proteomics. *Nat. Rev. Mol. Cell Biol.* 11, 427–439.

Giepmans BNG, Adams SR, Ellisman MH & Tsien RY (2006) The fluorescent toolbox for assessing protein location and function. *Science* 312, 217–224.

Golemis EA & Adams PD (eds) (2005) Protein-*Protein Interactions: A Molecular Cloning Manual*, 3rd ed. New York: Cold Spring Harbor Press.

Jones RB, Gordus A, Krall JA & MacBeath G (2006) A quantitative protein interaction network for the ErbB receptors using protein microarrays. *Nature* 439, 168–174.

Lippincott-Schwartz J & Patterson GH (2003) Development and use of fluorescent protein markers in living cells. *Science* 300, 87–91.

Michnick SW (2003) Protein fragment complementation strategies for biochemical network mapping. *Curr. Opin. Biotechnol.* 14, 610–617.

Wu JQ & Pollard TD (2005) Counting cytokinesis proteins globally and locally in fission yeast. *Science* 310, 310–314.

METHODS TO PERTURB CELL SIGNALING NETWORKS AND MONITOR CELLULAR RESPONSES

Barrios-Rodiles M, Brown KR, Ozdamar B, et al. (2005) High-throughput mapping of a dynamic signaling network in mammalian cells. *Science* 307, 1621–1625.

Bishop A, Buzko O, Heyeck-Dumas S, et al. (2000) Unnatural ligands for engineered proteins: New tools for chemical genetics. *Annu. Rev. Biophys. Biomol. Struct.* 29, 577–606.

Fujioka A, Terai K, Itoh RE, et al. (2006) Dynamics of the Ras/ERK MAPK cascade as monitored by fluorescent probes. *J. Biol. Chem.* 281, 8917–8926.

Hwang B, Lee JH & Bang D (2018) Single-cell RNA sequencing technologies and bioinformatics pipelines. *Exp. Mol. Med.* 50, 1–14.

Levskaya A, Weiner OD, Lim WA & Voigt CA (2009) Spatiotemporal control of cell signalling using a light-switchable protein interaction. *Nature* 461, 997–1001.

Pertz OD & Hahn KM (2004) Designing biosensors for Rho family proteins-deciphering the dynamics of Rho family GTPase activation in living cells. *J. Cell Sci.* 117, 1313–1318.

Perez OD & Nolan GP (2002) Simultaneous measurement of multiple active kinase states using polychromatic flow cytometry. *Nat. Biotechnol.* 20, 155–162.

Toettcher JE, Voigt CA, Weiner OD, & Lim WA (2011) The promise of optogenetics in cell biology: interrogating molecular circuits in space and time. *Nat. Methods* 8, 35–38.

Young JW, Locke JC, Altinok A, et al. (2011) Measuring single-cell gene expression dynamics in bacteria using fluorescence time-lapse microscopy. *Nat. Protoc.* 7, 80–88.

Glossary

14-3-3 proteins A class of small proteins that bind specifically to target proteins that are phosphorylated on serine or threonine residues. 14-3-3 binding can regulate the activity, conformation, and/or subcellular localization of its targets.

activation loop An important regulatory element for protein kinases that undergoes dramatic conformational changes between the inactive and active forms of the catalytic domain. The active conformation of the kinase is typically promoted by phosphorylation of the activation loop.

adaptation A property of a signaling system in which, in response to a signaling input, there is an initial burst of output and then output returns to basal levels even if the input stimulus remains.

adaptive immune response Response of the immune system to a specific antigen that typically generates an immunological memory.

adaptor A protein with multiple protein-binding domains that mediates the assembly of complexes of three or more components. Grb2, a canonical adaptor, binds to tyrosine-phosphorylated proteins such as growth factor receptors via its SH2 domain, and binds downstream effectors such as the Ras activator Sos via its SH3 domains.

ADP-ribosylation The enzymatic addition of an ADP-ribose group to proteins. ADP-ribose can be added to a variety of amino acid side chains including Glu, Asp, and Lys. Modification may consist of a single ADP-ribose group, or of long and sometimes branched chains (poly-ADP-ribosylation).

affinity Strength of a noncovalent binding interaction; the higher the affinity, the more likely two binding partners will exist in a complex.

agonist A ligand that binds a receptor and stimulates its activity.

AKAPs (A-kinase anchoring proteins) Scaffold proteins with specific binding sites for protein kinase A regulatory subunits and other proteins, and which localize to specific subcellular compartments such as the plasma membrane, mitochondrion, or centrosome.

Akt A family of serine/threonine kinases activated by phosphoinositides that play a role in signaling pathways that control cell growth, proliferation, and survival. Also known as protein kinase B (PKB).

allosteric switch proteins Modular enzymes in which the activity of catalytic domains is coupled to conformational changes induced by upstream inputs.

allostery The property of being able to exist in two or more structural states of differing activity. The equilibrium between these states is modulated by ligand binding or covalent modification.

alpha helix (α helix) Common secondary structural element in proteins in which a linear sequence of amino acids adopts a right-handed helical conformation stabilized by internal hydrogen bonds between backbone atoms.

amphipathic Molecules containing both hydrophilic and hydrophobic portions.

amplification—*see* **signal amplification**

anaphase-promoting complex (APC) A large, multisubunit ubiquitin ligase complex that regulates cell-cycle progression by ubiquitylating specific proteins and targeting them for proteasomal degradation.

angiogenesis The generation of new blood vessels.

antagonists Compounds that bind to a receptor but which fail to evoke an activating response.

antigen Any molecule or part of a molecule recognized by the variable antigen receptors of lymphocytes.

antigen-presenting cell (APC) Cell that displays a foreign antigen complexed with MHC at its cell surface for presentation to T cells.

apoptosis A highly programmed form of cell death initiated by specific signals that activate caspases and lead to characteristic biochemical and morphological changes.

apoptosome A large cytosolic complex that serves as a scaffold for the recruitment and activation of caspases.

arginine demethylase (RDM) The enzyme that removes methyl groups from arginine residues of proteins that had been modified by protein arginine methyltransferases.

autophagy A highly regulated process whereby cells engulf their own cytosol and organelles and target them for degradation in the lysosome, allowing the raw materials to be recycled.

avidity Increased apparent affinity of a molecule for its ligand due to the presence of multiple binding sites on both partners.

Bcl2 family Family of proteins that either promote or inhibit apoptosis by regulating the permeability of mitochondria.

beta sheet (β sheet) Common secondary structural element in proteins in which individual polypeptide chains are held together by hydrogen bonds between peptide bonds in adjacent strands. Strands can run in the same or opposite directions.

binding isotherm Curve in which binding (fractional occupancy) of one component of a binding reaction is plotted as a function of concentration of the other component, under conditions of constant temperature.

binding The relatively stable association of two components.

biofilm A microbial community that forms on surfaces, containing microbes embedded in a secreted extracellular polymer matrix. A biofilm may consist entirely of the same species of organism, or of a mixture of species that make complementary contributions to the community.

biomolecular condensates Regions of cytosol in which some components are highly enriched, while others are largely excluded. Condensates have many properties of liquid droplets, and their formation is typically driven by multivalent interactions between proteins, or between proteins and RNA. Condensates can enhance the rate and specificity of reactions by concentrating reactants and excluding potential competitors.

biosensor Molecular device, often a fluorescent small molecule or protein, that can be used to monitor specific changes in the physiological state of a cell.

bistability A property of signaling systems that exist in either of two distinct output states, in the absence of stable intermediate states.

bromodomain Modular protein domain that recognizes peptide motifs containing an acetylated lysine.

buried surface area The total surface area (in Å^2) at a macromolecular interface that is exposed to solvent in the uncomplexed proteins but is buried in the complex.

Ca_2^+/CaM-dependent protein kinases (CaMKs) A family of serine/threonine kinases that are regulated by the binding of Ca^{2+}-bound calmodulin (CaM).

CAAX box motif A motif found at the C-terminus of proteins destined for prenylation. During prenylation, the C-terminal three residues (AAX, where A is an aliphatic amino acid and X is any amino acid) are proteolytically removed and the isoprenyl group is added to the cysteine by a thioether linkage.

calmodulin (CaM) A small calcium-binding protein that confers calcium regulation on cellular signaling and effector molecules, including protein kinases and phosphatases, to which it binds.

cAMP Cyclic nucleotide synthesized from ATP and used as a signaling mediator.

CaM—*see* **calmodulin**

cascade A signal transduction pathway in which multiple enzymes are linked in series, such that output of one enzyme directly or indirectly regulates the next enzyme in the cascade.

caspases A specialized group of cysteine proteases, activated in apoptosis, which specifically cleave peptide bonds after aspartic acid residues in their protein targets.

catalytic domain Part of an enzyme that is responsible for promoting a specific chemical reaction.

catalytic rate constant (k_{cat}) The rate constant for the catalytic step carried out by the enzyme when saturated with substrate.

caveolae Small, cup-shaped patches of plasma membrane enriched in cholesterol and the protein caveolin.

CDK—*see* **cyclin-dependent kinase**

cDNA expression libraries Large collections of cDNAs reverse-transcribed from total cell mRNA.

cDNA sequencing analysis (RNA-seq) High-throughput sequencing of cDNA fragments generated from a cell sample, providing detailed information about the expression levels of all mRNA species present in the sample.

cell cycle The cyclical series of events encompassing duplication of the genomic DNA, mitosis, and cell division.

cell lysate A cell extract usually made by solubilizing cells with detergent-containing buffer solution.

cGMP Cyclic nucleotide synthesized from GTP and used as a signaling mediator.

chemotaxis The directed movement of a cell either up or down a concentration gradient of a chemical stimulus.

cholesterol A component of vertebrate membranes that is highly hydrophobic and has a rigid, polycyclic structure. It can profoundly affect membrane fluidity and lateral diffusion through its interaction with the fatty acid chains of other membrane lipids.

chromatin immunoprecipitation (ChIP) A method in which antibodies to specific modified histones or chromatin-associated proteins are used to purify DNA and other proteins associated with that protein. Analysis of associated DNA provides a measure of which genes and sequences are associated with particular chromatin modifications or chromatin-associated proteins.

chromodomain Modular protein domain that recognizes peptide motifs containing a methylated lysine.

clathrin A structural protein that self-assembles into a hollow, spherical lattice surrounding a membrane vesicle during the process of clathrin-mediated endocytosis.

coherent feed-forward—*see* **feed-forward, coherent**

co-immunoprecipitation (co-IP) Isolation of a protein and its binding partners from a cell extract through the use of an antibody specific for one of the proteins, usually bound to beads.

co-IP—*see* **co-immunoprecipitation**

combinatorial complexity For protein modifications, refers to the fact that multiple independent modifications lead to an exponential increase in the number of possible states of the protein. This greatly expands the coding capacity of the genome.

conformation Three-dimensional shape of a molecule.

cooperativity When binding of one ligand alters the binding affinity of an additional ligand (or additional ligands). Thermodynamically, cooperativity is observed when the free energy of two ligands binding simultaneously differs from the sum of the free energies of the two ligands binding individually.

cryo-electron microscopy (cryo-EM) A method in which samples of proteins or other large macromolecules are immobilized in a thin layer of vitreous ice at very low temperature, and imaged using an electron microscope. Images of many individual particles are collected and analyzed by computational methods to generate an average image, which has a much higher resolution than each of the individual images.

cyclic nucleotides Small signaling mediators synthesized from either ATP or GTP by cyclase enzymes.

cyclin The regulatory subunit of a cyclin-dependent kinase (CDK). Cyclins are necessary for activity of the associated catalytic subunit, and also contribute to substrate specificity. Because cyclins are essential for activity of CDKs, modulation of cyclin abundance by transcription and regulated proteolysis is used to regulate CDK activity in the cell cycle.

cyclin-dependent kinases (CDKs) Serine/threonine protein kinases that require binding to a regulatory cyclin subunit for activity. CDKs control progression through the cell cycle by phosphorylating specific substrate proteins.

cytokine Polypeptide signaling molecule that participates in immune responses. Cytokines often act locally but can act systemically.

cytokinesis Division of the cytoplasm of a eukaryotic cell into two daughter cells.

cytosol Contents of a cell contained within the plasma membrane, excluding the nucleus of eukaryotic cells.

DAG—*see* **diacylglycerol**

death domain (DD) Modular domain that mediates homotypic interactions, such as between death receptors and downstream effectors.

death effector domain (DED) Modular domain that mediates homotypic interactions, such as between the adaptor protein FADD and initiator caspases.

death receptors A related family of transmembrane receptors that induce apoptosis when bound to their ligands.

death-inducing signaling complex (DISC) Supra-molecular complex containing death receptors, adaptor proteins, and initiator caspases that induces apoptosis.

desensitization The process by which receptors respond to sustained activation by becoming less responsive to input signal.

deubiquitinase (DUB) A specialized type of protease catalyzing cleavage of the isopeptide bond between a lysine amino group and the C-terminus of ubiquitin, thereby removing the ubiquitin group from a protein.

diacylglycerol (DAG) A membrane lipid consisting of a glycerol backbone linked to two fatty acid chains. DAG is generated when phospholipase C cleaves the phosphorylated head group from a phospholipid such as PIP_2.

digital response—*see* **switchlike response**

dissociation constant (K_d) Quantitative measure of binding affinity, reflecting the concentrations of free components and complex at equilibrium. The lower the dissociation constant, the higher the affinity of the interaction.

distributive (multisite phosphorylation) Independent phosphorylation of multiple sites within a protein by the same kinase, where each phosphorylation is the result of a distinct binding event.

DNA tumor viruses Viruses whose genomes consist of double-stranded DNA, which depend on the host cell for most of the machinery required for their replication. Such viruses typically manipulate the host cell cycle to provide appropriate conditions for genome replication.

docking sites In protein kinases, substrate-binding sites distant from the catalytic cleft that participate in determining substrate specificity.

domain Compact unit of protein structure that is usually capable of folding stably as an independent entity in solution.

dwell time The amount of time a species or molecule remains in the same state or location; in signaling, dwell time often refers to the time that signaling molecules spend associated with a structure (such as the plasma membrane) before dissociating away.

dynamic encoding In cell signaling, refers to signaling mechanisms where the timing, duration, and/or frequency of input signals affects the output of the system.

E1 ubiquitin activating enzyme The first step of the protein-ubiquitylation reaction. Uses energy of ATP hydrolysis to couple the C-terminus of ubiquitin to a cysteine residue on the E1.

E2 ubiquitin conjugating enzyme The second step of the protein-ubiquitylation reaction. Ubiquitin is transferred from the E1 to a cysteine on the E2 enzyme. E2 enzymes generally determine the linkage between ubiquitin units in a polyubiquitin chain.

E3 ubiquitin ligase The third step of the protein-ubiquitylation reaction. The E3 binds to specific substrates and to E2 enzymes, facilitating transfer of ubiquitin from the E2 to the substrate.

ECM—*see* **extracellular matrix**

EF hand Conserved protein motif containing acidic residues that can chelate Ca^{2+}.

effector caspase A capase that is activated by cleavage by upstream initiator caspases. Effector caspases cleave cell proteins to execute the apoptotic program of cell death. Also known as executioner caspases.

eicosanoids A large family of bioactive lipids derived from arachidonic acid that includes the prostaglandins and leukotrienes. Eicosanoids signal through G-protein-coupled receptors to regulate physiological processes such as inflammation.

electron microscopy (EM) Use of a microscope that uses a beam of electrons to create the image.

endocrine Relating to hormones or the glands and tissues that secrete them.

endocytosis A process in which a portion of the plasma membrane invaginates and pinches off into the cytosol, generating a free cytosolic vesicle or endosome.

endosome A cytosolic vesicle generated by endocytosis.

enthalpy A form of energy, equivalent to work, that can be released or absorbed as heat at constant pressure.

entropy A measure of the disorder or randomness of a molecule or system.

enzymes Proteins (or other biological macromolecules) that greatly increase the rate of a chemical reaction without altering the thermodynamic equilibrium between the reactants and the products.

epitope tag Short peptide which, when fused to a protein, allows that protein to be specifically bound by an antibody to that peptide.

ESCRT (endosomal sorting complex required for transport) machinery Series of large multiprotein complexes that mediates sorting of endocytosed proteins for recycling or lysosomal degradation.

executioner caspase—*see* **effector caspase**

exportin A transport protein that functions to ferry cargo proteins out of the nucleus.

extracellular matrix The meshwork of proteins and carbohydrates that forms between cells in tissues; it can include fibronectin, collagen, vitronectin, and other components.

extrinsic apoptotic pathway The apoptotic cell death pathway induced by extracellular ligands that bind to death receptors on the cell surface.

FAK—*see* **focal adhesion kinase**

far-Western blotting Method of detecting proteins that can bind directly to a protein of interest. Proteins in a cell lysate are separated by gel electrophoresis and transferred to a membrane, and the membrane is probed with a labeled purified protein of interest.

feedback When the output from a signaling node follows a path of links that returns to regulate the original node.

feed-forward Distinct paths fanning out from an upstream signaling node that reconverge on another downstream node.

feed-forward, coherent A feed-forward loop with two divergent branches with the same overall sign.

feed-forward, incoherent A feed-forward loop with two divergent branches with opposite signs.

flow cytometry High-throughput analysis that assays the fluorescence and optical properties of many individual cells in a population. Cells are typically incubated with multiple antibodies, each conjugated to a different fluorescent dye, before analysis.

fluorescence resonance energy transfer (FRET) Method for detecting the physical proximity of two different fluorescent molecules on the basis of the nonradiative transfer of energy from one fluorescent molecule to the other.

focal adhesion kinase (FAK) A nonreceptor tyrosine kinase that is activated upon integrin engagement. FAK plays a critical role in formation and turnover of focal adhesions.

focal adhesions Highly complex cellular structures that couple sites of cell-matrix adhesion to intracellular actin cables or stress fibers.

fractional occupancy The fraction of the total amount of A that is complexed with B for the reaction: A + B ↔ AB.

free energy The energy that can be extracted from a system to drive reactions.

free-energy barrier The difference in free energy between the initial state (ground state) of reactants and the high-energy transition state that must be passed through for the reaction to proceed to completion. The higher the free-energy barrier, the slower the reaction will be, even if the overall reaction is highly favored (ΔG is negative).

FRET—*see* **fluorescence resonance energy transfer**

G domain A 20 kD domain that binds guanine nucleotides and can adopt alternative conformations depending on whether GDP or GTP is bound. The small G proteins essentially consist of a single G domain, while the heterotrimeric G proteins contain a G domain in their α subunits.

G protein Any of a large class of GTPases that act as molecular switches that are active when bound to GTP and inactive when bound to GDP. They may be heterotrimeric G proteins with α, β, and γ subunits, which typically signal from seven-transmembrane receptors, or small G proteins of the Ras superfamily. G proteins are also known as GTPases, GTP-binding proteins, or guanine-nucleotide-binding proteins.

GAP—*see* **GTPase-activator protein**

gated ion channels Ion channels that undergo regulated opening and closing in response to a stimulus such as ligand binding or change in membrane potential.

GDF—*see* **GDI displacement factor**

GDI displacement factor (GDF) A protein that facilitates the dissociation of a GDI from a GTPase, leading to delivery of the GTPase and insertion of its lipid group into the target membrane.

GDI—*see* **guanine nucleotide dissociation inhibitor**

GEF—*see* **guanine nucleotide exchange factor**

gel electrophoresis Method of separating macromolecules based on their migration through a porous gel under an electric current. In most experimental conditions, smaller molecules migrate more rapidly than larger ones.

GFP—*see* **green fluorescent protein**

glycerophospholipid A phospholipid in which two of the hydroxyl groups of glycerol are linked to fatty acids and the third to a phosphate group.

glycosylation The addition of carbohydrate groups (sugars) to proteins. In the case of cell-surface or secreted proteins, complex chains of carbohydrate are added in the endoplasmic reticulum and Golgi apparatus either to the hydroxyl groups of serine or threonine (*O*-glycosylation) or to the amino group of asparagine (*N*-glycosylation). Single *N*-acetyl glucosamine groups can also be added to cytosolic proteins.

glycosylphosphatidylinositol anchor (GPI anchor) A complex structure consisting of lipids and carbohydrates that is attached to some proteins to target them to the outer leaflet of the plasma membrane.

GPCRs—*see* **G-protein-coupled receptors**

G-protein-coupled receptors (GPCRs) Any of a diverse class of cell-surface receptors with seven membrane-spanning segments that, upon activation, serves as a guanine nucleotide exchange factor to activate heterotrimeric G proteins.

graded response—*see* **linear response**

green fluorescent protein (GFP) Fluorescent protein first isolated from a marine jellyfish; it is widely used as a fusion partner to visualize a protein of interest in cells by fluorescence microscopy.

ground-state energy The state of the lowest energy for a molecule or system.

growth factors Inducers of cell growth and cell bulk (the term is also sometimes used more broadly to include mitogens).

GTPase-activator protein (GAP) Protein that interacts with G proteins and accelerates their rate of GTP hydrolysis, leading to inactivation of the G protein.

GTPase—*see* **G protein**

guanine nucleotide dissociation inhibitor (GDI) A protein that can bind specific G proteins and shield their lipid groups, thus sequestering the G protein in the cytoplasm. GDIs lock G proteins in their GDP-bound (inactive) state and prevent their localization to the membrane.

guanine nucleotide exchange factor (GEF) Protein that interacts with G proteins and catalyzes the exchange of bound GDP for GTP, leading to activation of the G protein.

half-life For a binding reaction, the time it takes for half of the complex to dissociate (or the time at which there is a 50% chance that an individual complex will have dissociated).

hedgehog (Hh) signaling pathway Signaling pathway in which ligand binding causes a change in the processing of Gli, a protein that acts as a transcriptional activator, controlling developmental processes.

heterotrimeric G protein A G protein composed of three different subunits: an α subunit with GTPase activity, and associated β and γ subunits. Exchange of bound GDP for GTP on the α subunit causes dissociation of the heterotrimer into a free subunit and a βγ heterodimer; hydrolysis of the bound GTP causes reassociation of the subunits.

histidine kinases Protein kinases, primarily found in prokaryotes, that transfer the terminal phosphate group from ATP to one of their own histidine residues via a phosphoramidate linkage. In two-component signaling systems, the phosphate is then rapidly transferred to the carboxyl group of an aspartate side chain on a response regulator protein.

histone acetyl transferase (HAT) Enzyme that catalyzes the *N*-acetylation of lysine residues in histones (or other proteins).

histone deacetylase (HDAC) Enzyme that catalyzes the removal of *N*-acetyl groups from lysine residues in histones (or other proteins).

histones The major protein component of nucleosomes, which package genomic DNA into chromatin. Histone modification is a major mode of regulating chromatin structure and thus gene expression at different sites on the genome.

homeostasis The ability of living systems to adjust their behavior spontaneously to maintain a stable intracellular environment, despite varying environmental conditions.

hormone A soluble signaling molecule that induces physiological effects at a distance by binding to a specific receptor present on target cells.

hysteresis Describing a system where the input level at which the system switches between the two output states will be different if one is moving up from low input to high input, or moving down from high input to low input.

immune checkpoint proteins Proteins expressed by cells that activate T cell immune checkpoints, a natural "braking system" used to dampen T cell immune responses. Expression of immune checkpoint proteins allows tumor cells to evade attack by T cells.

immunoblotting Method of detecting and quantifying a protein of interest in a sample. Proteins are separated by gel electrophoresis and transferred to a membrane, and the membrane is probed with a specific antibody to the protein of interest (also termed Western blotting).

immunofluorescence Technique in which fluorescently labeled antibodies are used to determine the location of their corresponding antigens in fixed cells or tissues.

immunoreceptor tyrosine-based activating motif (ITAM) Two tyrosines, separated by approximately nine amino acids, that when phosphorylated serve to recruit and activate kinases of the ZAP-70 family.

importin A transport protein that functions to ferry cargo proteins into the nucleus.

incoherent feed-forward—*see* **feed-forward, incoherent**

inflammation Physiological response to infection, allergens, or trauma involving localized swelling, redness, and pain.

initiator caspase A caspase that is directly activated by apoptotic signals. Upon activation, it cleaves and activates effector caspases.

inositol 1,4,5-trisphosphate (IP$_3$) A soluble second messenger molecule generated by cleavage of PIP$_2$ by phospholipase C. IP$_3$ binds to and activates calcium channels on the endoplasmic reticulum, thereby releasing intracellular calcium stores.

input stimulus A substance or change in state that evokes a response in a cell.

integrins A family of cell-surface adhesion receptors that bind to cell-matrix- or cell-surface-associated proteins, such as fibronectin, laminin, and fibrinogen.

interaction domain Part of a protein that mediates interactions with other molecules.

intracellular receptors Sensor molecules within cells that bind to signaling molecules and transmit a response.

intrinsic apoptotic pathway The apoptotic cell death pathway induced by signals generated within the cell, such as stress responses. Involves permeabilization of mitochondrial outer membranes and release of cytochrome c and other components that lead to assembly of the apoptosome.

intrinsically disordered regions (IDRs) Regions of protein structure that do not fold into globular domains, but instead are flexible and sample many different conformational states. IDRs typically are enriched in small linear binding motifs, and in sites of post-translational modification.

IP$_3$—*see* **inositol 1,4,5-trisphosphate**

isothermal calorimetry (ITC) Analytical method using a highly sensitive calorimeter that provides thermodynamic parameters of the binding of two molecules.

ITAM—*see* **immunoreceptor tyrosine-based activating motif**

ITC—*see* **isothermal calorimetry**

JAK-STAT pathway The signaling pathway activated by the cytokine/hematopoietin receptor family, in which receptor engagement leads to activation of JAK family tyrosine kinases and the subsequent phosphorylation and activation of STAT family transcription factors.

juxtacrine signaling Signaling that involves direct contact between two adjacent cells or a cell and the extracellular matrix.

karyopherin A class of transport proteins that function to transport cargo proteins into or out of the nucleus; cargo binding is regulated by the Ran G protein.

k$_{cat}$—*see* **catalytic rate constant**

K$_d$—*see* **dissociation constant**

kinetic proofreading A signaling mechanism that involves a reaction with multiple steps, so the output is dependent on relatively sustained input (such as ligand binding or membrane localization). Such systems suppress noise, because chance encounters or random activation events are unlikely to lead to output.

kinetochore Structure formed by proteins of the mitotic chromosome that attaches to the microtubules of the mitotic spindle.

K$_m$ The Michaelis constant, which is the substrate concentration at which the reaction rate is one-half of V_{max}. For many enzymes, K_m is similar or equal to K_d, the affinity of the enzyme for the substrate.

ligand Small molecule or macromolecule that recognizes and binds to a macromolecule.

ligand-gated channel A membrane channel whose opening is controlled by the binding of specific ligands.

linear response (graded response) A system where the response is proportional to the input signal.

link The regulatory relationships between individual components (nodes) in a signaling network. Links are positive when the action of the upstream node on the downstream node results in activation, and negative when the result is repression/inhibition.

lipid bilayer This consists of two layers of polar lipids, arranged in a sheet with the hydrophobic chains oriented inward and the polar head groups on the surface of the sheet facing the aqueous environment.

lipid raft Local lipid domains, enriched in sphingomyelin and cholesterol, that are highly ordered and yet allow a high degree of lateral mobility.

liquid–liquid phase separation (LLPS) The transition from a homogeneous liquid solution to a state with lower free energy, in which some components segregate into a distinct, more concentrated liquid phase that co-exists with a less dense dilute phase.

local concentration The effective concentration of a component at a specific site, such as in the immediate vicinity of a second component.

logic gate Device that specifies output depending on the combination of two inputs.

lysine acetyl transferase (KAT)—*see* **histone acetyl transferase (HAT)**

lysine deacetylase (KDAC)—*see* **histone deacetylase (HDAC)**

lysine demethylase (KDM) Enzyme that removes methyl groups from methylated lysine residues within a protein.

lysine methyl transferase (KMT) Enzyme that transfers a methyl group from *S*-adenosyl methionine (SAM) to the terminal amino group of lysine residues within a protein.

lysophospholipid A glycerophospholipid from which one of the two fatty acid chains has been cleaved.

lysosome A membrane-enclosed intracellular compartment where the protein and lipid components of endocytosed vesicles are broken down enzymatically by proteases and lipases.

major histocompatibility complex (MHC) Proteins that present antigen peptides on the cell surface, where they can be recognized by T cells. Most peptides displayed by the MHC are self peptides, which do not elicit an immune response. However, non-self peptides (for example, from pathogens or from tumor cells expressing mutant proteins) can strongly activate the adaptive immune system.

MAP kinase (mitogen-activated protein kinase) An important family of serine/threonine kinases that are activated by upstream signals and which phosphorylate targets such as transcription factors. MAP kinase is the third in a three-kinase series (the MAP kinase cascade).

MAP kinase (mitogen-activated protein kinase) cascade A pathway module, consisting of three kinases that act in series, utilized in a remarkably wide variety of cellular responses. The essential components are a MAP kinase kinase kinase (MAPKKK), which phosphorylates and activates a MAP kinase kinase (MAPKK), which in turn phosphorylates and activates a MAP kinase (MAPK). MAP kinases often phosphorylate nuclear targets such as transcription factors. The three kinases of the cascade are often co-localized in a single multiprotein complex by scaffold proteins.

mass spectrometry (MS) Analytical method that separates molecules such as peptides on the basis of their mass-to-charge ratio, providing extremely accurate information on their atomic masses.

mechanistic target of rapamycin (mTOR) A protein serine/threonine kinase that acts as a master regulator of cell growth, survival, and metabolism.

membrane channels Protein pores in a lipid membrane that allow passage of hydrophilic molecules such as ions through the membrane.

Michaelis–Menten constant—*see* K_m

Michaelis–Menten equation An equation describing the velocity of an enzymatic reaction as a function of enzyme and substrate concentration.

microarrays High-throughput analytic method in which many protein or nucleic acid samples are arrayed as tiny spots on a solid support, which can then be probed with labeled binding partners or cell lysates.

mitogen Extracellular molecule that stimulates cell proliferation.

mitogen-activated protein kinase—*see* MAP kinase

mitosis Division of the nucleus of eukaryotic cells, such that each resulting nucleus contains one copy of each chromosome.

modular domain Domain found in many different proteins in the same organism that confers a specific function or activity.

molecular memory The conversion of a transient input into a permanent (or semipermanent) change in output.

motif Conserved peptide sequence that is recognized specifically by interaction domains.

MS—*see* mass spectrometry

mTOR—*see* mechanistic target of rapamycin

multivesicular body (MVB) Organelle to which receptor–ligand complexes are transported for sorting to either lysosomes or the cell surface.

***N*-acetylation** The transfer of an acetyl group to the terminal amino group of lysine residues in proteins.

necrosis Cell death induced by physical trauma or acute stress, in which cells spill their contents into their surrounding tissues. This tends to evoke strong inflammatory responses, which may be harmful to surrounding cells.

negative feedback When the output from a signaling node follows a path of links that results in negative regulation of the original node.

neoantigens Immunogenic molecules not found in normal, unmutated cells. When displayed on the cell surface by MHC, they could be recognized by host T cells as foreign.

NES—*see* nuclear export signal

network architecture The topology of a signaling network, defined by the specific links between molecular nodes and the sign of these links (positive or negative).

network Linked system of multiple signaling molecules that regulate one another.

NF-κB Transcription factor that is present in a latent form in the cytosol of unstimulated cells and which is translocated to the nucleus upon activation.

nitric oxide (NO) A small diatomic gas that acts as a signaling mediator. It passively diffuses into responding cells and directly activates guanylyl cyclase. Nitric oxide plays an important role in regulating smooth muscle relaxation and blood vessel dilation.

NLS—*see* nuclear localization signal

***N*-methylation** The transfer of methyl groups to amino groups of proteins.

***N*-myristoylation** Irreversible attachment of a myristoyl group to the N-terminal glycine of a protein via an amide linkage.

node Individual signaling component in a signaling pathway or network.

Notch A family of cell-surface receptors that are proteolytically cleaved upon ligand binding and that often participate in cell-fate determination during development.

nuclear export signal (NES) A short modular peptide motif that mediates nuclear export by binding to an exportin.

nuclear localization signal (NLS) A short modular peptide motif that mediates nuclear import by binding to an importin.

nuclear magnetic resonance (NMR) spectroscopy Method for determining protein structure and conformation; based on the resonant absorption of electromagnetic radiation at a specific frequency by atomic nuclei in a magnetic field, due to flipping of the orientation of their magnetic dipole moments.

nuclear pore complex A very large multiprotein complex that regulates the passage of macromolecules through the nuclear pore, and thus into and out of the nucleus.

nuclear receptor (NR) superfamily Intracellular receptors for hydrophobic signaling molecules such as steroid hormones. The receptor–ligand complex acts as a transcription factor.

nucleation promoting factors (NPFs) Enzymes that catalyze the initial addition of actin subunits to the Arp2/3 complex, thus nucleating new branches on existing actin filaments. NPFs serve as the primary mechanism for regulating where and when in the cell dendritic actin networks will form.

nucleosome The basic unit of chromatin, consisting of eight histone subunits arranged in a disclike structure, around which is wrapped ~147 base pairs of DNA. A typical nucleosome contains two molecules each of histone 2A (H2A), histone 2B (H2B), histone 3 (H3), and histone 4 (H4).

nucleotide-binding and leucine-rich repeat receptors (NLRs) Key components of innate immunity in animals and plants, which recognize foreign molecules and assemble into multimeric structures to promote host defense. They typically consist of three domains: a leucine-rich repeat (LRR) region that recognizes specific substrates, a nucleotide-binding domain, and a more variable N-terminal domain, which can include coiled-coil or TIR domains.

***O*-methylation** The transfer of methyl groups to oxygen atoms of protein side chains such as that of glutamate. Important for regulating prokaryotic systems such as bacterial chemotaxis.

oncogene A gene that when mutated or disregulated can lead to the uncontrolled cell growth characteristic of cancer. Oncogenes can be activated by mutation (point mutation, deletion, or truncation) or by an increase in expression level. Oncogenes act dominantly (they exert their effect even when the normal version of the oncogene is present in the cell).

operon A locus in bacterial genomes where genes needed for a particular task are controlled by a single promoter and a set of regulatory elements, leading to expression of a polycistronic mRNA encoding multiple proteins.

optogenetic (methods) These take advantage of light-sensitive proteins and domains, originally found in plants or protists, that when illuminated with light of a specific wavelength, will undergo conformational changes that result in functional changes. The aim is to achieve local control over specific signaling inputs using light.

orphan receptors Receptors for which the relevant physiological ligand(s) has not yet been identified.

oscillation Fluctuation of output levels between states of high and low activity in a periodic manner.

output responses The responses of a cell to signaling input.

p53 A master regulator of cellular responses to a wide range of environmental stresses such as DNA damage. Depending on the specific stress and cellular context, p53 can induce momentary cell-cycle arrest, or permanent cell-cycle arrest and apoptosis. Prevents cells from replicating inappropriately and passing on damaged genomic DNA. Most frequently mutated gene in human cancers.

paracrine Local signaling to nearby or adjacent cells.

pathway A linear chain of interactions where the output of each node serves as the input for the next downstream node.

PCA—*see* **protein-fragment complementation assay**

PH domain Modular protein domain, many of which bind to specific phosphoinositol-derived lipids.

phase separation Segregation of a homogeneous mixture into two distinct physical phases, such as when oil and water separate into distinct layers after mixing.

phase transition The physical process by which matter changes from one physical state to another, such as water changing from solid to liquid, or from liquid to vapor.

phorbol esters Organic compounds that mimic the structure of diacylglycerol (DAG) and thereby promote the activation of PKC *in vivo*.

phosphatidic acid (PA) A glycerophospholipid in which the head group consists only of phosphate. Can be generated either from other glycerophospholipids by the action of phospholipase D or from diacylglycerol by the action of diacylglycerol kinase.

phosphatidylinositol (PI) A membrane phospholipid with the six-membered sugar inositol as its head group.

phosphatidylinositol 3,4,5-trisphosphate (PIP_3) A membrane phospholipid in which the inositol head group is phosphorylated on positions 3, 4, and 5.

phosphatidylinositol 3-kinase (PI3K) A signaling enzyme that adds a phosphate to the 3 position of PIP_2 to generate the membrane-bound second messenger PIP_3.

phosphatidylinositol 3-phosphate [PI(3)P] A membrane phospholipid in which the inositol head group is phosphorylated on position 3.

phosphatidylinositol 4,5-bisphosphate (PIP_2) A membrane phospholipid in which the inositol head group is phosphorylated on positions 4 and 5.

phosphoinositides Phospholipids where the head group is the six-member sugar inositol. Phosphates can be added and removed from specific positions on the ring via the action of specific lipid kinases and phosphatases.

phospholipase A_2 (PLA_2) An enzyme that cleaves at the *sn*-2 (middle) position of the glycerol backbone of a glycerophospholipid, generating a free fatty acid and a lysophospholipid.

phospholipase C (PLC) An enzyme that cleaves the phosphorylated head group from a phospholipid, generating diacylglycerol and the phosphorylated head group.

phospholipase D (PLD) An enzyme that cleaves the head group from a phospholipid, generating phosphatidic acid and the unphosphorylated head group.

phosphorylation The transfer of the terminal phosphate group from ATP to proteins or other molecules.

photoreceptor Cell or molecule that is sensitive to light.

phytohormones Plant hormones.

PI3K—*see* **phosphatidylinositol 3-kinase**

PIP_2—*see* **phosphatidylinositol 4,5-bisphosphate**

PIP_3—*see* **phosphatidylinositol 3,4,5-trisphosphate**

PKA—*see* **protein kinase A**

PKC—*see* **protein kinase C**

plasmodesmata Membranous tubules that connect the cytosol of adjacent plant cells. These structures, consisting of an outer membrane continuous with the plasma membrane of adjacent cells, and an inner core continuous with their endoplasmic reticulum, allow proteins, nucleic acids, and other molecules up to ~70 kDa in size to move freely between cells.

podosome Actin-rich cell-surface protrusion that mediates adhesion and invasion in normal cells (a similar structure found in tumor cells is termed an invadopodium).

polarity Functional or structural asymmetry in a cell.

polyproline type II helix (PPII helix) A left-handed helical structure with three amino acids per turn that forms spontaneously in proline-rich peptides.

positive feedback When the output from a signaling node follows a path of links that results in positive regulation of the original node.

postsynaptic density (PSD) A specialized cellular substructure on the postsynaptic side of a neuronal junction, containing neurotransmitter receptors and other signaling proteins.

post-translational modifications Covalent modifications of proteins that are added and removed by specific enzymes after synthesis of the protein.

prenylation Irreversible attachment of either a farnesyl or geranylgeranyl lipid group to a protein via a thioether linkage.

primary cilium A specialized filamentous organelle constructed from microtubules. In many cells, it acts as a signaling center.

primary structure The simple linear sequence of amino acids in the polypeptide chain.

priming (multisite phosphorylation) Phosphorylation of a substrate by one kinase makes it a better substrate for phosphorylation by a second, distinct kinase.

processive (multisite phosphorylation) Phosphorylation of a protein on multiple sites by an enzyme that remains associated with its substrate.

proline hydroxylation Hydroxylation of the 4 position of the proline ring, generating 4-OH-proline. Proline hydroxylation of the transcription factor HIF-1α is an important component of the oxygen-sensing mechanism in metazoans.

prolyl *cis-trans* **isomerization** A switch in the conformation of a proline residue caused by rotation around the peptide bond. Spontaneously, this reaction proceeds very slowly, but it can be speeded up greatly through the action of peptidyl prolyl *cis-trans* isomerase (PPIase).

proteases The enzymes that cleave the peptide bonds of proteins.

proteasome A large multiprotein structure that mediates the proteolysis of cytosolic proteins. It consists of a hollow cylinder, lined on the inside with proteases, capped at each end by a "lid" structure that controls access to the proteolytic machinery of the inner chamber.

protein arginine methyl transferase (PRMT) Enzyme that transfers a methyl group from *S*-adenosyl methionine (SAM) to arginine residues within a protein.

protein kinase A (PKA) A protein serine/threonine kinase that is activated by cyclic AMP.

protein kinase C (PKC) A family of related serine/threonine protein kinases, whose activation is variously dependent on calcium and diacylglycerol (DAG).

protein kinase Enzyme that covalently modifies proteins by the addition of a phosphate group.

protein phosphatase Enzyme that removes phosphate groups from proteins.

protein trafficking The targeted transport of proteins from one subcellular location to another within the cell, generally by membrane vesicles.

protein-fragment complementation assay (PCA) Technique for detecting protein interactions in living cells, based on the reconstitution of a functional reporter molecule (such as green fluorescent protein) when two proteins bind to each other.

proteolysis The cleavage of the peptide bonds that link individual amino acid residues in a protein.

proteostasis The dynamic process whereby rates of protein synthesis, folding, modification, and degradation are regulated in order to maintain homeostasis of the proteome.

PTEN A lipid phosphatase that opposes the activity of PI 3-kinase. PTEN converts phosphatidylinositol 3,4,5-trisphosphate (PIP_3) to phosphatidylinositol 4,5-bisphosphate (PIP_2). PTEN counteracts the growth and survival signaling induced by PIP_3, and thus acts as a potent tumor suppressor.

PTP (protein tyrosine phosphatase or phosphotyrosine phosphatase)—*see* **tyrosine phosphatase**

pull-down assay A protein or protein fragment is produced in quantity and bound to beads, and then used to detect binding partners in cell extracts.

quaternary structure The arrangement of multiple subunits within a protein complex.

quorum sensing The process by which bacteria exchange information with other bacteria in a community such as a biofilm. Individual bacteria secrete signaling molecules called autoinducers (AI) that are bound by receptors and can induce transcriptional responses, both in the secreting cell and in other cells in the community. Quorum sensing typically regulates responses that would be costly for an individual cell, but can be beneficial to a larger population when it responds collectively.

Ras A small G protein that functions as a central regulator of cell proliferation and differentiation. Ras was first discovered as a viral oncogene, and it is frequently mutated and activated in human cancers.

reactive oxygen species (ROS) Highly reactive forms of oxygen (such as superoxide, hydrogen peroxide, and hydroxyl radical) that have the potential to cause oxidative damage to the cell.

receptor A protein that, when bound to a specific signaling molecule (ligand), undergoes a change in activity that transmits a signal.

receptor tyrosine kinases (RTKs) Single-pass transmembrane receptors with intracellular protein tyrosine kinase domains.

regulated cell death A process of cell death that is actively regulated by signaling pathways, which respond to specific triggers and then execute an orderly program of self-destruction. Regulated cell death includes diverse processes such as apoptosis, necroptosis, pyroptosis, and autophagy-dependent cell death.

regulated intramembrane proteolysis (RIP) Sequential proteolytic cleavage of a membrane protein, first by ADAM-mediated ectodomain cleavage, followed by further processing within the membrane by the γ-secretase complex.

regulator of G protein signaling protein (RGS protein) Protein that binds to the free GTP-bound α subunit of a heterotrimeric G protein and stimulates its GTPase activity.

response regulator (RR) The second (effector) component of two-component systems in bacteria and simple eukaryotes. After activation of a histidine kinase by incoming signals, phosphate is transferred to aspartate side chains of the response regulator, inducing conformational changes. Many response regulators are DNA-binding proteins that regulate transcription.

RGS protein—*see* **regulator of G protein signaling protein**

RNA-seq—*see* **cDNA sequencing analysis**

robustness The ability of a biological system or network to maintain its performance under a wide range of conditions and to be relatively insensitive to perturbations of the system.

S-acylation Reversible attachment of a fatty acid group to a protein via a thioester linkage; palmitoylation is an example of *S*-acylation.

scaffold proteins Proteins that bind multiple proteins, such as enzymes and their substrates, involved in a single process.

Scatchard analysis Method of quantifying the binding of a labeled, soluble analyte to an immobilized binding partner. Binding data can be plotted in the form of a straight line, the slope of which reveals the K_d of the interaction. This graph is called a Scatchard plot.

SCF complex A multisubunit ubiquitin ligase complex consisting of an E3 ubiquitin ligase, a Skp1 adaptor, a cullin, and an F-box specificity factor. Among other roles, the SCF complex targets phosphorylated substrates for degradation during cell-cycle transitions.

second messengers—*see* **small signaling mediators**

secondary structure Local, regular structural elements in proteins, primarily α helices and β strands.

serine/threonine kinase A protein kinase that phosphorylates serine or threonine residues on its substrates.

serine/threonine phosphatases A protein phosphatase that dephosphorylates phosphoserine or phosphothreonine residues on its substrates.

SH2 domain Modular protein domain that binds to tyrosine-phosphorylated peptides.

SH3 domain Modular protein domain that binds to proline-rich peptides that adopt a specific helical secondary structure.

signal amplification A property of signaling mechanisms where a single activated enzyme molecule can generate many molecules of product, thereby amplifying the output of the system.

signal integration The integration of multiple signaling inputs to produce a single output.

single-cell RNA-seq (scRNA-seq) A method in which single cells or nuclei are isolated from a population, and the mRNA from each individual cell is reverse-transcribed, amplified, and sequenced. Bioinformatic analysis can reveal different sub-populations of cells based on their RNA expression patterns.

sister chromatids A pair of identical chromosomes generated by DNA replication during S phase. One of the pair is segregated to each daughter cell during mitosis.

small G protein The large and diverse family of G proteins that consist entirely of a G domain, in the absence of other domains or subunits. Small G proteins are distinct from the other major class of G proteins, the heterotrimeric G proteins.

small signaling mediators Small, highly diffusible molecules that carry signaling information within cells.

soluble guanylyl cyclase (sGC) An enzyme that converts GTP into cGMP and pyrophosphate; the primary intracellular effector of nitric oxide signaling.

S-palmitoylation Transfer of the 16-carbon fatty acid palmitic acid to cysteine residues of a target protein. Unlike most other lipid modifications, S-palmitoylation is relatively dynamic.

specificity constant A measure of the efficiency of an enzyme for a particular substrate, defined as k_{cat}/K_m.

specificity The degree of selectivity for one partner or class of partners relative to a set of competing interactions.

sphingomyelin An abundant sphingolipid consisting of ceramide linked to a phosphorylcholine head group. A major component of the outer leaflet of the plasma membrane, it can be metabolized to generate a variety of bioactive lipids.

SPR—see surface plasmon resonance

Src family kinases A group of nonreceptor tyrosine kinases that regulate processes such as adhesion and lymphocyte activation.

standard free energy ($\Delta G°$) The change in free energy associated with formation of a compound, under standard conditions of concentration, temperature, and pressure.

STAT A family of transcription factors that are rapidly activated by cytokine and growth factor receptors. Phosphorylation by JAK family kinases leads to dimerization, nuclear localization, and DNA-binding activity.

state machines Devices that can exist in multiple discrete states and can change their state in response to specific instructional inputs.

stoichiometry the relative amounts of each of the individual components of a molecule or complex, for example, the number of each type of subunit present in a multimeric complex.

substrate specificity The selectivity of an enzyme to react with certain substrates over others. The degree of specificity of an enzyme for two substrates can be described quantitatively by the relative values of k_{cat}/K_m for two different substrates.

surface plasmon resonance (SPR) Analytical method for determining quantitative binding parameters. Uses an instrument that can monitor over time the extent and kinetics of binding and dissociation of a macromolecule to a second macromolecule fixed to a surface.

switch I and II regions Regions of a G protein that alter their conformation upon binding to the γ-phosphate of GTP.

switchlike response A nonlinear or all-or-none response in which the system does not respond to a stimulus until it reaches a threshold value, at which point the system responds with maximum output.

synthetic biology The use of natural biological components to build and engineer novel functional systems.

targeted therapies In cancer treatment, pharmaceuticals that specifically target known molecules or pathways that contribute to the disease. Examples include inhibitors of constitutively active oncogenic kinases, and immune checkpoint inhibitors.

tertiary structure Three-dimensional folded structure of a polymer chain such as a protein or RNA.

Toll-like receptors (TLRs) A family of receptors that bind pathogen-specific ligands and indirectly activate the NF-κB family of transcription factors.

transcription factor A protein or protein complex that binds to a gene near its promoter and regulates transcription of that gene.

transcription The highly regulated process that generates messenger RNA (mRNA) from the corresponding genomic DNA.

transformation (of cells in culture) Transformation is a term used to describe a cancer-like phenotype in cultured cells. Compared to normal cells, transformed cells typically are less dependent on serum and other mitogens for proliferation; are less adherent and more rounded in shape (more refractile); and have the ability to grow under conditions where normal cells typically cannot, such as when they are crowded and touching other cells, and when they are unattached to a solid substrate.

transition state The species with the highest free energy either in a reaction or in a step of a reaction.

transmembrane receptor A transmembrane protein that binds to extracellular ligands and transmits information into the cell by ligand-induced changes in conformation and/or enzymatic activity.

transphosphorylation The phosphorylation of one kinase molecule by another in a dimer or multiprotein complex.

tumor suppressor Gene product that helps prevent the formation of cancer by antagonizing cell proliferation and survival pathways; loss of both copies of the tumor suppressor can lead to cancer.

two-component system A common signaling system in prokaryotes, also found in some eukaryotes such as plants and fungi. Two-component systems are composed of a histidine kinase linked to a receptor and a response regulator. Activation of the histidine kinase by the receptor leads to its autophosphorylation on histidine. Phosphate is then transferred to aspartate on the response regulator, leading to conformational changes that transmit the signal.

type III secretion system (T3SS) A plasmid-encoded mechanism used by some gram-negative bacteria to transfer bioactive molecules such as effector proteins into host cells. Transfer is mediated by a hollow syringe-like apparatus structurally related to the flagellum, which spans the two bacterial membranes and the plasma membrane or endocytic vesicle of the host cell.

tyrosine kinase A protein kinase that phosphorylates tyrosine residues on its substrates.

tyrosine phosphatase A protein phosphatase that dephosphorylates phosphotyrosine residues on its substrates.

ubiquitin A 76-residue protein that is enzymatically added via its C-terminus to lysine side chains in a target protein, generating an isopeptide bond linking the two. Additional ubiquitin units can be added to lysine side chains or the N-terminus of each ubiquitin, generating long chains (polyubiquitylation).

ubiquitin-binding domain (UBD) One of several structurally distinct small binding domains and motifs that bind specifically to ubiquitin, usually to a hydrophobic patch of ubiquitin centered around Ile44.

ultrasensitivity When a relatively small change in input leads to a much larger than proportional change in output.

voltage-gated ion channel Ion channel across a membrane that only allows the passage of ions in response to a change in voltage across the membrane.

Western blotting—*see* **immunoblotting**

Wnt signaling pathway A conserved pathway regulating key developmental events. Activation of canonical Wnt pathway results in transcription mediated by β-catenin. Transduction mechanism involves protein phosphorylation and regulated protein degradation.

X-ray crystallography Technique for determining the three-dimensional arrangement of atoms in a molecule based on the diffraction pattern of X-rays passing through a crystal of the molecule.

Y2H—*see* **yeast two-hybrid assay**

yeast two-hybrid (Y2H) assay Molecular genetic technique for finding proteins that interact with a protein or protein fragment of interest.

zero-order ultrasensitivity When an increase in the amount or activity of the forward modifying enzyme leads to an ultrasensitive increase in the amount of the modified target protein. Occurs when both forward and reverse enzymes are fully saturated with substrate.

zymogen An inactive precursor form of an enzyme (usually a protease), which must be processed by proteolytic cleavage in order to be activated.

Index

Note: **Bold** page numbers refer to tables and *italic* page numbers refer to figures.

14-3-3 proteins **235**
  control of nuclear transport 112
  dimerization 247, *247*
  prevention of dephosphorylation 91
  recognition of phosphorylated sites 89, 242
  regulation of kinase activity 254, *254*
  regulation of subcellular localization 243, *243*
19S regulatory particle 212–213, *212*
26S proteasome *212*, 213
53BP1 371

Abl
  association with membrane 116
  oncogenic fusion proteins 257–258, *257*
accessory domains/subunits
  regulating kinase substrate specificity 51, *51*, 52
  regulating phosphatase activity 58
acetylation 78–79, *78*
  chemical effects 81, *82*
  histone *see under* histone(s)
  p53 87
  protein domains recognizing 243–244, *244*
N-Acetylation 78, *78*
N-Acetyl glucosamine (GlcNAc) 78, *79*
ActA *313*
actin cytoskeleton
  integrin-mediated adhesion *181*, 182–183
  regulation by phospholipase D 159
  studying changes in *400*
actin polymerization by nucleation. *310*
actin propulsion *314*
actin-related protein 2/3 (Arp2/3) complex 271, *272*
activation loop, protein kinase 47, *48*
  phosphorylation 49–50, *49*
  transphosphorylation 171–172, *172*
active state 14, *14*
S-Acylation **114**, 115, *115*
acyl-homoserine lactones (AHLs) 329
ADAM-10 208–209, *209*, 211
ADAM-17 209, 211–212
ADAM metalloproteases 208–210
  domain structure *209*

Eph-ephrin signaling 209, *209*
  Notch signaling 187
  regulated intramembrane proteolysis 211–212, *212*
ADAMTS proteins *209*
adaptability 2–3
adaptation 5, 283–285, *283*, *285*
  visual system 353
adaptive immune response 374
adaptor proteins 251, *251*
adenosine triphosphate (ATP)
  analogs 403–404, *403*
  driving phosphorylation/dephosphorylation 41, *41*
  hydrolysis 323
adenoviruses 316
adenylyl cyclase 130, 134–135, *134*, *135*
affinity 22–23; *see also* dissociation constant
  classification 29–31
  cooperative binding and 32–33, *35*
  effect of local concentration 29, *29*
  effect of multiple binding sites 28–29, *28*
  factors determining ideal 29
  functional constraints 29–31, *30*
  independent modulation 31–32, *32*
  range 29–31
affinity tags 398
agonists 180
AI receptors 329
A-kinase anchoring proteins (AKAPs) 137, 143–145, *144*
Akt kinases
  activation by membrane recruitment 117–118, *118*
  mTOR pathway 160
  PH domain function 248
all-or-none responses *see* switchlike responses
allosteric conformational changes 43–44, 392
  as consequence of binding 20–21
  diversity of types 44, *44*
  lipid-modified proteins 116
  as mechanism of signal transmission 40
  regulation 44, *44*
allosteric regulation 44, *44*
allosteric switch proteins 253–256, *253*

engineered 258–258, *259*
  signaling functions 267–268, *268*
α helices 393, *394*
  armadillo repeats 233, *234*
  effects of phosphorylation 45, *45*
  gated ion channels 188–189, *189*
  structure 393, *393*
  transmembrane receptors 170, *170*
Alzheimer's disease 187, 212
amino acid sequence 393, *393*
amphipathic molecules 148, 150
amplification, signal *see* signal amplification
amyloid β 211–212
amyloid precursor protein (APP) 187, 211, *212*
analytical methods 389–391
anaphase-promoting complex (APC) 214, 215–217, 369
  regulation by phosphorylation 215–217, *216*
  structure *214*, 215
  Xenopus cell cycle 286, *287*
AND gates, biological 269–270, *270*
  coherent feed-forward loops 279–280, *280*
  cyclin-dependent kinases 271
  modular signaling proteins 271–272, *272*
  transcriptional promoters 273, *273*
AND NOT gate, biological *270*
androgens 86, *87*
angiogenesis 195
animals, emergence of 340–342, *341*
ankyrin (ANK) repeats **235**
  NF-κB proteins 217, *217*
antagonists 180
ANTH/CALM domain **235**
anti-apoptotic factors, overexpression of 300
antibodies
  avidity of binding 28–29, *28*
  recognizing post-translational modifications 405–406, *407*
  specificity of phosphotyrosine binding 23, *23*
antigen 374; *see also* peptide-MHC complexes
  antibody binding 28, *28*
antigen-presenting cells 374, 375–376, 380
AP-1 273, *273*, 378

Apaf1 227, *227*
APC *see* anaphase-promoting complex (APC)
Apc2 *214*
APC protein 184, *184*
apoptosis 86, 219–220, 295
    extrinsic pathway 222–225, *223*
    intrinsic pathway 225–228, *226*
    irreversibility 206
    morphological features 219–220, *220*
    phosphatidylserine (PS) signaling 150
    receptor signaling via proteolysis
        185–186, *186*
    signals initiating 219, 220
apoptosis inducing factor (AIF) 228
apoptosome 221, 225, 227, *227*
AQUA (absolute quantification) 409
Arabidopsis 334
AraC 326, *326*
arachidonic acid (AA) 153, 163
    metabolites 163–164, *164*
archaea 325
Arf proteins 61, **61**
arginine finger 66–67, *67*
arginine methylation 100–101
armadillo repeats 233, *234*
ARM repeat **235**
Arp2/3 complex 271, *272*
arrestins
    GPCR desensitization 199–200, *200*,
        201, *201*
    hedgehog signaling 185, *186*
    rhodopsin desensitization 354
ASC 224, *224*
aspartate phosphorylation 93–94, *94*
aspartic acid, phosphorylation of 322
assemblies, protein *see* complexes, protein
association constant (Ka) 25
ATM/ATR kinases 371
ATP *see* adenosine triphosphate (ATP)
atrial natriuretic peptide (ANP) receptor
    179, *180*
aurora-B kinase 270, *270*
autocrine signaling 381
autoimmune disorders 293
autoinducers (AI) 329
automatic door analogy 264–266, *265*
auxin signaling 335, *335*
avidin-biotin interaction 27
avidity 28–29, *28*
axin 184, *184*, 185

B7 376, 383
bacteria 325
    chemotaxis *see* chemotaxis, bacterial
    populations 329–330, *330*
    two-component systems 93–94, *93*
    types of 329
BAD *225*, 226
BAK 225–226, *225*
BAR domains **235**, 249–250, *249*
BAX 225–226, *225*, *226*, *227*
Bcl2 226, *226*
Bcl2 family proteins 225, *225*
    activation 226–227, *227*
    anti-apoptotic 225–226, *225*
    induction of apoptosis 225, *226*, *227*

    pro-apoptotic 225–226, *225*
Bcl homology domains *see* BH1-BH4
    domains
Bcl-XL 225, *225*, 243
BCR-Abl fusion protein 257–258, *257*
BEACH domain **235**
Bem1 71, *72*
β-adrenergic receptor kinase (β-ARK; GRK2)
    199–200
β-adrenergic receptors *175*
beta blockers 180
β-catenin
    armadillo repeat domain 233, *234*
    Wnt signaling 184–185, *184*
β sheets 393, *393*
β strands 393, *393*
    WD40 repeats 233, *234*
β-TrCP 184, *186*, 218
BH1-BH4 domains 225, *225*, **235**
BH3-only proteins 225, *225*
    activation 226–227
    induction of apoptosis 226, *227*, 228
BID 225, *225*, 226
BIM 225, *225*, 226
binding 19, 20–21
    affinity *see* affinity
    cooperative *see* cooperativity
    direct and indirect consequences 20–21
    effects of post-translational modifications
        81–82, *82*
    interaction surfaces 21–22, *21*
    K_d *see* dissociation constant
    parameters, determination 389–391, *390*
    specificity *see under* specificity
    thermodynamics 25–26, *33*
binding isotherm 24, *24*, 276
binding probability 24
binding rate constant *see* on-rate
binding sites, total number of 390, *390*
biochemical analysis 389–397
biofilms 329
bioinformatics, identifying protein domains
    232–233, *233*
biophysical analysis 389–392
biosensors 411–412
    calcium-binding fluorescent dyes 141, 411
    fluorescent protein fusions 401, 411–412
    FRET-based 397, 410–411, *411*
BIR domains 222, **235**
bispecific engager 306
bistable systems 288, *289*
    flow cytometric analysis 413–414, *413*
black death *308*
blood coagulation *see* coagulation, blood
BRCA1 98
BRCT domain **235**
bromodomains **235**, *237*, 243, *244*
BTB/POZ domain **235**
buried surface area 21, *21*

C1 domain **235**
C2 domain **235**, 242
CAAX box 115
Caenorhabditis elegans 7, 219
    programmed cell death 219
    vulval development 7–8, *8*

calcineurin (PP2B) 54, *56*
    T cell signaling 273, *273*, 378
calcium (Ca^{2+}) **133**
    biosensors 141, 411
    clotting cascade 206–207, *207*
    influx into cells 140–141
    protein kinase C activation *138*, 139
    signaling 130, 139–143, *140*
    visual transduction cascade 354, 355
    waves 142–143, *142*
calcium (Ca^{2+})-binding proteins 141
calcium (Ca^{2+})/calmodulin 141, *141*; *see also*
    calmodulin
    control of cAMP degradation 134
    nitric oxide release 133
calcium/calmodulin-dependent protein kinase
    (CaMK) 53, 142
    II isoform (CaMKII) 252
calcium (Ca^{2+}) channels 139–140, *140*
calcium (Ca^{2+}-ATPase) pumps *140*, 141
calmodulin (CaM) *140*, 141–142, *141*; *see*
    also calcium (Ca^{2+})/calmodulin
    fused Ca^{2+} biosensors 141
    NMR spectroscopy *396*
cAMP *see* cyclic AMP
cancer
    biology *6*, 7–9
    cell cycle regulators 298–299, *299*
    engineered T cells 305–307, *305*
    forms of 293
    immune checkpoint inhibitors
        304–305, *304*
    immune evasion by 301, *301*
    immune surveillance by immune
        checkpoint activation 301–302, *301*
    loss of tumor suppressors-signaling
        molecules 300, *300*
    malfunctions 294–296, *295*, **296**
    mitogenic and growth 296–298, *297*
    mitogenic signaling proteins 297
    modular domain rearrangements 257–258,
        257
    signaling defects 295, **296**
    therapies 302–304, *303*
carbon monoxide (CO) 194
CARD domains 223, 224, *224*, **235**
CAR T cells *305*, 306
cascades, signaling enzyme 68–72
    signal amplification 281, *281*
    ultrasensitive switchlike responses
        277, *278*
casein kinase 1 (CK1) *53*
    hedgehog signaling 185, *186*
    Wnt signaling 184, *184*, 185
catabolite activator protein (CAP), regulation
    of 328
caspase-activated DNase (CAD) 222
catalysis, enzymatic 40–43
    driving reactions in one direction
        40–41, *41*
    quantitative analysis 391–392, *392*
    thermodynamic mechanisms 41–43, *41*
catalytic domains 231
    allosteric switch proteins 253–256, *253*
    GAPs 64–65, *65*
    GEFs 63–67, *64*, *65*

protein kinases 47–48, *48*
protein phosphatases 54, 59, *59*
structure 231
catalytic rate constant (k$_{cat}$) 42, 392, *392*
Caulobacter crescentus 94
caveolin-mediated endocytosis 119, *119*
inhibiting TGFβ signaling 120–121, *120*
Cbl 96, *250*
oncogenic forms 199
phosphotyrosine-binding domains 241, *241*
receptor down-regulation 197, *198*
CD3 379
CD4 co-receptor 180, 377
CD8 co-receptor 376
CD28 co-stimulatory receptor 376, 380
CD45 377
Cdc4, WD40 repeats *234*
Cdc20
complex with APC (APC^{Cdc20}) 215–216, *216*, 369
robust oscillator 286, *287*
spindle assembly checkpoint 372–373
Cdc24 71, *72*
Cdc25 56
cell cycle regulation *271*, 286, *287*, 369
homology domain 64–65
targeting for proteolysis 217, 371
Cdc42 272, *272*
CDCP1 242
Cdh1-APC complex (APCC^{dh1}) 215–216, *216*, 217
Cdk1 369
Cdk7 102
CDKs *see* cyclin-dependent kinases
cDNA expression libraries 399
cDNA microarrays 405–406, *406*
cDNA sequencing analysis (RNAseq) 405, *406*
cell(s); *see also* living cells
adhesion to extracellular matrix 182–183, *182*
environmental cues received by 168, *168*
extracts, isolation of proteins 398, *398*
polarity 107
as state machines 265, *265*
cell-cell adhesion 179
cell-cell communication 331
cell-cell signaling complexes 252–253, *252*
cell cycle 363–373
checkpoints 371
control mechanisms *315*
inhibitors 299
irreversible commitment 206, 364
molecular network 365
oscillations 286–287, *287*
phases 364
phase transitions 364, 366
regulators, mutation of *299*
ubiquitylating complexes controlling 214, *214*, 369
cell death
necrotic 220, *220*
programmed *see* apoptosis
cell division 363
cell lysate 398–399

cell motility, small G proteins controlling 62, *62*
cell signaling 1–18, 293–317
biological functions 6–10
in cancer and cancer therapy 294–299
challenges 4–5, **5**
defects in 293, *294*
diversity of inputs and outputs 3–4, *3*
hierarchical organization 16–17, *16*, 265
information processing perspective 2–3, *2*
length and timescales of *322*
mechanism of vision 325
molecular basis 10–14
pathogenic microbes 307
spatial and temporal scales 9–10
cells, size of 322
ceramidase (CDase) 161, *162*
ceramide **133**, 138
downstream targets 163
metabolism 161–163, *162*
structure 148–149
ceramide 1-phosphate (C1P) 161, *162*
cGMP *see* cyclic GMP
channels, membrane 169
gated *see* gated ion channels
chaperones, affinity and specificity of binding 30, 31
CH domain **235**
CheA/CheB/CheR/CheY 284, *284*
checkpoint inhibitors 304
chemical dimerizers 404–405, *404*
chemotaxis, bacterial 93
negative feedback adaptation 284–286
receptors 172
Chk2 371
chloroplast to nucleus *332*
choanoflagellates 339
cholesterol 150
chemical structure *150*
membrane fluidity and 151, *151*
choline 159
chromatin 98–99
nuclear receptor binding 195–196
remodeling 98, *101*, 102
structure 98, *98*
chromatin immunoprecipitation (ChIP) 405, *406*
chromodomains **235**, *238*, 243
chromo-shadow domain **235**
chromosomal translocations 257–258, *257*
chronic myelogenous leukemia (CML) 257–258
cilium, primary 185, *186*
CK1 *see* casein kinase 1
Claritin® (loratadine) 170
clathrin-mediated endocytosis 119–120, *120*
GPCR desensitization 200
mediating TGFβ signaling 120–121, *120*
receptor down-regulation 197–198
cluster analysis *406*
coagulation, blood 206–208
clotting cascade *207*
physiological role 207, *207*
cofilin 210, *210*
coherent feed-forward loops 274–275, 279–281, *280*

phospholipase D 156, 161
coiled coil (CC) domains **235**
co-immunoprecipitation (co-IP) 398, *398*
combinatorial complexity 84–85
comparative genomics 324–325, *325*
complexes, protein 20; *see also* dimerization; oligomerization; protein-protein interactions
cooperative binding 32–33, *35*
dynamic molecular assemblies 35
electron microscopy 396–397
isolation from cell extracts 398–399, *398*
variability of stability and homogeneity 34–35
concentrations, protein 13
binding affinities and specificities and 29–31, *30*
local *see* local concentration
cones (photoreceptor) 347
conformational changes 12, 40, 43–44
allosteric *see* allosteric conformational changes
binding-induced 20–21, *20*
diversity of mechanisms 44, *44*
ligand-transmembrane receptor binding-induced 170–172, *171*
light-induced 350–351
methods of determining/analyzing 395–396
multiple protein substates 43–44, *44*
phosphorylation-induced 45, *45*
post-translational modification induced 81, *82*
regulating nuclear import of STATs 112, *112*
regulation of stability 44, *44*
conformation, protein 12, 20, 392–395; *see also* protein structure
methods of determining/analyzing 395–396
cool temperatures, receptors responding to 190
cooperativity 33–34
cAMP binding to protein kinase A 135
coupled SH2 domains 239–240, *241*
enzyme activation 392, *392*
functional consequences 33–34
homotypic/heterotypic 34, *35*
inducing switchlike behavior 34, 277, *277*
membrane binding 116, *117*
molecular mechanisms 33, *33*
negative 33
positive 33–35, *35*
cortactin *210*
CRKL adaptor *400*
cryo-electron microscopy (cryo-EM) 396–397
Csk *232*, 377
C-terminal peptide motifs, recognition by PDZ domains 246, *246*
CTLA-4 383
CUE domain **235**
Cul1 *214*, 215
cullins 214–215, *214*
cyan fluorescent protein (CFP) 397, 401, *401*

cyclic AMP (cAMP) 130–131, **133**, 134–136
  AKAP-mediated regulation of signaling 143–145, *144*
  diffusion rate 131, *131*
  downstream targets 130–131, 135, *135*
  regulation of protein kinase A 136–137, *136*
  synthesis and degradation 134–135, *134, 135*
cyclic AMP-dependent protein kinase *see* protein kinase A
cyclic AMP (cAMP) phosphodiesterase 134–135, *134, 135*
  AKAP interactions 143–145, *144*
cyclic GMP (cGMP) **133**, 134–135
  downstream targets 135, *135*
  nitric oxide signaling 193–194, *193*
  synthesis and degradation 134–135, *134, 135*
  visual transduction cascade 349–351
cyclic GMP (cGMP)-dependent protein kinases (cGK) 136
cyclic GMP (cGMP) phosphodiesterase 134–135, *134, 135*, 136
cyclic nucleotide binding domains (CNB) 135, *135*
cyclic-nucleotide-gated (CNG) channel 349–351
cyclic nucleotides **133**, 135, *135*
cyclin-dependent kinase (CDK) 299, 365–369
  allosteric signal integration 271, *271*
  phylogenetic relationships *53*
  regulation of catalytic activity 47, *48*, 366
  substrate specificity 51, *52*
  targeting proteins for destruction 85, *86*, 212–213, *213*
  Xenopus cell cycle 286, *287*
cyclin-dependent kinase (CDK)
  inhibitors 366
  proteolytic destruction 215, 368
cyclins 365–369
  allosteric activation 47, *48*
  docking sites 51
  regulation of CDK activity 366
  ubiquitin-mediated regulation 214, 215–217, *216*
cyclooxygenase (cox) 163, *164*
Cys-loop superfamily of ion channels *189*, 191
cytochrome *c*
  apoptosome assembly 226, *227*
  induction of release 225, *225*
cytokine receptors
  activation of coupled tyrosine kinases 179–180, *181*
  nuclear import of activated STATs 112
cytokines 168, 223–224
cytokinins 332
  signaling *333*
cytosol 4
cytotoxic (killer) T cells 374, 376

death domain (DD) 183, 187, 222, **235**
death effector domain (DED) 188, *188*, 223, **235**

death-inducing signaling complex (DISC) 188, *188*, 221
  activation of apoptosis 222
death receptors
  activation of apoptosis 222–225, *223*
  signaling via proteolysis 185–187, *186*
delta, regulated intramembrane proteolysis 211–212, *212*
delta-serrate-Lag2 (DSL) *187*
dendritic cells 374–376
DEP domain **235**
dephosphorylation
  membrane lipids 154
  protein *see* protein dephosphorylation
desensitization, receptor 197, *197*
deubiquitinases (DUBs) 96, 213
developmental biology *6, 7*
diabetes 293
DIABLO/Smac 222, 228
diacylglycerol (DAG) **133**, 137, 138–139
  biophysical properties 152
  biosynthesis 138–139, *138*, 153–154, *154*
  phosphorylation 154–155
  regulation of protein kinase C activation 138–139, *138*
  T cell activation 378
diacylglycerol (DAG) kinases 153–154
diffusion
  membrane fluidity and 151
  within membranes 151–152, *152*
  small signaling mediators 131–132, *131*
digital responses *see* switchlike responses
dimerization
  chemical induction 404, *404*
  protein interaction domains 247, *248*
  transmembrane receptors 171–173, *171*
disheveled (DVL) 184, *184*
disorder/order transitions 44, *44*
dissociation constant (K_d) 25–27; *see also* affinity
  binding energy and 25–26
  calculation methods 389, 390, *390*
  effect of multiple binding sites 28–29, *29*
  factors determining ideal 29
  functional constraints 29–31, *30*
  local concentration effects 29
  physiological examples **24**
  rates of binding and dissociation and 26–27, *27*
dissociation rate constant *see* off-rate
distributive protein phosphorylation 91–92, *92*, 277
diversity in signaling across phylogeny 321–343
  constraints on 321
  differences in physical properties 321–323, *322*
  lifestyle 323–324
  mechanisms in multicellular plants 330
  themes in prokaryotic organisms 325
  toolkit for transition to multicellular life 318
DNA
  damage 98, 371
  degradation during apoptosis 219
  tumor viruses 315, *315*

Doc1 *214*
docking sites
  protein kinases 51, 52, *52*
  protein phosphatases 54
domains, protein 22, 231–260
  bioinformatics for identifying 232–233, *233*
  catalytic *see* catalytic domains
  combinations (multidomain proteins) 250–256
  containing smaller repeats 233, *234*
  interaction *see* interaction domains, protein
  recognition functions 234–237
  recombinations 256–259
  structure 232, *233*
dose-response curves 276–277, *276*
double-strand breaks 98
DR4 222, *223*
DR5 223, *223*
Drk 8
Drosophila melanogaster 7, 185
DSL (Delta-Serrate-Lag2) *187*
dual-specificity phosphatases (DSPs) 45, *55*, 56, 58
dynamic molecular assemblies 35

E1 ubiquitin activating enzymes 95, *95*
E2 ubiquitin conjugating enzymes 95, *95*, 96, *96*
E3 ubiquitin ligases 95, *95*, 96, *96*; *see also* Cbl; RING E3 ligases
  cell cycle regulation 214, *214*
E. coli 326, 328–329
ectodomain shedding 208–209
EDG receptors 163
effectors 336
effector-triggered immunity (ETI) 337
EF-hand domains **235**
  Ca^{2+}-binding proteins 141, 143
  SH2 domain interactions 239, *240*
EH domain **235**
eicosanoids **133**, 138, 153, 163–164
electron microscopy (EM) 396–397
endocrine glands/tissues 168
endocrinology *6, 7*
endocytosis 83, 119
  mechanisms 119, *119*
  Notch signaling 187, *187*
  phospholipid-protein interactions 249, *249*
  receptor down-regulation 197–200, *198*
  receptor signaling after 119–120, *120*
  role of ubiquitylation 97, 119
  signaling cascades 71, *72*
EndoG 228
endoplasmic reticulum, Ca^{2+} storage 140, *140*
endosomes 119–120
  FYVE domain proteins 249
  signaling 120
engineered signaling proteins 259, *259*
enterohemorrhagic *E. coli* (EHEC)
  actin pedestals 310, *310*
  attachment and pedestal formation by *311*
enteropathogenic *E. coli* (EPEC) 312
  attachment and pedestal formation by *311*

enthalpy of binding 26, 390, *391*
ENTH domain **235**
entropy of binding 26, 390, *391*
environment
  ability to respond to changes in 2–3
  types of signals from 168–169, *168*
enzymes (signaling) 11, 12, 39–72
  allosteric changes 20–21, 40–41,
    43–44, 392
  allosteric regulation 44, *44*
  as allosteric switches *253*, 254
  binding to substrates 20
  cascades 67–72
  cooperative activation 392, *392*
  cooperative switches 277, *277*
  dose-response curves 276–277, *276*
  hallmark 40, *40*
  principles of catalysis 40–43
  quantifying catalytic power
    391–392, *392*
  signal amplification 40–41, 281, *281*
Epac 65, 135, *135*
ephrin (Eph) receptors *177*, 209, *209*
epidermal growth factor receptor (EGFR)
  ADAM-mediated cleavage of ligands
    208, 210
  domain structure *177*
  internalization 119–120, 197, *198*
  signaling *8*, 177
epinephrine 4
epitope tags 398
ErbB4/HER4 211–212, *212*
Erbin *246*
Erk MAP kinase 69; *see also* Raf-MEK-Erk
  kinase pathway
  Fos interactions 280, *280*
  methods for studying *402*, *407*
  PDGF signaling 358, 362
  subcellular localization 112–113, *113*
  T cell signaling 273, *273*, 381–382, 385
ESCRT complexes
  receptor down-regulation 198, *199*
  ubiquitin-binding domains 97
ethylene 333
ethylene signaling *334*
euchromatin 99
EVH1 domains 234, **235**, *237*
evolution, recombination of protein domains
  250–251
exportins 110–111, *111*
extracellular matrix (ECM) 168, 342
  integrin-mediated adhesion 182–183, *182*
  regulated proteolysis 208, *209*
eye
  development, *Drosophila* 7–8, *8*
  vertebrate 348

factor V 207
factor X 207, *207*
factor XIII 207, *207*
FADD 187–188, *188*, 222, 223, *223*
fan-in network architecture 274, *274*
fan-out network architecture 274, *274*
farnesyl groups 115
far-Western blotting 399, *399*
Fas 187–188, *188*

activation of apoptosis 222–223, *223*
Fas ligand (FasL) 222, *223*
fatty acids
  enzymatic cleavage 154, *154*
  phospholipids 148–149, *149*
  saturated *151*
  unsaturated 148, 151, *151*
F-box domain **235**
F-box proteins *214*, 215, *215*
FCH domain **235**
FCP phosphatases 54, *55*
feedback 4, 5, 274; *see also* negative feedback
  loops; positive feedback loops
    cell cycle phase transitions 357
feed-forward 275; *see also* coherent
  feed-forward loops; incoherent feed-forward
  loops
FERM domain **235**
FF domain **235**
FH2 domain **235**
FHA domain **235**
fibrin 207, *207*
fibrinogen 207, *207*
fibroblast growth factor (FGF) receptor
  177, *177*
fibroblast growth factor receptor substrate 2
  (FRS2) 177
fibroblasts 356
  detection of wounds 359
  molecular control of proliferation 356,
    357–319
  responses to wounding 356
  termination of proliferative response 362
fibrosis 293
FLIP 223
flow cytometry **367**, 413–414, *413*
fluorescence resonance energy transfer
  (FRET)
  biosensors based on 397, 410–411, *412*
  detecting conformational changes
    397, *397*
  detecting protein interactions 401,
    401, *402*
  visualizing Ca²⁺ changes 140–141
fluorescent dyes, calcium binding
  140–141, 411
fluorescent protein tags 400–401, *400*,
  411–412
focal adhesion kinase (FAK) *180*, *181*, 182
focal adhesions 182, *400*
Fos 280, *280*
FOXO 243, *243*
fractional occupancy 24, 390
free energy 25–26, *26*
  barrier, effect of enzymes 41, *41*
  conformational substates 43, 44, *44*
FRET *see* fluorescence resonance energy
  transfer
frizzled 184, *184*
Fura-2 141, 411
Fus3 70, 71
FYVE domains **235**, 249, *249*

G₁/S transition 364, 367
G₂/M transition 364
GADS 246, *246*, 377

Gα subunit 62
  activity cycle 62, *62*
  control of activation 68
  families and their effectors 62, **62**
γ isoform (PLC-γ)
  activation by ZAP-70 253
  domain structure 251, *251*
  SH2 domain 239, *240*
  T cell activation 377–378
γ-secretase 187, *187*, 211–212, *212*
GAPs 64–65, *65*; *see* GTPase-activator
  proteins
GAT domain **235**
gated ion channels 169, *169*, 188–192
  ligand-induced conformational changes
    170–171
  structure 188–189, *189*
Gcn5 102, *238*
GDI displacement factors (GDFs) 68, 116, *117*
GDIs (guanine nucleotide dissociation
  inhibitors) 68, 116, *117*
G domain 60, *61*, 62
GDP *see* guanosine diphosphate
GEFs *see* guanine nucleotide exchange
  factors
GEL domain **235**
gel electrophoresis 398, *398*
gene expression
  changes in 9–10, *9*
  methods of analyzing 405, *406*
genetic deletion mutations (knockouts) 403
genomic analysis 339
geranylgeranyl groups, modification by 115, *117*
GK domain **235**
GlcNAc (N-acetyl glucosamine) 79, *79*
GLIC *192*
Gli transcription factors 185, *186*
glucocorticoid receptor (GR) 195, *196*
glucose, levels of *327*
GLUE domain **236**
glutamate 348
glycerophospholipids 148–149, *149*
glycogen synthase kinase 3 (GSK-3) 92
  hedgehog signaling 185, *186*
  Wnt signaling 184, *184*, 185
glycosylation 79, *79*
  N-Glycosylation 79
  O-Glycosylation 79
glycosylphosphatidylinositol (GPI) anchor
  114, **114**, *115*
GoLoco motifs (or domains) 68
GPCRs *see* G-protein-coupled receptors
G-protein-coupled receptor kinases (GRKs)
  199–200, *200*
  rhodopsin desensitization 354
G-protein-coupled receptors (GPCRs) 62,
  174–176
  desensitization 200–201, *200*
  diversity 63–64, *64*, 174
  as drug targets 170
  GEF activity *62*, 64–65, 175
  ligand-induced conformational changes
    170–171
  protease-activated receptors 211, *211*
  recycling (resensitization) 200
  signaling 174–175, *174*

G proteins 7–8, 39, 58–63
  biosensors 411, *412*
  classification 60
  as conformational switches 58–59, *59*
  downstream signaling 61–62
  heterotrimeric G proteins 60, 62–63
  molecular basis of conformational change 60, *60*
  regulation of activity 58–59, *58*, 63–68
  signaling cascades regulating 71–72, *72*
  small *see* small G proteins
  structure 60, *61*
  switch I/II regions 60, *60*
graded (linear) responses 275–277, *276*
  flow cytometric analysis 413–414, *413*
GRAM domain **246**
granzyme B 222
Grb2
  domain structure *250*
  flexibility of adaptor function 251, *251*
  homologs 8, *8*
  PDGF signaling 358, 360
  receptor down-regulation 198, *199*
  SH2 domain 239, *239*, *240*
  Sos recruitment 65, *118*
green fluorescent protein (GFP) 400–401, 413
  biosensors *158*
  negative feedback loops 282–283, *282*, *289*
  protein-fragment complementation assay 402, *402*
Grip1 *252*
GRIP domain **236**
GRK2 199–200
GRKs *see* G-protein-coupled receptor kinases
ground-state energies 41
growth factors 168
growth hormone, human (hGH) 21, 30, *30*
growth hormone receptor, human 21, 30, *30*
GRP1 *237*
GTP *see* guanosine triphosphate
GTPase-activator proteins (GAPs) 59, 61, 63–67
  domains, small G proteins 64, *65*, **68**
  GPCR signaling 174, *174*, 176
  heterotrimeric G proteins 68
  mechanisms of action 67, *67*
  mechanisms of regulation 65–66, *65*
  small G proteins 64–65, *64*, *65*
guanine nucleotide dissociation inhibitors (GDIs) 68, 116, *117*
guanine nucleotide exchange factors (GEFs) 59, *59*, 63–67, *64*, *65*
  control of membrane association 117, *117*
  different organisms 7, *8*
  domains 64–65, *65*, **68**
  heterotrimeric G proteins *62*, 64, 174–175
  mechanism of action 66, *66*
  mechanisms of regulation 64–65, *65*
  rhodopsin function 350–351
  small G proteins 64–65, *74*, *75*
guanosine diphosphate (GDP) 59, *59*
guanosine triphosphate (GTP) 59, *59*
  γ-phosphate group 60, *60*
guanylyl cyclase 135, *135*
  membrane (mGCs) 179, *180*

soluble (sGC) 193–194, *193*
visual transduction cascade 349, 351–352
GYF domain **236**

half-life, biological complex 26–27
Hck *50*
heat, receptors responding to 190
HEAT repeat **236**
HECT domain **236**
HECT domain E3 ligases 96
hedgehog (Hh) 185–187, *186*, 341
helper T cells 374, 375
hemoglobin, conformational changes 44
heptahelical receptors 174
heterochromatin 99
heterotrimeric G proteins 60, 62–63
  activation of cAMP 134–135, *134*, *135*
  activity cycle 62, *62*
  control of activation 64, 68
  GPCR signaling 174, *174*
  limited numbers 62, 174–175
  phosphatidylinositol 3-kinase activation 157
  protein kinase C activation *138*
hierarchical organization, information-processing systems 16–17, *17*, 264–266, *266*
HIF-1α 194, *194*
high-throughput sequencing 405, *406*
histamine H1 receptor antagonists 170
histidine kinase 52, 93–94, *93*
histidine phosphorylation
  bacteria 93–94, *94*
  eukaryotes 94–95, *95*
histidine protein (HPr) 327
histone(s) 98–103
  acetylation 79
  androgen-mediated changes 85, *86*
  antagonistic modifications 270, *270*
  methylation 79, 99–101, *100*
  nucleosome structure 98–99, *98*
  post-translational modification 98–99
histone acetyl transferase (HAT) 79, 102
  chromatin modification *101*, 102
  nuclear receptor interactions 195, *196*
histone deacetylase (HDAC) 79, 102
  chromatin modification *101*, 102
  nuclear receptor interactions 195, *196*
homeostasis 4, 326
homology domains 232–233
hormones 4, 7, 168, 323–324
  mechanisms in multicellular plants 332–336, *333*, *334*, *335*, *336*, *338*
host actin cytoskeleton 309–316, *310*, *311*, *313–315*
host T cells *301*
HP1
  antagonism by Aurora-B kinase 270, *270*
  chromodomain *238*, 243
Hrs
  FYVE domain 249
  UIM domain 97, *97*
Hsp90 195, *196*
human papilloma virus (HPV) 314–315
hydrophobic motif pocket (PIF pocket) 51
hyperpolarization, light-induced 348

hypersensitive response (HR) 337
hypoxia 194, *194*
hysteresis 288, *289*

IAP antagonists 228
IAPs (inhibitor of apoptosis proteins) 222–223
IκB
  NF-κB regulation 217–218, *218*
  Toll-like receptor signaling 183, *183*
IKK (IκB kinase) complex 98
  NF-κB regulation 217–218, *218*
  Toll-like receptor signaling 183, *183*
  tumor necrosis factor receptor signaling 222, *223*
imatinib 258, 302
immobilized metal affinity chromatography (IMAC) 409
immune checkpoint proteins 301
immune response, adaptive 375
immune synapse 180, 380
immune tyrosine inhibitory motifs (ITIMs) 302
immunoblotting 398–399
immunofluorescence 400
immunology 6, 7
immunoreceptor tyrosine-based activating motifs (ITAMs)
  phosphorylation 379
  SH2 domain interactions 35, 239–240, *241*
  T cell activation 377
  ZAP-70 recognition 180
immunoreceptor tyrosine-based inhibition motifs (ITIMs) 302
importins 110, *111*
inactive state 14, *14*
incoherent feed-forward loops 275
  adaptive responses 283–284, *283*
infections, adaptive immune response 375
infectious diseases 293
inflammasome 221, 223–224, *224*
inflammation 163
inflammatory mediators 163–164, *164*
information processing devices/systems 2–3, *2*, 263–290
  hierarchical organization 16–17, *16*, 264–266, *266*
  input detection 266–267, *266*
  inputs and outputs 3–4, *3*
  integrating multiple inputs 269–273, *269*
  modifying strength/duration of output 281–289
  molecular currencies 10–14, *12*, **12**
  proteins as 267–268, *268*
  responding to strength/duration of input 273–280
  state machines 264–265, *265*
inhibitor titration *278*, 279
inositol trisphosphate (IP₃) **133**, 137
  biophysical properties 152
  biosynthesis 138–139, *138*, 154
  control of protein kinase C activation 138–139, *139*
  T cell activation 378
inositol trisphosphate (IP₃)-gated Ca²⁺ channels (IP₃ receptors)

activation by Ca²⁺ 140–141, *140*
    propagation of Ca²⁺ waves 142–143, *142*
inputs, signal 3–4, *3*, 263–264
    detection 267–268, *268*
    generating changes in state 11–12, *12*
    integrating multiple 269–273, *269*
    responding to strength or duration of
        273–280
inside-out signaling 182
insulin 3–4
insulin-like growth factor-1 (IGF-1)
    receptor *177*
insulin receptors 7
    activation 171
    domain structure *177*
    signaling 89–90, 177
insulin receptor substrate 1 (IRS1)
    PTB domain *237*, 241–242, *242*
    scaffold function 89–90, 177
insulin receptor tyrosine kinase (IRK)
    49–50, *49*
integration, signal 5, 263–269, *263*
integrins 180–183, *182*, *183*, 342
interaction domains, protein 231, 232–237,
    235–236
    combination 250–256
    detection of modified binding sites
        405–406
    dimer and oligomer formation 247, *248*
    recognition functions 237–245
    structure 232, *233*
interleukin-2 (IL-2)
    autocrine and paracrine signaling 381
    expression by T cells 273, *273*, 378
interlocking regulatory mechanisms 328–329
internal state, monitoring 4
internalization, receptors
    as means of down-regulation 197, *197*
    mechanisms 119, *119*
    modulating signal transduction 119–120
    retrograde signaling in neurons 121, *121*
    TGFβ signaling output after 120–121, *120*
intracellular domain (ICD) 212, *212*
intracellular receptors 169
invadopodia 210, *210*
ion channels, gated *see* gated ion channels
IQ domain **236**
IRAKs 183, *183*
irreversible systems 288, *289*
IRS1 *see* insulin receptor substrate 1 (IRS1)
ISG15 95
isothermal calorimetry (ITC) 391, *391*
isotherm, binding 24, *24*, 276
ITAMs *see* immunoreceptor tyrosine-based
    activating motifs
ITIMs (immune tyrosine inhibitory motifs)
    383
iTRAQ 409

JAKs (Janus kinases) 112, *112*, 179–180
JAK-STAT pathway 112, 179–180, 342
JNK MAP kinase pathway 69, *70*
juxtacrine signaling 209

Kₐ (association constant) 25
Kalirin *252*

karyopherins 110, *111*
KCa3.1 potassium (K+) channel 95, *95*
k_cat (catalytic rate constant) 42, 392, *392*
k_cat/K_m (specificity constant) 393
KcsA potassium channel *190*
K_d *see* dissociation constant
KHD (kinase homology domain) 179, *180*
killer T cells 374, 375
kinetochores, detection of unattached
    372–373
K_m (Michaelis-Menten constant) 42, 392, *392*
knockout/knockdown 403, *403*
k_off *see* off-rate
k_on *see* on-rate
KSR 71
Kv channels *see* voltage-gated potassium
    channels

lac operon 326, *327*, *328*
    regulation of *327*
lac repressor (LacI) *289*, 327
lactose, levels of *327*
lac transcription 328
LAT 253, 377–378
Lck
    phosphorylation of adjacent receptors 381
    Shp1 phosphorylation 384–385
    T cell receptor interactions 180, *181*, 377,
        379
lectins 23
leucine-rich repeats (LRR) 309
leukemia 257
leukotrienes 163, 164, *164*
ligand-gated calcium (Ca²⁺) channels 139
ligand-gated ion channels
    desensitization 199–200
    open and closed conformations 192–193,
        192
    overall topology *189*, 191–192
ligands 19, 167
    allosteric regulation 44, *44*
    effect of local concentration on binding
        29, *29*
    polymeric, avidity 28–29, *28*
    receptor internalization and 120–121
light
    adaptation 353
    mechanism of detection 350–351
    signal amplification 351
    signal transduction cascade 349
    visual processing 347–355
light-gated domains
    engineered 259, *259*
    regulating plant protein kinases 254, *254*
LIM domain **236**
linear responses *276*, 277, *277*
links, network *274*, 287
lipid(s) 147–165
    membrane *see* membrane(s)
    modification of proteins 78–79
    mediating membrane association
        114–115, **114**, *115*
    reversible membrane association and
        115, *115*
lipid bilayer 147–148, *148*, 150
    organization states 151, *151*

lipid-binding domains 22
    mediating membrane associations
        115–116
    phosphoinositide signaling *155*, 157
lipid-derived signaling mediators 137–138,
    153–155
    biosynthesis 153–155, *155*
    classes **133**, 137–138
    difficulties in studying 153
    diversity of biophysical properties
        152–153
    major signaling pathways 155–165
    phosphorylation/dephosphorylation *155*,
        156–157
lipid kinases 154–155
lipid-modifying enzymes 153–155
    complexity of regulation 156
lipid phosphatases 154–155
lipid rafts 151, *151*
liposomes *148*
5-lipoxygenase (5-LO) 164, *164*
Listeria monocytogenes 312, *313*
living cells
    biosensors 411–412, *412*
    localization and tracking proteins
        400–401, *401*
    time-lapse microscopy 410–411, *410*
    visualizing protein-protein interactions
        401–402, *401*
living organisms 321
local concentration 13, 107–108
    effect of membrane localization on
        108–109, *109*
    effect on binding 29, *29*
localization of proteins 108, 114–118
    coupling with activation 117, *118*
    effect on local concentration
        108–109, *108*
    mechanisms 114, *114*
    reversibility 116–117, *117*
    subcellular *see* subcellular localization
logic gates 269–270, *270*
Loratadine (Claritin®) 170
LOV domains 255, *255*
    Rac recombination 259, *259*
LRP 184, *184*
LRR repeat **236**
LUBAC 98
lysine acetyl transferase (KAT) 79
lysine deacetylase (KDAC) 78–79
lysine demethylases (KDMs) 79, 100
lysine methylation 101
lysine methyl transferases (KMTs) 79, 100
lysophosphatidic acid (LPA) 153, 154
lysophosphatidylcholine 164, *164*
lysophospholipid 153
lysosomes 97, 119–120, 212

macrophage-colony-stimulating factor
    (M-CSF) receptor *177*
macropinocytosis 119, *119*
Mad1 372–373
Mad2 372–373
Mae 247, *248*
major histocompatibility complex
    (MHC) 301

major histocompatibility complex (MHC)-peptide complexes *see* peptide-MHC complexes

mammalian target of rapamycin *see* mTOR

MAP kinase (MAPK) 68–69, *69*; *see also* Erk MAP kinase
  control of nuclear localization 112–113, *113*

MAP kinase cascades 15, 68–71, *69*
  chimeric engineered scaffolds 258–259, *258*
  in different eukaryotes 7–8, *8*, 69, *70*
  in mammalian cells 69, *70*
  scaffold proteins 20, 70–71, *70*
  switchlike activation 277
  T cell activation 378, 381

MAP kinase kinase (MAPKK) 68–69, *69*

MAP kinase kinase kinase (MAPKKK) 68–69, *69*

MAP kinase phosphatases (MKPs) 362

mass spectrometry (MS) 399, 408–409, *408*

matrix metalloproteases (MMPs) 208, 210–211, *210*

MBT repeat **236**

Mdm2 87, *87*

mechanistic target of rapamycin *see* mTOR

medicine 6, *6*

MEK 69, *70*; *see also* Raf-MEK-Erk kinase pathway
  control of Erk2 localization 113, *113*
  visualizing Erk association *402*

membrane(s) 147–165; *see also* plasma membrane
  binding by BAR domains 249–250, *249*
  effects of phosphatidic acid 159, *160*
  fluidity 151, *151*
  functional domains 108, *108*, 151–152
  lipid bilayer 147–148, *147*, 150
  lipid components 147–148, *149*
  localization of proteins to 108, 114–118
  molecular interactions within 151–152, *152*
  physical properties and lipid composition 150–151, *151*
  trafficking, modulating signaling 118–122

membrane channels 169; *see also* gated ion channels

membrane guanylyl cyclases (mGCs) 179, *180*

membrane lipids *see* lipid(s), membrane

membrane proteins 114, *114*

membrane-type matrix metalloproteases (MT-MMPs) *209*

memory, molecular *287*, 288

Mena *237*

metabolic activity, transcriptional regulation with 326–329, *326–329*

metabolic diseases 293

metalloenzymes 54, *56*

metalloproteases, extracellular 208–211, *209*

metaphase-to-anaphase transition 215, *216*, 364

metastasis, tumor 210

metazoan transition *341*

methods, study 389–414

N-methylation 78–79, *78*, *100*, 101–102

O-methylation 79

methylation, protein 78–79, *78*
  histone *see under* histone(s)
  protein domains recognizing 101–102, 243–244, *244*

MH1/MH2 domains **236**

MHC-peptide complexes *see* peptide-MHC complexes

micelles 148, 150

Michaelis-Menten analysis 391–392, *392*

Michaelis-Menten constant (Km) 42, 392, *392*

Michaelis-Menten equation 391–392

microarrays 399, 405, *406*

microbial pathogens 293, 330–331

microfluidic devices 410, *410*

microsatellite instability (MSI) 305

mitochondria, intrinsic apoptotic pathway 225–228

mitochondrial outer membrane permeabilization (MOMP) 225, 226, *226*, 227–228

mitogenic signaling proteins *303*

mitogens 168

mitosis 4, 364
  regulation 369
  ubiquitin-mediated regulation 215–217, *216*

MIU domain **236**, *244*

modular protein domains 231, 232–237; *see also* domains, protein

modular proteins, signal integration 271–272, *272*

molecular currencies, cell signaling 10–14

molecular memory *287*, 288

monoubiquitylation 95

motifs, peptide 234
  integrating multiple posttranslational modifications 270, *271*
  modified *see* post-translational modifications
  unmodified, recognition by protein domains 245–247

mRNA, microarray analysis 405, *406*

mTORC1 160, *160*, 161

mTORC2 160, *160*, 161

mTOR pathway, regulation by phospholipase D 156, 160–161, *160*

multicellularity 323

multicellular organisms 329–330, *330*

multicellular plants, mechanisms in 330
  hormones 332–336, *333*, *334*, *335*, *336*, *338*
  needs of plants 330–332, *332*

multidomain proteins *see under* domains, protein

multiple reaction monitoring (MRM) 409

multivesicular body 198, *199*

mutation of growth suppressors *300*

mutual inhibition, lipid-modifying enzymes 156–157

Myc 358

MyD88 183, *183*

MYPT1 *56*

N-Myristoylation **114**, 115, *115*
  control of membrane association and 116

Myt1 271

NADPH oxidase, allosteric regulation 255–256, *256*

NALP1 inflammasome 224, *224*

Nck 232, *234*, 272, *272*

NDPK-B 95, *95*

necrosis, cell 219, *220*

Nedd8 95

negative discrimination, modulating affinity and specificity 31–32, *32*

negative feedback loops 274
  adaptive responses 283, *283*
  bacterial chemotaxis 284–286, *284*
  bistable systems *289*
  controling output level and precision 281–283, *282*
  oscillators 286, *287*
  PDGF signaling 362
  T cell signaling 383–384
  visual system 353

NEMO
  NF-κB regulation 218, *218*
  recognition of polyubiquitin linkages 97–98, *98*

nerve growth factor (NGF) receptor *177*

networks 16; *see also* feedback; feed-forward
  architectures 273–275, **275**
  artificial activation methods 404–405, *405*
  detecting sustained input 279–281, *280*
  methods for studying 403–404
  perturbation methods 402–403, *403*
  switchlike activation 277–279, *278*

neuregulin-1 212

neurobiology 6, 7

neurons
  Ca²⁺ signaling 141
  mechanisms of signaling 130–131
  retrograde signaling 121, *121*

neurotransmission 191–192

neurotrophins, retrograde signaling 121, *121*

neutrophils, NADPH oxidase 255, *256*

NFAT 273, *273*, 378

NF-κB *see* nuclear factor κB (NF-κB)

nicotinic acetylcholine receptor (nAChR) 191

nitric oxide (NO) **133**
  activation of guanylyl cyclase 135, *135*
  signaling 193–194, *193*
  speed of signaling 131
  transfer across plasma membrane 169

nodes, signaling 10–11, *12*, 274, *274*

nonlinear responses *see* switchlike responses

Noonan syndrome 255

NOR gate, biological 271

notch 340
  nuclear localization 82, 114, *114*
  proteolytic processing 187, *188*, 212
  signaling pathway 114, 187, *188*

notch extracellular domain (NECD) 187

notch intracellular domain (NICD) 113, *113*, 187, *188*, 211

novel elements, combination of 339–340

Noxa 225, *225*, 226

nuclear export signals (NES) 110, *111*

nuclear factor κB (NF-κB) 98–99, 217–218
  canonical pathway of activation 218, *218*

family 217–218, *217*
noncanonical pathway of activation 218, *218*
regulation of apoptosis 222–223, *223*
T cell activation 378
Toll-like receptor signaling 183–184, *183*
nuclear localization signals (NLS) 110, *111*
nuclear magnetic resonance (NMR) spectroscopy 396, *396*
nuclear pore complex 109–110
nuclear receptor (NR) superfamily 195–196, *196*, 341–342
nucleation promoting factors (NPFs) 310
nucleosome 98–88, *98*
nucleotide-binding and leucine-rich repeat receptors (NLRs) 336
    structures *338*
    system 337–338
nucleus
    localization of molecules to 107, 109–113
    transport to/from, control of 110–113
nutrients, as signals 3
NZF domain **236**

off-rate (k_{off}) 24, 26–28, *27*
    measurement 390, *390*
okadaic acid 57, 403
olfaction 175
oligomerization
    protein interaction domains 247, *247*, *248*
    transmembrane receptors 171–173, *171*
oncogenes 7, *8*, 61, 257, 295
on-rate (k_{on}) 23, 26–28, *27*
    measurement 390, *390*
operon 326
opsin 350
    photosensitive proteins of 325
optogenetic systems 404–405, *405*
optogenetic tools 331
OR gate, biological 272, *272*
orphan receptors 195
oscillation 286–287
outputs, signal *3*, 263–264
    changes in state generating 11–12, *12*
    modifying strength or duration 5, 281–290
oxygen (O_2) signaling 194, *194*

p21 215, 371
p27 215, 271, *271*
p38 MAP kinase pathway 69, *69*
p53 86–88
    post-translational modifications 86–87, *87*
    regulation of activity 87–88, *87*
p65 217, *217*
p100/p52
    activation pathway 217–218, *218*
    domain structure 217–218, *217*
p105/p50
    activation pathway 217–218, *218*
    domain structure 217–218, *217*
p300/CBP complex *87*, 88, 102
Pak1 393, *394*
S-Palmitoylation 80, 115
    Ras isoforms 80, 115
paracrine signaling 193, 381
PAS domain **236**

Patched (Ptc) 185, *186*
pathogenic microbes 307
    host actin cytoskeleton 309–316, *310*, *311*, *313–315*
    Yersinia pestis 307–309, *308*, **308**
pathways, signaling 15, *15*; *see also* networks
    artificial activation 404–405, *405*
pattern-recognition receptors (PRRs) 336
PB1 domain **236**
Pbs2 *70*, 71
PCAF/Gcn5 102, *238*
PDGF *see* platelet-derived growth factor
PDK1 kinase
    Akt activation 117–118, *118*
    PH domain 248
PDZ domains **236**
    recognition of C-terminal motifs 246–247, *246*
    scaffold proteins with 252–253, *253*
peptide binding 22, *22*, 237–242, *238*
    adjacent domains influencing 239, *240*
    cooperativity 34, *34*, 239–240, *240*
    recognition function 234
    specificity 239–240, *240*
peptide-MHC complexes 374, 375, 376
    binding properties 30, *30*
    discrimination of non-antigenic 384
peptide-protein interactions 22, *22*
peroxisome proliferator activated receptors (PPARs) 164–165
pharmacological small-molecule inhibitors 403, *403*
PH domains *see* pleckstrin homology domains (PH domains)
Philadelphia chromosome 257, *257*
Pho4 111–112, *111*
Pho80/85 111–112, *111*
phorbol esters 139
phosphatase and tensin homolog (PTEN) 298
phosphatidic acid (PA) 159
    biosynthesis 138–139, *154*, 155, 159
    effect on membrane curvature 159, *160*
    regulation of mTOR pathway 160–160, *160*
phosphatidylcholine (PC) 148, *149*
    enzymatic cleavage *154*, 159
    position in membrane 150
phosphatidylethanolamine (PE) 148, *149*
phosphatidylinositol (PI) 137–138, 148
    chemical structure *149*, 155
    position in membrane 150
phosphatidylinositol 3,4,5-trisphosphate (PI(3,4,5)P3) 156
    Akt regulation 117–118, *118*
    biosynthesis 137, 156, 158
    chemical structure *165*
    dephosphorylation 158–159
phosphatidylinositol 3-kinase (PI3K) 117, *118*
    isoforms 158–159
    p85 adaptor subunit, SH2 domains 239
    PDGF signaling 359
    phospholipase C interactions 156–157
phosphatidylinositol 3-phosphate (PI(3)P)
    binding by FYVE domains 249, *249*
    biosynthesis 157, *158*
    subcellular localization 157, *158*

TGFβ receptor internalization 120–121, *120*
phosphatidylinositol 4,5-bisphosphate (PI(4,5)P2) 116
    enzymatic cleavage 138, *138*, 156
    as membrane-localized binding site 156–157
    N-WASP regulation 271–272, *272*
    phospholipase Cδ interaction 248, *249*
    phosphorylation/dephosphorylation 156, 158–159
    subcellular localization 157, *158*
phosphatidylinositol 4-phosphate (PI(4)P) 157, *158*
phosphatidylinositol 4-phosphate 5-kinases (PIP5K) 159
phosphatidylinositol 5-kinase (PI5K) 156
phosphatidylinositol 5-phosphatases 157
phosphatidylinositol-specific phospholipase C (PI-PLC) 156–157
phosphatidylserine (PS) 148, *149*
phosphodiesterase 6 (PDE6)
    GTPase activating protein (GAP) activity 354, 355
    signal amplification 351
    speed of response 353
    visual transduction cascade 349–351
phosphodiesterases (PDEs) 134–135, *135*; *see also* cyclic AMP; cyclic AMP phosphodiesterase; cyclic GMP phosphodiesterase
    AKAP interactions 143–145, *144*
phosphoenolpyruvate (PEP) 327
phosphoinositides 156–159
    biosensors *158*, 411
    as membrane-localized binding sites 109, 157–159
    phosphorylation/dephosphorylation 155, 156
    protein domains recognizing 248–2
    recognition by PH domains 248–249, *248*
    as source of signaling mediators **133**, 138, *138*, 156–157, *156*
    subcellular distribution 157–158, *158*
phospholipase(s) 153–154, *154*
phospholipase A₂ (PLA₂) 153, *154*, 163–164
    cytosolic α isoform (cPLA₂α) 163–164, *164*
phospholipase C (PLC) 153, 155
    β isoform (PLC-β) 138, 156–157
    δ isoform (PLC-δ), PH domain 249, *249*
    γ isoform (PLC-γ) 251, *251*, *240*, 253
    hydrolysis of PIP₂ *137*, 138, 156
    phosphatidylinositol-specific (PI-PLC) 156–157
    protein kinase C activation 138–139, *138*
    regulation of activity 156–157
phospholipase D (PLD) 154, 159–160
    isoforms 159
    mTOR pathway regulation 156, 159–161
    phosphatidylcholine cleavage *153*, 159
    phosphatidylinositol 5-kinase interactions 156
    regulation of membrane curvature 159, *160*
    responses to environmental signals 161

phospholipids
  bilayer formation *148*
  chemical structure 148–148, *148*
  enzymatic modification 153–154, *154*
  position in membrane 150
  protein domains recognizing 248–249, *248*
phosphorylation
  membrane lipids 154–155
  protein *see* protein phosphorylation
phosphoserine/phosphothreonine
    14-3-3 protein interactions 242–243, *242*
  protein domains recognizing 242
phosphotransferase system (PTS) 327
  phosphorylation of 328–329
  system *328*
phosphotyrosine
  motifs, scaffold proteins with multiple
    253, *253*
  other domains recognizing 241–242, *242*
  recognition by SH2 domains 238–241
  specificity of antibodies binding 23, *23*
photoreceptor cells 347–355
  adaptation 353
  light-induced response 348
  mechanism of light detection 350–351
  signal amplification 351
  signal transduction cascade 349
  speed of response 353
phox proteins 255–256, *256*
phytohormones 332
PICK1 *252*
PIDD-osome 221
PIF pocket 51
PINS protein 68
plant innate immunity *336*
plants
  activation of kinases by light 255, *255*
  MAP kinase cascades 69, *69*
  serine/threonine kinases 177–178
plasma membrane 107; *see also*
  transmembrane signaling
    effect on local concentration 107–108, *108*
    information transfer across 167–201
    localization to 109
plasmin 208
plasmodesmata 331, *331*
platelet-activating factor (PAF) 164, *164*
platelet-derived growth factor (PDGF)
    356–362
  actin cytoskeletal rearrangements *400*
  secretion in response to injury 357
  signaling 358–360
platelet-derived growth factor receptors
    (PDGFR) 358–359
  domain structure *177*
  down-regulation 362
  ligand-induced dimerization 171
  mechanism of activation 359
  responses to autophosphorylation 360
  tyrosine kinase activity 171
platelets 211, 356, 357
pleckstrin homology domains (PH domains)
    22, **236**
  biosensors based on 412
  GPCR desensitization 199–200, *200*
  identification by database analysis 232, *233*

mediating membrane association 115,
    118, *118*
  phosphoinositide recognition 248–249, *248*
  recognition functions 234–237, *237*
podosomes 210
polarity, cell 107
polo-box domain **236**
polyacrylamide-gel electrophoresis
  (PAGE) *408*
polyhomeotic 247
polyomaviruses 316
polyproline type II (PPII) helix 245, *245*
polyubiquitylation 95, *96*
positive discrimination, modulating affinity
  and specificity 32–33, *33*
positive feedback loops 274
  bistable systems 288, *289*
  leading to switchlike activation *278*, 279
  lipid-modifying enzymes 156, 159
  robust oscillators 286, *287*
  T cell activation 381–382, 384
postsynaptic density 252–253, *252*
post-translational modifications (PTMs)
    12–13, 77–103
  chemical effects 81–87, *82*
  control by writer/eraser/reader systems
    83–84
  interplay between 84–85, *86*
  methods of detection *340*, 405–405, *407*
  multiplicity of sites and types 84–85, *85*
  p53 86–88, *87*
  peptide motifs integrating multiple
    270, *271*
  protein interaction domains recognizing
    238–241, *238*
  regulating membrane associations
    114–115
  reversibility 205
  role in subcellular localization 82, *82*,
    109, 114–115
  simple functional groups 78–79, *78*
  speed of signaling 84
  sugars, lipids and other proteins 79–81,
    79, *80*
  switching between distinct states
    85–86, *86*
  transmitting spatial information 84
potassium (K⁺) channels
  KCa3.1 95, *95*
  KcsA 1 *90*
  Kv (voltage-gated) *see* voltage-gated
    potassium channels
PP1 phosphatase 54, *56*
PP2B phosphatase *see* calcineurin
PPII (polyproline type II) helix 245, *245*
PPM phosphatases 54, *55*
PPP phosphatases 54, *55*
prenylation **114**, 115, *115*
  membrane association and 115
primary cilium 185, *186*
primary structure 393, *393*
priming, of phosphorylation 92, *92*
processive protein phosphorylation 92–93,
  92, 277
programmed cell death *see* apoptosis
prokaryotes 323; *see also* bacteria

evolutionary transition from 323
  extraordinary diversity of 325
  protein kinase-like genes 52
proline *cis-trans* isomerization 79, *79*
proline hydroxylase domain (PHD) proteins
    194, *194*
proline hydroxylation *78*, 79
proline-rich peptide motifs, recognition
    245, *245*
promoters, transcriptional, signal integration
    273, *273*
propeller structure 233, *234*
prostacyclin 164, *164*
prostaglandins 163, 164, *164*
protease-activated receptors (PARs)
    211–212, *211*
proteases 81, 206–208, 219–229; *see also*
  proteolysis
  activation by death receptors
    185–187, *188*
  classification 206
  domain structure 220, *221*
  effector (executioner) 220–221, *221*
  extrinsic apoptotic pathway 222–225, *223*
  inflammasome activation 223–224, *223*
  inflammatory 221
  initiator 187–188, 220–221, *221*
  intrinsic apoptotic pathway 226, *227*
  receptors coupled to 185–188
  receptors indirectly activating 183
  regulation of activation 220–222, *221*
  substrate specificity 206, 222
proteasome 212, *213*
  cell cycle regulation 215, *215*
  NF-κB regulation 218, *218*
  p53 targeting to 95, *96*
  protein targeting to 85, *86*, 212
protein(s)
  biochemical and biophysical analysis
    389–397
  degradation *see* proteolysis
  dynamics, spectroscopic analysis 397, *397*
  identification 398–399, 408–409
  internalization of cell surface
    119–120, *119*
  membrane 114, *114*
  methods for studying 389–414
  modular architecture 231–260
  as signaling devices 267–268, *268*
  structure *see* protein structure
  trafficking 82
protein arginine methyl transferases (PRMTs)
    79, 100
protein-complex allosteric switches 267, *268*
protein complexes *see* complexes, protein
protein dephosphorylation; *see also* protein
  phosphatases
    energetics driving 43, *43*
protein domains *see* domains, protein
protein-fragment complementation assay
  (PCA) 402–403
protein histidine phosphatase (PHPT-1) 95, *95*
protein interactions 19–37
  intermolecular *see* protein-protein
    interactions
  intramolecular 237–238

methods of mapping 399–402
protein kinase(s) 11, 39, 47–52; *see also*
  serine/threonine kinases; tyrosine kinases;
  specific kinases
    activation loop *see* activation loop, protein
      kinase
    allosteric switches 254
    biosensors 412
    C-helix 48, *48*
    dimerized/oligomerized receptors
      171–172, *172*
    families 52, *53*
    inhibitors 89
    key conserved catalytic residues 47–48,
      48
    pairing with phosphatases 41, *41*
    plant, regulation by light 255, *255*
    reaction energetics 41–42, *41*
    regulation of catalytic activity *48*, 49–51,
      49, *50*
    as state machines 264, *265*, *266*
    structure 47–48, *48*
    substrate specificity 51–52, *51*, 91, *91*
    transmembrane receptors covalently
      linked to 179–182
    transmembrane receptors indirectly
      activating 183
protein kinase A (PKA)
    activation by cAMP 130–131, *135*
    AKAP interactions 143–145, *144*
    allosteric switch function 254
    cooperative switchlike activation 277
    coupling Rac activation to 259, *259*
    diffusion rate *131*
    hedgehog signaling 185, *186*
    key catalytic residues *47*, 48
    myristoylation 115
    phylogenetic relationships *53*
    regulation by cAMP binding 136–137, *136*
protein kinase B (PKB) *see* Akt kinases
protein kinase C (PKC)
    activation 138–139, *138*
    AKAP-mediated regulation 145
    δ isoform (PKCδ) 242
protein kinase G (PKG) 135
protein kinase-like (PKL) genes 52
protein phosphatases 11, 52–58
    classes and reaction mechanisms 53–56, *55*
    pairing with kinases 41, *41*
    reaction energetics 43, *43*
    regulation of activity 58–59
    substrate specificity 91
protein phosphorylation 88–95; *see also*
  protein kinase(s); protein phosphatases
    allosteric regulation 44, *44*
    amino acid residues involved 45, *45*
    chemical effects 81, *82*
    coupling with protein interactions 89–91
    distributive 92–93, *92*, 277
    E3 ligase affinity and 96
    energetics 42–43, *42*
    induced conformational changes 45, *45*
    methods of studying 405–407, *407*,
      409, 413
    at multiple sites, diverse modes 92–93, *93*
    p53 87

pairing of kinases and phosphatases 41, *41*
    as post-translational modification 78, *78*
    priming *92*, 93
    processive 92–93, *92*, 277
    protein domains recognizing 237–245
    regulatory role 39, 45–47
protein-protein interactions 12, 19–37; *see
  also* complexes, protein
    cellular and molecular context 28–36
    coupling with phosphorylation 89–91
    effects of post-translational modifications
      81–82, *82*
    identifying binding partners 398
    methods of detection/mapping 399–402
    properties 20–28
    regulation by ubiquitylation 244–245
    solid-phase screening 399, *399*
protein stability
    allosteric regulation 44, *44*
    post-translational modifications and
      82–83, *82*
protein structure 392–395
    methods of determining/analyzing
      395–396
    primary 393, *393*
    quaternary *see* quaternary structure
    secondary 393, *393*
    tertiary *see* tertiary structure
protein tyrosine phosphatases (PTPs) 54–57,
  55, 339
    acting as receptors 58, 178–179, *179*
    cysteine (Cys)-based 54–58
    domains 58
    modular domain structure *58*, 59
    oxidative regulation 57
    reaction mechanism 56, *57*
proteolysis 12, 81, 205
    irreversibility 205, *206*
    nuclear localization of Notch 113, *113*
    post-translational modifications and
      81–82, *82*
    receptors coupled to 209–211
    receptors indirectly activating 183
    regulated 205–229
    regulated intramembrane (RIP)
      211–212, *211*
proteostasis 331
prothrombin 207–208, *207*
PSD93 *252*
PSD95 252–253, *252*
pseudo-histidine phosphotransferases
  (PHPs) 333
PTB domains 90, **236**
    binding phosphotyrosine sites
      241–242, *242*
    recognition functions 234–235, *237*
PTEN 158–159, *158*
PTP1B 57, *57*
PTPs *see* protein tyrosine phosphatases
pTyr-dependent signaling 339
pull-down assays 398
PUMA 225, *225*, 226
pumilio repeat **236**
PWWP domain **236**
PX domain **236**
pyrin (PYD) domains 222, *223*

quaternary structure 393, *393*
    phosphorylation-induced changes
      46–47, *47*
    rearrangements 44, *44*
quorum sensing 329
    in bacterial populations *330*
    mechanisms 329

Rab
    control of membrane association 116, *116*
    GEF binding *66*
    signaling cascade regulating 70–71, *71*
    subfamily 62, **68**
RabGDI 116, *117*
Rac1-related G proteins 62, *62*
Rac fusion proteins, engineered 259, *259*
Rad9 371
Raf
    in different organisms 8, 9
    PDGF-mediated activation 361
Raf-MEK-Erk kinase pathway *8*, 69, *69*; *see
  also* Erk MAP kinase; MEK
    EGF receptor signaling 15
    regulation by 14-3-3 protein 255, *255*
    scaffold protein *70*, 71
RalA 161
Ran proteins 61, **61**
    control of nuclear transport 110, *111*
    GEF binding *66*
Rap1/Rap2 134, *135*
Rap80 97–98, *97*
rapamycin 160, *161*
raptor 160, *161*
Ras 61
    activation by Sos 117, *118*
    different organisms 7–8, *8*
    family proteins 61, **61**
    mechanisms of regulation 64–65, *64*, *65*
    palmitoylation 122
    PDGF signaling 358, *359*, 360
    prenylation 115
    structure *61*
    subcellular localization and signal outputs
      122–123, *122*
    T cell activation 378
    H-Ras 121–122, *395*
    K-Ras 121–122
    N-Ras 121–122
RasGAP *250*
RasGRP 378, 381
Ras-MAP kinase pathway *8*, 15, 69; *see also*
    Raf-MEK-Erk kinase pathway
Rbx1 *214*, 215
Rcc1 110
reaction intermediates, enzyme catalyzed
    reactions 41, *41*
reactive oxygen species (ROS) 331
receptor protein tyrosine phosphatases
    (RPTPs) 57, 178–179, *179*
receptors 167–168
    desensitization 197, *197*
    dimerization/oligomerization 171–173,
      171, *172*
    down-regulation 119, 196–201, *197*, *199*
    as drug targets 170
    intracellular 168

receptors (*cont.*)
    ligand-induced conformational changes
        170–171, *171*, *172*
    transmembrane receptors 170–174,
        179, *179*
    up-regulation 119
receptor tyrosine kinases (RTKs) 8, 9,
    176–177, 296, 342
    activation 171–172, *172*, 177
    autophosphorylation 89, *90*
    as dynamically regulated scaffolds
        253, *253*
    major families *177*
    recruitment of downstream effectors
        89–91, *90*
red fluorescent protein (RFP) 400–401, *400*
regulated intramembrane proteolysis (RIP)
    211–212, *211*
regulators of G protein signaling (RGS)
    proteins 67, 176
regulatory particle 19S 212, *213*
c-Rel 217, *217*
RelA 217
RelB
    activation pathway 217–218, *218*
    domain structure 217, *217*
rel homology domain 217, *217*
repeated sequences, protein domains 233, *234*
resistosome 337
response regulator (RR) 93–94, *93*
retinal 350
retina, vertebrate 348
retrograde signaling 121, *121*, 332, *332*
RGS (regulators of G protein signaling)
    proteins 67, 176
RH (RGS homology) domain 199, *200*, **236**
Rheb 156, *160*, 161
Rho-associated kinase (ROCK) 222
rhodopsin
    desensitization 354
    light-induced conformational change
        350–351
    signal amplification 176, 351
    signaling 176, 349–351
    speed of signaling 176, 353
RhoGAPs 64–65, *64*, *65*
RhoGDI 116–117
RhoGEFs 64–65, *64*, *65*
Rho proteins 61, **61**
    biosensor *412*
    cell motility 61, *62*
    enzymes regulating 64–65, *64*, *65*
    membrane association 116–117
Rictor 160, 161
RING domain **236**
RING E3 ligases 96
    caspase inhibitors 222
    cell cycle regulation 214, *214*
    control of NF-κB activation 218
RNA interference (RNAi) 403
RNA polymerase 99–100, *101*, 102, 327
robust oscillators 286, *287*
rods 347–355; *see also* photoreceptor cells
RPTPs *see* receptor protein tyrosine
    phosphatases (RPTPs)
ryanodine receptors 142

Saccharomyces cerevisiae (budding yeast)
    control of nuclear transport 111–112, *111*
    MAP kinase cascades 69, *69*
    scaffold proteins 70–71
    SCF complex *215*
SAM domains **236**
    Abl activation, in leukemia 257–258, *257*
    oligomerization 247, *247*, *248*
SARA 120–121, *120*, 249
SBF 367, 368
scaffold complexes
    caspase activation 221, *221*
    T cell activation 377–378
scaffold proteins 20, *21*, 251
    chimeric, engineered 258–259, *258*
    determining substrate specificity 91
    G protein signaling 71, *72*
    MAP kinase cascades 20, 69, *69*
    mechanisms of action 70–71, *70*, *71*
    with multiple phosphotyrosine motifs 253,
        253
    PDZ domain-containing 252–253, *252*
    phosphorylation by receptor tyrosine
        kinases 89–90, *90*, 177
    small-mediator signaling 143–145, *144*
Scatchard analysis 390, *390*
SCF complex
    cell cycle control 215, *215*
    NF-κB regulation 218, *218*
    structure 214, *214*
Schizosaccharomyces pombe (fission yeast),
    Ras localization and signaling 122, *122*
secondary structure 393, *393*
second messengers *see* small signaling
    mediators
securin 216–217, *216*, 369
Sem-5 8
separase 369
serine phosphorylation 45, *45*; *see also*
    phosphoserine/phosphothreonine
        inducing protein interactions 89–90
serine/threonine (Ser/Thr) kinases 45
    docking sites 51, *51*
    families 44, 52, *53*
    plant, activation by light 255, *255*
    receptors with intrinsic activity
        177–178, *178*
    substrate specificity 51, *51*, 52
serine/threonine (Ser/Thr) phosphatases 45
    families 54–56, *55*
    holoenzyme complexes 56, *56*
    reaction mechanisms 52–53, *56*, *57*
    regulation 58
seven-transmembrane receptors (7-TMRs)
    174
SH2 domain-containing proteins
    combinatorial diversity 250–251, *250*
    recruitment by receptor tyrosine kinases
        89–90, *90*
SH2 domains **236**, 237–242, *238*, 240, *240*
    peptide binding 22, *22*, 237–242, *238*
    receptor tyrosine kinases 88–89, *90*
    regulation of kinase activity 50–51, *50*
    regulation of phosphatases *58*, 91
    regulation of substrate specificity 51, 91
    structure 237–239, *238*

SH3-containing proteins *232*
SH3 domains 22, *232*, **236**
    recognition of proline-rich motifs *245*, 246
    regulation of kinase activity 50–51, *50*
    regulation of NADPH oxidase 255, *256*
    regulation of substrate specificity 51
    specific binding to proline 32, *32*
Shank *252*
Shc 241
sheddases *see* ADAM metalloproteases
SHIP 159
Shp1 384–385
Shp2 *58*
Sic1 368–369
signal amplification 5, 281, *281*
    cleavage of zymogens 206
    enzymes 41, 281, *281*
    GPCRs 176
    small signaling mediators 131
    T cell activation 380–381
    visual transduction system 351
signal amplitude
    increasing *see* signal amplification
    responses to 275–277, *276*
signal duration
    measuring 279–280, *280*
    modifying 281, 283–284, *283*
    signaling, calcium 130, 139–143, *140*
    downstream effectors 130–131, 141, *141*
    negative feedback control 282, *282*
    speed 131, *131*
    T-cell activation 273, *273*
    types of 296
signal integration 5, 263–269, *263*
signals, diversity of 3–4, *3*
SILAC *408*, 409
sildenafil (Viagra®) 134, *135*
Simian Virus 40 (SV40) 315
single-cell analysis
    flow cytometry 413–414, *413*
    methods compared **367**
    time-lapse microscopy 410–411, *410*
single-celled organisms 338
Skp1 214–215, *215*
Skp2 *214*, 215
SLP-76
    GADS SH3 domain binding *245*, 246
    T cell receptor signaling 377–378
    transient SH2 binding sites 254
Smac/DIABLO 222, 227
SMAD2 120, *121*
SMAD3 120, *121*
SMAD4 120, 178, *178*
SMAD7 120, *121*
SMAD proteins 177–178, *178*
R-SMAD proteins 177, *178*
small G proteins (small GTPases) 61
    control of cell motility 61, *62*
    mechanisms of membrane association 116
    phosphoinositide signaling and 157
    regulation of activity 64–65, *65*, 68
    signaling cascades regulating 71–72, *72*
    structure 60, *61*
    subcellular locations and signal outputs
        121–122, *122*
    subfamilies 61, **61**

small-molecule inhibitors, pharmacological 403, *403*

small signaling mediators (second messengers) 13, 129–145; *see also* specific mediators
  classes 133, **133**
  complex spatiotemporal pattern generation 132–133, *132*
  mechanisms of downstream effects 130–131
  properties 129–132
  range of physical properties 133–134
  scaffold proteins 143–145, *144*
  signal amplification 132
  specificity and regulation 143–145, *143*
  speed of signaling 130–131
  synthesis-degradation balance 130, *130*

smell, sense of 175

Smoothened (Smo) 185–187, *186*, 200–201

SMURF ubiquitin ligases *120*, 121

SNARE domain **236**

SOCS box **236**

solid-phase screening 399, *399*

soluble guanylyl cyclases (sGC) 193–194, *193*

sonic hedgehog (Shh) 185, *186*

Sos
  activation of Ras 117, *118*
  different organisms 8
  Grb2 function 251, *251*
  mechanisms of regulation 64–65, *65*
  PDGF signaling 358, 359–360
  T cell activation 378, 381

spatial information, transmission of 84

spatial organization/localization 9–10; *see also* subcellular localization

specificity, binding 23, *23*
  cooperative binding and 33–35, *35*
  factors determining ideal 29
  functional constraints 29–31, *30*
  independent modulation 31–32, *32*
  quantification 25, *25*

specificity constant (k_{cat}/K_m) 392

spectroscopic analysis, protein dynamics 397, *397*

sphingolipids 150, 161–163

sphingomyelin 161–163
  chemical structure 148–149, *149*
  lipid rafts 151
  metabolites 138, 161–163, *162*

sphingomyelinase (SMase) 161–163, *162*

sphingosine
  generation/metabolism 161, *162*, 163
  phosphorylation 154–155, 161

sphingosine 1-phosphate (S1P) **133**, 161–163
  biophysical properties 152–153
  biosynthesis 138, 155, 161, *162*
  downstream signaling 163

sphingosine 1-phosphate receptors (S1PR) 163

sphingosine kinases (SKs) *162*, 163

spindle assembly checkpoint 217, 372–373

SPRY domain **236**

Src 13
  domain structure *250*

Src family kinases (SFK) 13–14, *14*, 181–182, integrin signaling *181*

as allosteric switches 254
  phosphorylation-mediated regulation 50–51, *50*

Src homology domains *see* SH2 domains; SH3 domains

stable-isotope labeling of amino acids in cell culture (SILAC) 408, *409*

standard free energy ($\Delta G°$) 25

START domain **236**

START phase, cell cycle 364, 367

state, changes in 10–14, *12*
  coupling between 13–14, *13*, *14*
  generating pathways and networks 14–17
  types *12*, **12**, 13–14

state machines 264, *265*
  hierarchical organization 264–266, *266*

statins 115

STAT proteins
  control of nuclear import 112, *112*
  cytokine receptor signaling 179, *181*
  domain structure *250*

Ste5
  membrane localization 116
  scaffold function 70–71, *70*

Ste7 70–71, *70*

Ste11 *70*, 71

Ste20 71, *72*

STE family kinases *53*

steroid hormones 195
  receptors 195–196, *196*
  transfer across plasma membrane 169

sterols, membrane 151

stoichiometry 34

stress responses
  activation of apoptosis 225–226
  sphingolipid signaling 163

subcellular localization 13, 107–126
  changes induced by binding 20, *20*
  important sites 108, *108*
  localization switches 267, *268*
  mechanisms of control 109–118
  methods of mapping 400–401, *400*
  modulating signaling 118–122
  post-translational modifications and 82, *82*, 109, 114–115
  regulation by 14-3-3 proteins 243, *243*
  regulation by PH domains 248–249, *248*
  transmitting information 108–109, *109*

substrates
  concentrations, Michaelis-Menten analysis 392, *392*
  enzyme binding to 20

SUMO 95

suppress proliferation 295

supramolecular activation complex (SMAC) 380

surface plasmon resonance (SPR) 390, *390*

surfaces, protein-protein interactions 21–22, *21*, *22*

sustained proliferative signaling 295

SWIRM domain **236**

switches, molecular 10, *11*
  allosteric switch proteins *see* allosteric switch proteins
  bistable 288, *289*
  G proteins 59, *59*

kinase/phosphatase systems 48–49
  post-translational modifications and 85–86, *86*

switchlike responses 276, *276*
  cooperativity leading to 35, 277, *277*
  networks yielding 277–279, *278*

synapses, scaffold proteins 252, *252*

synaptojanin 157

synaptotagmin 141

synthetic biology 258–259
  bistable systems 288, *289*

TAB2/TAB3 183, *183*

TACE metalloprotease 187, *187*, 209

TAK1 183, *183*

talin 182, *182*

T cell receptors (TCRs) 374
  activation of coupled tyrosine kinases 179, *181*
  down-regulation 383
  interaction with peptide-MHC complexes 379
  phosphorylation initiating activation 379
  signaling network 377–378
  ZAP-70 interaction 241, *241*

T cells 374–375
  activation 379–379
  antigen presentation to 376
  discrimination of non-antigenic peptides 384–385
  effector 375
  immune synapse formation 380
  interleukin-2 expression 273, *273*, 378
  naive 374, 375

TEL 247

TEL-Abl oncogene *257*, 258

temperature
  conformational changes and 43–44
  receptors responding to 190

tertiary structure 393, *393*
  phosphorylation-induced changes 46–47, *47*
  rearrangements 44, *44*

Tet repressor (TetR) *289*

TGFβ *see* transforming growth factor β

thermal breathing 43

thermodynamics
  binding interactions 25–26, *26*, 390–391, *392*
  enzymatic catalysis 41–42, *42*
  G protein regulation 59

threonine phosphorylation 45, *45*; *see also* phosphoserine/phosphothreonine
  inducing protein interactions 89, 90

thrombin
  activation of protease-activated receptors 211, *211*
  proteolytic activation 209, *209*

thrombolysis *207*, 208

thromboxane 164, *164*

thrombus formation 207, *207*

time-lapse microscopy, live cells 410–411, *410*

time scales 9–10
  signaling by second messengers 131–132, *131*

TIR domain 183, **236**
tissue inhibitors of metalloproteases (TIMPs) 211
tissue plasminogen activator (t-PA) 208
TKB domain 239, *240*
T lymphocytes *see* T cells
toll-like receptors (TLRs) 183–184, *183*
TPR repeat **236**
TRADD 223, *223*
TRAF-2 223, *223*
TRAF-6 183, *183*
TRAF domain **236**
TRAF proteins 218
TRAIL 187, 222, *223*
TRAIL receptors 187–188, 223, *223*
transcription
    changes in 9–10
    chromatin remodeling 98–99, *101*, 102–103
    monitoring changes in 405, *406*
transcription factors 99
    signal integration 273, *273*
transducin *174*, 349, 350
    inactivation 355
    signal amplification 351
    speed of response 353
transformation of mammalian cells 298
transforming growth factor β (TGFβ)
    receptors 177–178, 340
    modes of internalization 120–121, *120*
    SARA binding 249
    signaling 178, *178*
    wound healing 356
transistor 264, *265*
transition state, stabilization by enzymes 41–42, *41*
translocations, chromosomal 257–258, *257*
transmembrane receptors 170–174, 179, *179*
    activating both kinases and proteases 183
    coupled to proteolysis 185–188
    covalently linked to protein kinases 179–193
    with intrinsic enzymatic activity 176–179
    membrane-spanning segments *70*, 170–171
    with multiple membrane spanning segments 170–171, *171*
    with single membrane-spanning segment 171–172, *171*
    transduction strategies 170–173
transmembrane signaling 167–173; *see also* gated ion channels; receptors, transmembrane
    membrane-permeable 193–196
    three general strategies 169–170, *169*
transphosphorylation 171–172, *172*
TRP (transient receptor potential) channels 190, *190*
TRPV1 vanilloid receptor 190
TSC complex (TSC1/TSC2) 160, *160*
TUB domain **236**
tudor domains *101*, **237**, 243, *244*
tumor necrosis factor α (TNFα) 187, 222–223

proteolytic processing 209
    signaling pathway 223–224, *223*
tumor necrosis factor receptor (TNFR) 187, 222–223, *223*
tumors *301*
two-component systems 322, 324, 333
    bacteria 93–94, *93*
    eukaryotes 94
two-hybrid assay, yeast 398, *398*
type III secretion system (T3SS) 307–308, *308*
tyrosine (Tyr) kinases 45, 52, *53*
    effect of receptor clustering on activation 172–173, *173*
    oncogenic fusion proteins 257–258, *257*
    phosphatidylinositol 3-kinase activation 157
    phospholipase C activation 157
    receptor *see* receptor tyrosine kinases
    receptors covalently linked to 179–183
    substrate specificity 51, 52
tyrosine (Tyr) phosphatases 45, 54–58
    classes 54–58, *65*
    modular domain structure 58, *58*
    reaction mechanism 56, *57*
tyrosine phosphorylation 45, *45*; *see also* phosphotyrosine
    inducing protein interactions 89–91, *90*
    T cell activation 379

UBA domain **237**, *244*
UBAN domain *97*, 98
ubiquitin 81, 95
    enzymes adding and removing 95–96, *95*, *96*
    polyubiquitin formation 95, *96*
ubiquitin-binding domains (UBDs) 95, 244–245
    recognition of ubiquitin-mediated signals 97–98, *97*
    structural features 244, *244*
ubiquitin interactions 98–99, *98*, *244*
ubiquitin ligase complexes; *see also* E3 ubiquitin ligases
    cell cycle regulation 214, *214*, 369
ubiquitin-like (UBL) peptides 95
ubiquitylation 81, 95–97
    cell cycle control 214, *214*
    domains recognizing *see* ubiquitin binding domains
    functional consequences 97, *97*
    machinery 96–97, *96*
    methods of detection 409
    NF-κB regulation 218, *218*
    p53 87
    proteasome targeting 85, *86*, 213
    receptor down-regulation 197–199, *199*
    regulation of protein-protein interactions 244–245
    role in endocytosis 97, 119
    selection of substrates 96–97, *97*
UEV domain **237**
UIM domains **237**, *238*
ultrasensitive responses *276*, 277, *277*

oscillators 286, *287*
    signaling cascades 277, *278*
    zero order *278*, 279–281
unsaturated fatty acids 148, 151, *151*

vaccinia virus *314*
vanadate 57, *57*
vanilloid receptor 190
vascular endothelial growth factor (VEGF) receptor *177*
VAV *250*, 378
vesicles, membrane
    organization of lipids into *148*, 150
    regulation of intracellular trafficking 159, *160*
VH1-like phosphatases *see* dual specificity phosphatases
VHL-β protein *238*
VHL domain **237**
VHL protein 194, *194*
VHS domain **237**
Viagra® (sildenafil) 134, *135*
vision, vertebrate 347–355
V_{max} 392, *392*
voltage-gated calcium (Ca^{2+}) channels 139
voltage-gated potassium (Kv) channels 189–191, topology *190*
    gating mechanism 190, *191*
    selectivity filter 189–190, *190*
Von Hippel-Lindau syndrome 195
Vps27 *238*
Vps34 158

WASP proteins, signal integration 271–272, *272*
N-WASP, signal integration 271–272, *272*
WD40 repeat domains **237**
    binding modified histones 243, *244*
    diversity of functions 234
    F-box proteins 215
    structure 233, *234*
WDR5 243
Wee1
    cell cycle regulation 271, *271*, 286, *287*, 369
    targeting for proteolysis 215
Western blotting 399
Whi5 367–368
whole-genome sequencing 324
Wnt signaling pathway 184–185, *184*, 340
wound healing 356–361
writer/eraser/reader systems 11, *11*, 72
    G protein-regulating enzymes 59, *59*
    histone modifications 99–100
    kinase/phosphatase systems 41, *41*, 237
    linking together multiple 85, *86*
    phosphoinositide signaling 159
    photoreceptor signaling 351
    post-translational control machinery and 83–84, *83*
WW domain **237**

Xenopus laevis embryos 286–287, *287*
X-ray crystallography 395–396, *395*

Yan 247, *248*
yeast *see Saccharomyces cerevisiae;*
   Schizosaccharomyces pombe
yeast two-hybrid (Y2H) assay 101, *399*
yellow fluorescent protein (YFP) 397,
   400–401, *401*, *402*
Yersinia pestis 307–309, *308*, **308**

Yop effectors **309**
YopH 309
Yops 309
Ypt1 *117*

ZAP-70
   coupled SH2 domains 33, *34*, 240–241, *241*

phospholipase C-γ activation 253
   T cell activation 180, 378, 379
zero-order ultrasensitivity *278*, 279–281
zymogens 206, 208